Statistical Signal Processing of Complex-Valued Data

Complex-valued random signals are embedded into the very fabric of science and engineering, yet the usual assumptions made about their statistical behavior are often a poor representation of the underlying physics. This book deals with improper and noncircular complex signals, which do not conform to classical assumptions, and it demonstrates how correct treatment of these signals can have significant payoffs.

The book begins with detailed coverage of the fundamental theory and presents a variety of tools and algorithms for dealing with improper and noncircular signals. It provides a comprehensive account of the main applications, covering detection, estimation, and signal analysis of stationary, nonstationary, and cyclostationary processes.

Providing a systematic development from the origin of complex signals to their probabilistic description makes the theory accessible to newcomers. This book is ideal for graduate students and researchers working with complex data in a range of research areas from communications to oceanography.

PETER J. SCHREIER is an Associate Professor in the School of Electrical Engineering and Computer Science, The University of Newcastle, Australia. He received his Ph.D. in electrical engineering from the University of Colorado at Boulder in 2003. He currently serves on the Editorial Board of the *IEEE Transactions on Signal Processing*, and on the IEEE Technical Committee *Machine Learning for Signal Processing*.

LOUIS L. SCHARF is Professor of Electrical and Computer Engineering and Statistics at Colorado State University. He received his Ph.D. from the University of Washington at Seattle. He has since received numerous awards for his research contributions to statistical signal processing, including an IEEE Distinguished Lectureship, an IEEE Third Millennium Medal, and the Technical Achievement and Society Awards from the IEEE Signal Processing Society. He is a Life Fellow of the IEEE.

Statistical Signal Processing of Complex-Valued Data

The Theory of Improper and Noncircular Signals

PETER J. SCHREIER
University of Newcastle, New South Wales, Australia

LOUIS L. SCHARF
Colorado State University, Colorado, USA

CAMBRIDGE
UNIVERSITY PRESS

CAMBRIDGE
UNIVERSITY PRESS

University Printing House, Cambridge CB2 8BS, United Kingdom

Published in the United States of America by Cambridge University Press, New York

Cambridge University Press is part of the University of Cambridge.

It furthers the University's mission by disseminating knowledge in the pursuit of education, learning and research at the highest international levels of excellence.

www.cambridge.org
Information on this title: www.cambridge.org/9780521897723

First published 2010

A catalogue record for this publication is available from the British Library

ISBN 978-0-521-89772-3 Hardback

Contents

Preface

Complex-valued random signals are embedded into the very fabric of science and engineering, being essential to communications, radar, sonar, geophysics, oceanography, optics, electromagnetics, acoustics, and other applied sciences. A great many problems in detection, estimation, and signal analysis may be phrased in terms of two channels' worth of real signals. It is common practice in science and engineering to place these signals into the real and imaginary parts of a complex signal. Complex representations bring economies and insights that are difficult to achieve with real representations.

In the past, it has often been assumed – usually implicitly – that complex random signals are *proper* and *circular*. A *proper* complex random variable is uncorrelated with its complex conjugate, and a *circular* complex random variable has a probability distribution that is invariant under rotation in the complex plane. These assumptions are convenient because they simplify computations and, in many aspects, make complex random signals look and behave like real random signals. Yet, while these assumptions can often be justified, there are also many cases in which proper and circular random signals are very poor models of the underlying physics. This fact has been known and appreciated by oceanographers since the early 1970s, but it has only recently been accepted across disciplines by acousticians, optical scientists, and communication theorists.

This book develops the tools and algorithms that are necessary to deal with *improper* complex random variables, which are correlated with their complex conjugate, and with *noncircular* complex random variables, whose probability distribution varies under rotation in the complex plane. Accounting for the improper and noncircular nature of complex signals can have big payoffs. In digital communications, it can lead to a significantly improved tradeoff between spectral efficiency and power consumption. In array processing, it can enable us to estimate with increased accuracy the direction of arrival of one or more signals impinging on a sensor array. In independent component analysis, it may be possible to blindly separate Gaussian sources – something that is impossible if these sources are *proper*.

In the electrical engineering literature, the story of improper and noncircular complex signals began with Brown and Crane, Gardner, van den Bos, Picinbono, and their co-workers. They have laid the foundations for the theory we aim to review and extend in this research monograph, and to them we dedicate this book. The story is continuing, with work by a number of our colleagues who are publishing new findings as we write this preface. We have tried to stay up to date with their work by referencing it as carefully as we have been able. We ask their forbearance for results not included.

Outline of this book

The book can be divided into three parts. Part I (Chapters 1 and 2) gives an overview and introduction to complex random vectors and processes. In Chapter 1, we describe the origins and uses of complex signals. The chapter answers the following question: why do engineers and applied scientists represent real measurable effects by complex signals? Chapter 2 lays the foundation for the remainder of the book by introducing important concepts and definitions for complex random vectors and processes, such as widely linear transformations, complementary correlations, the multivariate improper Gaussian distribution, and complementary power spectra of wide-sense stationary processes. Chapter 2 should be read before proceeding to any of the later chapters.

Part II (Chapters 3–7) deals with complex random vectors and their application to correlation analysis, estimation, performance bounding, and detection. In Chapter 3, we discuss in detail the second-order description of a complex random vector. In particular, we are interested in those second-order properties that are invariant under either widely unitary or widely linear transformation. This leads us to a test for impropriety and applications in independent component analysis (ICA). Chapter 4 treats the assessment of multivariate association between two complex random vectors. We provide a unifying treatment of three popular correlation-analysis techniques: canonical correlation analysis, multivariate linear regression, and partial least squares. We also present several generalized likelihood-ratio tests for the correlation structure of complex Gaussian data, such as sphericity, independence within one data set, and independence between two data sets.

Chapter 5 is on estimation. Here we are interested in linear and widely linear least-squares problems, wherein parameter estimators are constrained to be linear or widely linear in the measurement and the performance criterion is mean-squared error or squared error under a constraint. Chapter 6 deals with performance bounds for parameter estimation. We consider quadratic performance bounds of the Weiss–Weinstein class, the most notable representatives of which are the Cramér–Rao and Fisher–Bayes bound. Chapter 7 addresses detection, where the problem is to determine which of two or more competing models best describes experimental measurements. In order to demonstrate the role of widely linear and widely quadratic forms in the theory of hypothesis testing, we concentrate on hypothesis testing within Gaussian measurement models.

Part III (Chapters 8–10) deals with complex random processes, both continuous- and discrete-time. Throughout this part, we focus on second-order spectral properties, and optimum linear (or widely linear) minimum mean-squared error filtering. Chapter 8 discusses wide-sense stationary (WSS) processes, with a focus on the role of the complementary power spectral density in rotary-component and polarization analysis. WSS processes admit a spectral representation in terms of the Fourier basis, which allows a frequency interpretation. The transform-domain description of a WSS signal is a spectral process with *orthogonal* increments. For nonstationary signals, we have to sacrifice either the Fourier basis and thus its frequency interpretation, or the orthogonality of the transform-domain representation. In Chapter 9, we will discuss both possibilities, which leads either to the Karhunen–Loève expansion or the Cramér–Loève spectral representation. The latter is the basis for bilinear time–frequency representations. Then, in Chapter 10 we treat cyclostationary processes. They are an important class of nonstationary processes that have periodically varying correlation properties. They can model periodic phenomena occurring in science and technology, including communications, meteorology, oceanography, climatology, astronomy, and economics.

Three appendices provide background material. Appendix 1 presents rudiments of matrix analysis. Appendix 2 introduces Wirtinger calculus, which enables us to compute generalized derivatives of a *real* function with respect to *complex* parameters. Finally, Appendix 3 discusses majorization, which is used at several places in this book. Majorization introduces a preordering of vectors, and it will allow us to optimize certain scalar real-valued functions with respect to real vector-valued parameters.

This book is mainly targeted at researchers and graduate students who rely on the theory of signals and systems to conduct their work in signal processing, communications, radar, sonar, optics, electromagnetics, acoustics, oceanography, geophysics, and geography. Although it is not primarily intended as a textbook, chapters of the book may be used to support a special-topics course at a second-year graduate level. We would expect readers to be familiar with basic probability theory, linear systems, and linear algebra, at a level covered in a typical first-year graduate course.

Acknowledgments

We would like to thank Dr. Patrik Wahlberg for giving us detailed feedback on many chapters of this book. We further thank Dr. Phil Meyler of Cambridge University Press for his support throughout the writing of this book. Peter Schreier acknowledges financial support from the Australian Research Council (ARC) under its Discovery Project scheme, and thanks Colorado State University, Ft. Collins, USA, for its hospitality during a five-month study leave in the winter and spring of 2008 in the northern hemisphere. Louis Scharf acknowledges years of research support by the Office of Naval Research and the National Science Foundation (NSF), and thanks the University of Newcastle, Australia, for its hospitality during a one-month study leave in the autumn of 2009 in the southern hemisphere.

Peter J. Schreier
Newcastle, New South Wales, Australia

Louis L. Scharf
Ft. Collins, Colorado, USA

Notation

Conventions

$\langle x, y\rangle$	inner product
$\|x\|$	norm (usually Euclidean)
$\hat{x}$	estimate of x
$\widetilde{x}$	complementary quantity to x
$x \perp y$	x is orthogonal to y

Vectors and matrices

$\mathbf{x}$	column-vector with components x_i
$\mathbf{x} \prec \mathbf{y}$	$\mathbf{x}$ is majorized by $\mathbf{y}$
$\mathbf{x} \prec_{\mathrm{w}} \mathbf{y}$	$\mathbf{x}$ is weakly majorized by $\mathbf{y}$
$\mathbf{X}$	matrix with components $(\mathbf{X})_{ij} = X_{ij}$
$\mathbf{X} > \mathbf{Y}$	$\mathbf{X} - \mathbf{Y}$ is positive definite
$\mathbf{X} \geq \mathbf{Y}$	$\mathbf{X} - \mathbf{Y}$ is positive semidefinite (nonnegative definite)
$\mathbf{X}^*$	complex conjugate
$\mathbf{X}^{\mathrm{T}}$	transpose
$\mathbf{X}^{\mathrm{H}} = (\mathbf{X}^{\mathrm{T}})^*$	Hermitian (conjugate) transpose
$\mathbf{X}^{\dagger}$	Moore–Penrose pseudo-inverse
$\langle\mathbf{X}\rangle$	subspace spanned by columns of $\mathbf{X}$
$\underline{\mathbf{x}} = \begin{bmatrix} \mathbf{x} \\ \mathbf{x}^* \end{bmatrix}$	augmented vector
$\underline{\mathbf{X}} = \begin{bmatrix} \mathbf{X}_1 & \mathbf{X}_2 \\ \mathbf{X}_2^* & \mathbf{X}_1^* \end{bmatrix}$	augmented matrix

Functions

$x(t)$	continuous-time signal
$x[k]$	discrete-time signal
$\hat{x}(t)$	Hilbert transform of $x(t)$; estimate of $x(t)$
$X(f)$	scalar-valued Fourier transform of $x(t)$
$\mathbf{X}(f)$	vector-valued Fourier transform of $\mathbf{x}(t)$
$\mathbb{X}(f)$	matrix-valued Fourier transform of $\mathbf{X}(t)$

$X(z)$	scalar-valued z-transform of $x[k]$
$\mathbf{X}(z)$	vector-valued z-transform of $\mathbf{x}[k]$
$\mathbb{X}(z)$	matrix-valued z-transform of $\mathbf{X}[k]$
$x(t) * y(t) = (x * y)(t)$	convolution of $x(t)$ and $y(t)$

Commonly used symbols and operators

$\arg(x) = \measuredangle x$	argument (phase) of complex x
$\mathbb{C}$	field of complex numbers
$\mathbb{C}_*^{2n}$	set of augmented vectors $\underline{\mathbf{x}} = [\mathbf{x}^{\mathrm{T}}, \mathbf{x}^{\mathrm{H}}]^{\mathrm{T}}$, $\mathbf{x} \in \mathbb{C}^n$
$\delta(x)$	Dirac δ-function (distribution)
$\det(\mathbf{X})$	matrix determinant
$\mathbf{diag}(\mathbf{X})$	vector of diagonal values $X_{11}, X_{22}, \ldots, X_{nn}$ of $\mathbf{X}$
$\mathbf{Diag}(x_1, \ldots, x_n)$	diagonal or block-diagonal matrix with diagonal elements $x_1, \ldots, x_n$
$\mathbf{e}$	error vector
$E(x)$	expectation of x
$\mathbf{ev}(\mathbf{X})$	vector of eigenvalues of $\mathbf{X}$, ordered decreasingly
$\mathbf{I}$	identity matrix
$\mathrm{Im}\ x$	imaginary part of x
$\mathbf{K}$	matrix of canonical/half-canonical correlations k_i
$\mathbf{\Lambda}$	matrix of eigenvalues λ_i
$\mathbf{m}_x$	sample mean vector of $\mathbf{x}$
$\boldsymbol{\mu}_x$	mean vector of $\mathbf{x}$
$p_x(x)$	probability density function (pdf) of x (often used without subscript)
$\mathbf{P}_{\mathbf{U}}$	orthogonal projection onto subspace $\langle \mathbf{U} \rangle$
$P_{xx}(f)$	power spectral density (PSD) of $x(t)$
$\widetilde{P}_{xx}(f)$	complementary power spectral density (C-PSD) of $x(t)$
$\underline{\mathbb{P}}_{xx}(f)$	augmented PSD matrix of $x(t)$
$\mathbf{Q}$	error covariance matrix
ρ	correlation coefficient; degree of impropriety; coherence
$\mathbb{R}$	field of real numbers
$\mathbf{R}_{xy}$	cross-covariance matrix of $\mathbf{x}$ and $\mathbf{y}$
$\widetilde{\mathbf{R}}_{xy}$	complementary cross-covariance matrix of $\mathbf{x}$ and $\mathbf{y}$
$\underline{\mathbf{R}}_{xy}$	augmented cross-covariance matrix of $\mathbf{x}$ and $\mathbf{y}$
$\mathbb{R}_{xy}$	covariance matrix of composite vector $[\mathbf{x}^{\mathrm{T}}, \mathbf{y}^{\mathrm{T}}]^{\mathrm{T}}$
$r_{xy}(t, \tau)$	cross-covariance function of $x(t)$ and $y(t)$
$\tilde{r}_{xy}(t, \tau)$	complementary cross-covariance function of $x(t)$ and $y(t)$
$\mathrm{Re}\ x$	real part of x
$\mathrm{sgn}(x)$	sign of x
$\mathbf{sv}(\mathbf{X})$	vector of singular values of $\mathbf{X}$, ordered decreasingly

$\mathbf{S}_{xx}$	sample covariance matrix of $\mathbf{x}$
$\widetilde{\mathbf{S}}_{xx}$	sample complementary covariance matrix of $\mathbf{x}$
$\underline{\mathbf{S}}_{xx}$	augmented sample covariance matrix of $\mathbf{x}$
$S_{xx}(\nu, f)$	(Loève) spectral correlation of $x(t)$
$\widetilde{S}_{xx}(\nu, f)$	(Loève) complementary spectral correlation of $x(t)$
$\mathbf{T} = \begin{bmatrix} \mathbf{I} & \mathrm{j}\mathbf{I} \\ \mathbf{I} & -\mathrm{j}\mathbf{I} \end{bmatrix}$	real-to-complex transformation
$\mathrm{tr}(\mathbf{X})$	matrix trace
$\mathbf{W}$	Wiener (linear or widely linear minimum mean-squared error) filter matrix
$\mathcal{W}^{m\times n}$	set of $2m \times 2n$ augmented matrices
$\mathbf{x} = \mathbf{u} + \mathrm{j}\mathbf{v}$	complex message/source
$\boldsymbol{\xi}$	internal (latent) description of $\mathbf{x}$
$x(t) = u(t) + \mathrm{j}v(t)$	complex continuous-time message/source signal
$\xi(f)$	spectral process corresponding to $x(t)$
$\mathbf{y} = \mathbf{a} + \mathrm{j}\mathbf{b}$	complex measurement/observation
$y(t)$	complex continuous-time measurement/observation signal
$\upsilon(f)$	spectral process corresponding to $y(t)$
$\boldsymbol{\omega}$	internal (latent) description of $\mathbf{y}$
Ω	sample space
$\mathbf{0}$	zero vector or matrix

Part I

Introduction

1 The origins and uses of complex signals

Engineering and applied science rely heavily on complex variables and complex analysis to model and analyze real physical effects. Why should this be so? That is, why should *real* measurable effects be represented by *complex* signals? The ready answer is that one complex signal (or channel) can carry information about two real signals (or two real channels), and the algebra and geometry of analyzing these two real signals as if they were one complex signal brings economies and insights that would not otherwise emerge. But ready answers beg for clarity. In this chapter we aim to provide it. In the bargain, we intend to clarify the language of engineers and applied scientists who casually speak of complex velocities, complex electromagnetic fields, complex baseband signals, complex channels, and so on, when what they are really speaking of is the x- and y-coordinates of velocity, the x- and y-components of an electric field, the in-phase and quadrature components of a modulating waveform, and the sine and cosine channels of a modulator or demodulator.

For electromagnetics, oceanography, atmospheric science, and other disciplines where two-dimensional trajectories bring insight into the underlying physics, it is the complex representation of an ellipse that motivates an interest in complex analysis. For communication theory and signal processing, where amplitude and phase modulations carry information, it is the complex baseband representation of a real bandpass signal that motivates an interest in complex analysis.

In Section 1.1, we shall begin with an elementary introduction to complex representations for Cartesian coordinates and two-dimensional signals. Then we shall proceed to a discussion of phasors and Lissajous figures in Sections 1.2 and 1.3. We will find that phasors are a complex representation for the motion of an undamped harmonic oscillator and Lissajous figures are a complex representation for polarized electromagnetic fields. The study of communication signals in Section 1.4 then leads to the Hilbert transform, the complex analytic signal, and various principles for modulating signals. Section 1.5 demonstrates how real signals can be loaded into the real and imaginary parts of a complex signal in order to make efficient use of the fast Fourier transform (FFT).

The second half of this chapter deals with complex *random* variables and signals. In Section 1.6, we introduce the univariate complex Gaussian probability density function (pdf) as an alternative parameterization for the bivariate pdf of two real correlated Gaussian random variables. We will see that the well-known form of the univariate complex Gaussian pdf models only a special case of the bivariate real pdf, where the two real random variables are independent and have equal variances. This special case is called *proper* or *circular*, and it corresponds to a uniform phase distribution of the complex random variable. In general, however, the complex Gaussian pdf depends not only on the variance but also on another term, which we will call the *complementary variance*. In Section 1.7, we extend this discussion to complex random signals. Using the polarization ellipse as an example, we will find an interplay of reality/complexity, propriety/impropriety, and wide-sense stationarity/nonstationarity. Section 1.8 provides a first glance at the mathematical framework that underpins the study of complex random variables in this book. Finally, Section 1.9 gives a brief survey of some recent papers that apply the theory of improper and noncircular complex random signals in communications, array processing, machine learning, acoustics, optics, and oceanography.

1.1 Cartesian, polar, and complex representations of two-dimensional signals

It is commonplace to represent two Cartesian coordinates (u, v) in their two polar coordinates (A, θ), or as the single complex coordinate $x = u + \mathrm{j}v = A\mathrm{e}^{\mathrm{j}\theta}$. The real coordinates $(u, v) \longleftrightarrow (A, \theta)$ are thus equivalent to the complex coordinates $u + \mathrm{j}v \longleftrightarrow A\mathrm{e}^{\mathrm{j}\theta}$. The virtue of this complex representation is that it leads to an economical algebra and an evocative geometry, especially when polar coordinates A and θ are used. This virtue extends to vector-valued coordinates $(\mathbf{u}, \mathbf{v})$, with complex representation $\mathbf{x} = \mathbf{u} + \mathrm{j}\mathbf{v}$. For example, $\mathbf{x}$ could be a mega-vector composed by stacking scan lines from a stereoscopic image, in which case $\mathbf{u}$ would be the image recorded by camera one and $\mathbf{v}$ would be the image recorded by camera two. In oceanographic applications, $\mathbf{u}$ and $\mathbf{v}$ could be the two orthogonal components of surface velocity and $\mathbf{x}$ would be the complex velocity. Or $\mathbf{x}$ could be a window's worth of a discrete-time communications signal. In the context of communications, radar, and sonar, $\mathbf{u}$ and $\mathbf{v}$ are called the *in-phase* and *quadrature* components, respectively, and they are obtained as sampled-data versions of a continuous-time signal that has been demodulated with a quadrature demodulator. The quadrature demodulator itself is designed to extract a baseband information-bearing signal from a passband carrying signal. This is explained in more detail in Section 1.4.

The virtue of complex representations extends to the analysis of time-varying coordinates $(u(t), v(t))$, which we call two-dimensional signals, and which we represent as the complex signal $x(t) = u(t) + \mathrm{j}v(t) = A(t)\mathrm{e}^{\mathrm{j}\theta(t)}$. Of course, the next generalization of this narrative would be to vector-valued complex signals $\mathbf{x}(t) = [x_1(t), x_2(t), \ldots, x_N(t)]^{\mathrm{T}}$, a generalization that produces technical difficulties, but not conceptual ones. The two best examples are complex-demodulated signals in a multi-sensor antenna array, in which case $x_k(t)$ is the complex signal recorded at sensor k, and complex-demodulated signals in spectral subbands of a wideband communication signal, in which case $x_k(t)$ is the complex signal recorded in subband k. When these signals are themselves sampled in time, then the vector-valued discrete-time sequence is $\mathbf{x}[n]$, with $\mathbf{x}[n] = \mathbf{x}(nT)$ a sampled-data version of $\mathbf{x}(t)$.

This introductory account of complex signals gives us the chance to remake a very important point. In engineering and applied science, measured signals are *real*. Correspondingly, in all of our examples, the components u and v are *real*. It is only our representation x that is complex. Thus one channel's worth of complex signal serves to represent two channels' worth of real signals. There is no fundamental reason why this would have to be done. We aim to make the point in this book that the algebraic economies, probabilistic computations, and geometrical insights that accrue to complex representations justify their use. The examples of the next several sections give a preview of the power of complex representations.

1.2 Simple harmonic oscillator and phasors

The damped harmonic oscillator models damped pendulums and second-order electrical and mechanical systems. A measurement (of position or voltage) in such a system obeys the second-order, homogeneous, linear differential equation

$$\frac{\mathrm{d}^2}{\mathrm{d}t^2}u(t) + 2\xi\omega_0\frac{\mathrm{d}}{\mathrm{d}t}u(t) + \omega_0^2 u(t) = 0. \tag{1.1}$$

The corresponding characteristic equation is

$$s^2 + 2\xi\omega_0 s + \omega_0^2 = 0. \tag{1.2}$$

If the damping coefficient ξ satisfies $0 \le \xi < 1$, the system is called *underdamped*, and the quadratic equation (1.2) has two complex conjugate roots $s_1 = -\xi\omega_0 + \mathrm{j}\sqrt{1-\xi^2}\omega_0$ and $s_2 = s_1^*$. The real homogeneous response of the damped harmonic oscillator is then

$$u(t) = A\mathrm{e}^{\mathrm{j}\theta}\mathrm{e}^{s_1 t} + A\mathrm{e}^{-\mathrm{j}\theta}\mathrm{e}^{s_1^* t} = \mathrm{Re}\{A\mathrm{e}^{\mathrm{j}\theta}\mathrm{e}^{s_1 t}\} = A\mathrm{e}^{-\xi\omega_0 t}\cos(\sqrt{1-\xi^2}\omega_0 t + \theta), \tag{1.3}$$

and A and θ may be determined from the initial values of $u(t)$ and $(\mathrm{d}/\mathrm{d}t)u(t)$ at $t = 0$. The real response (1.3) is the sum of two complex modal responses, or the real part of one of them. In anticipation of our continuing development, we might say that $A\mathrm{e}^{\mathrm{j}\theta}\mathrm{e}^{s_1 t}$ is a complex representation of the real signal $u(t)$.

For the *undamped* system with damping coefficient $\xi = 0$, we have $s_1 = \mathrm{j}\omega_0$ and the solution is

$$u(t) = \mathrm{Re}\{A\mathrm{e}^{\mathrm{j}\theta}\mathrm{e}^{\mathrm{j}\omega_0 t}\} = A\cos(\omega_0 t + \theta). \tag{1.4}$$

In this case, $A\mathrm{e}^{\mathrm{j}\theta}\mathrm{e}^{\mathrm{j}\omega_0 t}$ is the complex representation of the real signal $A\cos(\omega_0 t + \theta)$. The complex signal in its polar form

$$x(t) = A\mathrm{e}^{\mathrm{j}(\omega_0 t + \theta)} = A\mathrm{e}^{\mathrm{j}\theta}\mathrm{e}^{\mathrm{j}\omega_0 t}, \quad t \in \mathbb{R}, \tag{1.5}$$

is called a *rotating phasor*. The *rotator* $\mathrm{e}^{\mathrm{j}\omega_0 t}$ rotates the *stationary phasor* $A\mathrm{e}^{\mathrm{j}\theta}$ at the angular rate of ω_0 radians per second. The rotating phasor is periodic with period $2\pi/\omega_0$, thus overwriting itself every $2\pi/\omega_0$ seconds. Euler's identity allows us to express the rotating phasor in its Cartesian form as

$$x(t) = A\cos(\omega_0 t + \theta) + \mathrm{j}A\sin(\omega_0 t + \theta). \tag{1.6}$$

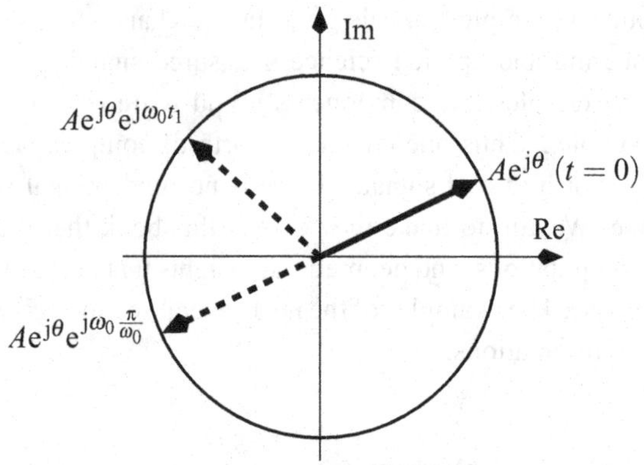

Figure 1.1 Stationary and rotating phasors.

Thus, the complex representation of the undamped simple harmonic oscillator turns out to be the trajectory in the complex plane of a rotating phasor of radian frequency ω_0, with starting point $Ae^{j\theta}$ at $t = 0$. The rotating phasor of Fig. 1.1 is illustrative.

The rotating phasor is one of the most fundamental complex signals we shall encounter in this book, as it is a basic building block for more complicated signals. As we build these more complicated signals, we will allow A and θ to be correlated random processes.

1.3 Lissajous figures, ellipses, and electromagnetic polarization

We might say that the circularly rotating phasor $x(t) = Ae^{j\theta}e^{j\omega_0 t} = A\cos(\omega_0 t + \theta) + jA\sin(\omega_0 t + \theta)$ is the simplest of Lissajous figures, consisting of real and imaginary parts that are $\pi/2$ radians out of phase. A more general Lissajous figure allows complex signals of the form

$$x(t) = u(t) + jv(t) = A_u\cos(\omega_0 t + \theta_u) + jA_v\cos(\omega_0 t + \theta_v). \tag{1.7}$$

Here the real part $u(t)$ and the imaginary part $v(t)$ can be mismatched in amplitude and phase. This Lissajous figure overwrites itself with period $2\pi/\omega_0$ and turns out an ellipse in the complex plane. (This is still not the most general Lissajous figure, since Lissajous figures generally also allow different frequencies in the u- and v-components.)

In electromagnetic theory, this complex signal would be the time-varying position of the electric field vector in the (u, v)-plane perpendicular to the direction of propagation. Over time, as the electric field vector propagates, it turns out an elliptical corkscrew in three-dimensional space. But in the two-dimensional plane perpendicular to the direction of propagation, it turns out an ellipse, so the electric field is said to be elliptically polarized. As this representation shows, the elliptical polarization may be modeled, and in fact produced, by the superposition of a one-dimensional, *linearly polarized*,

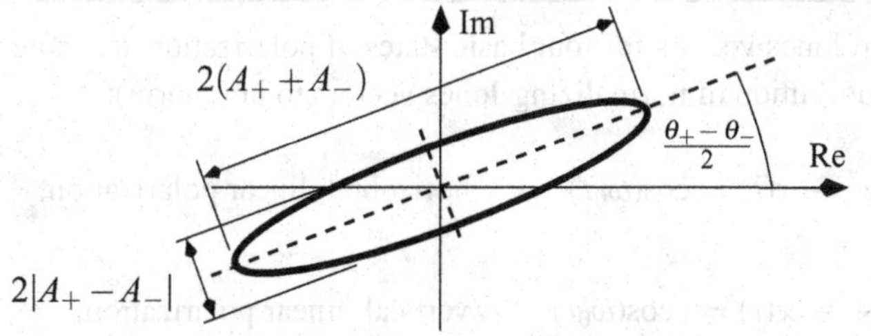

Figure 1.2 A typical polarization ellipse.

component of the form $A_u \cos(\omega_0 t + \theta_u)$ in the u-direction and another of the form $A_v \sin(\omega_0 t + \theta_v)$ in the v-direction.

But there is more. Euler's identity may be used to write the electric field vector as

$$\begin{aligned} x(t) &= \tfrac{1}{2}A_u e^{j\theta_u} e^{j\omega_0 t} + \tfrac{1}{2}A_u e^{-j\theta_u} e^{-j\omega_0 t} + \tfrac{1}{2}jA_v e^{j\theta_v} e^{j\omega_0 t} + \tfrac{1}{2}jA_v e^{-j\theta_v} e^{-j\omega_0 t} \\ &= \underbrace{\tfrac{1}{2}\left(A_u e^{j\theta_u} + jA_v e^{j\theta_v}\right)}_{A_+ e^{j\theta_+}} e^{j\omega_0 t} + \underbrace{\tfrac{1}{2}\left(A_u e^{-j\theta_u} + jA_v e^{-j\theta_v}\right)}_{A_- e^{-j\theta_-}} e^{-j\omega_0 t}. \end{aligned} \tag{1.8}$$

This representation of the two-dimensional electric field shows it to be the superposition of a two-dimensional, circularly polarized, component of the form $A_+ e^{j\theta_+} e^{j\omega_0 t}$ and another of the form $A_- e^{-j\theta_-} e^{-j\omega_0 t}$. The first rotates *counterclockwise* (CCW) and is said to be *left-circularly polarized.* The second rotates *clockwise* (CW) and is said to be *right-circularly polarized.* In this representation, the complex constants $A_+ e^{j\theta_+}$ and $A_- e^{-j\theta_-}$ fix the amplitude and phase of their respective circularly polarized components.

The circular representation of the ellipse makes it easy to determine the orientation of the ellipse and the lengths of the major and minor axes. In fact, by noting that the magnitude-squared of $x(t)$ is $|x(t)|^2 = A_+^2 + 2A_+A_- \cos(\theta_+ + \theta_- + 2\omega_0 t) + A_-^2$, it is easy to see that $|x(t)|^2$ has a maximum value of $(A_+ + A_-)^2$ at $\theta_+ + \theta_- + 2\omega_0 t = 2k\pi$, and a minimum value of $(A_+ - A_-)^2$ at $\theta_+ + \theta_- + 2\omega_0 t = (2k+1)\pi$. This orients the major axis of the ellipse at angle $(\theta_+ - \theta_-)/2$ and fixes the major and minor axis lengths at $2(A_+ + A_-)$ and $2|A_+ - A_-|$. A typical polarization ellipse is illustrated in Fig. 1.2.

Jones calculus

It is clear that the polarization ellipse $x(t)$ may be parameterized either by four real parameters $(A_u, A_v, \theta_u, \theta_v)$ or by two complex parameters $(A_+ e^{j\theta_+}, A_- e^{-j\theta_-})$. In the first case, we modulate the real basis $(\cos(\omega_0 t), \sin(\omega_0 t))$, and in the second case, we modulate the complex basis $(e^{j\omega_0 t}, e^{-j\omega_0 t})$. If we are interested only in the path that the electric field vector describes, and do not need to evaluate $x(t_0)$ at a particular time t_0, knowing the phase *differences* $\theta_u - \theta_v$ or $\theta_+ - \theta_-$ rather than the phases themselves is sufficient. The choice of parameterization – whether real or complex – is somewhat arbitrary, but it is common to use the *Jones vector* $[A_u, A_v e^{j(\theta_u - \theta_v)}]$ to describe the state of polarization. This is illustrated in the following example.

Example 1.1. The Jones vectors for four basic states of polarization are (note that we do not follow the convention of normalizing Jones vectors to unit norm):

$$\begin{bmatrix}1\\0\end{bmatrix} \longleftrightarrow x(t) = \cos(\omega_0 t) \qquad \text{horizontal, linear polarization,}$$

$$\begin{bmatrix}0\\1\end{bmatrix} \longleftrightarrow x(t) = \mathrm{j}\cos(\omega_0 t) \qquad \text{vertical, linear polarization,}$$

$$\begin{bmatrix}1\\j\end{bmatrix} \longleftrightarrow x(t) = \mathrm{e}^{\mathrm{j}\omega_0 t} \qquad \text{CCW (left-) circular polarization,}$$

$$\begin{bmatrix}1\\-j\end{bmatrix} \longleftrightarrow x(t) = \mathrm{e}^{-\mathrm{j}\omega_0 t} \qquad \text{CW (right-) circular polarization.}$$

Various polarization filters can be coded with two-by-two complex matrices that selectively pass components of the polarization. For example, consider these two polarization filters, and their corresponding Jones matrices:

$$\begin{bmatrix}1 & 0\\0 & 0\end{bmatrix} \qquad \text{horizontal, linear polarizer,}$$

$$\frac{1}{2}\begin{bmatrix}1 & -\mathrm{j}\\ \mathrm{j} & 1\end{bmatrix} \qquad \text{CCW (left-)circular polarizer.}$$

The first of these passes horizontal linear polarization and rejects vertical linear polarization. Such polarizers are used to reduce vertically polarized glare in Polaroid sunglasses. The second passes CCW circular polarization and rejects CW circular polarization. And so on.

1.4 Complex modulation, the Hilbert transform, and complex analytic signals

When analyzing the damped harmonic oscillator or the elliptically polarized electric field, the appropriate complex representations present themselves naturally. We now establish that this is so, as well, in the theory of modulation. Here the game is to modulate a *baseband*, information-bearing, signal onto a *passband* carrier signal that can be radiated from a real antenna onto a real channel. When the aim is to transmit information from here to there, then the channel may be "air," cable, or fiber. When the aim is to transmit information from now to then, then the channel may be a magnetic recording channel.

Actually, since a sinusoidal carrier signal can be modulated in amplitude and phase, the game is to modulate *two* information-bearing signals onto a carrier, suggesting again that *one complex* signal might serve to represent these two real signals and provide insight into how they should be designed. In fact, as we shall see, without the notion of a *complex analytic signal*, electrical engineers might never have discovered the Hilbert transform and single-sideband (SSB) modulation as the most spectrally efficient way to modulate *one* real channel of baseband information onto a passband carrier. Thus modulation theory provides the proper context for the study of the Hilbert transform and complex analytic signals.

1.4.1 Complex modulation using the complex envelope

Let us begin with two *real* information-bearing signals $u(t)$ and $v(t)$, which are combined in a *complex baseband signal* as

$$\begin{aligned} x(t) &= u(t) + \mathrm{j}v(t) \\ &= A(t)\mathrm{e}^{\mathrm{j}\theta(t)} = A(t)\cos\theta(t) + \mathrm{j}A(t)\sin\theta(t). \end{aligned} \quad (1.9)$$

The amplitude $A(t)$ and phase $\theta(t)$ are real. We take $u(t)$ and $v(t)$ to be lowpass signals with Fourier transforms supported on a baseband interval of $-\Omega < \omega < \Omega$. The representation $A(t)\mathrm{e}^{\mathrm{j}\theta(t)}$ is a generalization of the stationary phasor, wherein the fixed radius and angle of a phasor are replaced by a time-varying radius and angle. It is a simple matter to go back and forth between $x(t)$ and $(u(t), v(t))$ and $(A(t), \theta(t))$.

From $x(t)$ we propose to construct the *real* passband signal

$$p(t) = \mathrm{Re}\,\{x(t)\mathrm{e}^{\mathrm{j}\omega_0 t}\} = A(t)\cos(\omega_0 t + \theta(t)) = u(t)\cos(\omega_0 t) - v(t)\sin(\omega_0 t). \quad (1.10)$$

In accordance with standard communications terminology, we call $x(t)$ the complex baseband signal or *complex envelope* of $p(t)$, $A(t)$ and $\theta(t)$ the amplitude and phase of the complex envelope, and $u(t)$ and $v(t)$ the *in-phase* and *quadrature(-phase) components*. The term "quadrature component" refers to the fact that it is in phase quadrature ($+\pi/2$ out of phase) with respect to the in-phase component.

We say the complex envelope $x(t)$ complex-modulates the *complex carrier* $\mathrm{e}^{\mathrm{j}\omega_0 t}$, when what we really mean is that the *real* amplitude and phase $(A(t), \theta(t))$ real-modulate the amplitude and phase of the real carrier $\cos(\omega_0 t)$; or the in-phase and quadrature signals $(u(t), v(t))$ real-modulate the *real* in-phase carrier $\cos(\omega_0 t)$ and the *real* quadrature carrier $\sin(\omega_0 t)$. These are three equivalent ways of saying exactly the same thing. Figure 1.3(a) suggests a diagram for complex modulation. In Fig. 1.3(b), we stress the point that complex channels are actually two parallel real channels.

It is worth noting that when $\theta(t)$ is constant (say zero), then modulation is amplitude modulation only. In the complex plane, the complex baseband signal $x(t)$ writes out a trajectory $x(t) = A(t)$ that does not leave the real line. When $A(t)$ is constant (say 1), then modulation is phase modulation only. In the complex plane, the complex baseband signal $x(t)$ writes out a trajectory $x(t) = \mathrm{e}^{\mathrm{j}\theta(t)}$ that does not leave the unit circle. In general quadrature modulation, both $A(t)$ and $\theta(t)$ are time-varying, and they combine to write out quite arbitrary trajectories in the complex plane. These trajectories are composed of real part $u(t) = A(t)\cos\theta(t)$ and imaginary part $v(t) = A(t)\sin\theta(t)$, or of amplitude $A(t)$ and phase $\theta(t)$.

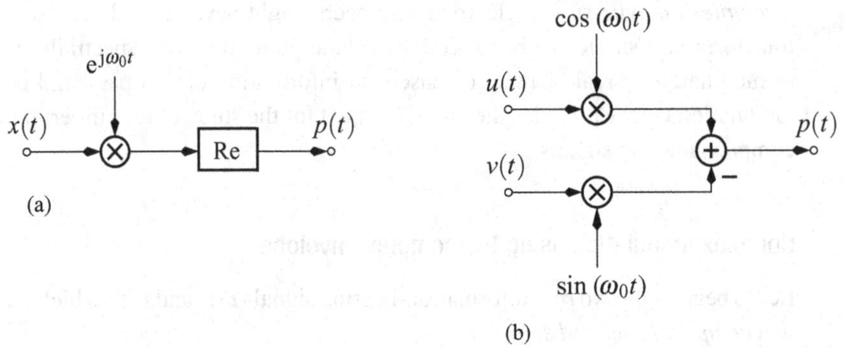

Figure 1.3 (a) Complex and (b) quadrature modulation.

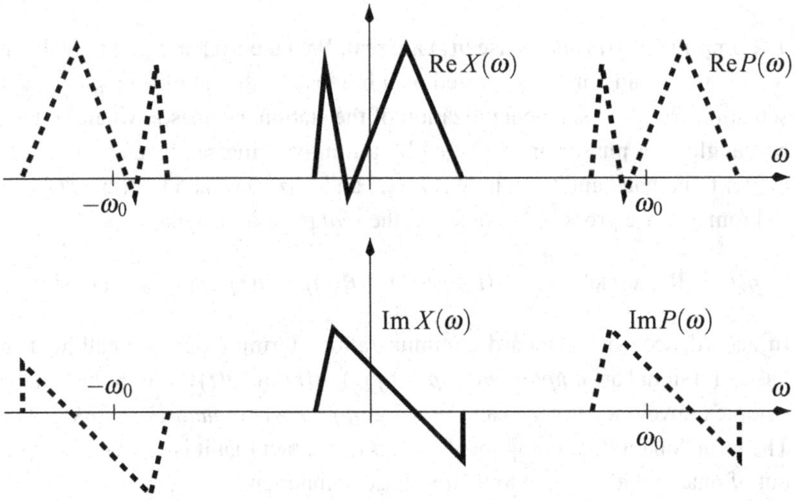

Figure 1.4 Baseband spectrum $X(\omega)$ (solid line) and passband spectrum $P(\omega)$ (dashed line).

If the complex signal $x(t)$ has Fourier transform $X(\omega)$, denoted $x(t) \longleftrightarrow X(\omega)$, then $x(t)e^{j\omega_0 t} \longleftrightarrow X(\omega - \omega_0)$ and $x^*(t) \longleftrightarrow X^*(-\omega)$. Thus,

$$p(t) = \operatorname{Re}\{x(t)e^{j\omega_0 t}\} = \tfrac{1}{2}x(t)e^{j\omega_0 t} + \tfrac{1}{2}x^*(t)e^{-j\omega_0 t} \tag{1.11}$$

has Hermitian-symmetric Fourier transform

$$P(\omega) = \tfrac{1}{2}X(\omega - \omega_0) + \tfrac{1}{2}X^*(-\omega - \omega_0). \tag{1.12}$$

Because $p(t)$ is real its Fourier transform satisfies $P(\omega) = P^*(-\omega)$. Thus, the real part of $P(\omega)$ is even, and the imaginary part is odd. Moreover, the magnitude $|P(\omega)|$ is even, and the phase $\measuredangle P(\omega)$ is odd. Fanciful spectra $X(\omega)$ and $P(\omega)$ are illustrated in Fig. 1.4.

1.4.2 The Hilbert transform, phase splitter, and analytic signal

If the complex baseband signal $x(t)$ can be recovered from the passband signal $p(t)$, then the two real channels $u(t)$ and $v(t)$ can be easily recovered as $u(t) = \text{Re } x(t) = \frac{1}{2}[x(t) + x^*(t)]$ and $v(t) = \text{Im } x(t) = [1/(2\text{j})][x(t) - x^*(t)]$. But how is $x(t)$ to be recovered from $p(t)$?

The *real* operator Re in the definition of $p(t)$ is applied to the complex signal $x(t)\text{e}^{\text{j}\omega_0 t}$ and returns the real signal $p(t)$. Suppose there existed an inverse operator Φ, i.e., a linear, convolutional, *complex* operator, that could be applied to the real signal $p(t)$ and return the complex signal $x(t)\text{e}^{\text{j}\omega_0 t}$. Then this complex signal could be complex-demodulated for $x(t) = \text{e}^{-\text{j}\omega_0 t}\text{e}^{\text{j}\omega_0 t}x(t)$. The complex operator Φ would have to be defined by an impulse response $\phi(t) \longleftrightarrow \Phi(\omega)$, whose Fourier transform $\Phi(\omega)$ were zero for negative frequencies and 2 for positive frequencies, in order to return the signal $x(t)\text{e}^{\text{j}\omega_0 t} \longleftrightarrow X(\omega - \omega_0)$.

This brings us to the Hilbert transform, the phase splitter, and the complex analytic signal. The Hilbert transform of a signal $p(t)$ is denoted $\hat{p}(t)$, and defined as the linear shift-invariant operation

$$\hat{p}(t) = (h * p)(t) \triangleq \int_{-\infty}^{\infty} h(t-\tau)p(\tau)\text{d}\tau \longleftrightarrow (HP)(\omega) \triangleq H(\omega)P(\omega) = \hat{P}(\omega). \tag{1.13}$$

The impulse response $h(t)$ and complex frequency response $H(\omega)$ of the Hilbert transform are defined to be, for $t \in \mathbb{R}$ and $\omega \in \mathbb{R}$,

$$h(t) = \frac{1}{\pi t} \longleftrightarrow -\text{j}\ \text{sgn}(\omega) = H(\omega). \tag{1.14}$$

Here $\text{sgn}(\omega)$ is the function

$$\text{sgn}(\omega) = \begin{cases} 1, & \omega > 0, \\ 0, & \omega = 0, \\ -1, & \omega < 0. \end{cases} \tag{1.15}$$

So $h(t)$ is real and odd, and $H(\omega)$ is imaginary and odd. From the Hilbert transform $h(t) \longleftrightarrow H(\omega)$ we define the *phase splitter*

$$\phi(t) = \delta(t) + \text{j}h(t) \longleftrightarrow 1 - \text{j}^2\ \text{sgn}(\omega) = 2\Gamma(\omega) = \Phi(\omega). \tag{1.16}$$

The complex frequency response of the phase splitter is $\Phi(\omega) = 2\Gamma(\omega)$, where $\Gamma(\omega)$ is the standard unit-step function. The convolution of the complex filter $\phi(t)$ and the real signal $p(t)$ produces the *analytic signal* $y(t) = p(t) + \text{j}\hat{p}(t)$, with Fourier transform identity

$$y(t) = (\phi * p)(t) = p(t) + \text{j}\hat{p}(t) \longleftrightarrow P(\omega) + \text{sgn}(\omega)P(\omega) = 2(\Gamma P)(\omega) = Y(\omega). \tag{1.17}$$

Recall that the Fourier transform $P(\omega)$ of a real signal $p(t)$ has Hermitian symmetry $P(-\omega) = P^*(\omega)$, so $P(\omega)$ for $\omega < 0$ is redundant. In the polar representation

$P(\omega) = B(\omega)e^{j\Psi(\omega)}$, we have $B(\omega) = B(-\omega)$ and $\Psi(-\omega) = -\Psi(\omega)$. Therefore, the corresponding spectral representations for real $p(t)$ and complex analytic $y(t)$ are

$$p(t) = \int_{-\infty}^{\infty} P(\omega)e^{j\omega t}\frac{d\omega}{2\pi} = 2\int_{0}^{\infty} B(\omega)\cos(\omega t + \Psi(\omega))\frac{d\omega}{2\pi}, \tag{1.18}$$

$$y(t) = \int_{0}^{\infty} 2P(\omega)e^{j\omega t}\frac{d\omega}{2\pi} = 2\int_{0}^{\infty} B(\omega)e^{j(\omega t+\Psi(\omega))}\frac{d\omega}{2\pi}. \tag{1.19}$$

The analytic signal replaces the redundant two-sided spectral representation (1.18) by the efficient one-sided representation (1.19). Equivalently, the analytic signal uses a linear combination of amplitude- and phase-modulated complex exponentials to represent a linear combination of amplitude- and phase-modulated cosines. We might say the analytic signal $y(t)$ is a bandwidth-efficient representation for its corresponding real signal $p(t)$.

Example 1.2. Begin with the real signal $\cos(\omega_0 t)$. Its Hilbert transform is the real signal $\sin(\omega_0 t)$ and its analytic signal is the complex exponential $e^{j\omega_0 t}$. This is easily established from the Fourier-series expansions for $e^{j\omega_0 t}$, $\cos(\omega_0 t)$, and $\sin(\omega_0 t)$. This result extends to the complex Fourier series

$$x(t) = \sum_{m=1}^{M} A_m e^{j\theta_m} e^{j2\pi\frac{m}{T}t}. \tag{1.20}$$

This complex signal is analytic, with spectral lines at positive frequencies of $2\pi m/T$ only. Its real and imaginary parts are

$$u(t) = \sum_{m=1}^{M} A_m \cos\left(2\pi\frac{m}{T}t + \theta_m\right) = \sum_{m=-M}^{M} \frac{A_m}{2} e^{j\theta_m} e^{j2\pi\frac{m}{T}t}, \tag{1.21}$$

$$\hat{u}(t) = \sum_{m=1}^{M} A_m \sin\left(2\pi\frac{m}{T}t + \theta_m\right) = \sum_{m=-M}^{M} \frac{\operatorname{sgn}(m)A_m}{2j} e^{j\theta_m} e^{j2\pi\frac{m}{T}t}, \tag{1.22}$$

where $A_0 = 0$, $A_{-m} = A_m$, and $\theta_{-m} = -\theta_m$. Of course $\hat{u}(t)$ is the Hilbert transform of $u(t)$.

But caution is required. Define the complex signal $e^{j\psi(t)} = \cos\psi(t) + j\sin\psi(t)$, with $\psi(t)$ real. But $\sin\psi(t)$ is the Hilbert transform of $\cos\psi(t)$ *if and only if* $e^{j\psi(t)}$ is analytic, meaning that its Fourier transform is one-sided. This means that $\sin\psi(t)$ is the Hilbert transform of $\cos\psi(t)$ *only in special cases.*

The imaginary part of a complex analytic signal is the Hilbert transform of its real part, and its Fourier transform is causal (zero for negative frequencies). There is also a dual for an analytic Fourier transform. Its imaginary part is the Hilbert transform of its real part, and its inverse Fourier transform is a causal signal (zero for negative time). This is called the *Kramers–Kronig relation*, after the physicists who first answered the question of what could be said about the spectrum of a causal signal.

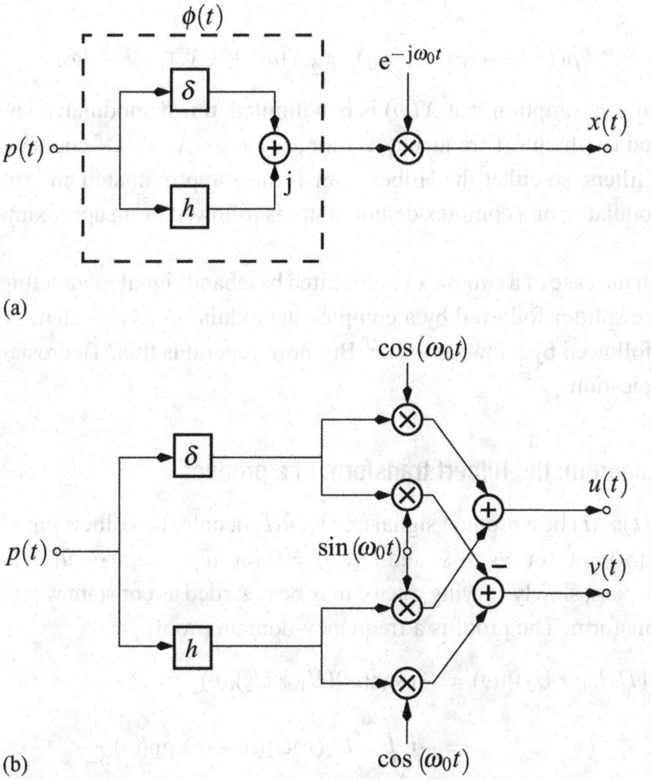

Figure 1.5 Complex demodulation with the phase splitter: (a) one complex channel and (b) two real channels.

1.4.3 Complex demodulation

Let's recap. The output of the phase splitter Φ applied to a passband signal $p(t) = \mathrm{Re}\{x(t)\mathrm{e}^{\mathrm{j}\omega_0 t}\} \longleftrightarrow \frac{1}{2}X(\omega-\omega_0)+\frac{1}{2}X^*(-\omega-\omega_0) = P(\omega)$ is the analytic signal

$$y(t) = (\phi * p)(t) \longleftrightarrow (\Phi P)(\omega) = 2(\Gamma P)(\omega) = Y(\omega). \tag{1.23}$$

Under the assumption that the bandwidth Ω of the spectrum $X(\omega)$ satisfies $\Omega < \omega_0$, we have

$$Y(\omega) = \begin{cases} X(\omega-\omega_0), & \omega > 0, \\ 0, & \omega \leq 0. \end{cases} \tag{1.24}$$

From here, we only need to shift $y(t)$ down to baseband to obtain the complex baseband signal

$$x(t) = y(t)\mathrm{e}^{-\mathrm{j}\omega_0 t} = (\phi * p)(t)\mathrm{e}^{-\mathrm{j}\omega_0 t} \longleftrightarrow 2(\Gamma P)(\omega+\omega_0) = X(\omega). \tag{1.25}$$

Two diagrams of a complex demodulator are shown in Fig. 1.5.

The Hilbert transform is an idealized convolution operator that can only be approximated in practice. The usual approach to approximating it is to complex-demodulate

$p(t)$ as

$$e^{-j\omega_0 t}p(t) \longleftrightarrow P(\omega+\omega_0) = \tfrac{1}{2}X(\omega) + \tfrac{1}{2}X^*(-\omega-2\omega_0). \tag{1.26}$$

Again, under the assumption that $X(\omega)$ is bandlimited, this demodulated signal may be lowpass-filtered (with cutoff frequency Ω) for $x(t) \longleftrightarrow X(\omega)$. Of course there are no ideal lowpass filters, so either the Hilbert transform is approximated and followed by a complex demodulator, or a complex demodulator is followed by an approximate lowpass filter.

Evidently, in the case of a complex bandlimited baseband signal modulating a complex carrier, a phase splitter followed by a complex demodulator is equivalent to a complex demodulator followed by a lowpass filter. But how general is this? Bedrosian's theorem answers this question.

1.4.4 Bedrosian's theorem: the Hilbert transform of a product

Let $u(t) = u_1(t)u_2(t)$ be a product signal and let $\hat{u}(t)$ denote the Hilbert transform of this product. If $U_1(\omega) = 0$ for $|\omega| > \Omega$ and $U_2(\omega) = 0$ for $|\omega| < \Omega$, then $\hat{u}(t) = u_1(t)\hat{u}_2(t)$. That is, the lowpass, slowly varying, factor may be regarded as constant when calculating the Hilbert transform. The proof is a frequency-domain proof:

$$\begin{aligned}
(H(U_1 * U_2))(\omega) &= -j\,\mathrm{sgn}(\omega)(U_1 * U_2)(\omega) \\
&= -j\int_{-\infty}^{\infty} U_1(\nu)U_2(\omega-\nu)\mathrm{sgn}(\omega)\frac{d\nu}{2\pi} \\
&= -j\int_{-\infty}^{\infty} U_1(\nu)U_2(\omega-\nu)\mathrm{sgn}(\omega-\nu)\frac{d\nu}{2\pi} \\
&= (U_1 * (HU_2))(\omega).
\end{aligned} \tag{1.27}$$

Actually, this proof is not as simple as it looks. It depends on the fact that $\mathrm{sgn}(\omega-\nu) = \mathrm{sgn}(\omega)$ over the range of values ν for which the integrand is nonzero.

Example 1.3. If $a(t)$ is a real lowpass signal with bandwidth $\Omega < \omega_0$, then the Hilbert transform of $u(t) = a(t)\cos(\omega_0 t)$ is $\hat{u}(t) = a(t)\sin(\omega_0 t)$. Hence, the analytic signal $x(t) = u(t) + j\hat{u}(t)$ computed from the real amplitude-modulated signal $u(t)$ is $x(t) = a(t)(\cos(\omega_0 t) + j\sin(\omega_0 t)) = a(t)e^{j\omega_0 t}$.

1.4.5 Instantaneous amplitude, frequency, and phase

So far, we have spoken rather loosely of amplitude and phase modulation. If we modulate two real signals $a(t)$ and $\psi(t)$ onto a cosine to produce the real signal $p(t) = a(t)\cos(\omega_0 t + \psi(t))$, then this language seems unambiguous: we would say the respective signals amplitude- and phase-modulate the cosine. But is it really unambiguous? The following example suggests that the question deserves thought.

Example 1.4. Let's look at a "purely amplitude-modulated" signal

$$p(t) = a(t)\cos(\omega_0 t). \tag{1.28}$$

Assuming that $a(t)$ is bounded such that $0 \le a(t) \le A$, there is a well-defined function

$$\psi(t) = \cos^{-1}\left(\frac{1}{A}p(t)\right) - \omega_0 t. \tag{1.29}$$

We can now write $p(t)$ as

$$p(t) = a(t)\cos(\omega_0 t) = A\cos(\omega_0 t + \psi(t)), \tag{1.30}$$

which makes it look like a "purely phase-modulated" signal.

This example shows that, for a given real signal $p(t)$, the factorization $p(t) = a(t)\cos(\omega_0 t + \psi(t))$ is not unique. In fact, there is an infinite number of ways for $p(t)$ to be factored into "amplitude" and "phase."

We can resolve this ambiguity by resorting to the complex envelope of $p(t)$. The complex envelope $x(t) = \mathrm{e}^{-\mathrm{j}\omega_0 t}(p(t) + \mathrm{j}\hat{p}(t))$, computed from the real bandpass signal $p(t)$ for a given carrier frequency ω_0, is uniquely factored as

$$x(t) = A(t)\mathrm{e}^{\mathrm{j}\theta(t)}. \tag{1.31}$$

We call $A(t) = |x(t)|$ the *instantaneous amplitude* of $p(t)$, $\theta(t) = \measuredangle x(t)$ the *instantaneous phase* of $p(t)$, and the derivative of the instantaneous phase $(\mathrm{d}/\mathrm{d}t)\theta(t)$ the *instantaneous frequency* of $p(t)$. This argument works also for $\omega_0 = 0$.

1.4.6 Hilbert transform and SSB modulation

There is another important application of the Hilbert transform, again leading to the definition of a complex signal from a real signal. In this case the aim is to modulate *one* real channel $u(t)$ onto a carrier. The direct way would be to use double-sideband suppressed carrier (DSB-SC) modulation of the form $u(t)\cos(\omega_0 t)$, whose spectrum is shown in Fig. 1.6(a). However, since $u(t)$ is real, its Fourier transform satisfies $U(-\omega) = U^*(\omega)$, so half the bandwidth is redundant.

The alternative is to Hilbert-transform $u(t)$, construct the analytic signal $x(t) = u(t) + \mathrm{j}\hat{u}(t)$, and complex-modulate with it to form the real passband signal $p(t) = u(t)\cos(\omega_0 t) - \hat{u}(t)\sin(\omega_0 t)$. This signal, illustrated in Fig. 1.6(b), is bandwidth-efficient and it is said to be single-sideband (SSB) modulated for obvious reasons. Without the notion of the Hilbert transform and the complex analytic signal, no such construction would have been possible.

1.4.7 Passband filtering at baseband

Consider the problem of linear shift-invariant filtering of the real passband signal $p(t)$ with a filter whose real-valued passband impulse response is $g(t)$ and whose passband

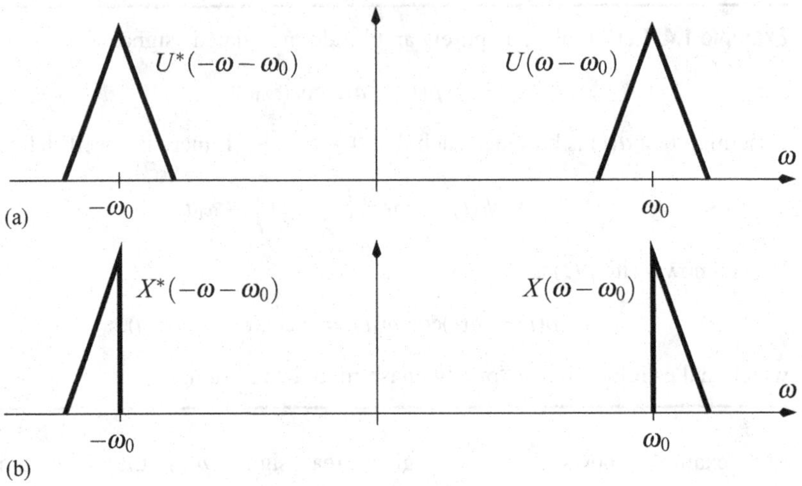

Figure 1.6 DSB-SC (a) and SSB modulation (b).

frequency response is $G(\omega)$. The filter output is

$$(g * p)(t) \longleftrightarrow (GP)(\omega). \tag{1.32}$$

Instead of filtering at passband, we can filter at baseband. Similarly to the definition of the complex baseband signal $x(t)$ in (1.25), we define the complex baseband impulse response and frequency response as

$$g_{\mathrm{b}}(t) = \tfrac{1}{2}(\phi * g)(t)\mathrm{e}^{-\mathrm{j}\omega_0 t} \longleftrightarrow G_{\mathrm{b}}(\omega) = (\Gamma G)(\omega + \omega_0), \tag{1.33}$$

where Γ again denotes the unit-step function. We note that the definition of a complex baseband impulse response includes a factor of $1/2$ that is not included in the definition of the complex baseband signal $x(t)$. Therefore,

$$G(\omega) = G_{\mathrm{b}}(\omega - \omega_0) + G_{\mathrm{b}}^*(-\omega - \omega_0), \tag{1.34}$$

whereas

$$P(\omega) = \tfrac{1}{2}X(\omega - \omega_0) + \tfrac{1}{2}X^*(-\omega - \omega_0). \tag{1.35}$$

The filter output is

$$(GP)(\omega) = \tfrac{1}{2}[G_{\mathrm{b}}(\omega - \omega_0) + G_{\mathrm{b}}^*(-\omega - \omega_0)][X(\omega - \omega_0) + X^*(-\omega - \omega_0)]. \tag{1.36}$$

If $G_{\mathrm{b}}(\omega)$ and $X(\omega)$ are bandlimited with bandwidth $\Omega < \omega_0$, then $G_{\mathrm{b}}(\omega - \omega_0)X^*(-\omega - \omega_0) \equiv 0$ and $G_{\mathrm{b}}^*(-\omega - \omega_0)X(\omega - \omega_0) \equiv 0$. This means that

$$\begin{aligned}(GP)(\omega) &= \tfrac{1}{2}[G_{\mathrm{b}}(\omega - \omega_0)X(\omega - \omega_0) + G_{\mathrm{b}}^*(-\omega - \omega_0)X^*(-\omega - \omega_0)]\\ &= \tfrac{1}{2}[(G_{\mathrm{b}}X)(\omega - \omega_0) + (G_{\mathrm{b}}X)^*(-\omega - \omega_0)],\end{aligned} \tag{1.37}$$

and in the time domain,

$$(g * p)(t) = \mathrm{Re}\,\{(g_{\mathrm{b}} * x)(t)\mathrm{e}^{\mathrm{j}\omega_0 t}\}. \tag{1.38}$$

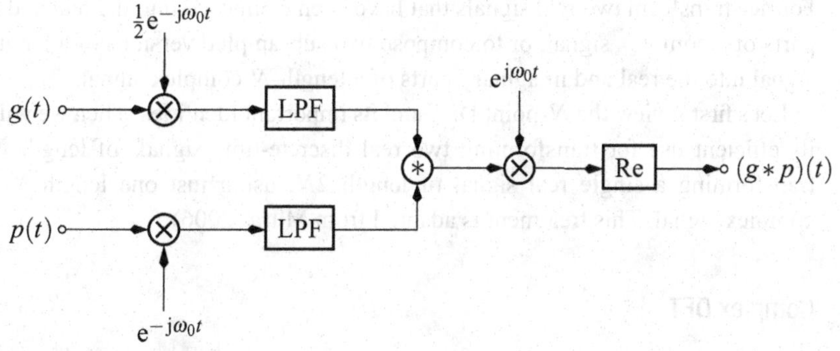

Figure 1.7 Passband filtering at baseband.

Hence, passband filtering can be performed at baseband. In the implementation shown in Fig. 1.7, the passband signal $p(t)$ is complex-demodulated and lowpass-filtered to produce the complex baseband signal $x(t)$ (as discussed in Section 1.4.3), then filtered using the complex baseband impulse response $g_b(t)$, and finally modulated back to passband. This is what is done in most practical applications.

This concludes our discussion of complex signals for general modulation of the amplitude and phase of a sinusoidal carrier. The essential point is that, once again, the representation of two real modulating signals as a complex signal leads to insights and economies of reasoning that would not otherwise emerge. Moreover, complex signal theory allows us to construct bandwidth-efficient versions of amplitude modulation, using the Hilbert transform and the complex analytic signal.

1.5 Complex signals for the efficient use of the FFT

There are four Fourier transforms: the continuous-time Fourier transform, the continuous-time Fourier series, the discrete-time Fourier transform (DTFT), and the discrete-time Fourier series, usually called the discrete Fourier transform (DFT). In practice, the DFT is always computed using the fast Fourier transform (FFT). All of these transforms are applied to a complex signal and they return a complex transform, or spectrum. When they are applied to a real signal, then the returned complex spectrum has Hermitian symmetry in frequency, meaning the negative-frequency half of the spectrum has been (inefficiently) computed when it could have been determined by simply complex conjugating an efficiently computed positive-frequency half of the spectrum.

One might be inclined to say that the analytic signal solves this problem by placing a real signal in the real part and the real Hilbert transform of this signal in the imaginary part to form a complex analytic signal, whose spectrum no longer has Hermitian symmetry (being zero for negative frequencies). However, again, this special non-Hermitian spectrum is known to be zero for negative frequencies and therefore the negative-frequency part of the spectrum has been computed inefficiently for its zero values. So, the only way to exploit the Fourier transform efficiently is to use it to simultaneously

Fourier-transform two real signals that have been composed into the real and imaginary parts of a complex signal, or to compose two subsampled versions of a length-$2N$ real signal into the real and imaginary parts of a length-N complex signal.

Let's first review the N-point DFT and its important identities. Then we will illustrate its efficient use for transforming two real discrete-time signals of length N, and for transforming a single real signal of length $2N$, using just one length-N DFT of a complex signal. This treatment is adapted from Mitra (2006).

1.5.1 Complex DFT

We shall denote the DFT of the length-N sequence $\{x[n]\}_{n=0}^{N-1}$ by the length-N sequence $\{X[m]\}_{m=0}^{N-1}$ and establish the shorthand notation $\{x[n]\}_{n=0}^{N-1} \longleftrightarrow \{X[m]\}_{m=0}^{N-1}$. The mth DFT coefficient $X[m]$ is computed as

$$X[m] = \sum_{n=0}^{N-1} x[n] W_N^{-mn}, \qquad W_N = \mathrm{e}^{\mathrm{j}2\pi/N}, \tag{1.39}$$

and these DFT coefficients are inverted for the original signal as

$$x[n] = \frac{1}{N} \sum_{m=0}^{N-1} X[m] W_N^{mn}. \tag{1.40}$$

The complex number $W_N = \mathrm{e}^{\mathrm{j}2\pi/N}$ is an Nth root of unity. When raised to the powers $n = 0, 1, \ldots, N-1$, it visits all the Nth roots of unity.

An important symmetry of the DFT is $\{x^*[n]\}_{n=0}^{N-1} \longleftrightarrow \{X^*[(N-m)_N]\}_{m=0}^{N-1}$, which reads "the DFT of a complex-conjugated sequence is the DFT of the original sequence, complex-conjugated and cyclically reversed in frequency." The notation $(N-m)_N$ stands for "$N - m$ modulo N" so that $X[(N)_N] = X[0]$.

Now consider a length-$2N$ sequence $\{x[n]\}_{n=0}^{2N-1}$ and its length-$2N$ DFT sequence $\{X[m]\}_{m=0}^{2N-1}$. Call $\{e[n] = x[2n]\}_{n=0}^{N-1}$ the *even polyphase* component of x, and $\{o[n] = x[2n+1]\}_{n=0}^{N-1}$ the *odd polyphase* component of x. Their DFTs are $\{e[n]\}_{n=0}^{N-1} \longleftrightarrow \{E[m]\}_{m=0}^{N-1}$ and $\{o[n]\}_{n=0}^{N-1} \longleftrightarrow \{O[m]\}_{m=0}^{N-1}$. The DFT of $\{x[n]\}_{n=0}^{2N-1}$ is, for $m = 0, 1, \ldots, 2N-1$,

$$\begin{aligned} X[m] = \sum_{n=0}^{2N-1} x[n] W_{2N}^{-mn} &= \sum_{n=0}^{N-1} e[n] W_N^{-mn} + W_N^{-m} \sum_{n=0}^{N-1} o[n] W_N^{-mn} \\ &= E[(m)_N] + W_N^{-m} O[(m)_N]. \end{aligned} \tag{1.41}$$

That is, the $2N$-point DFT is computed from two N-point DFTs. In fact, this is the basis of the decimation-in-time FFT.

1.5.2 Twofer: two real DFTs from one complex DFT

Begin with the two real length-N sequences $\{u[n]\}_{n=0}^{N-1}$ and $\{v[n]\}_{n=0}^{N-1}$. From them form the complex signal $\{x[n] = u[n] + \mathrm{j}v[n]\}_{n=0}^{N-1}$. DFT this complex sequence

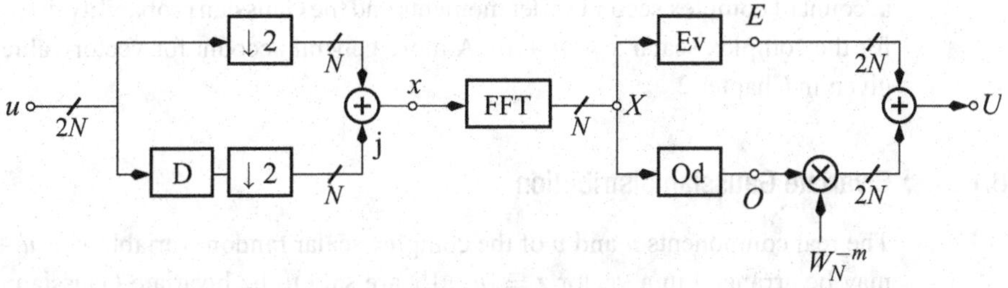

Figure 1.8 Using one length-N complex DFT to compute the DFT for a length-$2N$ real signal.

for $\{x[n]\}_{n=0}^{N-1} \longleftrightarrow \{X[m]\}_{m=0}^{N-1}$. Now note that $u[n] = \frac{1}{2}(x[n] + x^*[n])$, and $v[n] = [1/(2\text{j})](x[n] - x^*[n])$. So for $m = 0, 1, \ldots, N-1$,

$$U[m] = \frac{1}{2}(X[m] + X^*[(N-m)_N]), \tag{1.42}$$

$$V[m] = \frac{1}{2\text{j}}(X[m] - X^*[(N-m)_N]). \tag{1.43}$$

In this way, the N-point DFT is applied to a complex N-sequence x and efficiently returns an N-point DFT sequence X, from which the DFTs for u and v are extracted frequency-by-frequency.

1.5.3 Twofer: one real 2*N*-DFT from one complex *N*-DFT

Begin with the real length-$2N$ sequence $\{u[n]\}_{n=0}^{2N-1}$ and subsample it on its even and odd integers to form the real polyphase length-N sequences $\{e[n] = u[2n]\}_{n=0}^{N-1}$ and $\{o[n] = u[2n+1]\}_{n=0}^{N-1}$. From these form the complex N-sequence $\{x[n] = e[n] + \text{j}o[n]\}_{n=0}^{N-1}$. DFT this for $\{x[n]\}_{n=0}^{N-1} \longleftrightarrow \{X[m]\}_{m=0}^{N-1}$. Extract the DFTs for $\{e[n]\}_{n=0}^{N-1} \longleftrightarrow \{E[m]\}_{m=0}^{N-1}$ and $\{o[n]\}_{n=0}^{N-1} \longleftrightarrow \{O[m]\}_{m=0}^{N-1}$, according to (1.42) and (1.43). Then for $m = 0, 1, \ldots, 2N-1$, construct the $2N$-DFT $\{u[n]\}_{n=0}^{2N-1} \longleftrightarrow \{U[m]\}_{m=0}^{2N-1}$ according to (1.41). In this way, the DFT of a real $2N$-sequence is efficiently computed with the DFT of one complex N-sequence, followed by simple frequency-by-frequency computations. A hardware diagram is given in Fig. 1.8.

The two examples in this section show that the *complex* FFT can be made efficient for the Fourier analysis of *real* signals by constructing one complex signal from two real signals, providing one more justification for the claim that complex representations of two real signals bring efficiencies not otherwise achievable.

1.6 The bivariate Gaussian distribution and its complex representation

How is a complex random variable x, a complex random vector $\mathbf{x}$, a complex random signal $x(t)$, or a complex vector-valued random signal $\mathbf{x}(t)$ statistically described using probability distributions and moments? This question will occupy many of the ensuing chapters in this book. However, as a preview of our methods, we will offer a sketchy account of complex second-order moments and the Gaussian probability density function for the complex scalar $x = u + jv$. A more general account for vector-valued $\mathbf{x}$ will be given in Chapter 2.

1.6.1 Bivariate Gaussian distribution

The real components u and v of the complex scalar random variable $x = u + jv$, which may be arranged in a vector $\mathbf{z} = [u, v]^{\mathrm{T}}$, are said to be bivariate Gaussian distributed, with mean zero and covariance matrix $\mathbf{R}_{zz}$, if their joint probability density function (pdf) is

$$\begin{aligned} p_{uv}(u, v) &= \frac{1}{2\pi \det^{1/2} \mathbf{R}_{zz}} \exp\left\{ -\tfrac{1}{2} \begin{bmatrix} u & v \end{bmatrix} \mathbf{R}_{zz}^{-1} \begin{bmatrix} u \\ v \end{bmatrix} \right\} \\ &= \frac{1}{2\pi \det^{1/2} \mathbf{R}_{zz}} \exp\left\{ -\tfrac{1}{2} q_{uv}(u, v) \right\}. \end{aligned} \tag{1.44}$$

Here the quadratic form $q_{uv}(u, v)$ and the covariance matrix $\mathbf{R}_{zz}$ of the composite vector $\mathbf{z}$ are defined as follows:

$$q_{uv}(u, v) = \begin{bmatrix} u & v \end{bmatrix} \mathbf{R}_{zz}^{-1} \begin{bmatrix} u \\ v \end{bmatrix}, \tag{1.45}$$

$$\mathbf{R}_{zz} = E(\mathbf{z}\mathbf{z}^{\mathrm{T}}) = \begin{bmatrix} E(u^2) & E(uv) \\ E(vu) & E(v^2) \end{bmatrix} = \begin{bmatrix} R_{uu} & \sqrt{R_{uu}}\sqrt{R_{vv}}\rho_{uv} \\ \sqrt{R_{uu}}\sqrt{R_{vv}}\rho_{uv} & R_{vv} \end{bmatrix}. \tag{1.46}$$

In the right-most parameterization of $\mathbf{R}_{zz}$, the terms are

$R_{uu} = E(u^2)$ — variance of the random variable u,

$R_{vv} = E(v^2)$ — variance of the random variable v,

$R_{uv} = \sqrt{R_{uu}}\sqrt{R_{vv}}\rho_{uv} = E(uv)$ — correlation of the random variables u, v,

$\rho_{uv} = \dfrac{R_{uv}}{\sqrt{R_{uu}}\sqrt{R_{vv}}}$ — correlation coefficient of the random variables u, v.

As in (A1.38), the inverse of the covariance matrix $\mathbf{R}_{zz}^{-1}$ may be factored as

$$\mathbf{R}_{zz}^{-1} = \begin{bmatrix} 1 & 0 \\ -(\sqrt{R_{uu}}/\sqrt{R_{vv}})\rho_{uv} & 1 \end{bmatrix} \begin{bmatrix} 1/[R_{uu}(1-\rho_{uv}^2)] & 0 \\ 0 & 1/R_{vv} \end{bmatrix} \begin{bmatrix} 1 & -(\sqrt{R_{uu}}/\sqrt{R_{vv}})\rho_{uv} \\ 0 & 1 \end{bmatrix}. \tag{1.47}$$

Using (A1.3) we find $\det \mathbf{R}_{zz} = R_{uu}(1-\rho_{uv}^2)R_{vv}$, and from here the bivariate pdf $p_{uv}(u, v)$ may be written as

$$\begin{aligned} p_{uv}(u, v) = {} & \frac{1}{(2\pi R_{uu}(1-\rho_{uv}^2))^{1/2}} \exp\left\{ -\frac{1}{2R_{uu}(1-\rho_{uv}^2)} \left(u - \frac{\sqrt{R_{uu}}}{\sqrt{R_{vv}}}\rho_{uv} v \right)^2 \right\} \\ & \times \frac{1}{(2\pi R_{vv})^{1/2}} \exp\left\{ -\frac{1}{2R_{vv}} v^2 \right\}. \end{aligned} \tag{1.48}$$

The term $(\sqrt{R_{uu}}/\sqrt{R_{vv}})\rho_{uv}v$ is the conditional mean estimator of u from v and $e = u - (\sqrt{R_{uu}}/\sqrt{R_{vv}})\rho_{uv}v$ is the error of this estimator. Thus the bivariate pdf $p(u, v)$ factors into a zero-mean Gaussian pdf for the error e, with variance $R_{uu}(1 - \rho_{uv}^2)$, and a zero-mean Gaussian pdf for v, with variance R_{vv}. The error e and v are independent.

From $u = r\cos\theta$, $v = r\sin\theta$, $\mathrm{d}u\,\mathrm{d}v = r\,\mathrm{d}r\,\mathrm{d}\theta$, it is possible to change variables and obtain the pdf for the polar coordinates (r, θ)

$$p_{r\theta}(r, \theta) = r \cdot p_{uv}(u, v)|_{u=r\cos\theta, v=r\sin\theta}. \tag{1.49}$$

From here it is possible to integrate over r to obtain the marginal pdf for θ and over θ to obtain the marginal pdf for r. But this sequence of steps is so clumsy that it is hard to find formulas in the literature for these marginal pdfs. There is an alternative, which demonstrates again the power of complex representations.

1.6.2 Complex representation of the bivariate Gaussian distribution

Let's code the real random variables u and v as

$$\begin{bmatrix} u \\ v \end{bmatrix} = \frac{1}{2}\begin{bmatrix} 1 & 1 \\ -j & j \end{bmatrix}\begin{bmatrix} x \\ x^* \end{bmatrix}. \tag{1.50}$$

Then the quadratic form $q_{uv}(u, v)$ in the definition of the bivariate Gaussian distribution (1.45) may be written as

$$\begin{aligned} q_{uv}(u, v) &= \frac{1}{4}\begin{bmatrix} x^* & x \end{bmatrix}\begin{bmatrix} 1 & j \\ 1 & -j \end{bmatrix}\mathbf{R}_{zz}^{-1}\begin{bmatrix} 1 & 1 \\ -j & j \end{bmatrix}\begin{bmatrix} x \\ x^* \end{bmatrix} \\ &= \begin{bmatrix} x^* & x \end{bmatrix}\underline{\mathbf{R}}_{xx}^{-1}\begin{bmatrix} x \\ x^* \end{bmatrix}, \end{aligned} \tag{1.51}$$

where the covariance matrix $\underline{\mathbf{R}}_{xx}$ and its inverse $\underline{\mathbf{R}}_{xx}^{-1}$ are

$$\underline{\mathbf{R}}_{xx} = E\begin{bmatrix} x \\ x^* \end{bmatrix}\begin{bmatrix} x^* & x \end{bmatrix} = \begin{bmatrix} R_{xx} & \widetilde{R}_{xx} \\ \widetilde{R}_{xx}^* & R_{xx} \end{bmatrix} = \begin{bmatrix} 1 & j \\ 1 & -j \end{bmatrix}\mathbf{R}_{zz}\begin{bmatrix} 1 & 1 \\ -j & j \end{bmatrix}, \tag{1.52}$$

$$\underline{\mathbf{R}}_{xx}^{-1} = \frac{1}{4}\begin{bmatrix} 1 & j \\ 1 & -j \end{bmatrix}\mathbf{R}_{zz}^{-1}\begin{bmatrix} 1 & 1 \\ -j & j \end{bmatrix} = \frac{1}{R_{xx}^2 - |\widetilde{R}_{xx}|^2}\begin{bmatrix} R_{xx} & -\widetilde{R}_{xx} \\ -\widetilde{R}_{xx}^* & R_{xx} \end{bmatrix}. \tag{1.53}$$

The new terms in this representation of the quadratic form $q_{uv}(u, v)$ bear comment. So let's consider the elements of $\underline{\mathbf{R}}_{xx}$. The variance term R_{xx} is

$$R_{xx} = E|x|^2 = E[(u + \mathrm{j}v)(u - \mathrm{j}v)] = R_{uu} + R_{vv} + \mathrm{j}0. \tag{1.54}$$

This variance alone is an incomplete characterization for the bivariate pair (u, v), and it carries no information at all about ρ_{uv}, the correlation coefficient between the random variables u and v. But $\underline{\mathbf{R}}_{xx}$ contains another complex second-order moment

$$\widetilde{R}_{xx} = Ex^2 = E[(u + \mathrm{j}v)(u + \mathrm{j}v)] = R_{uu} - R_{vv} + \mathrm{j}2\sqrt{R_{uu}}\sqrt{R_{vv}}\rho_{uv}, \tag{1.55}$$

which we will call the *complementary variance*. The complementary variance is the correlation between x and its conjugate x^*. It is zero if and only if $R_{uu} = R_{vv}$ and $\rho_{uv} = 0$. This is the so-called *proper* case. All others are *improper*.

Now let's introduce the *complex correlation coefficient* ρ between x and x^* as

$$\rho = \frac{\widetilde{R}_{xx}}{R_{xx}}. \tag{1.56}$$

Thus, we may write $\widetilde{R}_{xx} = R_{xx}\rho$ and $R_{xx}^2 - |\widetilde{R}_{xx}|^2 = R_{xx}^2(1 - |\rho|^2)$. The complex correlation coefficient $\rho = |\rho|e^{j\psi}$ satisfies $|\rho| \leq 1$. If $|\rho| = 1$, then $x = \widetilde{R}_{xx}R_{xx}^{-1}x^* = \rho x^* = e^{j\psi}x^*$ with probability 1. Equivalent conditions for $|\rho| = 1$ are $R_{uu} = 0$, or $R_{vv} = 0$, or $\rho_{uv} = \pm 1$. The first of these conditions makes the complex signal x purely imaginary and the second makes it real. The third condition means $v = \tan(\psi/2)u$ and $x = [1 + j\tan(\psi/2)]u$. All these cases with $|\rho| = 1$ are called *maximally improper* because the support of the pdf for the complex random variable x degenerates into a line in the complex plane.

There are three real parameters R_{uu}, R_{vv}, and ρ_{uv} required to determine the bivariate pdf for (u, v), and these may be obtained from the three real values R_{xx}, $\operatorname{Re}\rho$, and $\operatorname{Im}\rho$ (or alternatively, R_{xx}, $\operatorname{Re}\widetilde{R}_{xx}$, and $\operatorname{Im}\widetilde{R}_{xx}$) using the following inverse formulas of (1.54) and (1.55):

$$R_{uu} = \tfrac{1}{2}R_{xx}(1 + \operatorname{Re}\rho), \tag{1.57}$$

$$R_{vv} = \tfrac{1}{2}R_{xx}(1 - \operatorname{Re}\rho), \tag{1.58}$$

$$\rho_{uv} = \frac{\operatorname{Im}\rho}{\sqrt{1 - (\operatorname{Re}\rho)^2}}. \tag{1.59}$$

Now, by replacing the quadratic form $q_{uv}(u, v)$ in (1.44) with the expression (1.51), and noting that $\det \underline{\mathbf{R}}_{xx} = 4\det \mathbf{R}_{zz}$, we may record the complex representation of the pdf for the bivariate Gaussian distribution or, equivalently, the pdf for complex x:

$$\begin{aligned} p_x(x) \triangleq p_{uv}(u, v) &= \frac{1}{\pi \det^{1/2}\underline{\mathbf{R}}_{xx}} \exp\left\{-\tfrac{1}{2}\begin{bmatrix} x^* & x \end{bmatrix} \underline{\mathbf{R}}_{xx}^{-1} \begin{bmatrix} x \\ x^* \end{bmatrix}\right\} \\ &= \frac{1}{\pi R_{xx}\sqrt{1 - |\rho|^2}} \exp\left\{-\frac{|x|^2 - \operatorname{Re}(\rho x^{*2})}{R_{xx}(1 - |\rho|^2)}\right\}. \end{aligned} \tag{1.60}$$

This shows that the bivariate pdf for the real pair (u, v) can be written in terms of the complex variable x. Yet the formula (1.60) is not what most people expect to see when they talk about the pdf of a complex Gaussian random variable x. In fact, it is often implicitly assumed that x is *proper*, i.e., $\rho = 0$ and x is not correlated with x^*. Then the pdf takes on the simple and much better-known form

$$p_x(x) = \frac{1}{\pi R_{xx}} \exp\left(-\frac{|x|^2}{R_{xx}}\right). \tag{1.61}$$

But it is clear from our development that (1.61) models only a special case of the bivariate pdf for (u, v) where $R_{uu} = R_{vv}$ and $\rho_{uv} = 0$. In general, we need to incorporate both the variance R_{xx} and the complementary variance $\widetilde{R}_{xx} = R_{xx}\rho$. That is, even in this very simple bivariate case, we need to take into account the correlation between x and its complex conjugate x^*. In Chapter 2, this general line of argumentation is generalized to derive the complex representation of the *multivariate* pdf $p_{uv}(\mathbf{u}, \mathbf{v})$.

1.6.3 Polar coordinates and marginal pdfs

What could be the virtue of the complex representation for the real bivariate pdf? One answer is this: with the change from Cartesian to polar coordinates, the bivariate pdf takes the simple form

$$p_{r\theta}(r, \theta) = \frac{r}{\pi R_{xx}\sqrt{1 - |\rho|^2}} \exp\left\{-\frac{r^2[1 - |\rho|\cos(2\theta - \psi)]}{R_{xx}(1 - |\rho|^2)}\right\}, \tag{1.62}$$

where x and ρ have been given their polar representations $x = r\mathrm{e}^{\mathrm{j}\theta}$ and $\rho = |\rho|\mathrm{e}^{\mathrm{j}\psi}$. It is now a simple matter to integrate this bivariate pdf to obtain the marginal pdfs for the radius r and the angle θ. The results, to be explored more fully in Chapter 2, are

$$p_r(r) = \frac{2r}{R_{xx}\sqrt{1 - |\rho|^2}} \exp\left\{-\frac{r^2}{R_{xx}(1 - |\rho|^2)}\right\} I_0\left\{\frac{r^2|\rho|}{R_{xx}(1 - |\rho|^2)}\right\}, \quad r > 0, \tag{1.63}$$

$$p_\theta(\theta) = \frac{\sqrt{1 - |\rho|^2}}{2\pi[1 - |\rho|\cos(2\theta - \psi)]}, \quad -\pi < \theta \leq \pi. \tag{1.64}$$

Here I_0 is the modified Bessel function of the first kind of order 0, defined as

$$I_0(z) = \frac{1}{\pi}\int_0^\pi \mathrm{e}^{z\cos\theta}\, \mathrm{d}\theta. \tag{1.65}$$

These results show that the parameters $(R_{xx}, |\rho|, \psi)$ for complex $x = u + \mathrm{j}v$, rather than the parameters $(R_{uu}, R_{vv}, \rho_{uv})$ for real (u, v), are the most natural parameterization for the joint and marginal pdfs of the polar coordinates r and θ. These marginals are illustrated in Fig. 1.9 for $\psi = \pi/2$ and various values of $|\rho|$. In the *proper* case $\rho = 0$, we see that the marginal pdf for r is Rayleigh and the marginal pdf for θ is uniform. Because of the uniform phase distribution, a proper Gaussian random variable is also called *circular*. The larger $|\rho|$ the more improper (or noncircular) x becomes, and the marginal distribution for θ develops two peaks at $\theta = \psi/2 = \pi/4$ and $\theta = \psi/2 - \pi$. At the same time, the maximum of the pdf for r is shifted to the left. However, the change in the marginal for r is not as dramatic as the change in the marginal for θ.

1.7 Second-order analysis of the polarization ellipse

In the previous section we found that the second-order description of a complex Gaussian random variable needs to take into account complementary statistics. The same holds for the second-order description of complex random *signals*. As an obvious extension of our previous definition, we shall call a zero-mean complex signal $x(t)$, with correlation function $r_{xx}(t, \tau) = E[x(t + \tau)x^*(t)]$, *proper* if its *complementary correlation function*

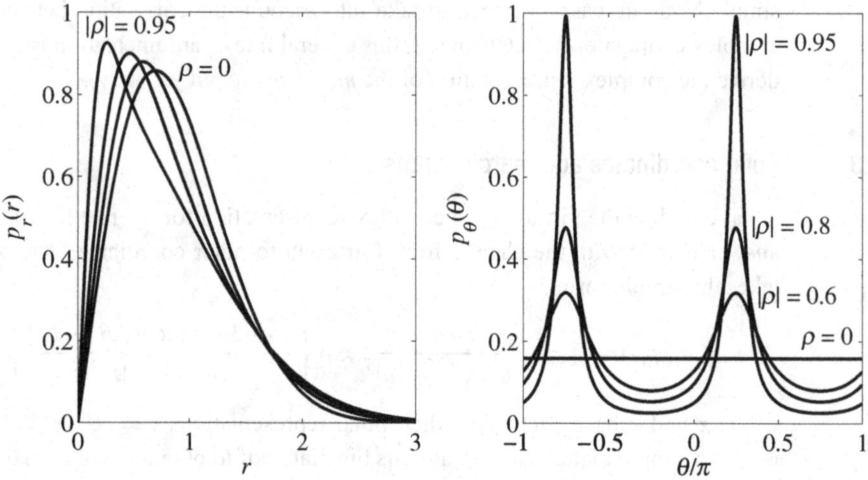

Figure 1.9 Marginal pdfs for magnitude r and angle θ in the general bivariate Gaussian distribution for $R_{xx} = 1$, $|\rho| = \{0, 0.6, 0.8, 0.95\}$, and $\psi = \pi/2$.

$\tilde{r}_{xx}(t,\tau) = E[x(t+\tau)x(t)] \equiv 0$. The following example is a preview of results that will be explored in much more detail in Chapters 8 and 9.

Example 1.5. Our discussion of the polarization ellipse in Section 1.3 has said nothing about the second-order statistical behavior of the complex coefficients $C_+ = A_+ e^{j\theta_+}$ and $C_- = A_- e^{-j\theta_-}$. We show here how the second-order moments of these coefficients determine the second-order behavior of the complex signal

$$x(t) = C_+ e^{j\omega_0 t} + C_- e^{-j\omega_0 t}.$$

We begin with the second-order Hermitian correlation function

$$\begin{aligned} r_{xx}(t,\tau) &= E[x(t+\tau)x^*(t)] \\ &= E[C_+ e^{j\omega_0(t+\tau)} + C_- e^{-j\omega_0(t+\tau)}][C_+^* e^{-j\omega_0 t} + C_-^* e^{j\omega_0 t}] \\ &= E[C_+C_+^*]e^{j\omega_0\tau} + 2\,\mathrm{Re}\,\{E[C_+C_-^*]e^{j2\omega_0 t}e^{j\omega_0\tau}\} + E[C_-C_-^*]e^{-j\omega_0\tau} \end{aligned}$$

and the second-order complementary correlation function

$$\begin{aligned} \tilde{r}_{xx}(t,\tau) &= E[x(t+\tau)x(t)] \\ &= E[C_+ e^{j\omega_0(t+\tau)} + C_- e^{-j\omega_0(t+\tau)}][C_+ e^{j\omega_0 t} + C_- e^{-j\omega_0 t}] \\ &= E[C_+C_+]e^{j2\omega_0 t}e^{j\omega_0\tau} + 2E[C_+C_-]\cos(\omega_0\tau) + E[C_-C_-]e^{-j2\omega_0 t}e^{-j\omega_0\tau}. \end{aligned}$$

Several observations are in order:

- the signal $x(t)$ is real if and only if $C_+^* = C_-$
- the signal is wide-sense stationary (WSS), that is $r_{xx}(t,\tau) = r_{xx}(0,\tau)$ *and* $\tilde{r}_{xx}(t,\tau) = \tilde{r}_{xx}(0,\tau)$, if and only if $E[C_+C_-^*] = 0$, $E[C_+C_+] = 0$, and $E[C_-C_-] = 0$

- the signal is proper, that is $\tilde{r}_{xx}(t,\tau) = 0$, if and only if $E[C_+C_+] = 0$, $E[C_+C_-] = 0$, and $E[C_-C_-] = 0$
- the signal is proper and WSS if and only if $E[C_+C_-^*] = 0$, $E[C_+C_+] = 0$, $E[C_+C_-] = 0$, and $E[C_-C_-] = 0$
- the signal cannot be nonzero, real, and proper, because $C_+^* = C_-$ makes $E[C_+C_-] = E[C_+C_+^*] > 0$

So a signal can be proper and WSS, proper and nonstationary, improper and WSS, or improper and nonstationary. If the signal is real then it is improper, but it can be WSS or nonstationary.

1.8 Mathematical framework

This chapter has shown that there are conceptual differences between *one complex signal* and *two real signals*. In this section, we will explain why there are also mathematical differences. This will allow us to provide a first glimpse at the mathematical framework that underpins much of what is to come later in this book. Consider the simple $\mathbb{C}$-linear relationship between two complex scalars $x = u + jv$ and $y = a + jb$:

$$y = kx. \tag{1.66}$$

We may write this relationship in terms of real and imaginary parts as

$$\begin{bmatrix} a \\ b \end{bmatrix} = \underbrace{\begin{bmatrix} \operatorname{Re} k & -\operatorname{Im} k \\ \operatorname{Im} k & \operatorname{Re} k \end{bmatrix}}_{\mathbf{M}} \begin{bmatrix} u \\ v \end{bmatrix}. \tag{1.67}$$

The 2×2 matrix $\mathbf{M}$ has a special structure and is determined by two real numbers $\operatorname{Re} k$ and $\operatorname{Im} k$. On the other hand, a general linear transformation on $\mathbb{R}^2$ is

$$\begin{bmatrix} a \\ b \end{bmatrix} = \underbrace{\begin{bmatrix} M_{11} & M_{12} \\ M_{21} & M_{22} \end{bmatrix}}_{\mathbf{M}} \begin{bmatrix} u \\ v \end{bmatrix}, \tag{1.68}$$

where all four elements of $\mathbf{M}$ may be chosen freely. This $\mathbb{R}^2$-linear expression is $\mathbb{C}$-linear, i.e., it can be expressed as $y = kx$, if and only if $M_{11} = M_{22}$ and $M_{21} = -M_{12}$. In the more general case, where $M_{11} \neq M_{22}$ and/or $M_{21} \neq -M_{12}$, the complex equivalent of (1.68) is the linear–conjugate-linear, or *widely linear*, transformation

$$y = k_1 x + k_2 x^*. \tag{1.69}$$

Widely linear transformations depend linearly on x and its conjugate x^*. The two complex coefficients k_1 and k_2 have a one-to-one correspondence to the four real coefficients $M_{11}, M_{12}, M_{21}, M_{22}$, which we derive in Section 2.1.

Traditionally, widely linear transformations have been employed only reluctantly. If the $\mathbb{R}^2$-linear transformation (1.68) does not satisfy $M_{11} = M_{22}$ and $M_{21} = -M_{11}$, most people would prefer the $\mathbb{R}^2$-linear representation (1.68) over the $\mathbb{C}$-widely linear representation (1.69). As a matter of fact, in classical complex analysis, functions that depend on x^* are not even considered differentiable *even if* they are differentiable when expressed as two-dimensional functions in terms of real and imaginary parts (see Appendix 2). In this book, we aim to make the point that a complex representation is not only feasible but also can be much more powerful and elegant, even if it leads to expressions involving complex conjugates.

A key tool will be the *augmented representation* of widely linear transformations, where we express (1.69) as

$$\begin{bmatrix} y \\ y^* \end{bmatrix} = \begin{bmatrix} k_1 & k_2 \\ k_2^* & k_1^* \end{bmatrix} \begin{bmatrix} x \\ x^* \end{bmatrix},$$

$$\underline{\mathbf{y}} = \underline{\mathbf{K}}\,\underline{\mathbf{x}}. \tag{1.70}$$

This representation utilizes the *augmented vectors*

$$\underline{\mathbf{y}} = \begin{bmatrix} y \\ y^* \end{bmatrix} \quad \text{and} \quad \underline{\mathbf{x}} = \begin{bmatrix} x \\ x^* \end{bmatrix} \tag{1.71}$$

and the *augmented matrix*

$$\underline{\mathbf{K}} = \begin{bmatrix} k_1 & k_2 \\ k_2^* & k_1^* \end{bmatrix}. \tag{1.72}$$

Augmented vectors and matrices are underlined. The augmented representation obviously has some built-in redundancy but it will turn out to be very useful and convenient as we develop the first- and second-order theories of improper signals. The space of augmented complex vectors $[x, y]^{\mathrm{T}}$, where $y = x^*$, is denoted $\mathbb{C}_*^2$, and it is *isomorphic* to $\mathbb{R}^2$. However, $\mathbb{C}_*^2$ is only an $\mathbb{R}$-linear (or $\mathbb{C}$-widely linear), but not a $\mathbb{C}$-linear, subspace of $\mathbb{C}^2$. It satisfies all properties of a linear subspace except that it is not closed under multiplication with a complex scalar α because $\alpha x^* \neq (\alpha x)^*$. Similarly, augmented matrices form a matrix algebra that is closed under addition, multiplication, inversion, and multiplication with a *real* scalar but not under multiplication with a *complex* scalar.

We have already had a first exposure to augmented vectors and matrices in our discussion of the complex Gaussian distribution in Section 1.6. In fact, we now recognize that the covariance matrix defined in (1.52) is

$$\underline{\mathbf{R}}_{xx} = E\underline{\mathbf{x}}\,\underline{\mathbf{x}}^{\mathrm{H}} = \begin{bmatrix} R_{xx} & \widetilde{R}_{xx} \\ \widetilde{R}_{xx}^* & R_{xx} \end{bmatrix}. \tag{1.73}$$

This is the covariance matrix of the augmented vector $\underline{\mathbf{x}} = [x, x^*]^{\mathrm{T}}$, and $\underline{\mathbf{R}}_{xx}$ itself is an augmented matrix because $R_{xx} = R_{xx}^*$. This is why we call $\underline{\mathbf{R}}_{xx}$ the *augmented covariance matrix* of x. The augmented covariance matrix $\underline{\mathbf{R}}_{xx}$ is more than simply a convenient way of keeping track of both the variance $R_{xx} = E|x|^2$ and the complementary variance $\widetilde{R}_{xx} = Ex^2$. By combining R_{xx} and $\widetilde{R}_{xx}$ into $\underline{\mathbf{R}}_{xx}$, we gain access to the large number of results on 2×2 matrices. For instance, we know that any covariance matrix, including $\underline{\mathbf{R}}_{xx}$, must be positive semidefinite and thus have nonnegative determinant, $\det \underline{\mathbf{R}}_{xx} \geq 0$.

This immediately leads to $|\widetilde{R}_{xx}|^2 \leq R_{xx}^2$, a simple upper bound on the magnitude of the complementary variance.

1.9 A brief survey of applications

The following is a brief survey of a few applications of improper and noncircular complex random signals, without any attempt at a complete bibliography. Our aim here is to indicate the breadth of applications spanning areas as diverse as communications and oceanography. We apologize in advance to authors whose work has not been included.

Currently, much of the research utilizing improper random signals concerns applications in **communications**. So what has sparked this recent interest in impropriety? There is an important result (cf. Results 2.15 and 2.16) stating that wide-sense stationary analytic signals, and also complex baseband representations of wide-sense stationary real bandpass signals, must be proper. On the other hand, nonstationary analytic signals and complex baseband representations of nonstationary real bandpass signals can be improper. In digital communications, thermal noise is assumed to be wide-sense stationary, but the transmitted signals are nonstationary (in fact, as we will discuss in Chapter 9, they are cyclostationary). This means that the analytic and complex baseband representations of thermal noise are always proper, whereas the analytic and complex baseband representations of the transmitted data signal are potentially improper.

In optimum maximum-likelihood detection, only the noise is assigned statistical properties. The likelihood function, which is the probability density function of the received signal conditioned on the transmitted signal, is proper because the noise is proper. Therefore, in maximum-likelihood detection, it is irrelevant whether or not the transmitted signal is improper. The communications research until the 1990s focused on optimal detection strategies based on maximum likelihood and was thus prone to overlook the potential impropriety of the data signal.

However, when more complicated scenarios are considered, such as multiuser or space–time communications, maximum likelihood is no longer a viable detection strategy because it is computationally too expensive. In these scenarios, detection is usually based on suboptimum algorithms that are less complex to implement. Many of these suboptimum detection algorithms do assign statistical properties to the signal. Hence, the potentially improper nature of signals must be taken into account when designing these detection algorithms. In mobile multiuser communications, this leads to a significantly improved tradeoff between spectral efficiency and power consumption. Important examples of digital modulation schemes that produce improper complex baseband signals are Binary Phase Shift Keying (BPSK), Pulse Amplitude Modulation (PAM), Gaussian Minimum Shift Keying (GMSK), Offset Quaternary Phase Shift Keying (OQPSK), and *baseband* (but not passband) Orthogonal Frequency Division Multiplexing (OFDM), which is commonly called Discrete Multitone (DMT). A small sample of papers addressing these issues is Yoon and Leib (1997), Gelli *et al.* (2000),

Lampe *et al.* (2002), Gerstacker *et al.* (2003), Nilsson *et al.* (2003), Napolitano and Tanda (2004), Witzke (2005), Buzzi *et al.* (2006), Jeon *et al.* (2006), Mirbagheri *et al.* (2006), Chevalier and Pipon (2006), Tauböck (2007), and Cacciapuoti *et al.* (2007).

Improper baseband communication signals can also arise due to imbalance between their in-phase and quadrature (I/Q) components. This can be caused by amplifier or receiver imperfections, or by communication channels that are not rotationally invariant. If the in-phase and quadrature components of a signal are subject to different gains, or if the phase-offset between them is not exactly 90°, even rotationally invariant modulation schemes such as Quaternary Phase Shift Keying (QPSK) become improper at the receiver. I/Q imbalance degrades the signal-to-noise ratio and thus bit error rate performance. Some papers proposing ways of compensating for I/Q imbalance in various types of communication systems include Anttila *et al.* (2008), Rykaczewski *et al.* (2008), and Zou *et al.* (2008). Morgan (2006) and Morgan and Madsen (2006) present techniques for wideband system identification when the system (e.g., a wideband wireless communication channel) is not rotationally invariant.

Array processing is the generic term applied to processing the output of an array of sensors. An important example of array processing is *beamforming*, which allows directional signal transmission or reception, either for radio or for sound waves. Besides the perhaps obvious applications in radar, sonar, and wireless communications, array processing is also employed in fields such as seismology, radio astronomy, and biomedicine. Often the aim is to estimate the direction of arrival (DOA) of one or more signals of interest impinging on a sensor array. If the signals of interest or the interference are improper (as they would be if they originated, e.g., from a BPSK transmitter), this can be exploited to achieve higher DOA resolution. Some papers addressing the adaptation of array-processing algorithms to improper signals include Charge *et al.* (2001), McWhorter and Schreier (2003), Haardt and Roemer (2004), Delmas (2004), Chevalier and Blin (2007), and Römer and Haardt (2009).

Another area where the theory of impropriety has led to important advances is **machine learning**. Much interest is centered around independent component analysis (ICA), which is a technique for separating a multivariate signal into additive components that are as independent as possible. A typical application of ICA is to functional magnetic resonance imaging (fMRI), which measures neural activity in the brain or spinal cord. The fMRI signal is naturally modeled as a complex signal (see Adali and Calhoun (2007) for an introduction to complex ICA applied to fMRI data). In the past, it has been assumed that the fMRI signal is proper, when in fact it isn't. As we will discuss in Section 3.5, impropriety is even a desirable property because it enables the separation of signals that would otherwise not be separable. The impropriety of the fMRI signal is now recognized, and techniques for ICA of complex signals that exploit impropriety have been proposed by DeLathauwer and DeMoor (2002), Eriksson and Koivunen (2006), Adali *et al.* (2008), Novey and Adali (2008a, 2008b), Li and Adali (2008), and Ollila and Koivunen (2009), amongst others.

Machine-learning techniques are often implemented using neural networks. Examples of neural-network implementations of signal processing algorithms utilizing complementary statistics are given by Goh and Mandic (2007a, 2007b). A comprehensive account of complex-valued nonlinear adaptive filters is provided in the research monograph by Mandic and Goh (2009).

The theory of impropriety has also found recent applications in **acoustics** (e.g., Rivet *et al.* (2007)) and **optics**. In optics, the standard correlation function is called the *phase-insensitive correlation*, and the complementary correlation function is called the *phase-sensitive correlation*. In a recent paper, Shapiro and Erkmen (2007) state that "Optical coherence theory for the complex envelopes of passband fields has been concerned, almost exclusively, with correlations that are all phase insensitive, despite decades of theoretical and experimental work on the generation and applications of light with phase-sensitive correlations. This paper begins the process of remedying that deficiency . . . ." More details on the work with phase-sensitive light can be found in Erkmen and Shapiro (2006) and the Ph.D. dissertation by Erkmen (2008). Maybe our discussion of polarization analysis in Sections 8.4 and 9.4 can make a contribution to this topic.

It is interesting to note that perhaps the first fields of research that recognized the importance of the complementary correlation and made consistent use of complex representations are **oceanography** and **geophysics**. The seminal paper by Mooers (1973), building upon prior work by Gonella (1972), presented techniques for the cross-spectrum analysis of bivariate time series by modeling them as complex-valued time series. Mooers realized that the information in the standard correlation function – which he called the *inner-cross correlation* – must be complemented by the complementary correlation function – which he called the *outer-cross correlation* – to fully model the second-order behavior of bivariate time series. He also recognized that the complex-valued description yields the desirable property that coherences are invariant under coordinate rotation. Moreover, testing for coherence between a pair of complex-valued time series is significantly simplified compared with the real-valued description. Early examples using this work are the analysis of wind fields by Burt *et al.* (1974) and the interpretation of ocean-current spectra by Calman (1978). Mooers' work is still frequently cited today, and it provides the basis for our discussion of polarization analysis in Section 8.4.

2 Introduction to complex random vectors and processes

This chapter lays the foundation for the remainder of the book by introducing key concepts and definitions for complex random vectors and processes. The structure of this chapter is as follows.

In Section 2.1, we relate descriptions of complex random vectors to the corresponding descriptions in terms of their real and imaginary parts. We will see that operations that are linear when applied to real and imaginary parts generally become widely linear (i.e., linear–conjugate-linear) when applied to complex vectors. We introduce a matrix algebra that enables a convenient description of these widely linear transformations.

Section 2.2 introduces a complete second-order statistical characterization of complex random vectors. The key finding is that the information in the standard, Hermitian, covariance matrix must be complemented by a second, complementary, covariance matrix. We establish the conditions that a pair of Hermitian and complementary covariance matrices must satisfy, and show what role the complementary covariance matrix plays in power and entropy.

In Section 2.3, we explain that probability distributions and densities for complex random vectors must be interpreted as joint distributions and densities of their real and imaginary parts. We present two important distributions: the complex multivariate Gaussian distribution and its generalization, the complex multivariate elliptical distribution. These distributions depend both on the Hermitian covariance matrix and on the complementary covariance matrix, and their well-known versions are obtained for the zero complementary covariance matrix. In Section 2.4, we establish that the Hermitian sample covariance and complementary sample covariance matrices are maximum-likelihood estimators and sufficient statistics for the Hermitian covariance and complementary covariance matrices. The sample covariance matrix is complex Wishart distributed.

In Section 2.5, we introduce characteristic and cumulant-generating functions, and use these to derive higher-order moments and cumulants of complex random vectors. We then discuss circular random vectors whose probability distributions are invariant under rotation. Circular random vectors may be regarded as an extension of proper random vectors, for which rotation invariance holds only for second-order moments.

In Section 2.6, we extend some of these ideas and concepts to continuous-time complex random processes. However, we treat only second-order properties of wide-sense stationary processes and widely linear shift-invariant filtering of them, postponing more advanced topics, such as higher-order statistics and circularity, to Chapter 8.

2.1 Connection between real and complex descriptions

Let Ω be the sample space of a random experiment, and $\mathbf{u}: \Omega \longrightarrow \mathbb{R}^n$ and $\mathbf{v}: \Omega \longrightarrow \mathbb{R}^n$ be two real random vectors defined on Ω. From $\mathbf{u}$ and $\mathbf{v}$ we construct three closely related vectors. The first is the *real composite* random vector $\mathbf{z}: \Omega \longrightarrow \mathbb{R}^{2n}$, obtained by stacking $\mathbf{u}$ on $\mathbf{v}$:

$$\mathbf{z} = \begin{bmatrix} \mathbf{u} \\ \mathbf{v} \end{bmatrix}. \tag{2.1}$$

The second vector is the *complex* random vector $\mathbf{x}: \Omega \longrightarrow \mathbb{C}^n$, obtained by composing $\mathbf{u}$ and $\mathbf{v}$ into its real and imaginary parts:

$$\mathbf{x} = \mathbf{u} + \mathrm{j}\mathbf{v}. \tag{2.2}$$

The third vector is the *complex augmented* random vector $\underline{\mathbf{x}}: \Omega \longrightarrow \mathbb{C}_*^{2n}$, obtained by stacking $\mathbf{x}$ on top of its complex conjugate $\mathbf{x}^*$:

$$\underline{\mathbf{x}} = \begin{bmatrix} \mathbf{x} \\ \mathbf{x}^* \end{bmatrix}. \tag{2.3}$$

The space of complex augmented vectors, whose bottom n entries are the complex conjugates of the top n entries, is denoted by $\mathbb{C}_*^{2n}$. Augmented vectors will always be underlined.

The complex augmented vector $\underline{\mathbf{x}}$ is related to the real composite vector $\mathbf{z}$ as

$$\underline{\mathbf{x}} = \mathbf{T}_n\mathbf{z} \iff \mathbf{z} = \tfrac{1}{2}\mathbf{T}_n^{\mathrm{H}}\underline{\mathbf{x}}, \tag{2.4}$$

where the real-to-complex transformation

$$\mathbf{T}_n = \begin{bmatrix} \mathbf{I} & \mathrm{j}\mathbf{I} \\ \mathbf{I} & -\mathrm{j}\mathbf{I} \end{bmatrix} \in \mathbb{C}^{2n\times 2n} \tag{2.5}$$

is unitary up to a factor of 2:

$$\mathbf{T}_n\mathbf{T}_n^{\mathrm{H}} = \mathbf{T}_n^{\mathrm{H}}\mathbf{T}_n = 2\mathbf{I}. \tag{2.6}$$

The complex augmented random vector $\underline{\mathbf{x}}: \Omega \longrightarrow \mathbb{C}_*^{2n}$ is obviously an equivalent redundant representation of $\mathbf{z}: \Omega \longrightarrow \mathbb{R}^{2n}$. But far from being a vice, this redundancy will be turned into an evident virtue as we develop the algebra of improper complex random vectors. Whenever the size of $\mathbf{T}_n$ is clear, we will drop the subscript n for economy.

2.1.1 Widely linear transformations

If a real linear transformation $\mathbf{M} \in \mathbb{R}^{2m\times 2n}$ is applied to the composite real vector $\mathbf{z}: \Omega \longrightarrow \mathbb{R}^{2n}$, it yields a real composite vector $\mathbf{w}: \Omega \longrightarrow \mathbb{R}^{2m}$,

$$\mathbf{w} = \begin{bmatrix} \mathbf{a} \\ \mathbf{b} \end{bmatrix} = \begin{bmatrix} \mathbf{M}_{11} & \mathbf{M}_{12} \\ \mathbf{M}_{21} & \mathbf{M}_{22} \end{bmatrix} \begin{bmatrix} \mathbf{u} \\ \mathbf{v} \end{bmatrix} = \mathbf{M}\mathbf{z}, \tag{2.7}$$

where $\mathbf{M}_{ij} \in \mathbb{R}^{m\times n}$. The augmented complex version of $\mathbf{w}$ is

$$\underline{\mathbf{y}} = \begin{bmatrix} \mathbf{y} \\ \mathbf{y}^* \end{bmatrix} = \mathbf{T}_m \begin{bmatrix} \mathbf{a} \\ \mathbf{b} \end{bmatrix} = \left(\tfrac{1}{2}\mathbf{T}_m \mathbf{M} \mathbf{T}_n^{\mathrm{H}}\right)(\mathbf{T}_n \mathbf{z}) = \underline{\mathbf{H}}\,\underline{\mathbf{x}}, \tag{2.8}$$

with $\mathbf{y} = \mathbf{a} + \mathrm{j}\mathbf{b}$. The matrix $\underline{\mathbf{H}} \in \mathbb{C}^{2m\times 2n}$ is called an *augmented matrix* because it satisfies a particular block pattern, where the southeast block is the conjugate of the northwest block, and the southwest block is the conjugate of the northeast block:

$$\underline{\mathbf{H}} = \tfrac{1}{2}\mathbf{T}_m \mathbf{M} \mathbf{T}_n^{\mathrm{H}} = \begin{bmatrix} \mathbf{H}_1 & \mathbf{H}_2 \\ \mathbf{H}_2^* & \mathbf{H}_1^* \end{bmatrix}, \tag{2.9}$$

$$\mathbf{H}_1 = \tfrac{1}{2}[\mathbf{M}_{11} + \mathbf{M}_{22} + \mathrm{j}(\mathbf{M}_{21} - \mathbf{M}_{12})],$$

$$\mathbf{H}_2 = \tfrac{1}{2}[\mathbf{M}_{11} - \mathbf{M}_{22} + \mathrm{j}(\mathbf{M}_{21} + \mathbf{M}_{12})].$$

Hence, $\underline{\mathbf{H}}$ is an augmented description of the *widely linear* or *linear–conjugate-linear* transformation[1]

$$\mathbf{y} = \mathbf{H}_1\mathbf{x} + \mathbf{H}_2\mathbf{x}^*. \tag{2.10}$$

Obviously, the set of complex linear transformations, $\mathbf{y} = \mathbf{H}_1\mathbf{x}$ with $\mathbf{H}_2 = \mathbf{0}$, is a subset of the set of widely linear transformations. A complex linear transformation (sometimes called *strictly linear* for emphasis) has the equivalent real representation

$$\begin{bmatrix} \mathbf{a} \\ \mathbf{b} \end{bmatrix} = \begin{bmatrix} \mathbf{M}_{11} & \mathbf{M}_{12} \\ -\mathbf{M}_{12} & \mathbf{M}_{11} \end{bmatrix} \begin{bmatrix} \mathbf{u} \\ \mathbf{v} \end{bmatrix}. \tag{2.11}$$

Even though the representation (2.9) contains some redundancy in that the northern blocks determine the southern blocks, it will prove to be very powerful in due course. For instance, it enables easy concatenation of widely linear transformations.

Let's recap. Linear transformations on $\mathbb{R}^{2n}$ are linear on $\mathbb{C}^n$ only if they have the particular structure (2.11). Otherwise, the equivalent operation on $\mathbb{C}^n$ is widely linear. Representing $\mathbb{R}$-linear operations as $\mathbb{C}$-widely linear operations often provides more insight. However, from a hardware implementation point of view, $\mathbb{R}$-linear transformations are usually preferable over $\mathbb{C}$-widely linear transformations because the former require fewer real operations (additions and multiplications) than the latter.

We will let $\mathcal{W}^{m\times n}$ denote the set of $2m \times 2n$ augmented matrices that satisfy the pattern (2.9). Elements of $\mathcal{W}^{m\times n}$ are always underlined. For $m = n$, the set $\mathcal{W}^{n\times n}$ is a *real* matrix algebra that is closed under addition, multiplication, inversion, and multiplication by a *real*, but not complex, scalar.

Example 2.1. Let us show that $\mathcal{W}^{n\times n}$ is closed under inversion. Using the matrix-inversion lemma (A1.42) in Appendix 1, the inverse of the block matrix

$$\underline{\mathbf{H}} = \begin{bmatrix} \mathbf{H}_1 & \mathbf{H}_2 \\ \mathbf{H}_2^* & \mathbf{H}_1^* \end{bmatrix}$$

can be calculated as

$$\underline{\mathbf{H}}^{-1} = \begin{bmatrix} (\mathbf{H}_1 - \mathbf{H}_2\mathbf{H}_1^{-*}\mathbf{H}_2^*)^{-1} & -(\mathbf{H}_1 - \mathbf{H}_2\mathbf{H}_1^{-*}\mathbf{H}_2^*)^{-1}\mathbf{H}_2\mathbf{H}_1^{-*} \\ -(\mathbf{H}_1^* - \mathbf{H}_2^*\mathbf{H}_1^{-1}\mathbf{H}_2)^{-1}\mathbf{H}_2^*\mathbf{H}_1^{-1} & (\mathbf{H}_1^* - \mathbf{H}_2^*\mathbf{H}_1^{-1}\mathbf{H}_2)^{-1} \end{bmatrix}$$

$$= \begin{bmatrix} \mathbf{N}_1 & \mathbf{N}_2 \\ \mathbf{N}_2^* & \mathbf{N}_1^* \end{bmatrix} = \underline{\mathbf{N}},$$

which has the block structure (2.9).

When working with the augmented matrix algebra $\mathcal{W}$ we often require that all factors in matrix factorizations represent widely linear transformations. If that is the case, we need to ensure that all factors satisfy the block pattern (2.9). If a factor $\mathbf{H} \notin \mathcal{W}$, it does not represent a widely linear transformation, since applying $\mathbf{H}$ to an augmented vector $\underline{\mathbf{x}}$ would yield a vector whose last n entries are not the conjugate of the first n entries.

Example 2.2. Consider the Cholesky factorization of a positive definite matrix $\underline{\mathbf{H}} \in \mathcal{W}^{n\times n}$ into a lower-triangular and upper-triangular factor. If we require that the Cholesky factors be widely linear transformations, then $\underline{\mathbf{H}} = \mathbf{X}\mathbf{X}^{\mathrm{H}}$ with $\mathbf{X}$ lower triangular and $\mathbf{X}^{\mathrm{H}}$ upper triangular will not work since generally $\mathbf{X} \notin \mathcal{W}^{n\times n}$. Instead, we determine the Cholesky factorization of the equivalent real matrix

$$\tfrac{1}{2}\mathbf{T}^{\mathrm{H}}\underline{\mathbf{H}}\mathbf{T} = \mathbf{L}\mathbf{L}^{\mathrm{H}}, \tag{2.12}$$

and transform $\mathbf{L}$ into the augmented complex notation as $\underline{\mathbf{N}} = \tfrac{1}{2}\mathbf{T}\mathbf{L}\mathbf{T}^{\mathrm{H}}$. Then

$$\underline{\mathbf{H}} = \underline{\mathbf{N}}\underline{\mathbf{N}}^{\mathrm{H}} \tag{2.13}$$

is the augmented complex representation of the Cholesky factorization of $\underline{\mathbf{H}}$ with $\underline{\mathbf{N}} \in \mathcal{W}^{n\times n}$. Note that (2.13) simply reexpresses (2.12) in the augmented algebra $\mathcal{W}^{n\times n}$ but $\underline{\mathbf{N}}$ itself is not generally lower triangular, and neither are its blocks $\mathbf{N}_1$ and $\mathbf{N}_2$. An exception is the block-diagonal case: if $\underline{\mathbf{H}}$ is block-diagonal, $\underline{\mathbf{N}}$ is block-diagonal and the diagonal block $\mathbf{N}_1$ (and $\mathbf{N}_1^*$) *is* lower triangular.

2.1.2 Inner products and quadratic forms

Consider the two $2n$-dimensional real composite vectors $\mathbf{w} = [\mathbf{a}^{\mathrm{T}}, \mathbf{b}^{\mathrm{T}}]^{\mathrm{T}}$ and $\mathbf{z} = [\mathbf{u}^{\mathrm{T}}, \mathbf{v}^{\mathrm{T}}]^{\mathrm{T}}$, the corresponding n-dimensional complex vectors $\mathbf{y} = \mathbf{a} + \mathrm{j}\mathbf{b}$ and $\mathbf{x} = \mathbf{u} + \mathrm{j}\mathbf{v}$, and their complex augmented descriptions $\underline{\mathbf{y}} = \mathbf{T}\mathbf{w}$ and $\underline{\mathbf{x}} = \mathbf{T}\mathbf{z}$. We may now relate the inner products defined on $\mathbb{R}^{2n}$, $\mathbb{C}_*^{2n}$, and $\mathbb{C}^n$ as

$$\mathbf{w}^{\mathrm{T}}\mathbf{z} = \tfrac{1}{2}\underline{\mathbf{y}}^{\mathrm{H}}\underline{\mathbf{x}} = \mathrm{Re}(\mathbf{y}^{\mathrm{H}}\mathbf{x}). \tag{2.14}$$

Thus, the usual inner product $\mathbf{w}^{\mathrm{T}}\mathbf{z}$ defined on $\mathbb{R}^{2n}$ equals (up to a factor of 1/2) the inner product $\underline{\mathbf{y}}^{\mathrm{H}}\underline{\mathbf{x}}$ defined on $\mathbb{C}_*^{2n}$, and also the real part of the usual inner product $\mathbf{y}^{\mathrm{H}}\mathbf{x}$ defined on $\mathbb{C}^n$. In this book, we will compute inner products on $\mathbb{C}_*^{2n}$ as well as inner products on $\mathbb{C}^n$. These inner products are discussed in more detail in Section 5.1.

Another common real-valued expression is the quadratic form $\mathbf{z}^{\mathrm{T}}\mathbf{M}\mathbf{z}$, which may be written as a (real-valued) *widely quadratic form* in $\mathbf{x}$:

$$\mathbf{z}^{\mathrm{T}}\mathbf{M}\mathbf{z} = \tfrac{1}{2}(\mathbf{z}^{\mathrm{T}}\mathbf{T}^{\mathrm{H}})\left(\tfrac{1}{2}\mathbf{T}\mathbf{M}\mathbf{T}^{\mathrm{H}}\right)(\mathbf{T}\mathbf{z}) = \tfrac{1}{2}\underline{\mathbf{x}}^{\mathrm{H}}\underline{\mathbf{H}}\,\underline{\mathbf{x}}. \tag{2.15}$$

The augmented matrix $\underline{\mathbf{H}}$ and the real matrix $\mathbf{M}$ are connected as before in (2.9). Thus, we obtain

$$\mathbf{z}^{\mathrm{T}}\mathbf{M}\mathbf{z} = \mathbf{x}^{\mathrm{H}}\mathbf{H}_1\mathbf{x} + \operatorname{Re}(\mathbf{x}^{\mathrm{H}}\mathbf{H}_2\mathbf{x}^*). \tag{2.16}$$

Widely quadratic forms are discussed in more detail in the context of widely quadratic estimation in Section 5.7.

2.2 Second-order statistical properties

In order to characterize the second-order statistical properties of $\mathbf{x} = \mathbf{u} + \mathrm{j}\mathbf{v}$, we consider the composite real random vector $\mathbf{z}$. Its mean vector is

$$\boldsymbol{\mu}_z = E\mathbf{z} = \begin{bmatrix} E\mathbf{u} \\ E\mathbf{v} \end{bmatrix} = \begin{bmatrix} \boldsymbol{\mu}_u \\ \boldsymbol{\mu}_v \end{bmatrix} \tag{2.17}$$

and its covariance matrix is

$$\mathbf{R}_{zz} = E(\mathbf{z} - \boldsymbol{\mu}_z)(\mathbf{z} - \boldsymbol{\mu}_z)^{\mathrm{T}} = \begin{bmatrix} \mathbf{R}_{uu} & \mathbf{R}_{uv} \\ \mathbf{R}_{uv}^{\mathrm{T}} & \mathbf{R}_{vv} \end{bmatrix} \tag{2.18}$$

with $\mathbf{R}_{uu} = E(\mathbf{u} - \boldsymbol{\mu}_u)(\mathbf{u} - \boldsymbol{\mu}_u)^{\mathrm{T}}$, $\mathbf{R}_{uv} = E(\mathbf{u} - \boldsymbol{\mu}_u)(\mathbf{v} - \boldsymbol{\mu}_v)^{\mathrm{T}}$, and $\mathbf{R}_{vv} = E(\mathbf{v} - \boldsymbol{\mu}_v)(\mathbf{v} - \boldsymbol{\mu}_v)^{\mathrm{T}}$. The *augmented mean vector* of $\mathbf{x}$ is

$$\underline{\boldsymbol{\mu}}_x = E\underline{\mathbf{x}} = \mathbf{T}\boldsymbol{\mu}_z = \begin{bmatrix} \boldsymbol{\mu}_x \\ \boldsymbol{\mu}_x^* \end{bmatrix} = \begin{bmatrix} \boldsymbol{\mu}_u + \mathrm{j}\boldsymbol{\mu}_v \\ \boldsymbol{\mu}_u - \mathrm{j}\boldsymbol{\mu}_v \end{bmatrix} \tag{2.19}$$

and the *augmented covariance matrix* of $\mathbf{x}$ is

$$\underline{\mathbf{R}}_{xx} = E(\underline{\mathbf{x}} - \underline{\boldsymbol{\mu}}_x)(\underline{\mathbf{x}} - \underline{\boldsymbol{\mu}}_x)^{\mathrm{H}} = \mathbf{T}\mathbf{R}_{zz}\mathbf{T}^{\mathrm{H}} = \begin{bmatrix} \mathbf{R}_{xx} & \widetilde{\mathbf{R}}_{xx} \\ \widetilde{\mathbf{R}}_{xx}^* & \mathbf{R}_{xx}^* \end{bmatrix} = \underline{\mathbf{R}}_{xx}^{\mathrm{H}}. \tag{2.20}$$

The augmented covariance matrix $\underline{\mathbf{R}}_{xx}$ is a member of the matrix algebra $\mathcal{W}^{n\times n}$. Its northwest block is the usual (Hermitian) covariance matrix

$$\mathbf{R}_{xx} = E(\mathbf{x} - \boldsymbol{\mu}_x)(\mathbf{x} - \boldsymbol{\mu}_x)^{\mathrm{H}} = \mathbf{R}_{uu} + \mathbf{R}_{vv} + \mathrm{j}(\mathbf{R}_{uv}^{T} - \mathbf{R}_{uv}) = \mathbf{R}_{xx}^{\mathrm{H}} \tag{2.21}$$

and its northeast block is the *complementary covariance matrix*

$$\widetilde{\mathbf{R}}_{xx} = E(\mathbf{x} - \boldsymbol{\mu}_x)(\mathbf{x} - \boldsymbol{\mu}_x)^{\mathrm{T}} = \mathbf{R}_{uu} - \mathbf{R}_{vv} + \mathrm{j}(\mathbf{R}_{uv}^{\mathrm{T}} + \mathbf{R}_{uv}) = \widetilde{\mathbf{R}}_{xx}^{\mathrm{T}}, \tag{2.22}$$

which uses a regular transpose rather than a Hermitian (conjugate) transpose. Other names for $\widetilde{\mathbf{R}}_{xx}$ include *pseudo-covariance matrix*, *conjugate covariance matrix*, and *relation matrix*.[2] It is important to note that *both* $\mathbf{R}_{xx}$ *and* $\widetilde{\mathbf{R}}_{xx}$ *are required for a complete second-order characterization of* $\mathbf{x}$. There is, however, an important special case in which the complementary covariance vanishes.

Definition 2.1. *If the complementary covariance matrix vanishes,* $\widetilde{\mathbf{R}}_{xx} = \mathbf{0}$, $\mathbf{x}$ *is called* proper, *otherwise* $\mathbf{x}$ *is called* improper.

The conditions for propriety on the covariance and cross-covariance of real and imaginary parts $\mathbf{u}$ and $\mathbf{v}$ are

$$\mathbf{R}_{uu} = \mathbf{R}_{vv}, \tag{2.23}$$

$$\mathbf{R}_{uv} = -\mathbf{R}_{uv}^{\mathrm{T}}. \tag{2.24}$$

The second condition, (2.24), requires $\mathbf{R}_{uv}$ to have zero diagonal elements, but its off-diagonal elements may be nonzero. When $x = u + \mathrm{j}v$ is scalar, then $R_{uv} = 0$ is necessary for propriety. If $\mathbf{x}$ is proper, its complementary covariance matrix $\widetilde{\mathbf{R}}_{xx} = \mathbf{0}$ and its Hermitian covariance matrix is

$$\mathbf{R}_{xx} = 2\mathbf{R}_{uu} - 2\mathrm{j}\mathbf{R}_{uv} = 2\mathbf{R}_{vv} + 2\mathrm{j}\mathbf{R}_{uv}^{\mathrm{T}}, \tag{2.25}$$

so its augmented covariance matrix $\underline{\mathbf{R}}_{xx}$ is block-diagonal. If complex x is proper and scalar, then $R_{xx} = 2R_{uu} = 2R_{vv}$. It is easy to see that propriety is preserved by strictly linear transformations, which are represented by block-diagonal augmented matrices.

2.2.1 Extending definitions from the real to the complex domain

A general question that has divided researchers is how to extend definitions from the real to the complex case. As an example, consider the definition of uncorrelatedness. Two real random vectors $\mathbf{z}$ and $\mathbf{w}$ are called uncorrelated if their cross-covariance matrix is zero:

$$\mathbf{R}_{zw} = E(\mathbf{z} - \boldsymbol{\mu}_z)(\mathbf{w} - \boldsymbol{\mu}_w)^{\mathrm{T}} = \mathbf{0}. \tag{2.26}$$

There are now two philosophies for a corresponding definition for two complex vectors $\mathbf{x}$ and $\mathbf{y}$. One could argue for the classical definition that calls $\mathbf{x}$ and $\mathbf{y}$ uncorrelated if

$$\mathbf{R}_{xy} = E(\mathbf{x} - \boldsymbol{\mu}_x)(\mathbf{y} - \boldsymbol{\mu}_y)^{\mathrm{H}} = \mathbf{0}. \tag{2.27}$$

This only considers the usual cross-covariance matrix but not the complementary cross-covariance matrix. On the other hand, if we consider $\mathbf{x}$ and $\mathbf{y}$ to be equivalent complex descriptions of real $\mathbf{z} = [\mathbf{u}^{\mathrm{T}}, \mathbf{v}^{\mathrm{T}}]^{\mathrm{T}}$ and $\mathbf{w} = [\mathbf{a}^{\mathrm{T}}, \mathbf{b}^{\mathrm{T}}]^{\mathrm{T}}$ as

$$\mathbf{x} = \mathbf{u} + \mathrm{j}\mathbf{v} \Leftrightarrow \underline{\mathbf{x}} = \mathbf{T}\mathbf{z}, \tag{2.28}$$

$$\mathbf{y} = \mathbf{a} + \mathrm{j}\mathbf{b} \Leftrightarrow \underline{\mathbf{y}} = \mathbf{T}\mathbf{w}, \tag{2.29}$$

then the condition equivalent to (2.26) is that the augmented cross-covariance matrix be zero:

$$\underline{\mathbf{R}}_{xy} = E(\underline{\mathbf{x}} - \underline{\boldsymbol{\mu}}_x)(\underline{\mathbf{y}} - \underline{\boldsymbol{\mu}}_y)^{\mathrm{H}} = \mathbf{0}. \tag{2.30}$$

Thus, (2.30) requires $\mathbf{R}_{xy} = \mathbf{0}$ but also $\widetilde{\mathbf{R}}_{xy} = \mathbf{0}$. If $\mathbf{x}$ and $\mathbf{y}$ have zero mean, an equivalent statement in the Hilbert space of second-order random variables is that the first definition (2.27) requires $\mathbf{x} \perp \mathbf{y}$, whereas the second definition (2.30) requires $\mathbf{x} \perp \mathbf{y}$ *and* $\mathbf{x} \perp \mathbf{y}^*$.

Which of these two definitions is more compelling? The first school of thought, which ignores complementary covariances in definitions, treats real and complex descriptions *differently*. This leads to unusual and counterintuitive results such as the following.

- Two uncorrelated Gaussian random vectors need not be independent (because their complementary covariance matrix need not be diagonal).
- A wide-sense stationary analytic signal may describe a nonstationary real signal (because the complementary covariance function of the analytic signal need not be shift-invariant).

We would like to avoid these displeasing results, and therefore *always* adhere to the following general principle in this book: *definitions and conditions derived for the real and complex domains must be equivalent.* This means that complementary covariances must be considered if they are nonzero.

2.2.2 Characterization of augmented covariance matrices

A matrix $\underline{\mathbf{R}}_{xx}$ is the augmented covariance matrix of a complex random vector $\mathbf{x}$ if and only if

(1) it satisfies the block pattern (2.9), i.e., $\underline{\mathbf{R}}_{xx} \in \mathcal{W}^{n\times n}$, and
(2) it is Hermitian and positive semidefinite.

Condition (1) needs to be enforced when factoring an augmented covariance matrix into factors that represent widely linear transformations. Then all factors must be members of $\mathcal{W}^{n\times n}$. A particularly important example is the eigenvalue decomposition of $\underline{\mathbf{R}}_{xx}$, which will be presented in Chapter 3 (Result 3.1).

Condition (2) leads to characterizations of the individual blocks of $\underline{\mathbf{R}}_{xx}$, i.e., the covariance matrix $\mathbf{R}_{xx}$ and the complementary covariance matrix $\widetilde{\mathbf{R}}_{xx}$.[3]

Result 2.1. *If* $\mathbf{R}_{xx}$ *is nonsingular, the following three conditions are necessary and sufficient for* $\mathbf{R}_{xx}$ *and* $\widetilde{\mathbf{R}}_{xx}$ *to be covariance and complementary covariance matrices of a complex random vector* $\mathbf{x}$.

1. *The covariance matrix* $\mathbf{R}_{xx}$ *is Hermitian and positive semidefinite.*
2. *The complementary covariance matrix is symmetric,* $\widetilde{\mathbf{R}}_{xx} = \widetilde{\mathbf{R}}_{xx}^{\mathrm{T}}$.
3. *The Schur complement of the augmented covariance matrix,* $\mathbf{R}_{xx} - \widetilde{\mathbf{R}}_{xx}\mathbf{R}_{xx}^{-*}\widetilde{\mathbf{R}}_{xx}^{*}$, *is positive semidefinite.*

   If $\mathbf{R}_{xx}$ *is singular, then condition 3 must be replaced with*

3a. *The generalized Schur complement of the augmented covariance matrix,* $\mathbf{R}_{xx} - \widetilde{\mathbf{R}}_{xx}(\mathbf{R}_{xx}^{*})^{\dagger}\widetilde{\mathbf{R}}_{xx}^{*}$, *where* $(\cdot)^{\dagger}$ *denotes the pseudo-inverse, is positive semidefinite.*
3b. *The null space of* $\mathbf{R}_{xx}$ *is contained in the null space of* $\widetilde{\mathbf{R}}_{xx}^{*}$.

This result says that (1) any given complex random vector $\mathbf{x}$ has covariance and complementary covariance matrices $\mathbf{R}_{xx}$ and $\widetilde{\mathbf{R}}_{xx}$ that satisfy conditions 1–3, and (2), given a pair of matrices $\mathbf{R}_{xx}$ and $\widetilde{\mathbf{R}}_{xx}$ that satisfies conditions 1–3, there exists a complex random vector $\mathbf{x}$ with covariance and complementary covariance matrices $\mathbf{R}_{xx}$ and $\widetilde{\mathbf{R}}_{xx}$. We will revisit the problem of characterizing augmented covariance matrices in Section 3.2.3, where we will develop an alternative point of view.

2.2.3 Power and entropy

The average power of complex $\mathbf{x}$ is defined as

$$P_x = \frac{1}{n}\sum_{i=1}^{n} E\,|x_i|^2. \tag{2.31}$$

It can be calculated as

$$P_x = \frac{1}{n}\operatorname{tr}\mathbf{R}_{xx} = \frac{1}{2n}\operatorname{tr}\underline{\mathbf{R}}_{xx} = \frac{1}{n}\operatorname{tr}\mathbf{R}_{zz}. \tag{2.32}$$

Hence, power is invariant under widely unitary transformation $\underline{\mathbf{U}}$, $\underline{\mathbf{U}}\,\underline{\mathbf{U}}^{\mathrm{H}} = \underline{\mathbf{U}}^{\mathrm{H}}\underline{\mathbf{U}} = \mathbf{I}$, and $\mathbf{x}' = \mathbf{U}_1\mathbf{x} + \mathbf{U}_2\mathbf{x}^*$ has the same power as $\mathbf{x}$.

The entropy of a complex random vector $\mathbf{x}$ is defined to be the entropy of the composite vector of real and imaginary parts $[\mathbf{u}^{\mathrm{T}}, \mathbf{v}^{\mathrm{T}}]^{\mathrm{T}} = \mathbf{z}$. If $\mathbf{u}$ and $\mathbf{v}$ are jointly Gaussian distributed, their differential entropy is

$$H(\mathbf{z}) = \tfrac{1}{2}\log\left[(2\pi\mathrm{e})^{2n}\det\mathbf{R}_{zz}\right]. \tag{2.33}$$

Since $\det\mathbf{T} = (-2\mathrm{j})^n$ and

$$\det\underline{\mathbf{R}}_{xx} = \det\mathbf{R}_{zz}|\det\mathbf{T}|^2 = 2^{2n}\det\mathbf{R}_{zz}, \tag{2.34}$$

we obtain the following result.

Result 2.2. *The differential entropy of a complex Gaussian random vector* $\mathbf{x}$ *with augmented covariance matrix* $\underline{\mathbf{R}}_{xx}$ *is*

$$H(\mathbf{x}) = \tfrac{1}{2}\log\left[(\pi\mathrm{e})^{2n}\det\underline{\mathbf{R}}_{xx}\right]. \tag{2.35}$$

The Fischer determinant inequality

$$\det\begin{bmatrix}\mathbf{R}_{xx} & \widetilde{\mathbf{R}}_{xx}\\ \widetilde{\mathbf{R}}^*_{xx} & \mathbf{R}^*_{xx}\end{bmatrix} \le \det\begin{bmatrix}\mathbf{R}_{xx} & \mathbf{0}\\ \mathbf{0} & \mathbf{R}^*_{xx}\end{bmatrix} \tag{2.36}$$

establishes the following classical result.

Result 2.3. *If* $\mathbf{x}$ *is Gaussian with given covariance matrix* $\mathbf{R}_{xx}$, *its differential entropy is maximized if* $\mathbf{x}$ *is proper. The differential entropy of a proper complex Gaussian* $\mathbf{x}$ *is*

$$H(\mathbf{x}) = \log[(\pi\mathrm{e})^n\det\mathbf{R}_{xx}]. \tag{2.37}$$

This formula for $H(\mathbf{x})$ is owed to $\det\underline{\mathbf{R}}_{xx} = \det^2\mathbf{R}_{xx}$ for block-diagonal $\underline{\mathbf{R}}_{xx}$. Like power, entropy is invariant under widely unitary transformation.

2.3 Probability distributions and densities

Rather than defining a complex random variable from first principles (where we would start with a probability measure on a sample space), we simply define a complex random variable $\mathbf{x}\colon \Omega \longrightarrow \mathbb{C}^n$ as $\mathbf{x} = \mathbf{u} + \mathrm{j}\mathbf{v}$, where $\mathbf{u}\colon \Omega \longrightarrow \mathbb{R}^n$ and $\mathbf{v}\colon \Omega \longrightarrow \mathbb{R}^n$ are a pair of real random variables. This pair $(\mathbf{u}, \mathbf{v})$ has the joint probability distribution

$$P(\mathbf{u}_0, \mathbf{v}_0) = \mathrm{Prob}(\mathbf{u} \le \mathbf{u}_0, \mathbf{v} \le \mathbf{v}_0) \tag{2.38}$$

and joint probability density function (pdf)

$$p(\mathbf{u}, \mathbf{v}) = \frac{\partial}{\partial \mathbf{u}} \frac{\partial}{\partial \mathbf{v}} P(\mathbf{u}, \mathbf{v}). \tag{2.39}$$

We will allow the use of Dirac delta functions in the pdf. When we write $P(\mathbf{x})$ or $p(\mathbf{x})$, we shall define this to mean

$$P(\mathbf{x}) = P(\mathbf{u} + \mathrm{j}\mathbf{v}) \triangleq P(\mathbf{u}, \mathbf{v}), \tag{2.40}$$

$$p(\mathbf{x}) = p(\mathbf{u} + \mathrm{j}\mathbf{v}) \triangleq p(\mathbf{u}, \mathbf{v}). \tag{2.41}$$

Thus, the probability distribution of a complex random vector is interpreted as the $2n$-dimensional joint distribution of its real and imaginary parts. The probability of $\mathbf{x}$ taking a value in the region $\mathcal{A} = \{\mathbf{u}_1 < \mathbf{u} \le \mathbf{u}_2;\ \mathbf{v}_1 < \mathbf{v} \le \mathbf{v}_2\}$ is thus

$$\mathrm{Prob}(\mathbf{x} \in \mathcal{A}) = \int_{\mathbf{v}_1}^{\mathbf{v}_2} \int_{\mathbf{u}_1}^{\mathbf{u}_2} p(\mathbf{x}) \mathrm{d}\mathbf{u}\, \mathrm{d}\mathbf{v}. \tag{2.42}$$

For a function $\mathbf{g}\colon \mathcal{D} \to \mathbb{C}^n$ whose domain $\mathcal{D}$ includes the range of $\mathbf{x}$, the expectation operator is defined accordingly as

$$\begin{aligned} E\{\mathbf{g}(\mathbf{x})\} &= E\{\mathrm{Re}[\mathbf{g}(\mathbf{x})]\} + \mathrm{j}E\{\mathrm{Im}[\mathbf{g}(\mathbf{x})]\} \\ &= \int_{\mathbb{R}^{2n}} \mathbf{g}(\mathbf{u} + \mathrm{j}\mathbf{v}) p(\mathbf{u} + \mathrm{j}\mathbf{v}) \mathrm{d}\mathbf{u}\, \mathrm{d}\mathbf{v}. \end{aligned} \tag{2.43}$$

In many cases, expressing $P(\mathbf{u}, \mathbf{v})$ or $p(\mathbf{u}, \mathbf{v})$ in terms of $\mathbf{x}$ requires the use of the complex conjugate $\mathbf{x}^*$. This has prompted many researchers to write $P(\mathbf{x}, \mathbf{x}^*)$ and $p(\mathbf{x}, \mathbf{x}^*)$, which raises the question of whether these are now the joint distribution and density for $\mathbf{x}$ and $\mathbf{x}^*$. This question is actually ill-posed since distributions and densities of complex random vectors are always interpreted in terms of (2.40) and (2.41) – whether we write this as $p(\mathbf{x})$ or $p(\mathbf{x}, \mathbf{x}^*)$ makes no difference. Nevertheless, the notation $p(\mathbf{x}, \mathbf{x}^*)$ does seem to carry potential for confusion since $\mathbf{x}$ perfectly determines $\mathbf{x}^*$, and vice versa. *It is not possible to assign densities to* $\mathbf{x}$ *and* $\mathbf{x}^*$ *independently*.

The advantage of expressing a pdf in terms of complex $\mathbf{x}$ lies not in the fact that $\mathrm{Prob}(\mathbf{x} \in \mathcal{A})$ becomes easier to evaluate – that is obviously not the case. However, direct calculations of $\mathrm{Prob}(\mathbf{x} \in \mathcal{A})$ via (2.42) are rare. In most practical cases, e.g., maximum-likelihood or minimum mean-squared error estimation, we can work directly with $p(\mathbf{x})$ since it contains all relevant information, conveniently parameterized in terms of the statistical properties of complex $\mathbf{x}$.

We will now take a look at two important complex distributions: the multivariate Gaussian distribution and its generalization, the multivariate elliptical distribution. We will be particularly interested in expressing these pdfs in terms of covariance and complementary covariance matrices.

2.3.1 Complex Gaussian distribution

In order to derive the general complex multivariate Gaussian pdf (proper or improper), we begin with the Gaussian pdf of the composite vector of real and imaginary parts $[\mathbf{u}^{\mathrm{T}}, \mathbf{v}^{\mathrm{T}}]^{\mathrm{T}} = \mathbf{z}: \Omega \longrightarrow \mathbb{R}^{2n}$:

$$p(\mathbf{z}) = \frac{1}{(2\pi)^{2n/2} \det^{1/2} \mathbf{R}_{zz}} \exp\left\{-\tfrac{1}{2}(\mathbf{z} - \boldsymbol{\mu}_z)^{\mathrm{T}} \mathbf{R}_{zz}^{-1} (\mathbf{z} - \boldsymbol{\mu}_z)\right\}. \tag{2.44}$$

Using

$$\mathbf{R}_{zz}^{-1} = \mathbf{T}^{\mathrm{H}} \underline{\mathbf{R}}_{xx}^{-1} \mathbf{T}, \tag{2.45}$$

$$\det \underline{\mathbf{R}}_{xx} = 2^{2n} \det \mathbf{R}_{zz} \tag{2.46}$$

in (2.44), we obtain

$$p(\mathbf{z}) = \frac{1}{\pi^n \det^{1/2} \underline{\mathbf{R}}_{xx}} \exp\left\{-\tfrac{1}{2}(\mathbf{z} - \boldsymbol{\mu}_z)^{\mathrm{T}} \mathbf{T}^{\mathrm{H}} \underline{\mathbf{R}}_{xx}^{-1} \mathbf{T} (\mathbf{z} - \boldsymbol{\mu}_z)\right\}. \tag{2.47}$$

With $\underline{\mathbf{x}} = \mathbf{T}\mathbf{z}$, we are now in a position to state the following.

Result 2.4. *The general pdf of a complex Gaussian random vector* $\mathbf{x}: \Omega \longrightarrow \mathbb{C}^n$ *is*

$$p(\mathbf{x}) = \frac{1}{\pi^n \det^{1/2} \underline{\mathbf{R}}_{xx}} \exp\left\{-\tfrac{1}{2}(\underline{\mathbf{x}} - \underline{\boldsymbol{\mu}}_x)^{\mathrm{H}} \underline{\mathbf{R}}_{xx}^{-1} (\underline{\mathbf{x}} - \underline{\boldsymbol{\mu}}_x)\right\}. \tag{2.48}$$

This pdf algebraically depends on $\underline{\mathbf{x}}$, i.e., $\mathbf{x}$ and $\mathbf{x}^*$, but is interpreted as the joint pdf of $\mathbf{u}$ and $\mathbf{v}$. It may be used for proper or improper $\mathbf{x}$. In the past, the term "complex Gaussian distribution" often implicitly assumed propriety. Therefore, some researchers call an improper complex Gaussian random vector "generalized complex Gaussian."[4] The simplification that occurs in the proper case, where $\widetilde{\mathbf{R}}_{xx} = \mathbf{0}$ and $\underline{\mathbf{R}}_{xx}$ is block-diagonal, is obvious and leads to the following classical result.

Result 2.5. *The pdf of a complex* proper *Gaussian random vector* $\mathbf{x}: \Omega \longrightarrow \mathbb{C}^n$ *is*

$$p(\mathbf{x}) = \frac{1}{\pi^n \det \mathbf{R}_{xx}} \exp\left\{-(\mathbf{x} - \boldsymbol{\mu}_x)^{\mathrm{H}} \mathbf{R}_{xx}^{-1} (\mathbf{x} - \boldsymbol{\mu}_x)\right\}. \tag{2.49}$$

Let's go back to the general Gaussian pdf in Result 2.4. As in (A1.38) of Appendix 1, $\underline{\mathbf{R}}_{xx}^{-1}$ may be factored as

$$\underline{\mathbf{R}}_{xx}^{-1} = \begin{bmatrix} \mathbf{R}_{xx} & \widetilde{\mathbf{R}}_{xx} \\ \widetilde{\mathbf{R}}_{xx}^* & \mathbf{R}_{xx}^* \end{bmatrix}^{-1} = \begin{bmatrix} \mathbf{I} & \mathbf{0} \\ -\mathbf{W}^{\mathrm{H}} & \mathbf{I} \end{bmatrix} \begin{bmatrix} \mathbf{P}^{-1} & \mathbf{0} \\ \mathbf{0} & \mathbf{R}_{xx}^{-*} \end{bmatrix} \begin{bmatrix} \mathbf{I} & -\mathbf{W} \\ \mathbf{0} & \mathbf{I} \end{bmatrix}, \tag{2.50}$$

where $\mathbf{P} = \mathbf{R}_{xx} - \widetilde{\mathbf{R}}_{xx} \mathbf{R}_{xx}^{-*} \widetilde{\mathbf{R}}_{xx}^*$ is the Schur complement of $\mathbf{R}_{xx}^*$ within $\underline{\mathbf{R}}_{xx}$. Furthermore, $\mathbf{W} = \widetilde{\mathbf{R}}_{xx} \mathbf{R}_{xx}^{-*}$ produces the *linear* minimum mean-squared error (LMMSE) estimate of

$\mathbf{x}$ from $\mathbf{x}^*$ as

$$\hat{\mathbf{x}} = \mathbf{W}(\mathbf{x} - \boldsymbol{\mu}_x)^* + \boldsymbol{\mu}_x, \tag{2.51}$$

and $\operatorname{tr}\mathbf{P} = E\|\mathbf{x} - \hat{\mathbf{x}}\|^2$ is the corresponding LMMSE. From (A1.3) we find $\det \underline{\mathbf{R}}_{xx} = \det \mathbf{R}_{xx}^* \det \mathbf{P} = \det \mathbf{R}_{xx} \det \mathbf{P}$, and, using (2.50), we may then factor the improper pdf $p(\mathbf{x})$ as

$$\begin{aligned} p(\mathbf{x}) = {} & \frac{1}{\pi^n} \times \frac{1}{\det^{1/2}\mathbf{R}_{xx}} \exp\left\{-\tfrac{1}{2}(\mathbf{x} - \boldsymbol{\mu}_x)^{\mathrm{H}}\mathbf{R}_{xx}^{-1}(\mathbf{x} - \boldsymbol{\mu}_x)\right\} \\ & \times \frac{1}{\det^{1/2}\mathbf{P}} \exp\left\{-\tfrac{1}{2}(\mathbf{x} - \hat{\mathbf{x}})^{\mathrm{H}}\mathbf{P}^{-1}(\mathbf{x} - \hat{\mathbf{x}})\right\}. \end{aligned} \tag{2.52}$$

This expresses the improper Gaussian pdf $p(\mathbf{x})$ in terms of two factors: the first factor involves only $\mathbf{x}$, its mean $\boldsymbol{\mu}_x$, and its covariance matrix $\mathbf{R}_{xx}$; and the second factor involves only the prediction error $\mathbf{x} - \hat{\mathbf{x}}$ and its covariance matrix $\mathbf{P}$. These two factors are "*almost*" proper Gaussian pdfs, albeit with incorrect normalization constants and a factor of $1/2$ in the quadratic form.

In Section 1.6.1, we found that the *real* bivariate Gaussian pdf $p(u, v)$ in (1.48) does indeed factor into a Gaussian pdf for the prediction error $u - \hat{u}$ and a Gaussian pdf for v. Importantly, in the real case, the error $u - \hat{u}$ and v are independent. The difference in the complex case is that, although $\mathbf{x} - \hat{\mathbf{x}}$ and $\mathbf{x}^*$ are uncorrelated, they cannot be independent because $\mathbf{x}^*$ perfectly determines $\mathbf{x}$ (through complex conjugation). If $\mathbf{x}$ is proper, then $\hat{\mathbf{x}} = \boldsymbol{\mu}_x$, so that $\mathbf{W} = \mathbf{0}$ and $\mathbf{P} = \mathbf{R}_{xx}$, and the two factors in (2.52) are identical. This makes the factor of $1/2$ in the quadratic form disappear.

By employing the Woodbury identity (cf. (A1.43) in Appendix 1)

$$\mathbf{P}^{-1} = \mathbf{R}_{xx}^{-1} + \mathbf{W}^{\mathrm{T}}\mathbf{P}^{-*}\mathbf{W}^* \tag{2.53}$$

and

$$\det \mathbf{P} = \det \mathbf{R}_{xx} \det(\mathbf{I} - \mathbf{W}\mathbf{W}^*), \tag{2.54}$$

we may find the following alternative expressions for $p(\mathbf{x})$:

$$\begin{aligned} p(\mathbf{x}) = {} & \frac{\det^{1/2}(\mathbf{I} - \mathbf{W}\mathbf{W}^*)}{\pi^n \det \mathbf{P}} \\ & \times \exp\left\{-(\mathbf{x} - \boldsymbol{\mu}_x)^{\mathrm{H}}\mathbf{P}^{-1}(\mathbf{x} - \boldsymbol{\mu}_x) + \operatorname{Re}\left((\mathbf{x} - \boldsymbol{\mu}_x)^{\mathrm{T}}\mathbf{P}^{-*}\mathbf{W}^*(\mathbf{x} - \boldsymbol{\mu}_x)\right)\right\}, \end{aligned} \tag{2.55}$$

$$\begin{aligned} p(\mathbf{x}) = {} & \frac{1}{\pi^n(\det \mathbf{R}_{xx} \det \mathbf{P})^{1/2}} \exp\left\{-(\mathbf{x} - \boldsymbol{\mu}_x)^{\mathrm{H}}\mathbf{R}_{xx}^{-1}(\mathbf{x} - \boldsymbol{\mu}_x)\right\} \\ & \times \exp\left\{-(\mathbf{x} - \boldsymbol{\mu}_x)^{\mathrm{H}}\mathbf{W}^{\mathrm{T}}\mathbf{P}^{-*}\mathbf{W}^*(\mathbf{x} - \boldsymbol{\mu}_x) + \operatorname{Re}\left((\mathbf{x} - \boldsymbol{\mu}_x)^{\mathrm{T}}\mathbf{P}^{-*}\mathbf{W}^*(\mathbf{x} - \boldsymbol{\mu}_x)\right)\right\}. \end{aligned} \tag{2.56}$$

Since the complex Gaussian pdf is simply a convenient way of expressing the joint pdf of real and imaginary parts, many results valid for the real case translate straightforwardly to the complex case. In particular, a linear or widely linear transformation of a Gaussian random vector (proper or improper) is again Gaussian (proper or improper). We note, however, that a *widely linear* transformation of a proper Gaussian will generally produce an *improper* Gaussian, and a widely linear transformation of an improper Gaussian may produce a proper Gaussian.

2.3.2 Conditional complex Gaussian distribution

If two real random vectors $\mathbf{z} = [\mathbf{u}^{\mathrm{T}}, \mathbf{v}^{\mathrm{T}}]^{\mathrm{T}} : \Omega \longrightarrow \mathbb{R}^{2n}$ and $\mathbf{w} = [\mathbf{a}^{\mathrm{T}}, \mathbf{b}^{\mathrm{T}}]^{\mathrm{T}} : \Omega \longrightarrow \mathbb{R}^{2m}$ are jointly Gaussian, then the conditional density for $\mathbf{z}$ given $\mathbf{w}$ is Gaussian,

$$p(\mathbf{z}|\mathbf{w}) = \frac{1}{(2\pi)^{2n/2} \det^{1/2} \mathbf{R}_{zz|w}} \exp\left\{-\tfrac{1}{2}(\mathbf{z} - \boldsymbol{\mu}_{z|w})^{\mathrm{T}} \mathbf{R}_{zz|w}^{-1} (\mathbf{z} - \boldsymbol{\mu}_{z|w})\right\} \tag{2.57}$$

with conditional mean vector

$$\boldsymbol{\mu}_{z|w} = \boldsymbol{\mu}_z + \mathbf{R}_{zw}\mathbf{R}_{ww}^{-1}(\mathbf{w} - \boldsymbol{\mu}_w) \tag{2.58}$$

and conditional covariance matrix

$$\mathbf{R}_{zz|w} = \mathbf{R}_{zz} - \mathbf{R}_{zw}\mathbf{R}_{ww}^{-1}\mathbf{R}_{zw}^{\mathrm{T}}. \tag{2.59}$$

This result easily generalizes to the complex case. Let $\mathbf{x} = \mathbf{u} + \mathrm{j}\mathbf{v}: \Omega \longrightarrow \mathbb{C}^n$ and $\mathbf{y} = \mathbf{a} + \mathrm{j}\mathbf{b}: \Omega \longrightarrow \mathbb{C}^m$, and $\underline{\mathbf{y}} = \mathbf{T}\mathbf{w}$. Then the augmented conditional mean vector is

$$\begin{aligned} \underline{\boldsymbol{\mu}}_{x|y} = \begin{bmatrix} \boldsymbol{\mu}_{x|y} \\ \boldsymbol{\mu}_{x|y}^* \end{bmatrix} &= \mathbf{T}\boldsymbol{\mu}_{z|w} = \mathbf{T}\boldsymbol{\mu}_z + (\mathbf{T}\mathbf{R}_{zw}\mathbf{T}^{\mathrm{H}})(\mathbf{T}^{-\mathrm{H}}\mathbf{R}_{ww}^{-1}\mathbf{T}^{-1})\mathbf{T}(\mathbf{w} - \boldsymbol{\mu}_w) \\ &= \underline{\boldsymbol{\mu}}_x + \underline{\mathbf{R}}_{xy}\underline{\mathbf{R}}_{yy}^{-1}(\underline{\mathbf{y}} - \underline{\boldsymbol{\mu}}_y). \end{aligned} \tag{2.60}$$

The augmented conditional covariance matrix is

$$\begin{aligned} \underline{\mathbf{R}}_{xx|y} = \mathbf{T}\mathbf{R}_{zz|w}\mathbf{T}^{\mathrm{H}} &= \mathbf{T}\mathbf{R}_{zz}\mathbf{T}^{\mathrm{H}} - (\mathbf{T}\mathbf{R}_{zw}\mathbf{T}^{\mathrm{H}})(\mathbf{T}^{-\mathrm{H}}\mathbf{R}_{ww}^{-1}\mathbf{T}^{-1})(\mathbf{T}\mathbf{R}_{zw}^{\mathrm{T}}\mathbf{T}^{\mathrm{H}}) \\ &= \underline{\mathbf{R}}_{xx} - \underline{\mathbf{R}}_{xy}\underline{\mathbf{R}}_{yy}^{-1}\underline{\mathbf{R}}_{xy}^{\mathrm{H}}. \end{aligned} \tag{2.61}$$

Therefore, the conditional pdf takes the general form

$$p(\mathbf{x}|\mathbf{y}) = \frac{1}{\pi^n \det^{1/2} \underline{\mathbf{R}}_{xx|y}} \exp\left\{-\tfrac{1}{2}(\underline{\mathbf{x}} - \underline{\boldsymbol{\mu}}_{x|y})^{\mathrm{H}} \underline{\mathbf{R}}_{xx|y}^{-1} (\underline{\mathbf{x}} - \underline{\boldsymbol{\mu}}_{x|y})\right\}. \tag{2.62}$$

Using the matrix inversion lemma for $\underline{\mathbf{R}}_{xx|y}^{-1}$, it is possible to derive an expression that explicitly shows the dependence of $p(\mathbf{x}|\mathbf{y})$ on $\mathbf{y}$ and $\mathbf{y}^*$. However, we shall postpone this until our discussion of widely linear estimation in Section 5.4.

Definition 2.2. *Two complex random vectors* $\mathbf{x}$ *and* $\mathbf{y}$ *are called* jointly proper *if the composite vector* $[\mathbf{x}^{\mathrm{T}}, \mathbf{y}^{\mathrm{T}}]^{\mathrm{T}}$ *is proper. This means they must be individually proper,* $\widetilde{\mathbf{R}}_{xx} = \mathbf{0}$ *and* $\widetilde{\mathbf{R}}_{yy} = \mathbf{0}$*, and also cross-proper,* $\widetilde{\mathbf{R}}_{xy} = \mathbf{0}$*.*

If $\mathbf{x}$ and $\mathbf{y}$ are jointly proper, the conditional Gaussian density for $\mathbf{x}$ given $\mathbf{y}$ is

$$p(\mathbf{x}|\mathbf{y}) = \frac{1}{\pi^n \det \mathbf{R}_{xx|y}} \exp\left\{-(\mathbf{x} - \boldsymbol{\mu}_{x|y})^{\mathrm{H}} \mathbf{R}_{xx|y}^{-1} (\mathbf{x} - \boldsymbol{\mu}_{x|y})\right\} \tag{2.63}$$

with mean

$$\boldsymbol{\mu}_{x|y} = \boldsymbol{\mu}_x + \mathbf{R}_{xy}\mathbf{R}_{yy}^{-1}(\mathbf{y} - \boldsymbol{\mu}_y) \tag{2.64}$$

and covariance matrix

$$\mathbf{R}_{xx|y} = \mathbf{R}_{xx} - \mathbf{R}_{xy}\mathbf{R}_{yy}^{-1}\mathbf{R}_{xy}^{\mathrm{H}}. \tag{2.65}$$

2.3.3 Scalar complex Gaussian distribution

The scalar complex Gaussian distribution is important enough to revisit in detail. Consider a *zero-mean* scalar Gaussian random variable $x = u + \mathrm{j}v$ with variance $R_{xx} = E|x|^2$ and complementary variance $\widetilde{R}_{xx} = E\,x^2 = \rho R_{xx}$ with $|\rho| < 1$. The complex correlation coefficient ρ between x and x^* is a measure for the degree of impropriety of x. From Result 2.4, the pdf of x is

$$p(x) = \frac{1}{\pi R_{xx}\sqrt{1-|\rho|^2}} \exp\left\{-\frac{|x|^2 - \mathrm{Re}(\rho x^{*2})}{R_{xx}(1-|\rho|^2)}\right\}. \tag{2.66}$$

Let R_{uu} and R_{vv} be the variances of the real part u and imaginary part v, and R_{uv} their cross-covariance. The correlation coefficient between u and v is

$$\rho_{uv} = \frac{R_{uv}}{\sqrt{R_{uu}}\sqrt{R_{vv}}}. \tag{2.67}$$

From (2.21) and (2.22) we know that

$$R_{uu} + R_{vv} = R_{xx}, \tag{2.68}$$

$$R_{uu} - R_{vv} + 2\mathrm{j}\sqrt{R_{uu}}\sqrt{R_{vv}}\rho_{uv} = \rho R_{xx}. \tag{2.69}$$

So the complementary variance ρR_{xx} carries information about the variance mismatch $R_{uu} - R_{vv}$ in its real part and about the correlation between u and v in its imaginary part. There are now four different cases.

1. If u and v have identical variances, $R_{uu} = R_{vv} = R_{xx}/2$, and are independent, $\rho_{uv} = 0$, then x is proper, i.e., $\rho = 0$. Its pdf is

$$p(x) = \frac{1}{\pi}\mathrm{e}^{-|x|^2}. \tag{2.70}$$

2. If u and v have different variances, $R_{uu} \neq R_{vv}$, but u and v are still independent, $\rho_{uv} = 0$, then ρ is *real*, $\rho = (R_{uu} - R_{vv})/R_{xx}$, and x is improper.
3. If u and v have identical variances, $R_{uu} = R_{vv} = R_{xx}/2$, but u and v are correlated, $\rho_{uv} \neq 0$, then ρ is *purely imaginary*, $\rho = \mathrm{j}\rho_{uv}$, and x is improper.
4. We can combine these two possible sources of impropriety so that u and v have different variances, $R_{uu} \neq R_{vv}$, and are correlated, $\rho_{uv} \neq 0$. Then ρ is generally complex.

With $x = r\mathrm{e}^{\mathrm{j}\theta}$ and $\rho = |\rho|\mathrm{e}^{\mathrm{j}\psi}$, we see that the pdf $p(x)$ is constant on the contour (or level curve) $r^2[1 - |\rho|\cos(2\theta - \psi)] = K^2$. This contour is an ellipse, and r is

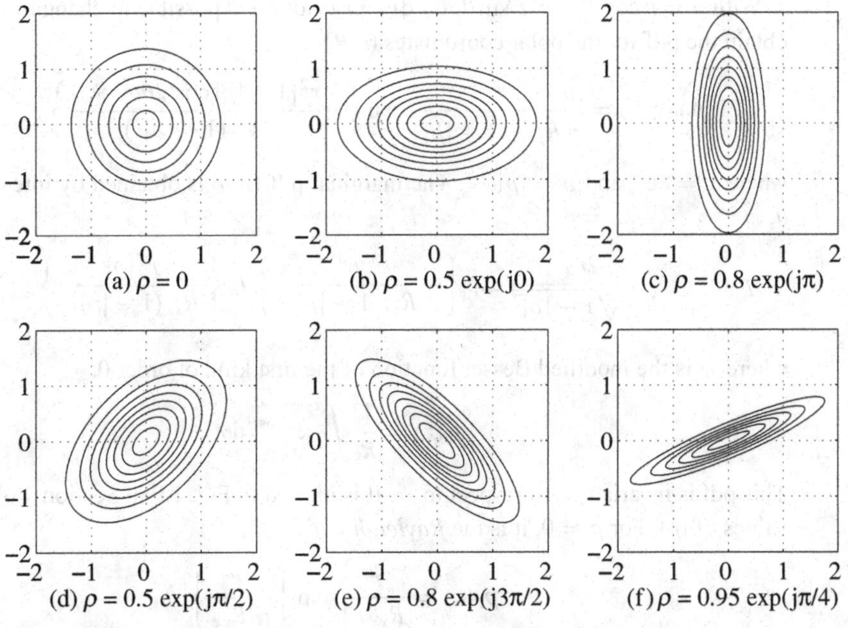

Figure 2.1 Probability-density contours of complex Gaussian random variables with different ρ.

maximum when $\cos(2\theta - \psi)$ is minimum. This establishes that the ellipse orientation (the angle between the u-axis and the major ellipse axis) is $\theta = \psi/2$, which is half the angle of the complex correlation coefficient $\rho = |\rho|e^{j\psi}$. It is also not difficult to show (see Ollila (2008)) that $|\rho|$ is the square of the ellipse eccentricity. This is compelling evidence for the usefulness of the complex description. The real description – in terms of R_{uu}, R_{vv}, and the correlation coefficient ρ_{uv} between u and v – is not nearly as insightful.

Example 2.3. Figure 2.1 shows contours of constant probability density for cases 1–4 listed above. In plot (a), we see the proper case with $\rho = 0$, which exhibits circular contour lines. All remaining plots are improper, with elliptical contour lines. We can make two observations. First, increasing the degree of impropriety of the signal by increasing $|\rho|$ leads to ellipses with greater eccentricity. Secondly, the angle of the ellipse orientation is half the angle of ρ, as proved above.

In plots (b) and (c), we have case 2: u and v have different variances but are still independent. In this situation, the ellipse orientation is either 0° or 90°, depending on whether u or v has greater variance. Plots (d) and (e) show case 3: u and v have the same variance but are now correlated. In this situation, the ellipse orientation is either 45° or 135°. The general case, case 4, is depicted in plot (f). Now the ellipse can have an arbitrary orientation $\psi/2$, which is controlled by the angle of $\rho = |\rho|e^{j\psi}$.

With $u = r\cos\theta$, $v = r\sin\theta$, $du\,dv = r\,dr\,d\theta$, it is possible to change variables and obtain the pdf for the polar coordinates (r, θ)

$$p_{r\theta}(r,\theta) = \frac{r}{\pi R_{xx}\sqrt{1-|\rho|^2}} \exp\left\{-\frac{r^2[1-|\rho|\cos(2\theta-\psi)]}{R_{xx}(1-|\rho|^2)}\right\}, \tag{2.71}$$

where $x = re^{j\theta}$ and $\rho = |\rho|e^{j\psi}$. The marginal pdf for r is obtained by integrating over θ,

$$p_r(r) = \frac{2r}{R_{xx}\sqrt{1-|\rho|^2}} \exp\left\{-\frac{r^2}{R_{xx}(1-|\rho|^2)}\right\} I_0\left\{\frac{r^2|\rho|}{R_{xx}(1-|\rho|^2)}\right\}, \quad r > 0, \tag{2.72}$$

where I_0 is the modified Bessel function of the first kind of order 0:

$$I_0(z) = \frac{1}{\pi}\int_0^{\pi} e^{z\cos\theta}\,d\theta. \tag{2.73}$$

This pdf is invariant with respect to ψ. It is plotted in Fig. 1.9 in Section 1.6 for several values of $|\rho|$. For $\rho = 0$, it is the *Rayleigh pdf*

$$p_r(r) = \frac{r}{R_{xx}/2} \exp\left\{-\frac{r^2}{R_{xx}}\right\}. \tag{2.74}$$

This suggests that we call $p_r(r)$ in (2.72) the *improper Rayleigh pdf*.[5] Integrating $p_{r\theta}(r,\theta)$ over r yields the marginal pdf for θ:

$$p_\theta(\theta) = \frac{\sqrt{1-|\rho|^2}}{2\pi[1-|\rho|\cos(2\theta-\psi)]}, \quad -\pi < \theta \le \pi. \tag{2.75}$$

This pdf is shown in Fig. 1.9 for several values of $|\rho|$. For $\rho = 0$, the pdf is uniform. For larger $|\rho|$, the pdf develops two peaks at $\theta = \psi/2$ and $\theta = \psi/2 - \pi$.

If $|\rho| = 1$, x is a singular random variable because the support of the pdf $p(x)$ collapses to a line in the complex plane and the pdf (2.66) must be expressed using a Dirac δ-function. This case is called *maximally improper* (terms used by other researchers are *rectilinear* and *strict-sense noncircular*). If x is maximally improper, we can express it as $x = ae^{j\psi/2} = a\cos(\psi/2) + ja\sin(\psi/2)$, where a is a *real* Gaussian random variable with zero mean and variance R_{xx}. Hence, the radius-squared r^2 of $x/\sqrt{R_{xx}}$ is χ^2-distributed with one degree of freedom, and the angle θ takes on values $\psi/2$ and $\psi/2 - \pi$, each with probability equal to $1/2$.

2.3.4 Complex elliptical distribution

A generalization of the Gaussian distribution, which has found some interesting applications in communications, is the family of elliptical distributions. We could proceed as in the Gaussian case by starting with the pdf of an elliptical distribution for a composite real random vector $\mathbf{z}$ and then deriving an expression in terms of complex $\mathbf{x}$. Instead, we will directly modify the improper Gaussian pdf (2.48) by replacing the exponential function with a nonnegative function g: $[0,\infty) \longrightarrow [0,\infty)$, called the *pdf generator*, that satisfies $\int_0^\infty t^{n-1}g(t)dt < \infty$. This necessitates two changes: First, since we do not yet know the second-order moments of $\mathbf{x}$, the matrix used in the expression for the pdf may no longer be the augmented covariance matrix of $\mathbf{x}$. Hence, instead of $\underline{\mathbf{R}}_{xx}$, we use the *augmented generating matrix*

$$\underline{\mathbf{H}}_{xx} = \begin{bmatrix} \mathbf{H}_{xx} & \widetilde{\mathbf{H}}_{xx} \\ \widetilde{\mathbf{H}}^*_{xx} & \mathbf{H}^*_{xx} \end{bmatrix}$$

to denote an arbitrary augmented positive definite matrix of size $2n \times 2n$. Secondly, we need to introduce a normalizing constant c_n to ensure that $p(\mathbf{x})$ is a valid pdf that integrates to 1.

We now state the general form of the complex elliptical pdf, which is a straightforward generalization of the real elliptical pdf, due to Ollila and Koivunen (2004).

Definition 2.3. *The pdf of a complex elliptical random vector* $\mathbf{x}\colon \Omega \longrightarrow \mathbb{C}^n$ *is*

$$p(\mathbf{x}) = \frac{c_n}{\det^{1/2}\underline{\mathbf{H}}_{xx}} g\left\{(\underline{\mathbf{x}} - \underline{\boldsymbol{\mu}}_x)^{\mathrm{H}}\underline{\mathbf{H}}_{xx}^{-1}(\underline{\mathbf{x}} - \underline{\boldsymbol{\mu}}_x)\right\}. \tag{2.76}$$

The normalizing constant c_n *is given by*

$$c_n = \frac{(n-1)!}{\pi^n \int_0^\infty t^{2n-1} g(t^2)\mathrm{d}t}. \tag{2.77}$$

If the mean exists, the parameter $\underline{\boldsymbol{\mu}}_x$ in the pdf (2.76) is the augmented mean of $\mathbf{x}$. This indicates that the mean is independent of the choice of pdf generator. However, there are distributions for which some or all moments are undefined. For instance, none of the moments exist for the Cauchy distribution, which belongs to the family of elliptical distributions. In this case, $\underline{\boldsymbol{\mu}}_x$ should be treated simply as a parameter of the pdf but not its augmented mean.

Since the complex elliptical pdf (2.76) contains the same quadratic form as the complex Gaussian pdf (2.48), we can obtain straightforward analogs of (2.55) and (2.56), which are expressions in terms of $\mathbf{x}$ and $\mathbf{x}^*$. We can also write down the expression for the pdf of an elliptical random vector with zero complementary generating matrix.

Result 2.6. *The pdf of a complex elliptical random vector* $\mathbf{x}\colon \Omega \longrightarrow \mathbb{C}^n$ *with* $\widetilde{\mathbf{H}}_{xx} = \mathbf{0}$ *is*

$$p(\mathbf{x}) = \frac{c_n}{\det \mathbf{H}_{xx}} g\left\{2(\mathbf{x} - \boldsymbol{\mu}_x)^{\mathrm{H}}\mathbf{H}_{xx}^{-1}(\mathbf{x} - \boldsymbol{\mu}_x)\right\}, \tag{2.78}$$

with normalizing constant c_n *given by (2.77).*

A good overview of real and complex elliptical distributions is given by Fang *et al.* (1990). However, Fang *et al.* consider only complex elliptical distributions with zero complementary generating matrix. Thus, Ollila and Koivunen (2004) refer to the general complex elliptical pdf in Definition 2.3 as a "generalized complex elliptical" pdf.

The family of complex elliptical distributions contains some important subclasses of distributions.

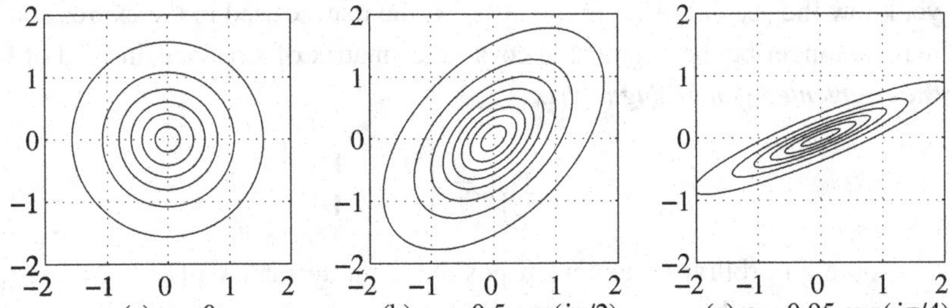

(a) $\rho = 0$ (b) $\rho = 0.5\exp(\mathrm{j}\pi/2)$ (c) $\rho = 0.95\exp(\mathrm{j}\pi/4)$

Figure 2.2 Probability-density contours of complex Cauchy random variables with different ρ.

- The complex multivariate Gaussian distribution, for pdf generator $g(t) = \exp(-t/2)$.
- The complex multivariate t-distribution, with pdf given by

$$p(\mathbf{x}) = \frac{2^n \Gamma(n + k/2)}{(\pi k)^n \Gamma(k/2) \det^{1/2} \underline{\mathbf{H}}_{xx}} \left\{1 + k^{-1}(\underline{\mathbf{x}} - \underline{\boldsymbol{\mu}}_x)^{\mathrm{H}} \underline{\mathbf{H}}_{xx}^{-1} (\underline{\mathbf{x}} - \underline{\boldsymbol{\mu}}_x)\right\}^{-n-k/2}, \tag{2.79}$$

  where k is an integer. We note that the Gamma function satisfies $\Gamma(n) = (n-1)!$ if n is a positive integer.
- The complex multivariate Cauchy distribution, which is a special case of the complex multivariate t-distribution with $k = 1$. Its pdf is

$$p(\mathbf{x}) = \frac{2^n \Gamma(n + 1/2)}{\pi^{n+1/2} \det^{1/2} \underline{\mathbf{H}}_{xx}} \left\{1 + (\underline{\mathbf{x}} - \underline{\boldsymbol{\mu}}_x)^{\mathrm{H}} \underline{\mathbf{H}}_{xx}^{-1} (\underline{\mathbf{x}} - \underline{\boldsymbol{\mu}}_x)\right\}^{-n-1/2}. \tag{2.80}$$

  None of the moments of the Cauchy distribution exist.

Example 2.4. Similarly to Example 2.3, consider a scalar complex Cauchy pdf with $\mu_x = 0$ (which is the median but not the mean, since the mean does not exist) and augmented generator matrix

$$\underline{\mathbf{H}}_{xx} = \begin{bmatrix} 1 & \rho \\ \rho^* & 1 \end{bmatrix}, \quad |\rho| < 1. \tag{2.81}$$

Its pdf is

$$p(x) = \frac{1}{\pi\sqrt{1 - |\rho|^2}} \left\{1 + \frac{|x|^2 - \mathrm{Re}(\rho x^{*2})}{1 - |\rho|^2}\right\}^{-3/2}. \tag{2.82}$$

If $\rho = 0$, then the pdf is $p(x) = (1/\pi)(1 + |x|^2)^{-3/2}$.

Just as for the scalar complex Gaussian pdf, the scalar complex Cauchy pdf is constant on elliptical contours whose major axis is $\psi/2$, half the angle of $\rho = |\rho|e^{\mathrm{j}\psi}$. Figure 2.2 shows contours of constant probability density for three Cauchy random variables. Plots (a), (b), and (c) in this figure should be compared with plots (a), (d), and (f) in Fig. 2.1, respectively, for the Gaussian case. The main difference is that Cauchy random variables have much heavier tails than Gaussian random variables. From the expression (2.82), it is straightforward to write down the joint pdf in polar coordinates (r, θ), and not quite as straightforward to integrate with respect to r or θ to obtain the marginal pdfs.

Complex elliptical distributions have a number of desirable properties.

- The mean of $\mathbf{x}$ (if it exists) is independent of the choice of pdf generator g.
- The augmented covariance matrix of $\mathbf{x}$ (if it exists) is proportional to the augmented generating matrix $\underline{\mathbf{H}}_{xx}$. The proportionality factor depends on g, and is most easily determined using the characteristic function of $\mathbf{x}$. Therefore, if the second-order moments exist, the pdf (2.76) is the pdf of a complex *improper* elliptical random vector, and the pdf (2.78) for $\widetilde{\mathbf{H}}_{xx} = \mathbf{0}$ is the pdf of a complex *proper* elliptical random vector with $\widetilde{\mathbf{R}}_{xx} = \mathbf{0}$. This is due to Result 2.8, to be discussed in Section 2.5.
- All marginal distributions are also elliptical. If $\mathbf{y}: \Omega \longrightarrow \mathbb{C}^m$ contains components of $\mathbf{x}: \Omega \longrightarrow \mathbb{C}^n$, $m < n$, then $\mathbf{y}$ is elliptical with the same g as $\mathbf{x}$, and $\boldsymbol{\mu}_y$ contains the corresponding components of $\boldsymbol{\mu}_x$. The augmented generating matrix of $\mathbf{y}$, $\underline{\mathbf{H}}_{yy}$, is the sub-matrix of $\underline{\mathbf{H}}_{xx}$ that corresponds to the components that $\mathbf{y}$ extracts from $\mathbf{x}$.
- Let $\underline{\mathbf{y}} = \underline{\mathbf{M}}\,\underline{\mathbf{x}} + \underline{\mathbf{b}}$, where $\underline{\mathbf{M}}$ is a given $2m \times 2n$ augmented matrix and $\underline{\mathbf{b}}$ is a given $2m \times 1$ augmented vector. Then $\mathbf{y}$ is elliptical with the same g as $\mathbf{x}$ and $\underline{\boldsymbol{\mu}}_y = \underline{\mathbf{M}}\,\underline{\boldsymbol{\mu}}_x + \underline{\mathbf{b}}$ and $\underline{\mathbf{H}}_{yy} = \underline{\mathbf{M}}\,\underline{\mathbf{H}}_{xx}\,\underline{\mathbf{M}}^{\mathrm{H}}$.
- If $\mathbf{x}: \Omega \longrightarrow \mathbb{C}^n$ and $\mathbf{y}: \Omega \longrightarrow \mathbb{C}^m$ are jointly elliptically distributed with pdf generator g, then the conditional distribution of $\mathbf{x}$ given $\mathbf{y}$ is also elliptical with

$$\underline{\boldsymbol{\mu}}_{x|y} = \underline{\boldsymbol{\mu}}_x + \underline{\mathbf{H}}_{xy}\underline{\mathbf{H}}_{yy}^{-1}(\underline{\mathbf{y}} - \underline{\boldsymbol{\mu}}_y), \tag{2.83}$$

which is analogous to the Gaussian case (2.60), and augmented conditional generating matrix

$$\underline{\mathbf{H}}_{xx|y} = \underline{\mathbf{H}}_{xx} - \underline{\mathbf{H}}_{xy}\underline{\mathbf{H}}_{yy}^{-1}\underline{\mathbf{H}}_{xy}^{\mathrm{H}}, \tag{2.84}$$

which is analogous to the Gaussian case (2.61). However, the conditional distribution will in general have a different pdf generator than $\mathbf{x}$ and $\mathbf{y}$.

2.4 Sufficient statistics and ML estimators for covariances: complex Wishart distribution

Let's draw a sequence of M independent and identically distributed (i.i.d.) samples $\{\mathbf{x}_i\}_{i=1}^M$ from a complex multivariate Gaussian distribution with mean $\boldsymbol{\mu}_x$ and augmented covariance matrix $\underline{\mathbf{R}}_{xx}$. We assemble these samples in a matrix $\mathbf{X} = [\mathbf{x}_1, \mathbf{x}_2, \ldots, \mathbf{x}_M]$, and let $\underline{\mathbf{X}} = [\underline{\mathbf{x}}_1, \underline{\mathbf{x}}_2, \ldots, \underline{\mathbf{x}}_M]$ denote the augmented sample matrix. Using the expression for the Gaussian pdf in Result 2.4, the joint pdf of the samples $\{\mathbf{x}_i\}_{i=1}^M$ is

$$p(\mathbf{X}) = \pi^{-Mn}(\det \underline{\mathbf{R}}_{xx})^{-M/2} \exp\left\{-\frac{1}{2}\sum_{m=1}^{M}(\underline{\mathbf{x}}_m - \underline{\boldsymbol{\mu}}_x)^{\mathrm{H}}\underline{\mathbf{R}}_{xx}^{-1}(\underline{\mathbf{x}}_m - \underline{\boldsymbol{\mu}}_x)\right\} \tag{2.85}$$

$$= \pi^{-Mn}(\det \underline{\mathbf{R}}_{xx})^{-M/2} \exp\left\{-\frac{M}{2}\operatorname{tr}(\underline{\mathbf{R}}_{xx}^{-1}\underline{\mathbf{S}}_{xx})\right\}. \tag{2.86}$$

In this expression, $\underline{\mathbf{S}}_{xx}$ is the *augmented sample covariance matrix*

$$\underline{\mathbf{S}}_{xx} = \begin{bmatrix} \mathbf{S}_{xx} & \widetilde{\mathbf{S}}_{xx} \\ \widetilde{\mathbf{S}}_{xx}^* & \mathbf{S}_{xx}^* \end{bmatrix} = \frac{1}{M}\sum_{m=1}^{M}(\underline{\mathbf{x}}_m - \underline{\mathbf{m}}_x)(\underline{\mathbf{x}}_m - \underline{\mathbf{m}}_x)^{\mathrm{H}} = \frac{1}{M}\underline{\mathbf{X}}\,\underline{\mathbf{X}}^{\mathrm{H}} - \underline{\mathbf{m}}_x\underline{\mathbf{m}}_x^{\mathrm{H}} \tag{2.87}$$

and $\underline{\mathbf{m}}_x$ is the augmented sample mean vector

$$\underline{\mathbf{m}}_x = \frac{1}{M}\sum_{m=1}^{M}\underline{\mathbf{x}}_m. \tag{2.88}$$

The augmented sample covariance matrix contains the sample Hermitian covariance matrix

$$\mathbf{S}_{xx} = \frac{1}{M}\sum_{m=1}^{M}(\mathbf{x}_m - \mathbf{m}_x)(\mathbf{x}_m - \mathbf{m}_x)^{\mathrm{H}} = \frac{1}{M}\mathbf{X}\mathbf{X}^{\mathrm{H}} - \mathbf{m}_x\mathbf{m}_x^{\mathrm{H}} \tag{2.89}$$

and the sample complementary covariance matrix

$$\widetilde{\mathbf{S}}_{xx} = \frac{1}{M}\sum_{m=1}^{M}(\mathbf{x}_m - \mathbf{m}_x)(\mathbf{x}_m - \mathbf{m}_x)^{\mathrm{T}} = \frac{1}{M}\mathbf{X}\mathbf{X}^{\mathrm{T}} - \mathbf{m}_x\mathbf{m}_x^{\mathrm{T}}. \tag{2.90}$$

We now appeal to the Fisher–Neyman factorization theorem to argue that $\underline{\mathbf{m}}_x$ and $\underline{\mathbf{S}}_{xx}$ are a pair of *sufficient statistics* for $\underline{\boldsymbol{\mu}}_x$ and $\underline{\mathbf{R}}_{xx}$, or, equivalently, $(\mathbf{m}_x, \mathbf{S}_{xx}, \widetilde{\mathbf{S}}_{xx})$ are a set of sufficient statistics for $(\boldsymbol{\mu}_x, \mathbf{R}_{xx}, \widetilde{\mathbf{R}}_{xx})$.

It is a straightforward consequence of the corresponding result for the real case that the sample mean $\mathbf{m}_x$ is also a *maximum-likelihood estimator* of the mean $\boldsymbol{\mu}_x$, and the sample covariances $(\mathbf{S}_{xx}, \widetilde{\mathbf{S}}_{xx})$ are *maximum-likelihood estimators* of the covariances $(\mathbf{R}_{xx}, \widetilde{\mathbf{R}}_{xx})$. Note that the sample mean is independent of the sample covariance matrices, but the Hermitian and complementary sample covariances are *not* independent.

Complex Wishart distribution

How are the sample covariance matrices distributed? Let $\mathbf{u}_i$ and $\mathbf{v}_i$ be the real and imaginary parts of the sample vector $\mathbf{x}_i = \mathbf{u}_i + \mathrm{j}\mathbf{v}_i$, and $\mathbf{U}$ and $\mathbf{V}$ be the real and imaginary parts of the sample matrix $\mathbf{X} = \mathbf{U} + \mathrm{j}\mathbf{V}$. Moreover, let $\mathbf{Z} = [\mathbf{U}^{\mathrm{T}}, \mathbf{V}^{\mathrm{T}}]^{\mathrm{T}}$. If the samples $\mathbf{x}_i$ are drawn from a *zero-mean* Gaussian distribution, then $\mathbf{W}_{zz} = \mathbf{Z}\mathbf{Z}^{\mathrm{T}} = M\mathbf{S}_{zz}$ is *Wishart distributed*:

$$p_{W_{zz}}(\mathbf{W}_{zz}) = \frac{[\det \mathbf{W}_{zz}]^{(M-1)/2-n}\exp\left\{-\frac{1}{2}\operatorname{tr}(\mathbf{R}_{zz}^{-1}\mathbf{W}_{zz})\right\}}{2^{Mn}\Gamma_{2n}(M/2)[\det \mathbf{R}_{zz}]^{M/2}}. \tag{2.91}$$

In this expression, $\Gamma_{2n}(M/2)$ is the multivariate Gamma function

$$\Gamma_{2n}(M/2) = \pi^{n(2n-1)/2}\prod_{i=1}^{2n}\Gamma[(M-i+1)/2]. \tag{2.92}$$

By following a path similar to the derivation of the complex Gaussian pdf in Section 2.3.1, it is possible to rewrite the pdf of real $\mathbf{W}_{zz}$ in terms of the complex augmented matrix $\underline{\mathbf{W}}_{xx} = \mathbf{T}\mathbf{W}_{zz}\mathbf{T}^{\mathrm{H}} = \underline{\mathbf{X}}\underline{\mathbf{X}}^{\mathrm{H}} = M\underline{\mathbf{S}}_{xx}$ and the augmented covariance matrix $\underline{\mathbf{R}}_{xx}$ as

$$p_{W_{zz}}(\mathbf{W}_{xx}, \widetilde{\mathbf{W}}_{xx}) = \frac{[\det \underline{\mathbf{W}}_{xx}]^{(M-1)/2-n} \exp\left\{-\frac{1}{2}\operatorname{tr}(\underline{\mathbf{R}}_{xx}^{-1}\underline{\mathbf{W}}_{xx})\right\}}{2^{n(M-2n-1)}\Gamma_{2n}(M/2)[\det \underline{\mathbf{R}}_{xx}]^{M/2}}. \tag{2.93}$$

If $\mathbf{x}$ is *proper*, then $\widetilde{\mathbf{R}}_{xx} = \mathbf{0}$ and $\underline{\mathbf{R}}_{xx} = \mathbf{Diag}(\mathbf{R}_{xx}, \mathbf{R}_{xx}^*)$. However, this does *not* imply that the sample complementary covariance matrix vanishes. In general, $\widetilde{\mathbf{W}}_{xx} = \mathbf{X}\mathbf{X}^{\mathrm{T}} \neq \mathbf{0}$. Thus, for proper $\mathbf{x}$ the pdf (2.93) simplifies, but it is still the pdf of real $\mathbf{W}_{zz}$, expressed as a function of both $\mathbf{W}_{xx} = \mathbf{X}\mathbf{X}^{\mathrm{H}}$ and $\widetilde{\mathbf{W}}_{xx} = \mathbf{X}\mathbf{X}^{\mathrm{T}}$:

$$p_{W_{zz}}(\mathbf{W}_{xx}, \widetilde{\mathbf{W}}_{xx}) = \frac{[\det \mathbf{W}_{xx} \det(\mathbf{W}_{xx}^* - \widetilde{\mathbf{W}}_{xx}^* \mathbf{W}_{xx}^{-1}\widetilde{\mathbf{W}}_{xx})]^{(M-1)/2-n} \exp\left\{-\operatorname{tr}(\mathbf{R}_{xx}^{-1}\mathbf{W}_{xx})\right\}}{2^{n(M-2n-1)}\Gamma_{2n}(M/2)[\det \mathbf{R}_{xx}]^{M}}. \tag{2.94}$$

In order to obtain the marginal pdf of $\mathbf{W}_{xx}$, we would need to integrate $p_{W_{zz}}(\mathbf{W}_{xx}, \widetilde{\mathbf{W}}_{xx})$ for given $\mathbf{W}_{xx}$ over all $\mathbf{W}_{xx}^* - \widetilde{\mathbf{W}}_{xx}^* \mathbf{W}_{xx}^{-1}\widetilde{\mathbf{W}}_{xx} > \mathbf{0}$. Alternatively, we can compute the characteristic function of $\mathbf{W}_{zz}$, set all complementary terms equal to zero, and then compute the corresponding pdf of $\mathbf{W}_{xx}$. The last of these three steps is the tricky one. In essence, this is the approach followed by Goodman (1963), who showed that

$$p(\mathbf{W}_{xx}) = \frac{[\det \mathbf{W}_{xx}]^{M-n} \exp\left\{-\operatorname{tr}(\mathbf{R}_{xx}^{-1}\mathbf{W}_{xx})\right\}}{\Gamma_n^{\mathrm{c}}(M)[\det \mathbf{R}_{xx}]^{M}}, \tag{2.95}$$

where $\Gamma_n^{\mathrm{c}}(M)$ is

$$\Gamma_n^{\mathrm{c}}(M) = \pi^{n(n-1)/2}\prod_{i=1}^{n}(M-i)! \tag{2.96}$$

The marginal pdf $p(\mathbf{W}_{xx})$ for *proper* $\mathbf{x}$ in (2.95) is what is commonly referred to as the *complex Wishart distribution*. Its support is $\mathbf{W}_{xx} > \mathbf{0}$, and it is interpreted as the joint pdf of real and imaginary parts of $\mathbf{W}_{xx}$.

An alternative, simpler, derivation of the complex Wishart distribution, not based on the characteristic function, was given by Srivastava (1965). The derivation of the marginal pdf $p(\mathbf{W}_{xx})$ for *improper* $\mathbf{x}$ is an unresolved problem.

2.5 Characteristic function and higher-order statistical description

The characteristic function is a characterization equivalent to the pdf of a random vector, yet working with the characteristic function can be more convenient. For the composite real random vector $\mathbf{z} = [\mathbf{u}^{\mathrm{T}}, \mathbf{v}^{\mathrm{T}}]^{\mathrm{T}}$, the characteristic function is defined in terms of $\mathbf{s}_u \in \mathbb{R}^n$ and $\mathbf{s}_v \in \mathbb{R}^n$ (which are *not* random) as

$$\psi(\mathbf{s}_u, \mathbf{s}_v) = E\left[\exp\left[\mathrm{j}(\mathbf{s}_u^{\mathrm{T}}\mathbf{u} + \mathbf{s}_v^{\mathrm{T}}\mathbf{v})\right]\right]. \tag{2.97}$$

This is the inverse Fourier transform of the pdf $p(\mathbf{u}, \mathbf{v})$. If we let $\mathbf{x} = \mathbf{u} + \mathrm{j}\mathbf{v}$ (random) and $\mathbf{s} = \mathbf{s}_u + \mathrm{j}\mathbf{s}_v$ (not random), we have

$$\mathbf{s}_u^{\mathrm{T}}\mathbf{u} + \mathbf{s}_v^{\mathrm{T}}\mathbf{v} = \tfrac{1}{2}\underline{\mathbf{s}}^{\mathrm{H}}\underline{\mathbf{x}} = \operatorname{Re}(\mathbf{s}^{\mathrm{H}}\mathbf{x}). \tag{2.98}$$

Thus, we may define the characteristic function of $\mathbf{x}$ as

$$\psi(\mathbf{s}) = E\left\{\exp\left(\frac{\mathrm{j}}{2}\underline{\mathbf{s}}^{\mathrm{H}}\underline{\mathbf{x}}\right)\right\} = E\left\{\exp[\mathrm{j}\,\mathrm{Re}(\mathbf{s}^{\mathrm{H}}\mathbf{x})]\right\}. \tag{2.99}$$

2.5.1 Characteristic functions of Gaussian and elliptical distributions

The characteristic function of a real elliptical random vector $\mathbf{z}$ (cf. Fang *et al.* (1990)) is

$$\psi(\mathbf{s}_u, \mathbf{s}_v) = \exp(\mathrm{j}\begin{bmatrix}\mathbf{s}_u^{\mathrm{T}} & \mathbf{s}_v^{\mathrm{T}}\end{bmatrix}\boldsymbol{\mu}_z)\phi\left(\begin{bmatrix}\mathbf{s}_u^{\mathrm{T}} & \mathbf{s}_v^{\mathrm{T}}\end{bmatrix}\mathbf{R}_{zz}\begin{bmatrix}\mathbf{s}_u \\ \mathbf{s}_v\end{bmatrix}\right), \tag{2.100}$$

where ϕ is a scalar function that determines the distribution. From this we obtain the following result, which was first published by Ollila and Koivunen (2004).

Result 2.7. *The characteristic function of a complex elliptical random vector* $\mathbf{x}$ *is*

$$\begin{aligned}\psi(\mathbf{s}) &= \exp\left(\frac{\mathrm{j}}{2}\underline{\mathbf{s}}^{\mathrm{H}}\underline{\boldsymbol{\mu}}_x\right)\phi\left(\tfrac{1}{4}\underline{\mathbf{s}}^{\mathrm{H}}\underline{\mathbf{H}}_{xx}\underline{\mathbf{s}}\right) \\ &= \exp[\mathrm{j}\,\mathrm{Re}(\mathbf{s}^{\mathrm{H}}\boldsymbol{\mu}_x)]\phi\left[\tfrac{1}{2}\mathbf{s}^{\mathrm{H}}\mathbf{H}_{xx}\mathbf{s} + \tfrac{1}{2}\,\mathrm{Re}(\mathbf{s}^{\mathrm{H}}\widetilde{\mathbf{H}}_{xx}\mathbf{s}^*)\right].\end{aligned} \tag{2.101}$$

If $\mathbf{x}$ *is complex Gaussian, then* $\mathbf{H}_{xx} = \mathbf{R}_{xx}$, $\widetilde{\mathbf{H}}_{xx} = \widetilde{\mathbf{R}}_{xx}$, *and* $\phi(t) = \exp(-t/2)$.

Both expressions in (2.101) – in terms of augmented vectors and in terms of unaugmented vectors – are useful. It is easy to see the simplification that occurs when $\widetilde{\mathbf{H}}_{xx} = \mathbf{0}$ or $\widetilde{\mathbf{R}}_{xx} = \mathbf{0}$, as when $\mathbf{x}$ is a proper Gaussian random vector.

The characteristic function has a number of useful properties. It exists even when the augmented covariance matrix $\underline{\mathbf{R}}_{xx}$ (or the covariance matrix $\mathbf{R}_{xx}$) is singular, a property not shared by the pdf. The characteristic function also allows us to state a simple connection between $\underline{\mathbf{R}}_{xx}$ and $\underline{\mathbf{H}}_{xx}$ for elliptical random vectors (consult Fang *et al.* (1990) for the expression in the real case).

Result 2.8. *Let* $\mathbf{x}$ *be a complex elliptical random vector with characteristic function (2.101). If the second-order moments of* $\mathbf{x}$ *exist, then*

$$\underline{\mathbf{R}}_{xx} = c\underline{\mathbf{H}}_{xx} \text{ with } c = -2\frac{\mathrm{d}}{\mathrm{d}t}\phi(t)|_{t=0}. \tag{2.102}$$

It follows that $\mathbf{R}_{xx} = c\mathbf{H}_{xx}$ and $\widetilde{\mathbf{R}}_{xx} = c\widetilde{\mathbf{H}}_{xx}$. In particular, we see that $\mathbf{x}$ is proper if and only if $\widetilde{\mathbf{H}}_{xx} = \mathbf{0}$. For Gaussian $\mathbf{x}$, $\phi(t) = \exp(-t/2)$ and therefore $c = 1$, as expected.

2.5.2 Higher-order moments

First- and second-order moments suffice for the characterization of many probability distributions. They also suffice for the solution of mean-squared-error estimation problems and Gaussian detection problems. Nevertheless, higher-order moments and cumulants carry important information that can be exploited when the underlying probability distribution is unknown. For a complex random variable x, there are $N + 1$ different Nth-order moments $E(x^q x^{*N-q})$, with $q = 0, 1, 2, \ldots, N$. It is immediately clear that there can be only a maximum of $\lfloor N/2 \rfloor + 1$ distinct moments, where $\lfloor \cdot \rfloor$ denotes the floor function. Since $E(x^q x^{*N-q}) = [E(x^{N-q} x^{*q})]^*$ we can restrict q to $q = 0, 1, 2, \ldots, \lfloor N/2 \rfloor$. Because moments become increasingly difficult to estimate in practice for larger N, it is rare to consider moments with $N > 4$.

Summarizing all Nth-order moments for a complex random *vector* $\mathbf{x}$ requires the use of tensor algebra, as developed in the complex case by Amblard *et al.* (1996a). We will avoid this by considering only moments of individual components of $\mathbf{x}$. These are of the form $E(x_{i_1}^{\diamond_1} x_{i_2}^{\diamond_2} \cdots x_{i_N}^{\diamond_N})$, where $1 \le i_j \le n$, $j = 1, \ldots, N$, and $\diamond_j$ indicates whether or not x_{i_j} is conjugated. Again, there are only a maximum of $\lfloor N/2 \rfloor + 1$ distinct moments since $E(x_{i_1}^{\diamond_1} x_{i_2}^{\diamond_2} \cdots x_{i_N}^{\diamond_N}) = [E((x_{i_1}^{\diamond_1})^* (x_{i_2}^{\diamond_2})^* \cdots (x_{i_N}^{\diamond_N})^*)]^*$.

As in the real case, the moments of $\mathbf{x}$ can be calculated from the characteristic function. To this end, we use two *generalized complex differential operators* for complex $s = s_u + \mathrm{j} s_v$, which are defined as

$$\frac{\partial}{\partial s} \triangleq \frac{1}{2}\left(\frac{\partial}{\partial s_u} - \mathrm{j}\frac{\partial}{\partial s_v}\right), \tag{2.103}$$

$$\frac{\partial}{\partial s^*} \triangleq \frac{1}{2}\left(\frac{\partial}{\partial s_u} + \mathrm{j}\frac{\partial}{\partial s_v}\right) \tag{2.104}$$

and discussed in Appendix 2. From the complex Taylor-series expansion of $\psi(\mathbf{s})$, we can obtain the Nth-order moment from the characteristic function. The following result is due to Amblard *et al.* (1996a).

Result 2.9. *For a random vector* $\mathbf{x}$ *with characteristic function* $\psi(\mathbf{s})$, *the* N*th-order moment can be computed as*

$$E(x_{i_1}^{\diamond_1} x_{i_2}^{\diamond_2} \cdots x_{i_N}^{\diamond_N}) = \frac{2^N}{\mathrm{j}^N} \left.\frac{\partial^N \psi(\mathbf{s})}{(\partial s_{i_1}^{\diamond_1})^* (\partial s_{i_2}^{\diamond_2})^* \cdots (\partial s_{i_N}^{\diamond_N})^*}\right|_{\mathbf{s}=\mathbf{0}}. \tag{2.105}$$

Example 2.5. Consider a scalar Gaussian random variable with zero mean, variance R_{xx}, and complementary variance $\widetilde{R}_{xx} = \rho R_{xx}$, whose pdf is

$$p(x) = \frac{1}{\pi R_{xx}\sqrt{1-|\rho|^2}} \exp\left\{-\frac{|x|^2 - \mathrm{Re}(\rho x^{*2})}{R_{xx}(1-|\rho|^2)}\right\}. \tag{2.106}$$

Its characteristic function is

$$\psi(s) = \exp\left\{-\tfrac{1}{4} R_{xx}[|s|^2 + \mathrm{Re}(\rho s^{*2})]\right\}. \tag{2.107}$$

The pdf $p(x)$ is not defined for $|\rho| = 1$ because the support of the pdf collapses to a line in the complex plane. The characteristic function, on the other hand, is defined for all $|\rho| \le 1$.

Using the differentiation rules from Appendix 2, we compute the first-order partial derivative

$$\frac{\partial \psi(s)}{\partial s^*} = -\tfrac{1}{4} R_{xx} \psi(s)(s + \rho s^*) \tag{2.108}$$

and the second-order partial derivatives

$$\frac{\partial^2\psi(s)}{\partial s\,\partial s^*} = -\tfrac{1}{4}R_{xx}\left[\frac{\partial\psi(s)}{\partial s}(s+\rho s^*)+\psi(s)\right], \tag{2.109}$$

$$\frac{\partial^2\psi(s)}{(\partial s^*)^2} = -\tfrac{1}{4}R_{xx}\left[\frac{\partial\psi(s)}{\partial s^*}(s+\rho s^*)+\rho\psi(s)\right]. \tag{2.110}$$

According to Result 2.9, the variance is obtained as

$$E(x^*x) = \frac{2^2}{\mathrm{j}^2}\left.\frac{\partial^2\psi(s)}{\partial s\,\partial s^*}\right|_{s=0} = R_{xx} \tag{2.111}$$

and the complementary variance as

$$E(xx) = \frac{2^2}{\mathrm{j}^2}\left.\frac{\partial^2\psi(s)}{(\partial s^*)^2}\right|_{s=0} = \rho R_{xx}. \tag{2.112}$$

2.5.3 Cumulant-generating function

The *cumulant-generating function* of $\mathbf{x}$ is defined as

$$\Psi(\mathbf{s}) = \log\psi(\mathbf{s}) = \log E\left[\exp\left[\mathrm{j}\,\mathrm{Re}(\mathbf{s}^{\mathrm{H}}\mathbf{x})\right]\right]. \tag{2.113}$$

The cumulants are found from the cumulant-generating function just like the moments are from the characteristic function:

$$\mathrm{Cum}(x_{i_1}^{\diamond_1}x_{i_2}^{\diamond_2}\cdots x_{i_N}^{\diamond_N}) = \frac{2^N}{\mathrm{j}^N}\left.\frac{\partial^N\Psi(\mathbf{s})}{(\partial s_{i_1}^{\diamond_1})^*(\partial s_{i_2}^{\diamond_2})^*\cdots(\partial s_{i_N}^{\diamond_N})^*}\right|_{\mathbf{s}=\mathbf{0}}. \tag{2.114}$$

The first-order cumulant is the mean,

$$\mathrm{Cum}\,x_i^{\diamond} = Ex_i^{\diamond}. \tag{2.115}$$

For a zero-mean random vector $\mathbf{x}$, the cumulants of second, third, and fourth order are related to moments through

$$\mathrm{Cum}(x_{i_1}^{\diamond_1}x_{i_2}^{\diamond_2}) = E(x_{i_1}^{\diamond_1}x_{i_2}^{\diamond_2}), \tag{2.116}$$

$$\mathrm{Cum}(x_{i_1}^{\diamond_1}x_{i_2}^{\diamond_2}x_{i_3}^{\diamond_3}) = E(x_{i_1}^{\diamond_1}x_{i_2}^{\diamond_2}x_{i_3}^{\diamond_3}), \tag{2.117}$$

$$\begin{aligned}\mathrm{Cum}(x_{i_1}^{\diamond_1}x_{i_2}^{\diamond_2}x_{i_3}^{\diamond_3}x_{i_4}^{\diamond_4}) &= E(x_{i_1}^{\diamond_1}x_{i_2}^{\diamond_2}x_{i_3}^{\diamond_3}x_{i_4}^{\diamond_4}) - E(x_{i_1}^{\diamond_1}x_{i_2}^{\diamond_2})E(x_{i_3}^{\diamond_3}x_{i_4}^{\diamond_4})\\ &\quad - E(x_{i_1}^{\diamond_1}x_{i_3}^{\diamond_3})E(x_{i_2}^{\diamond_2}x_{i_4}^{\diamond_4}) - E(x_{i_1}^{\diamond_1}x_{i_4}^{\diamond_4})E(x_{i_2}^{\diamond_2}x_{i_3}^{\diamond_3}).\end{aligned} \tag{2.118}$$

Thus, the second-order cumulant is the covariance or complementary covariance. Cumulants are sometimes preferred over moments because cumulants are additive for independent random variables. For Gaussian random vectors, all cumulants of order higher than two are zero. This leads to the following result for *proper Gaussians*, which is based on (2.118) and vanishing complementary covariance terms.

Result 2.10. *For a* proper *complex Gaussian random vector* $\mathbf{x}$*, only fourth-order moments with two conjugated and two nonconjugated terms are nonzero. Without loss of generality, assume that* x_{i_3} *and* x_{i_4} *are the conjugated terms. We then have*

$$E(x_{i_1}x_{i_2}x_{i_3}^*x_{i_4}^*) = E(x_{i_1}x_{i_3}^*)E(x_{i_2}x_{i_4}^*) + E(x_{i_1}x_{i_4}^*)E(x_{i_2}x_{i_3}^*). \tag{2.119}$$

2.5.4 Circularity

It is also possible to define a stronger version of propriety in terms of the probability distribution of a random vector. A vector is called *circular* if its probability distribution is *rotationally invariant*.

Definition 2.4. *A random vector* $\mathbf{x}$ *is called* circular *if* $\mathbf{x}$ *and* $\mathbf{x}' = \mathrm{e}^{\mathrm{j}\alpha}\mathbf{x}$ *have the same probability distribution for any given real* α.

Therefore, $\mathbf{x}$ must have zero mean in order to be circular. But circularity does not imply any condition on the standard covariance matrix $\mathbf{R}_{xx}$ because

$$\mathbf{R}_{x'x'} = E(\mathbf{x}'\mathbf{x}'^{\mathrm{H}}) = E(\mathrm{e}^{\mathrm{j}\alpha}\mathbf{x}\mathbf{x}^{\mathrm{H}}\mathrm{e}^{-\mathrm{j}\alpha}) = \mathbf{R}_{xx}. \tag{2.120}$$

On the other hand,

$$\widetilde{\mathbf{R}}_{x'x'} = E(\mathbf{x}'\mathbf{x}'^{\mathrm{T}}) = E(\mathrm{e}^{\mathrm{j}\alpha}\mathbf{x}\mathbf{x}^{\mathrm{T}}\mathrm{e}^{\mathrm{j}\alpha}) = \mathrm{e}^{\mathrm{j}2\alpha}\widetilde{\mathbf{R}}_{xx} \tag{2.121}$$

can be true for arbitrary α only if $\widetilde{\mathbf{R}}_{xx} = \mathbf{0}$. Because the Gaussian pdf is completely determined by $\boldsymbol{\mu}_x$, $\mathbf{R}_{xx}$, and $\widetilde{\mathbf{R}}_{xx}$, we obtain the following result, due to Grettenberg (1965).

Result 2.11. *A complex zero-mean Gaussian random vector* $\mathbf{x}$ *is proper if and only if it is circular.*

This result generalizes nicely to the elliptical pdf.

Result 2.12. *A complex elliptical random vector* $\mathbf{x}$ *is circular if and only if* $\boldsymbol{\mu}_x = \mathbf{0}$ *and the complementary generating matrix* $\widetilde{\mathbf{H}}_{xx} = \mathbf{0}$. *If the first- and second-order moments exist, then we may equivalently say that* $\mathbf{x}$ *is circular if and only if it has zero mean and is proper.*

The first half of this result is easily obtained by inspecting the elliptical pdf, and the second half is due to Result 2.8.

Propriety requires that second-order moments be rotationally invariant, whereas circularity requires that the pdf, and thus *all* moments (if they exist), be rotationally invariant. Therefore, circularity implies propriety, but not vice versa, and impropriety implies noncircularity, but not vice versa. By extending the reasoning of (2.120) and (2.121) to higher-order moments, we see that the following result holds.

Result 2.13. *If* $\mathbf{x}$ *is circular, an* N*th-order moment* $E\left(x_{i_1}^{\diamond_1}x_{i_2}^{\diamond_2}\cdots x_{i_N}^{\diamond_N}\right)$ *can be nonzero only if it has the same number of conjugated and nonconjugated terms. In particular, all odd moments must be zero. This holds for arbitrary* N.

We have already seen a forerunner of this fact in Result 2.10. We may also be interested in a term for a random vector that satisfies this condition up to some order N only.

Definition 2.5. *A vector* $\mathbf{x}$ *is called* Nth-order circular *or* Nth-order proper *if the only nonzero moments up to order* N *have the same number of conjugated and nonconjugated terms. In particular, all odd moments up to order* N *must be zero.*

Therefore, the terms *proper* and *second-order circular* are equivalent. We note that, while this terminology is most common, there is not uniform agreement in the literature. Some researchers use the terms "proper" and "circular" interchangeably. It is also possible to define stronger versions of circularity, as Picinbono (1994) has done.

Do circular random vectors have spherical pdf contours?

It is instructive to first take a closer look at a scalar circular random variable $x = A\mathrm{e}^{\mathrm{j}\phi}$. If x and $\mathrm{e}^{\mathrm{j}\alpha}x$ have the same probability distribution, then the phase ϕ of x must be uniformly distributed over $[0, 2\pi)$ and independent of the amplitude A, which may have an arbitrary distribution. This means that the pdf of x is complex elliptical with variance R_{xx} and zero complementary generator,

$$p(x) = \frac{c_1}{R_{xx}} g\left(2\frac{|x|^2}{R_{xx}}\right), \tag{2.122}$$

as developed in Section 2.3.4. The pdf (2.122) has circular contour lines of constant probability density. Does this result generalize to circular random *vectors*?

Unfortunately, nothing can be said in general about the contours of a circular random *vector* $\mathbf{x}$ – not even that they are elliptical. Elliptical random vectors have elliptical contours but circularity is not sufficient to make them spherical. In order to obtain spherical contours, $\mathbf{x}$ must be *spherically distributed*, which means that $\mathbf{x}$ is elliptical with $\boldsymbol{\mu}_x = \mathbf{0}$ and augmented generating matrix $\underline{\mathbf{H}}_{xx} = k\mathbf{I}$ for some positive constant k. Hence, only spherical circular random vectors indeed have spherical contours.

2.6 Complex random processes

In this section, we extend some of the ideas introduced so far to a continuous-time complex random process $x(t) = u(t) + \mathrm{j}v(t)$, which is built from two real-valued random processes $u(t)$ and $v(t)$ defined on $\mathbb{R}$. We restrict our attention to a second-order description of wide-sense stationary processes. Higher-order statistical characterizations and concepts such as circularity of random processes are more difficult to treat than in the vector case, and we postpone a discussion of more advanced topics to Chapters 8 and 9.[6]

To simplify notation in this section, we will assume that $x(t)$ has zero mean. The *covariance function* of $x(t)$ is denoted by

$$r_{xx}(t, \tau) = E[x(t + \tau)x^*(t)], \tag{2.123}$$

and the *complementary covariance function* of $x(t)$ is

$$\tilde{r}_{xx}(t,\tau) = E[x(t+\tau)x(t)]. \tag{2.124}$$

We also introduce the augmented signal $\underline{\mathbf{x}}(t) = \begin{bmatrix} x(t) & x^*(t) \end{bmatrix}^{\mathrm{T}}$, whose covariance matrix

$$\underline{\mathbf{R}}_{xx}(t,\tau) = E[\underline{\mathbf{x}}(t+\tau)\underline{\mathbf{x}}^{\mathrm{H}}(t)] = \begin{bmatrix} r_{xx}(t,\tau) & \tilde{r}_{xx}(t,\tau) \\ \tilde{r}^*_{xx}(t,\tau) & r^*_{xx}(t,\tau) \end{bmatrix} \tag{2.125}$$

is called the *augmented covariance function* of $x(t)$.

2.6.1 Wide-sense stationary processes

Definition 2.6. *A signal $x(t)$ is wide-sense stationary (WSS) if and only if $\underline{\mathbf{R}}_{xx}(t,\tau)$ is independent of t. That is, both the covariance function $r_{xx}(t,\tau)$ and the complementary covariance function $\tilde{r}_{xx}(t,\tau)$ are independent of t.*

This definition, in keeping with our general philosophy outlined in Section 2.2.1, calls $x(t)$ WSS if and only if its real and imaginary parts $u(t)$ and $v(t)$ are jointly WSS. We note that some researchers call a complex signal $x(t)$ WSS if $r_{xx}(t,\tau)$ alone is independent of t, and *second-order stationary* if both $r_{xx}(t,\tau)$ and $\tilde{r}_{xx}(t,\tau)$ are independent of t. If $x(t)$ is WSS, we drop the t-argument from the covariance functions. The covariance and complementary covariance functions then have the symmetries

$$r_{xx}(\tau) = r^*_{xx}(-\tau) \quad \text{and} \quad \tilde{r}_{xx}(\tau) = \tilde{r}_{xx}(-\tau). \tag{2.126}$$

The Fourier transform of $\underline{\mathbf{R}}_{xx}(\tau)$ is the *augmented power spectral density (PSD) matrix*

$$\underline{\mathbb{P}}_{xx}(f) = \begin{bmatrix} P_{xx}(f) & \widetilde{P}_{xx}(f) \\ \widetilde{P}^*_{xx}(-f) & P^*_{xx}(-f) \end{bmatrix}. \tag{2.127}$$

The augmented PSD matrix contains the PSD $P_{xx}(f)$, which is the Fourier transform of $r_{xx}(\tau)$, and the *complementary power spectral density* (C-PSD) $\widetilde{P}_{xx}(f)$, which is the Fourier transform of $\tilde{r}_{xx}(\tau)$. The augmented PSD matrix is positive semidefinite, which implies the following result.

Result 2.14. *There exists a WSS random process $x(t)$ with PSD $P_{xx}(f)$ and C-PSD $\widetilde{P}_{xx}(f)$ if and only if*

(1) the PSD is real and nonnegative (but not necessarily even),

$$P_{xx}(f) \geq 0; \tag{2.128}$$

(2) the C-PSD is even (but generally complex),

$$\widetilde{P}_{xx}(f) = \widetilde{P}_{xx}(-f); \tag{2.129}$$

(3) the PSD provides a bound on the magnitude of the C-PSD,

$$|\widetilde{P}_{xx}(f)|^2 \leq P_{xx}(f)P_{xx}(-f). \tag{2.130}$$

Condition (3) is due to det $\underline{\mathbf{P}}_{xx}(f) \geq 0$. The three conditions in this result correspond to the three conditions in Result 2.1 for a complex vector $\mathbf{x}$. Because of (2.128) and (2.129) the augmented PSD matrix simplifies to

$$\underline{\mathbf{P}}_{xx}(f) = \begin{bmatrix} P_{xx}(f) & \widetilde{P}_{xx}(f) \\ \widetilde{P}^*_{xx}(f) & P_{xx}(-f) \end{bmatrix}. \tag{2.131}$$

The time-invariant power of $x(t)$ is

$$P_x = r_{xx}(0) = \int_{-\infty}^{\infty} P_{xx}(f)\mathrm{d}f, \tag{2.132}$$

regardless of whether or not $x(t)$ is proper.

We now connect the complex description of $x(t) = u(t) + \mathrm{j}v(t)$ to the description in terms of its real and imaginary parts $u(t)$ and $v(t)$. Let $r_{uv}(\tau) = E[u(t+\tau)v(t)]$ denote the cross-covariance function between $u(t)$ and $v(t)$, and $P_{uv}(f)$ the Fourier transform of $r_{uv}(\tau)$, which is the cross-PSD between $u(t)$ and $v(t)$. Analogously to (2.20) for $n = 1$, there is the connection

$$\underline{\mathbf{P}}_{xx}(f) = \mathbf{T} \begin{bmatrix} P_{uu}(f) & P_{uv}(f) \\ P^*_{uv}(f) & P_{vv}(f) \end{bmatrix} \mathbf{T}^{\mathrm{H}}. \tag{2.133}$$

From it, we find that

$$P_{xx}(f) = P_{uu}(f) + P_{vv}(f) + 2\,\mathrm{Im}\,P_{uv}(f), \tag{2.134}$$

$$\widetilde{P}_{xx}(f) = P_{uu}(f) - P_{vv}(f) + 2\mathrm{j}\,\mathrm{Re}\,P_{uv}(f), \tag{2.135}$$

which are the analogs of (2.21) and (2.22). Note that, unlike the PSD of $x(t)$, the PSDs of $u(t)$ and $v(t)$ are even: $P_{uu}(f) = P_{uu}(-f)$ and $P_{vv}(f) = P_{vv}(-f)$. Propriety is now defined as the obvious extension from vectors to processes.

Definition 2.7. *A complex WSS random process $x(t)$ is called* proper *if $\tilde{r}_{xx}(\tau) = 0$ for all τ or, equivalently, $\widetilde{P}_{xx}(f) = 0$ for all f.*

Equivalent conditions on real and imaginary parts for propriety are

$$r_{uu}(\tau) = r_{vv}(\tau) \quad \text{and} \quad r_{uv}(\tau) = -r_{uv}(-\tau) \text{ for all } \tau \tag{2.136}$$

or

$$P_{uu}(f) = P_{vv}(f) \quad \text{and} \quad \mathrm{Re}\,P_{uv}(f) = 0 \text{ for all } f. \tag{2.137}$$

Therefore, if WSS $x(t)$ is proper, its PSD is

$$P_{xx}(f) = 2[P_{uu}(f) - \mathrm{j}P_{uv}(f)] = 2[P_{vv}(f) + \mathrm{j}P_{vu}(f)]. \tag{2.138}$$

The PSD of a proper $x(t)$ is even if and only if $P_{uv}(f) = 0$ because $P_{uv}(f)$ is purely imaginary and odd.

From (2.130), we obtain the following important result for WSS analytic signals, which have $P_{xx}(f) = 0$ for $f < 0$, and WSS anti-analytic signals, which have $P_{xx}(f) = 0$ for $f > 0$.

Result 2.15. *A WSS analytic (or anti-analytic) signal without a DC component, i.e., $P_{xx}(0) = 0$, is proper.*

Because the equivalent complex baseband signal of a real bandpass signal is a down-modulated analytic signal, and modulation keeps propriety intact, we also have the following result.

Result 2.16. *The equivalent complex baseband signal of a WSS real bandpass signal is proper.*

2.6.2 Widely linear shift-invariant filtering

Widely linear (linear–conjugate-linear) shift-invariant filtering is described in the time domain as

$$y(t) = \int_{-\infty}^{\infty} [h_1(t-\tau)x(\tau) + h_2(t-\tau)x^*(\tau)]\mathrm{d}\tau. \tag{2.139}$$

There is a slight complication with a corresponding frequency-domain expression: as we will discuss in Section 8.1, the Fourier transform of a WSS random process $x(t)$ does not exist. The way to deal with WSS processes in the frequency domain is to utilize the Cramér spectral representation for $x(t)$ and $y(t)$,

$$x(t) = \int_{-\infty}^{\infty} \mathrm{d}\xi(f)\mathrm{e}^{\mathrm{j}2\pi ft}, \tag{2.140}$$

$$y(t) = \int_{-\infty}^{\infty} \mathrm{d}\upsilon(f)\mathrm{e}^{\mathrm{j}2\pi ft}, \tag{2.141}$$

where $\xi(f)$ and $\upsilon(f)$ are spectral processes with orthogonal increments $\mathrm{d}\xi(f)$ and $\mathrm{d}\upsilon(f)$, respectively. This will be discussed in detail in Section 8.1. For now, we content ourselves with stating that

$$\mathrm{d}\upsilon(f) = H_1(f)\mathrm{d}\xi(f) + H_2(f)\mathrm{d}\xi^*(-f), \tag{2.142}$$

which may be written in augmented notation as

$$\begin{bmatrix} \mathrm{d}\upsilon(f) \\ \mathrm{d}\upsilon^*(-f) \end{bmatrix} = \underbrace{\begin{bmatrix} H_1(f) & H_2(f) \\ H_2^*(-f) & H_1^*(-f) \end{bmatrix}}_{\underline{\mathbb{H}}(f)} \begin{bmatrix} \mathrm{d}\xi(f) \\ \mathrm{d}\xi^*(-f) \end{bmatrix}. \tag{2.143}$$

The relationship between the PSDs of $x(t)$ and $y(t)$ is

$$\underline{\mathbb{P}}_{yy}(f) = \underline{\mathbb{H}}(f)\underline{\mathbb{P}}_{xx}(f)\underline{\mathbb{H}}^{\mathrm{H}}(f). \tag{2.144}$$

Strictly linear filters have $h_2(t) = 0$ for all t or, equivalently, $H_2(f) = 0$ for all f. It is clear that propriety is preserved under linear filtering, and it is also preserved under modulation.

Notes

1 The first account of widely linear filtering for complex signals seems to have been that by Brown and Crane (1969), who used the term "conjugate linear filtering." Gardner and his co-authors

have made extensive use of widely linear filtering in the context of cyclostationary signals, in particular for communications. See, for instance, Gardner (1993). The term "widely linear" was introduced by Picinbono and Chevalier (1995), who presented the widely linear minimum mean-squared error (WLMMSE) estimator for complex random vectors. Schreier and Scharf (2003a) revisited the WLMMSE problem using the augmented complex matrix algebra.

2 The terms "proper" and "pseudo-covariance" (for complementary covariance) were coined by Neeser and Massey (1993), who looked at applications of proper random vectors in communications and information theory. The term "complementary covariance" is used by Lee and Messerschmitt (1994), "relation matrix" by Picinbono and Bondon (1997), and "conjugate covariance" by Gardner. Both van den Bos (1995) and Picinbono (1996) utilize what we have called the augmented covariance matrix.

3 Conditions 1–3 in Result 2.1 for the nonsingular case were proved by Picinbono (1996). A rather technical proof of conditions 3a and 3b was presented by Wahlberg and Schreier (2008).

4 The complex *proper* multivariate Gaussian distribution was introduced by Wooding (1956). Goodman (1963) provided a more in-depth study, and also derived the proper complex Wishart distribution. The assumption of propriety when studying Gaussian random vectors was commonplace until van den Bos (1995) introduced the *improper* multivariate Gaussian distribution, which he called "generalized complex normal." Picinbono (1996) explicitly connected this distribution with the Hermitian and complementary covariance matrices.

5 There are other distributions thrown off by the scalar improper complex Gaussian pdf. For example, the conditional distribution $P_{\theta|r}(\theta|r) = p_{r\theta}(r,\theta)/P_r(r)$, with $p_{r\theta}(r,\theta)$ given by (2.71) and $p_r(r)$ by (2.72) is the *von Mises* distribution. The distribution of the radius-squared r^2 could be called the *improper* χ^2*-distribution with one degree of freedom*, and the sum of the radii-squared of k independent improper Gaussian random variables could be called the *improper* χ^2*-distribution with k degrees of freedom*. One could also derive improper extensions of the β- and exponential distributions.

6 Section 2.6 builds on results by Picinbono and Bondon (1997), who discussed second-order properties of complex signals, both stationary and nonstationary, and Amblard *et al.* (1996b), who developed higher-order properties of complex stationary signals. Rubin-Delanchy and Walden (2007) present an algorithm for the simulation of improper WSS processes having specified covariance and complementary covariance functions.

Part II

Complex random vectors

3 Second-order description of complex random vectors

In this chapter, we discuss in detail the second-order description of a complex random vector $\mathbf{x}$. We have seen in Chapter 2 that the second-order averages of $\mathbf{x}$ are completely described by the augmented covariance matrix $\underline{\mathbf{R}}_{xx}$. We shall now be interested in those second-order properties of $\mathbf{x}$ that are *invariant* under two types of transformations: widely unitary and nonsingular strictly linear.

The eigenvalues of the augmented covariance matrix $\underline{\mathbf{R}}_{xx}$ constitute a maximal invariant for $\underline{\mathbf{R}}_{xx}$ under widely unitary transformation. Hence, any function of $\underline{\mathbf{R}}_{xx}$ that is invariant under widely unitary transformation must be a function of these eigenvalues only. In Section 3.1, we consider the augmented eigenvalue decomposition (EVD) of $\underline{\mathbf{R}}_{xx}$ for a complex random vector $\mathbf{x}$. Since we are working with an augmented matrix algebra, this EVD looks somewhat different from what one might expect. In fact, because all factors in the EVD must be augmented matrices, widely unitary diagonalization of $\underline{\mathbf{R}}_{xx}$ is generally not possible. As an application for the augmented EVD, we discuss rank reduction and transform coding.

In Section 3.2, we introduce the canonical correlations between $\mathbf{x}$ and $\mathbf{x}^*$, which have been called the *circularity coefficients*. These constitute a maximal invariant for $\underline{\mathbf{R}}_{xx}$ under nonsingular strictly linear transformation. They are interesting and useful for a number of reasons.

- They determine the loss in entropy that an improper Gaussian random vector incurs compared with its proper version (see Section 3.2.1).
- They enable an easy characterization of the set of complementary covariance matrices for a fixed covariance matrix (Section 3.2.3) and can be used to quantify the degree of impropriety (Section 3.3).
- They define the test statistic in a generalized likelihood-ratio test for impropriety (Section 3.4).
- They play a key role in blind source separation of complex signals using independent component analysis (Section 3.5).

The connection between the circularity coefficients and the eigenvalues of $\underline{\mathbf{R}}_{xx}$ is explored in Section 3.3.1. We find that certain functions of the circularity coefficients (in particular, the degree of impropriety) are upper- and lower-bounded by functions of the eigenvalues.[1]

3.1 Eigenvalue decomposition

In this section, we develop the augmented eigenvalue decomposition (EVD), or spectral representation, of the augmented covariance matrix of a complex *zero-mean* random vector $\mathbf{x}: \Omega \longrightarrow \mathbb{C}^n$. Proceeding along the lines of Schreier and Scharf (2003a), the idea is to start with the EVD for the composite vector of real and imaginary parts of $\mathbf{x}$. Following the notation introduced in Chapter 2, let $\mathbf{x} = \mathbf{u} + \mathrm{j}\mathbf{v}$ and $\mathbf{z} = [\mathbf{u}^{\mathrm{T}}, \mathbf{v}^{\mathrm{T}}]^{\mathrm{T}}$. The covariance matrix of the composite vector $\mathbf{z}$ is

$$\mathbf{R}_{zz} = E\mathbf{z}\mathbf{z}^{\mathrm{T}} = \begin{bmatrix} E\mathbf{u}\mathbf{u}^{\mathrm{T}} & E\mathbf{u}\mathbf{v}^{\mathrm{T}} \\ E\mathbf{v}\mathbf{u}^{\mathrm{T}} & E\mathbf{v}\mathbf{v}^{\mathrm{T}} \end{bmatrix} = \begin{bmatrix} \mathbf{R}_{uu} & \mathbf{R}_{uv} \\ \mathbf{R}_{uv}^{\mathrm{T}} & \mathbf{R}_{vv} \end{bmatrix}. \tag{3.1}$$

Using the matrix

$$\mathbf{T} = \begin{bmatrix} \mathbf{I} & \mathrm{j}\mathbf{I} \\ \mathbf{I} & -\mathrm{j}\mathbf{I} \end{bmatrix} \tag{3.2}$$

we may transform $\mathbf{z}$ to $\underline{\mathbf{x}} = [\mathbf{x}^{\mathrm{T}}, \mathbf{x}^{\mathrm{H}}]^{\mathrm{T}} = \mathbf{T}\mathbf{z}$. The augmented covariance matrix of $\mathbf{x}$ is

$$\underline{\mathbf{R}}_{xx} = E\underline{\mathbf{x}}\,\underline{\mathbf{x}}^{\mathrm{H}} = \begin{bmatrix} E\mathbf{x}\mathbf{x}^{\mathrm{H}} & E\mathbf{x}\mathbf{x}^{\mathrm{T}} \\ E\mathbf{x}^*\mathbf{x}^{\mathrm{H}} & E\mathbf{x}^*\mathbf{x}^{\mathrm{T}} \end{bmatrix} = \begin{bmatrix} \mathbf{R}_{xx} & \widetilde{\mathbf{R}}_{xx} \\ \widetilde{\mathbf{R}}_{xx}^* & \mathbf{R}_{xx}^* \end{bmatrix}, \tag{3.3}$$

which can be connected with $\mathbf{R}_{zz}$ as

$$\underline{\mathbf{R}}_{xx} = E\underline{\mathbf{x}}\,\underline{\mathbf{x}}^{\mathrm{H}} = E[(\mathbf{T}\mathbf{z})(\mathbf{T}\mathbf{z})^{\mathrm{H}}] = \mathbf{T}\mathbf{R}_{zz}\mathbf{T}^{\mathrm{H}}. \tag{3.4}$$

Since $\mathbf{T}$ is unitary up to a factor of 2, the eigenvalues of $\underline{\mathbf{R}}_{xx}$ are the eigenvalues of $\mathbf{R}_{zz}$ multiplied by 2,

$$[\lambda_1, \lambda_2, \ldots, \lambda_{2n}]^{\mathrm{T}} = \mathbf{ev}(\underline{\mathbf{R}}_{xx}) = 2\,\mathbf{ev}(\mathbf{R}_{zz}). \tag{3.5}$$

We write the EVD of $\mathbf{R}_{zz}$ as

$$\mathbf{R}_{zz} = \mathbf{U} \begin{bmatrix} \frac{1}{2}\boldsymbol{\Lambda}^{(1)} & \mathbf{0} \\ \mathbf{0} & \frac{1}{2}\boldsymbol{\Lambda}^{(2)} \end{bmatrix} \mathbf{U}^{\mathrm{T}}, \tag{3.6}$$

where the eigenvalues $\lambda_1 \geq \lambda_2 \geq \cdots \geq \lambda_{2n} \geq 0$ are contained on the diagonals of the diagonal matrices

$$\boldsymbol{\Lambda}^{(1)} = \mathbf{Diag}(\lambda_1, \lambda_3, \ldots, \lambda_{2n-1}), \tag{3.7}$$

$$\boldsymbol{\Lambda}^{(2)} = \mathbf{Diag}(\lambda_2, \lambda_4, \ldots, \lambda_{2n}). \tag{3.8}$$

We will see shortly why we are distributing the eigenvalues over $\boldsymbol{\Lambda}^{(1)}$ and $\boldsymbol{\Lambda}^{(2)}$ in this fashion.

The orthogonal transformation $\mathbf{U}$ in (3.6) is sometimes called the Karhunen–Loève transform. On inserting (3.6) into (3.4), we obtain

$$\underline{\mathbf{R}}_{xx} = \left(\tfrac{1}{2}\mathbf{T}\mathbf{U}\mathbf{T}^{\mathrm{H}}\right)\left(\mathbf{T}\begin{bmatrix} \frac{1}{2}\boldsymbol{\Lambda}^{(1)} & \mathbf{0} \\ \mathbf{0} & \frac{1}{2}\boldsymbol{\Lambda}^{(2)} \end{bmatrix}\mathbf{T}^{\mathrm{H}}\right)\left(\tfrac{1}{2}\mathbf{T}\mathbf{U}\mathbf{T}^{\mathrm{H}}\right)^{\mathrm{H}}. \tag{3.9}$$

From this, we obtain the following expression for the EVD of $\underline{\mathbf{R}}_{xx}$.

Result 3.1. *The augmented EVD of the augmented covariance matrix* $\underline{\mathbf{R}}_{xx}$ *is*

$$\underline{\mathbf{R}}_{xx} = \underline{\mathbf{U}}\,\underline{\boldsymbol{\Lambda}}\,\underline{\mathbf{U}}^{\mathrm{H}} \tag{3.10}$$

with

$$\underline{\mathbf{U}} = \tfrac{1}{2}\mathbf{T}\mathbf{U}\mathbf{T}^{\mathrm{H}}, \tag{3.11}$$

$$\underline{\boldsymbol{\Lambda}} = \frac{1}{2}\begin{bmatrix}\boldsymbol{\Lambda}^{(1)} + \boldsymbol{\Lambda}^{(2)} & \boldsymbol{\Lambda}^{(1)} - \boldsymbol{\Lambda}^{(2)} \\ \boldsymbol{\Lambda}^{(1)} - \boldsymbol{\Lambda}^{(2)} & \boldsymbol{\Lambda}^{(1)} + \boldsymbol{\Lambda}^{(2)}\end{bmatrix}. \tag{3.12}$$

A crucial remark regarding this EVD is the following. When factoring augmented matrices, we need to make sure that all factors again have the block pattern required of an augmented matrix. In the EVD, $\underline{\mathbf{U}}$ is a *widely unitary* transformation, satisfying

$$\underline{\mathbf{U}} = \begin{bmatrix}\mathbf{U}_1 & \mathbf{U}_2 \\ \mathbf{U}_2^* & \mathbf{U}_1^*\end{bmatrix}, \tag{3.13}$$

and $\underline{\mathbf{U}}\,\underline{\mathbf{U}}^{\mathrm{H}} = \underline{\mathbf{U}}^{\mathrm{H}}\,\underline{\mathbf{U}} = \mathbf{I}$. The pattern of $\underline{\mathbf{U}}$ is necessary to ensure that, in the augmented internal description

$$\underline{\boldsymbol{\xi}} = \underline{\mathbf{U}}^{\mathrm{H}}\underline{\mathbf{x}} \Leftrightarrow \boldsymbol{\xi} = \mathbf{U}_1^{\mathrm{H}}\mathbf{x} + \mathbf{U}_2^{\mathrm{T}}\mathbf{x}^*, \tag{3.14}$$

the bottom half of $\underline{\boldsymbol{\xi}}$ is indeed the conjugate of the top half. This leads to the maybe unexpected result that the augmented eigenvalue matrix $\underline{\boldsymbol{\Lambda}}$ is generally not diagonal, but has diagonal blocks instead. It becomes diagonal if and only if all eigenvalues have even multiplicity.

In general, the internal description $\boldsymbol{\xi}$ has uncorrelated components:

$$E\,\xi_i\xi_j^* = 0 \quad \text{and} \quad E\,\xi_i\xi_j = 0 \quad \text{for } i \neq j. \tag{3.15}$$

The ith component has variance

$$E|\xi_i|^2 = \tfrac{1}{2}(\lambda_{2i-1} + \lambda_{2i}) \tag{3.16}$$

and complementary variance

$$E\xi_i^2 = \tfrac{1}{2}(\lambda_{2i-1} - \lambda_{2i}). \tag{3.17}$$

Thus, if the eigenvalues do not have even multiplicity, $\boldsymbol{\xi}$ is improper. If $\mathbf{x}$ is proper, the augmented covariance matrix is $\underline{\mathbf{R}}_{xx} = \underline{\mathbf{R}}_0 = \mathbf{Diag}(\mathbf{R}_{xx}, \mathbf{R}_{xx}^*)$. If we denote $\mathbf{ev}(\mathbf{R}_{xx}) = [\mu_1, \mu_2, \ldots, \mu_n]^{\mathrm{T}}$, then $\mathbf{ev}(\underline{\mathbf{R}}_0) = [\mu_1, \mu_1, \mu_2, \mu_2, \ldots, \mu_n, \mu_n]^{\mathrm{T}}$. Propriety of $\mathbf{x}$ is a sufficient but not necessary condition for the internal description $\boldsymbol{\xi}$ to be proper.

3.1.1 Principal components

Principal-component analysis (PCA) is a classical statistical tool for data analysis, compression, and prediction. PCA determines a unitary transformation into an internal coordinate system, where the ith component makes the ith largest possible contribution to the overall variance of $\mathbf{x}$. It is easy to show that the EVD achieves this objective, using results from majorization theory (see Appendix 3).

We shall begin with strictly unitary PCA, which computes an internal description as

$$\boldsymbol{\xi} = \mathbf{U}^{\mathrm{H}}\mathbf{x}, \tag{3.18}$$

restricted to *strictly unitary* transformations. The internal representation $\boldsymbol{\xi}$ has covariance matrix $\mathbf{R}_{\xi\xi} = \mathbf{U}^{\mathrm{H}}\mathbf{R}_{xx}\mathbf{U}$. The variance of component ξ_i is $d_i = E|\xi_i|^2 = (\mathbf{R}_{\xi\xi})_{ii}$. For simplicity and without loss of generality, we assume that we order the components ξ_i such that $d_1 \geq d_2 \geq \cdots \geq d_n$. PCA aims to maximize the sum over the r largest variances d_i, for each r:

$$\max_{\mathbf{U}} \sum_{i=1}^{r} d_i, \quad r = 1, \ldots, n. \tag{3.19}$$

An immediate consequence of the majorization relation (cf. Result A3.5)

$$[d_1, d_2, \ldots, d_n]^{\mathrm{T}} = \mathbf{diag}(\mathbf{R}_{\xi\xi}) \prec \mathbf{ev}(\mathbf{R}_{\xi\xi}) = \mathbf{ev}(\mathbf{R}_{xx}) \tag{3.20}$$

is that (3.19) is maximized if $\mathbf{U}$ is chosen from the EVD of $\mathbf{R}_{xx} = \mathbf{U}\mathbf{M}\mathbf{U}^{\mathrm{H}}$, where $\mathbf{M} = \mathbf{Diag}(\mu_1, \mu_2, \ldots, \mu_n)$ contains the eigenvalues of $\mathbf{R}_{xx}$. Therefore, for each r,

$$\max_{\mathbf{U}} \sum_{i=1}^{r} d_i = \sum_{i=1}^{r} \mu_i, \quad r = 1, \ldots, n. \tag{3.21}$$

The components of $\boldsymbol{\xi}$ are called the *principal components* and are Hermitian-uncorrelated.

The approach (3.18) is generally suboptimal because it ignores the complementary covariances. For improper complex vectors, the principal components $\boldsymbol{\xi}$ must instead be determined by (3.14), using the widely unitary transformation $\underline{\mathbf{U}}$ from the EVD $\underline{\mathbf{R}}_{xx} = \underline{\mathbf{U}}\,\underline{\boldsymbol{\Lambda}}\,\underline{\mathbf{U}}^{\mathrm{H}}$. This leads to an improved result

$$\max_{\underline{\mathbf{U}}} \sum_{i=1}^{r} d_i = \sum_{i=1}^{r} \tfrac{1}{2}(\lambda_{2i-1} + \lambda_{2i}) \geq \sum_{i=1}^{r} \mu_i, \quad r = 1, \ldots, n. \tag{3.22}$$

This maximization requires the arrangement of eigenvalues in (3.7) and (3.8). The inequality in (3.22) follows from the majorization

$$[\mu_1, \mu_1, \mu_2, \mu_2, \ldots, \mu_n, \mu_n]^{\mathrm{T}} = \mathbf{ev}(\underline{\mathbf{R}}_0) \prec \mathbf{ev}(\underline{\mathbf{R}}_{xx}) = [\lambda_1, \lambda_2, \ldots, \lambda_{2n}]^{\mathrm{T}}, \tag{3.23}$$

which in turn is a consequence of Result A3.7. The last result shows that the eigenvalues of the augmented covariance matrix of an improper vector are more spread out than those of a proper vector with augmented covariance matrix $\underline{\mathbf{R}}_0 = \mathbf{Diag}(\mathbf{R}_{xx}, \mathbf{R}_{xx}^*)$. We will revisit this finding later on.

3.1.2 Rank reduction and transform coding

The maximization property (3.19) makes PCA an ideal candidate for rank reduction and transform coding, as shown in Fig. 3.1. Example A3.5 in Appendix 3 considers the equivalent structure for real-valued vectors.

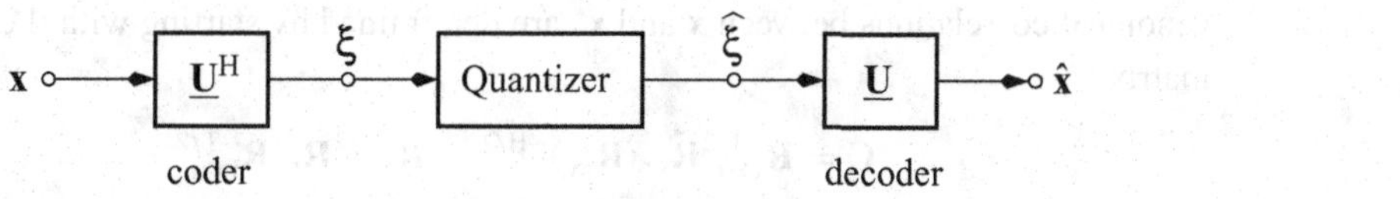

Figure 3.1 A widely unitary rank-reduction/transform coder.

The complex random vector $\mathbf{x} = [x_1, x_2, \ldots, x_n]^{\mathrm{T}}$, which is assumed to be zero-mean Gaussian, is passed through a widely unitary coder $\underline{\mathbf{U}}^{\mathrm{H}}$. The output of the coder is $\boldsymbol{\xi} = \mathbf{U}_1^{\mathrm{H}}\mathbf{x} + \mathbf{U}_2^{\mathrm{T}}\mathbf{x}^*$, which is subsequently processed by a bank of n scalar complex quantizers. The quantizer output $\widehat{\boldsymbol{\xi}}$ is then decoded as $\hat{\mathbf{x}} = \mathbf{U}_1\widehat{\boldsymbol{\xi}} + \mathbf{U}_2\widehat{\boldsymbol{\xi}}^*$. From (3.20) we know that any Schur-concave function of the variances $\{d_i\}_{i=1}^n$ will be minimized if the coder $\underline{\mathbf{U}}^{\mathrm{H}}$ decorrelates the quantizer input $\boldsymbol{\xi}$. In particular, consider the Schur-concave mean-squared error

$$\mathrm{MSE} = \tfrac{1}{2}E\|\underline{\mathbf{x}} - \hat{\underline{\mathbf{x}}}\|^2 = \tfrac{1}{2}E\|\underline{\mathbf{U}}^{\mathrm{H}}(\underline{\boldsymbol{\xi}} - \widehat{\underline{\boldsymbol{\xi}}})\|^2 = \tfrac{1}{2}E\|\underline{\boldsymbol{\xi}} - \widehat{\underline{\boldsymbol{\xi}}}\|^2 = E\|\boldsymbol{\xi} - \widehat{\boldsymbol{\xi}}\|^2. \tag{3.24}$$

A good model for the quantization error in the internal coordinate system is

$$E\|\boldsymbol{\xi} - \widehat{\boldsymbol{\xi}}\|^2 = \sum_{i=1}^{n} d_i f(b_i), \tag{3.25}$$

where $f(b_i)$ is a decreasing function of the number of bits b_i spent on quantizing ξ_i. We will spend more bits on components with higher variance, i.e., $b_1 \geq b_2 \geq \cdots \geq b_n$, and, therefore, $f(b_1) \leq f(b_2) \leq \cdots \leq f(b_n)$. For a fixed but arbitrary bit assignment, Result A3.4 in Appendix 3 shows that the linear function (3.25) is indeed a Schur-concave function of $\{d_i\}_{i=1}^n$. The minimum MSE is therefore

$$\mathrm{MMSE} = \tfrac{1}{2}\sum_{i=1}^{n}(\lambda_{2i-1} + \lambda_{2i})f(b_i). \tag{3.26}$$

There are two common choices for $f(b_i)$ that deserve explicit mention.

- To model rank-r reduction (sometimes called zonal sampling), we set $f(b_i) = 0$ for $i = 1, \ldots, r$ and $f(b_i) = 1$ for $i = r+1, \ldots, n$. Gaussianity need not be assumed for rank reduction.
- For fine quantizers with a large number of bits, we may employ the high-resolution assumption, as explained by Gersho and Gray (1992), and set $f(b_i) = c\,2^{-b_i}$. This assumes that, of the b_i bits spent on quantizing ξ_i, $b_i/2$ bits each go toward quantizing real and imaginary parts. The constant c is dependent on the distribution of ξ_i.

3.2 Circularity coefficients

We now turn to finding a maximal invariant (a complete set of invariants) for $\underline{\mathbf{R}}_{xx}$ under nonsingular strictly linear transformation. Such a set is given by the canonical correlations between $\mathbf{x}$ and its conjugate $\mathbf{x}^*$. Canonical correlations in the general case will be discussed in much more detail in Chapter 4. Assuming $\mathbf{R}_{xx}$ has full rank, the canonical correlations between $\mathbf{x}$ and $\mathbf{x}^*$ are determined by starting with the coherence matrix

$$\mathbf{C} = \mathbf{R}_{xx}^{-1/2}\widetilde{\mathbf{R}}_{xx}(\mathbf{R}_{xx}^*)^{-\mathrm{H}/2} = \mathbf{R}_{xx}^{-1/2}\widetilde{\mathbf{R}}_{xx}\mathbf{R}_{xx}^{-\mathrm{T}/2}. \tag{3.27}$$

Since $\mathbf{C}$ is complex symmetric, $\mathbf{C} = \mathbf{C}^{\mathrm{T}}$, yet not Hermitian symmetric, $\mathbf{C} \neq \mathbf{C}^{\mathrm{H}}$, there exists a special singular value decomposition (SVD), called the *Takagi factorization*, which is

$$\mathbf{C} = \mathbf{F}\mathbf{K}\mathbf{F}^{\mathrm{T}}. \tag{3.28}$$

The Takagi factorization is discussed more thoroughly in Section 3.2.2. The complex matrix $\mathbf{F}$ is unitary, and $\mathbf{K} = \mathbf{Diag}(k_1, k_2, \ldots, k_n)$ contains the canonical correlations $1 \geq k_1 \geq k_2 \geq \cdots \geq k_n \geq 0$ on its diagonal. The squared canonical correlations k_i^2 are the eigenvalues of the squared coherence matrix $\mathbf{C}\mathbf{C}^{\mathrm{H}} = \mathbf{R}_{xx}^{-1/2}\widetilde{\mathbf{R}}_{xx}\mathbf{R}_{xx}^{-*}\widetilde{\mathbf{R}}_{xx}^*\mathbf{R}_{xx}^{-\mathrm{H}/2}$, or equivalently, of the matrix $\mathbf{R}_{xx}^{-1}\widetilde{\mathbf{R}}_{xx}\mathbf{R}_{xx}^{-*}\widetilde{\mathbf{R}}_{xx}^*$ because

$$\mathbf{K}\mathbf{K}^{\mathrm{H}} = \mathbf{F}^{\mathrm{H}}\mathbf{R}_{xx}^{-1/2}\widetilde{\mathbf{R}}_{xx}\mathbf{R}_{xx}^{-*}\widetilde{\mathbf{R}}_{xx}^*\mathbf{R}_{xx}^{-\mathrm{H}/2}\mathbf{F}. \tag{3.29}$$

The canonical correlations between $\mathbf{x}$ and $\mathbf{x}^*$ are invariant to the choice of a square root for $\mathbf{R}_{xx}$, and they are a maximal invariant for $\underline{\mathbf{R}}_{xx}$ under nonsingular strictly linear transformation of $\mathbf{x}$. Therefore, any function of $\underline{\mathbf{R}}_{xx}$ that is invariant under nonsingular strictly linear transformation *must* be a function of these canonical correlations *only*.

The internal description

$$\boldsymbol{\xi} = \mathbf{F}^{\mathrm{H}}\mathbf{R}_{xx}^{-1/2}\mathbf{x} = \mathbf{A}\mathbf{x} \tag{3.30}$$

is said to be given in *canonical coordinates*. We will adopt the following terminology by Eriksson and Koivunen (2006).

Definition 3.1. *Vectors that are uncorrelated with unit variance, but possibly improper, are called* strongly uncorrelated. *The transformation* $\mathbf{A} = \mathbf{F}^{\mathrm{H}}\mathbf{R}_{xx}^{-1/2}$, *which transforms* $\mathbf{x}$ *into canonical coordinates* $\boldsymbol{\xi}$, *is called the* strong uncorrelating transform *(SUT). The canonical correlations* k_i *are referred to as* circularity coefficients, *and the set* $\{k_i\}_{i=1}^n$ *as the* circularity spectrum *of* $\mathbf{x}$.

The term "circularity coefficient" is not entirely accurate insofar as the circularity coefficients only characterize *second-order* circularity, or (im-)propriety. Thus, the name *impropriety coefficients* would have been more suitable. Moreover, the insight that the circularity coefficients are canonical correlations is critical because it enables us to utilize a wealth of results on canonical correlations in the literature.[2]

The canonical coordinates are strongly uncorrelated, i.e., they are uncorrelated,

$$E\,\xi_i\xi_j^* = E\,\xi_i\xi_j = 0 \quad \text{for } i \neq j, \tag{3.31}$$

and have unit variance, $E\,|\xi_i|^2 = 1$. However, they are generally improper as

$$E\,\xi_i^2 = k_i. \tag{3.32}$$

The circularity coefficients k_i measure the correlations between the white, unit-norm canonical coordinates $\boldsymbol{\xi}$ and their conjugates $\boldsymbol{\xi}^*$. More precisely, the circularity coefficients k_i are the cosines of the canonical angles between the linear subspaces spanned by $\boldsymbol{\xi}$ and the complex conjugate $\boldsymbol{\xi}^*$. If these angles are small, then $\mathbf{x}^*$ may be *linearly* estimated from $\mathbf{x}$, indicating that $\mathbf{x}$ is improper – obviously, $\mathbf{x}^*$ can always be perfectly estimated from $\mathbf{x}$ if *widely linear* operations are allowed. If these angles are large, then $\mathbf{x}^*$ may not be *linearly* estimated from $\mathbf{x}$, indicating a proper $\mathbf{x}$. These angles are invariant with respect to nonsingular linear transformation.

3.2.1 Entropy

Combining our results so far, we may factor $\underline{\mathbf{R}}_{xx}$ as

$$\underline{\mathbf{R}}_{xx} = \begin{bmatrix} \mathbf{R}_{xx} & \widetilde{\mathbf{R}}_{xx} \\ \widetilde{\mathbf{R}}^*_{xx} & \mathbf{R}^*_{xx} \end{bmatrix} = \begin{bmatrix} \mathbf{R}^{1/2}_{xx} & \mathbf{0} \\ \mathbf{0} & \mathbf{R}^{*/2}_{xx} \end{bmatrix} \begin{bmatrix} \mathbf{F} & \mathbf{0} \\ \mathbf{0} & \mathbf{F}^* \end{bmatrix} \begin{bmatrix} \mathbf{I} & \mathbf{K} \\ \mathbf{K} & \mathbf{I} \end{bmatrix} \begin{bmatrix} \mathbf{F}^{\mathrm{H}} & \mathbf{0} \\ \mathbf{0} & \mathbf{F}^{\mathrm{T}} \end{bmatrix} \begin{bmatrix} \mathbf{R}^{\mathrm{H}/2}_{xx} & \mathbf{0} \\ \mathbf{0} & \mathbf{R}^{\mathrm{T}/2}_{xx} \end{bmatrix}. \tag{3.33}$$

Note that each factor is an augmented matrix. This factorization establishes

$$\det \underline{\mathbf{R}}_{xx} = \det{}^2\mathbf{R}_{xx} \det(\mathbf{I} - \mathbf{K}\mathbf{K}^{\mathrm{H}}) = \det{}^2\mathbf{R}_{xx} \prod_{i=1}^{n}(1 - k_i^2). \tag{3.34}$$

This allows us to derive the following connection between the entropy of an improper Gaussian random vector with augmented covariance matrix $\underline{\mathbf{R}}_{xx}$ and the corresponding proper Gaussian random vector with covariance matrix $\mathbf{R}_{xx}$.

Result 3.2. *The entropy of a complex improper Gaussian random vector* $\mathbf{x}$ *is*

$$\begin{aligned} H_{\text{improper}} &= \tfrac{1}{2}\log[(\pi \mathrm{e})^{2n} \det \underline{\mathbf{R}}_{xx}] \\ &= \underbrace{\log[(\pi \mathrm{e})^{n} \det \mathbf{R}_{xx}]}_{H_{\text{proper}}} + \underbrace{\tfrac{1}{2}\log \prod_{i=1}^{n}(1 - k_i^2)}_{-I(\mathbf{x};\mathbf{x}^*)}, \end{aligned} \tag{3.35}$$

where H_{proper} *is the entropy of a proper Gaussian random vector with the same Hermitian covariance matrix* $\mathbf{R}_{xx}$ *(but* $\widetilde{\mathbf{R}}_{xx} = \mathbf{0}$*), and* $I(\mathbf{x};\mathbf{x}^*)$ *is the (nonnegative) mutual information between* $\mathbf{x}$ *and* $\mathbf{x}^*$*. Therefore, the entropy is maximized if and only if* $\mathbf{x}$ *is proper.*

If $\mathbf{x}$ is improper, the loss in entropy compared with the proper case is the mutual information between $\mathbf{x}$ and $\mathbf{x}^*$, which is a function of the circularity spectrum.

3.2.2 Strong uncorrelating transform (SUT)

The Takagi factorization is a special SVD for a complex symmetric matrix $\mathbf{C} = \mathbf{C}^{\mathrm{T}}$. But why do we need the Takagi factorization $\mathbf{C} = \mathbf{F}\mathbf{K}\mathbf{F}^{\mathrm{T}}$ rather than simply a regular SVD $\mathbf{C} = \mathbf{U}\mathbf{K}\mathbf{V}^{\mathrm{H}}$? The reason is that the canonical coordinates $\boldsymbol{\xi}$ and $\boldsymbol{\xi}^*$ are supposed to be complex conjugates of each other. If we used a regular SVD $\mathbf{C} = \mathbf{U}\mathbf{K}\mathbf{V}^{\mathrm{H}}$, then $\mathbf{V}^{\mathrm{H}}\mathbf{R}^{-*/2}_{xx}\mathbf{x}^*$ would not generally be the complex conjugate of $\boldsymbol{\xi} = \mathbf{U}^{\mathrm{H}}\mathbf{R}^{-1/2}_{xx}\mathbf{x}$. For this we need $\mathbf{V} = \mathbf{U}^*$. So how exactly is the Takagi factorization determined from the SVD? Because the matrix $\mathbf{C}$ is symmetric, it has two SVDs:

$$\mathbf{C} = \mathbf{U}\mathbf{K}\mathbf{V}^{\mathrm{H}} = \mathbf{V}^*\mathbf{K}\mathbf{U}^{\mathrm{T}} = \mathbf{C}^{\mathrm{T}}. \tag{3.36}$$

Now, since

$$\begin{aligned}\mathbf{C}\mathbf{C}^{\mathrm{H}} = \mathbf{U}\mathbf{K}^2\mathbf{U}^{\mathrm{H}} = \mathbf{V}^*\mathbf{K}^2\mathbf{V}^{\mathrm{T}} &\Leftrightarrow \mathbf{K}^2\mathbf{U}^{\mathrm{H}}\mathbf{V}^* = \mathbf{U}^{\mathrm{H}}\mathbf{V}^*\mathbf{K}^2 \\ &\Leftrightarrow \mathbf{K}\mathbf{U}^{\mathrm{H}}\mathbf{V}^* = \mathbf{U}^{\mathrm{H}}\mathbf{V}^*\mathbf{K},\end{aligned} \tag{3.37}$$

the unitary matrix $\mathbf{U}^{\mathrm{H}}\mathbf{V}^*$ commutes with $\mathbf{K}$. This is possible only if the (i, j)th element of $\mathbf{U}^{\mathrm{H}}\mathbf{V}^*$ is zero whenever $k_i \neq k_j$. Thus, every $\mathbf{C}$ can be expressed as

$$\mathbf{C} = \mathbf{U}\mathbf{K}\mathbf{D}\mathbf{U}^{\mathrm{T}} \tag{3.38}$$

with $\mathbf{D} = \mathbf{V}^{\mathrm{H}}\mathbf{U}^*$. Assume that among the circularity coefficients $\{k_i\}_{i=1}^n$ there are N distinct coefficients, denoted by $\sigma_1, \ldots, \sigma_N$. Their respective multiplicities are $m_1, \ldots, m_N$. Then we may write

$$\mathbf{D} = \mathbf{Diag}(\mathbf{D}_1, \mathbf{D}_2, \ldots, \mathbf{D}_N), \tag{3.39}$$

where $\mathbf{D}_i$ is an $m_i \times m_i$ unitary and symmetric matrix, and

$$\mathbf{K}\mathbf{D} = \mathbf{Diag}(\sigma_1\mathbf{D}_1, \sigma_2\mathbf{D}_2, \ldots, \sigma_N\mathbf{D}_N). \tag{3.40}$$

In the special case in which all circularity coefficients are distinct, we have $\sigma_i = k_i$ and $\mathbf{D}_i$ is a scalar with unit magnitude. Therefore,

$$\mathbf{D} = \mathbf{Diag}(\mathrm{e}^{\mathrm{j}\theta_1}, \mathrm{e}^{\mathrm{j}\theta_2}, \ldots, \mathrm{e}^{\mathrm{j}\theta_n}) \tag{3.41}$$

$$\mathbf{K}\mathbf{D} = \mathbf{Diag}(k_1\mathrm{e}^{\mathrm{j}\theta_1}, k_2\mathrm{e}^{\mathrm{j}\theta_2}, \ldots, k_n\mathrm{e}^{\mathrm{j}\theta_n}). \tag{3.42}$$

Then we have

$$\mathbf{C} = \mathbf{U}\mathbf{D}^{1/2}\mathbf{K}\mathbf{D}^{\mathrm{T}/2}\mathbf{U}^{\mathrm{T}} \tag{3.43}$$

with

$$\mathbf{D}^{1/2} = \mathbf{Diag}(\pm\mathrm{e}^{\mathrm{j}\theta_1/2}, \pm\mathrm{e}^{\mathrm{j}\theta_2/2}, \ldots, \pm\mathrm{e}^{\mathrm{j}\theta_n/2}) \tag{3.44}$$

with arbitrary signs, so that $\mathbf{F} = \mathbf{U}\mathbf{D}^{1/2}$ achieves the Takagi factorization $\mathbf{C} = \mathbf{F}\mathbf{K}\mathbf{F}^{\mathrm{T}}$. This shows how to obtain the Takagi factorization from the SVD. We note that, if only the circularity coefficients k_i need to be computed and the canonical coordinates $\boldsymbol{\xi}$ are not required, a regular SVD will suffice.

The SVD $\mathbf{C} = \mathbf{U}\mathbf{K}\mathbf{V}^{\mathrm{H}}$ of any matrix $\mathbf{C}$ with distinct singular values is unique up to multiplication of $\mathbf{U}$ from the right by a unitary diagonal matrix. If $\mathbf{C} = \mathbf{C}^{\mathrm{T}}$, the Takagi factorization determines this unitary diagonal factor in such a way that the matrix of left singular vectors is the conjugate of the matrix of right singular vectors. Hence, if all circularity coefficients are distinct, the Takagi factorization is unique up to the sign of the diagonal elements in $\mathbf{D}^{1/2}$. The following result then follows.

Result 3.3. *For distinct circularity coefficients, the strong uncorrelating transform* $\mathbf{A} = \mathbf{F}^{\mathrm{H}}\mathbf{R}_{xx}^{-1/2}$ *is unique up to the sign of its rows,* regardless *of the choice of the inverse square root of* $\mathbf{R}_{xx}$.

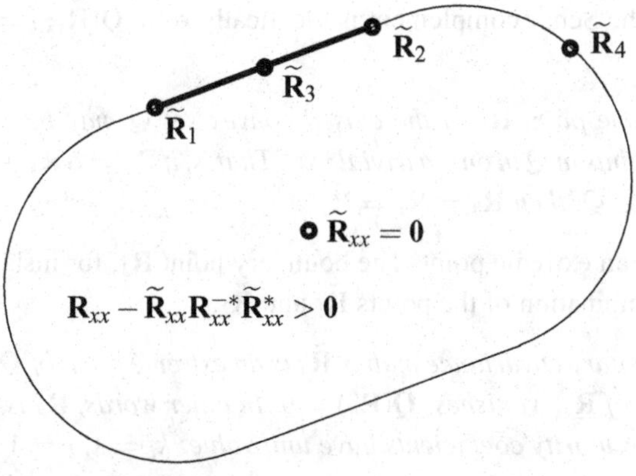

Figure 3.2 Geometry of the set $\mathcal{Q}$.

To see this, consider any inverse square root $\mathbf{G}\mathbf{R}_{xx}^{-1/2}$ for arbitrary choice of a unitary matrix $\mathbf{G}$. The Takagi factorization of the corresponding coherence matrix is $\mathbf{C} = \mathbf{G}\mathbf{R}_{xx}^{-1/2}\widetilde{\mathbf{R}}_{xx}\mathbf{R}_{xx}^{-\mathrm{T}/2}\mathbf{G}^{\mathrm{T}} = \mathbf{G}\mathbf{F}\mathbf{K}\mathbf{F}^{\mathrm{T}}\mathbf{G}^{\mathrm{T}}$. Hence, the strong uncorrelating transform $\mathbf{A} = \mathbf{F}^{\mathrm{H}}\mathbf{G}^{\mathrm{H}}\mathbf{G}\mathbf{R}_{xx}^{-1/2} = \mathbf{F}^{\mathrm{H}}\mathbf{R}_{xx}^{-1/2}$ is independent of $\mathbf{G}$.

On the other hand, if there are repeated circularity coefficients, the strong uncorrelating transform is no longer unique. If $\mathbf{D}_i$ in (3.39) is $m_i \times m_i$ with $m_i \geq 2$, there is an infinite number of ways of decomposing $\mathbf{D}_i = \mathbf{D}_i^{1/2}\mathbf{D}_i^{\mathrm{T}/2}$. For instance, any given decomposition of $\mathbf{D}_i$ can be modified by a *real* orthogonal $m_i \times m_i$ matrix $\mathbf{G}$ as $\mathbf{D}_i = (\mathbf{D}_i^{1/2}\mathbf{G})(\mathbf{G}^{\mathrm{T}}\mathbf{D}_i^{\mathrm{T}/2})$.

3.2.3 Characterization of complementary covariance matrices

We may now give two different characterizations of the set $\mathcal{Q}$ of complementary covariance matrices for a fixed covariance matrix $\mathbf{R}_{xx}$. When we say "characterization," we mean necessary and sufficient conditions for a matrix $\widetilde{\mathbf{R}}_{xx}$ to be a valid complementary covariance matrix for given covariance matrix $\mathbf{R}_{xx}$. The first characterization follows directly from the positive semidefinite property of the augmented covariance matrix $\underline{\mathbf{R}}_{xx}$, which implies a positive semidefinite Schur complement:

$$\mathbf{Q}(\widetilde{\mathbf{R}}_{xx}) = \mathbf{R}_{xx} - \widetilde{\mathbf{R}}_{xx}\mathbf{R}_{xx}^{-*}\widetilde{\mathbf{R}}_{xx}^{*} \geq \mathbf{0}. \tag{3.45}$$

This shows that $\mathcal{Q}$ is convex and compact. Alternatively, $\mathcal{Q}$ can be characterized by

$$\widetilde{\mathbf{R}}_{xx} = \mathbf{R}_{xx}^{1/2}\mathbf{F}\mathbf{K}\mathbf{F}^{\mathrm{T}}\mathbf{R}_{xx}^{\mathrm{T}/2}, \tag{3.46}$$

which follows from (3.27) and (3.28). Here, $\mathbf{F}$ is a unitary matrix, and $\mathbf{K}$ is a diagonal matrix of circularity coefficients $0 \leq k_i \leq 1$.

We will use the characterizations (3.45) and (3.46) to illuminate the structure of the convex and compact set $\mathcal{Q}$. This was first presented by Schreier *et al.* (2005). The geometry of $\mathcal{Q}$ is depicted in Fig. 3.2. The interior of $\mathcal{Q}$ contains complementary covariance matrices $\widetilde{\mathbf{R}}_{xx}$ for which the Schur complement is positive definite: $\mathbf{Q}(\widetilde{\mathbf{R}}_{xx}) > \mathbf{0}$. Boundary points are rank-deficient points. However, as we will show now, only boundary points where the Schur complement is identically zero, $\mathbf{Q}(\widetilde{\mathbf{R}}_{xx}) = \mathbf{0}$, are also extreme points.

Definition 3.2. *An extreme point $\widetilde{\mathbf{R}}_{\#}$ of the closed convex set $\mathcal{Q}$ may be written as a convex combination of points in $\mathcal{Q}$ in only a trivial way. That is, if $\widetilde{\mathbf{R}}_{\#} = \alpha\widetilde{\mathbf{R}}_1 + (1-\alpha)\widetilde{\mathbf{R}}_2$ with $0 < \alpha < 1$, $\widetilde{\mathbf{R}}_1, \widetilde{\mathbf{R}}_2 \in \mathcal{Q}$, then $\widetilde{\mathbf{R}}_{\#} = \widetilde{\mathbf{R}}_1 = \widetilde{\mathbf{R}}_2$.*

In Fig. 3.2, only $\widetilde{\mathbf{R}}_4$ is an extreme point. The boundary point $\widetilde{\mathbf{R}}_3$, for instance, can be obtained as a convex combination of the points $\widetilde{\mathbf{R}}_1$ and $\widetilde{\mathbf{R}}_2$.

Result 3.4. *A complementary covariance matrix $\widetilde{\mathbf{R}}_{\#}$ is an extreme point of $\mathcal{Q}$ if and only if the Schur complement of $\underline{\mathbf{R}}_{xx}$ vanishes: $\mathbf{Q}(\widetilde{\mathbf{R}}_{\#}) = \mathbf{0}$. In other words, $\widetilde{\mathbf{R}}_{\#}$ is an extreme point if and only if all circularity coefficients have unit value: $k_i = 1, i = 1, \ldots, n$.*

We prove this result by first showing that a point $\widetilde{\mathbf{R}}_{xx} \in \mathcal{Q}$ with nonzero Schur complement cannot be an extreme point. There exists at least one circularity coefficient, say k_j, with $k_j < 1$. Now choose $\varepsilon > 0$ such that $k_j + \varepsilon \leq 1$. Define $\mathbf{K}_1$ such that it agrees with $\mathbf{K}$ except that the jth entry, k_j, is replaced with $k_j + \varepsilon$. Similarly, define $\mathbf{K}_2$ such that it agrees with $\mathbf{K}$ except that the jth entry, k_j, is replaced with $k_j - \varepsilon$. Then, using (3.46), write

$$\widetilde{\mathbf{R}}_{xx} = \tfrac{1}{2}\mathbf{R}_{xx}^{1/2}\mathbf{F}(\mathbf{K}_1 + \mathbf{K}_2)\mathbf{F}^{\mathrm{T}}\mathbf{R}_{xx}^{\mathrm{T}/2} \tag{3.47}$$

$$= \tfrac{1}{2}\widetilde{\mathbf{R}}_1 + \tfrac{1}{2}\widetilde{\mathbf{R}}_2. \tag{3.48}$$

This expresses $\widetilde{\mathbf{R}}_{xx}$ as a nontrivial convex combination of $\widetilde{\mathbf{R}}_1 = \mathbf{R}_{xx}^{1/2}\mathbf{F}\mathbf{K}_1\mathbf{F}^{\mathrm{T}}\mathbf{R}_{xx}^{\mathrm{T}/2}$ and $\widetilde{\mathbf{R}}_2 = \mathbf{R}_{xx}^{1/2}\mathbf{F}\mathbf{K}_2\mathbf{F}^{\mathrm{T}}\mathbf{R}_{xx}^{\mathrm{T}/2}$, with $\alpha = 1/2$. Hence, $\widetilde{\mathbf{R}}_{xx}$ is not an extreme point.

We now assume that the Schur complement $\mathbf{Q}(\widetilde{\mathbf{R}}_{xx}) = \mathbf{R}_{xx} - \widetilde{\mathbf{R}}_{xx}\mathbf{R}_{xx}^{-*}\widetilde{\mathbf{R}}_{xx}^{*} = \mathbf{0}$ for some $\widetilde{\mathbf{R}}_{xx}$. We first note that

$$\alpha(1-\alpha)(\widetilde{\mathbf{R}}_1 - \widetilde{\mathbf{R}}_2)\mathbf{R}_{xx}^{-*}(\widetilde{\mathbf{R}}_1 - \widetilde{\mathbf{R}}_2)^{\mathrm{H}} \geq \mathbf{0} \tag{3.49}$$

with equality if and only if $\widetilde{\mathbf{R}}_1 = \widetilde{\mathbf{R}}_2$. This implies

$$(\alpha\widetilde{\mathbf{R}}_1 + (1-\alpha)\widetilde{\mathbf{R}}_2)\mathbf{R}_{xx}(\alpha\widetilde{\mathbf{R}}_1 + (1-\alpha)\widetilde{\mathbf{R}}_2)^{*} \leq \alpha\widetilde{\mathbf{R}}_1\mathbf{R}_{xx}\widetilde{\mathbf{R}}_1^{*} + (1-\alpha)\widetilde{\mathbf{R}}_2\mathbf{R}_{xx}\widetilde{\mathbf{R}}_2^{*}. \tag{3.50}$$

Thus, we find

$$\alpha\,\mathbf{Q}(\widetilde{\mathbf{R}}_1) + (1-\alpha)\mathbf{Q}(\widetilde{\mathbf{R}}_2) \leq \mathbf{Q}(\widetilde{\mathbf{R}}_{xx}) \tag{3.51}$$

for $\widetilde{\mathbf{R}}_{xx} = \alpha\widetilde{\mathbf{R}}_1 + (1-\alpha)\widetilde{\mathbf{R}}_2$, with equality if and only if $\widetilde{\mathbf{R}}_1 = \widetilde{\mathbf{R}}_2$. This makes $\mathbf{Q}(\widetilde{\mathbf{R}}_{xx})$ a strictly matrix-concave function. Since $\mathbf{Q}(\widetilde{\mathbf{R}}_1) \geq \mathbf{0}$ and $\mathbf{Q}(\widetilde{\mathbf{R}}_2) \geq \mathbf{0}$, a vanishing $\mathbf{Q}(\widetilde{\mathbf{R}}_{xx}) = \mathbf{0}$ requires $\widetilde{\mathbf{R}}_1 = \widetilde{\mathbf{R}}_2 = \widetilde{\mathbf{R}}_{xx}$. Thus, $\widetilde{\mathbf{R}}_{xx}$ is indeed an extreme point of $\mathcal{Q}$, which concludes the proof of the result.

3.3 Degree of impropriety

Building upon our insights from the previous section, we may now ask how to quantify the degree of impropriety of $\mathbf{x}$. Propriety is preserved by strictly linear (but not widely linear) transformation. This suggests that a measure for the degree of impropriety should also be invariant under linear transformation. Such a measure of impropriety must then be a function of the circularity coefficients k_i because the circularity coefficients constitute a maximal invariant for $\underline{\mathbf{R}}_{xx}$ under nonsingular linear transformation.[3]

There are several functions that are used to measure the multivariate association between two vectors, as discussed in more detail in Chapter 4. Applied to **x** and **x***, some examples are

$$\rho_1 = 1 - \prod_{i=1}^{r}(1 - k_i^2), \tag{3.52}$$

$$\rho_2 = \prod_{i=1}^{r} k_i^2, \tag{3.53}$$

$$\rho_3 = \frac{1}{n}\sum_{i=1}^{r} k_i^2. \tag{3.54}$$

These functions are defined for $r = 1, \ldots, n$. For full rank $r = n$, they can also be written in terms of $\mathbf{R}_{xx}$ and $\widetilde{\mathbf{R}}_{xx}$. Then, from (3.34), we obtain

$$\rho_1 = 1 - \frac{\det \underline{\mathbf{R}}_{xx}}{\det^2 \mathbf{R}_{xx}} = 1 - \frac{\det \mathbf{Q}}{\det \mathbf{R}_{xx}} \tag{3.55}$$

and from (3.29) we obtain

$$\rho_2 = \frac{\det(\widetilde{\mathbf{R}}_{xx}\mathbf{R}_{xx}^{-*}\widetilde{\mathbf{R}}_{xx}^{*})}{\det \mathbf{R}_{xx}} \tag{3.56}$$

$$\rho_3 = \frac{1}{n}\,\mathrm{tr}(\mathbf{R}_{xx}^{-1}\widetilde{\mathbf{R}}_{xx}\mathbf{R}_{xx}^{-*}\widetilde{\mathbf{R}}_{xx}^{*}). \tag{3.57}$$

These measures all satisfy $0 \le \rho_i \le 1$. However, only ρ_3 has the two properties that $\rho_3 = 0$ indicates the proper case, i.e., $k_i = 0$ for all $i = 1, \ldots, n$, and $\rho_3 = 1$ the maximally improper case, in the sense that $k_i = 1$ for all $i = 1, \ldots, n$. Measure ρ_2 is 0 if at least one k_i is 0, and ρ_1 is 1 if at least one k_i is 1.

While a case can be made for any of these measures, or many other functions of k_i, ρ_1 seems to be most compelling since it relates the entropy of an improper Gaussian random vector to that of the corresponding proper version through (3.35). Moreover, as we will see in Section 3.4, ρ_1 is also used as the test statistic in a generalized likelihood-ratio test for impropriety. For this reason, we will focus on ρ_1 in the remainder of this section.

Example 3.1. Figure 3.3 depicts a QPSK signalling constellation with I/Q imbalance characterized by gain imbalance (factor) $G > 0$ and quadrature skew ϕ. The four equally likely signal points are $\{\pm\mathrm{j}, \pm G\mathrm{e}^{\mathrm{j}\phi}\}$. We find

$$\widetilde{R}_{xx} = E\,x^2 = \tfrac{1}{4}(\mathrm{j}^2 + (-\mathrm{j})^2 + G^2\mathrm{e}^{2\mathrm{j}\phi} + (-G)^2\mathrm{e}^{2\mathrm{j}\phi}) = \tfrac{1}{2}(G^2\mathrm{e}^{2\mathrm{j}\phi} - 1), \tag{3.58}$$

$$R_{xx} = E|x|^2 = \tfrac{1}{2}(1 + G^2). \tag{3.59}$$

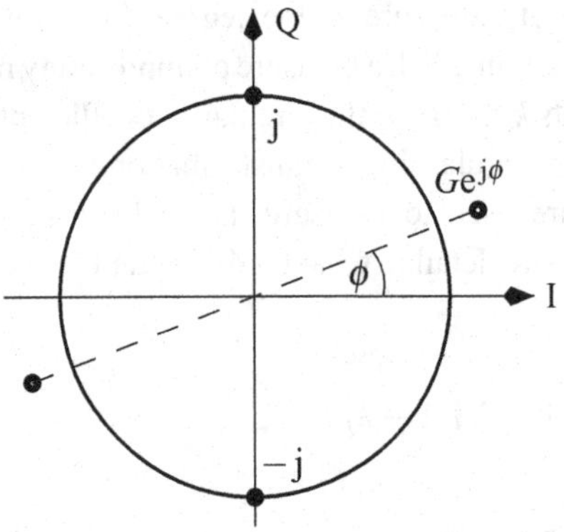

Figure 3.3 QPSK with I/Q imbalance.

Since x is *scalar*, with variance R_{xx} and complementary variance $\widetilde{R}_{xx}$, the degree of impropriety ρ_1 becomes particularly simple:

$$\rho_1 = \frac{|\widetilde{R}_{xx}|^2}{R_{xx}^2} = \frac{G^4 - 2G^2\cos(2\phi) + 1}{(1+G^2)^2} = \begin{cases} \dfrac{(G^2-1)^2}{(G^2+1)^2}, & \phi = 0 \\ 1, & \phi = \pi/2, \\ \frac{1}{2}(1-\cos(2\phi)), & G = 1. \end{cases} \tag{3.60}$$

Perfect I/Q balance is obtained with $G = 1$ and $\phi = 0$. QPSK with perfect I/Q balance is proper, i.e., $\rho_1 = 0$. The worst possible I/Q imbalance $\phi = \pi/2$ results in a maximally improper random variable, i.e., $\rho_1 = 1$, irrespective of G.

3.3.1 Upper and lower bounds

So far, we have developed two internal descriptions $\boldsymbol{\xi}$ for $\mathbf{x}$: the principal components (3.14), which are found by a widely unitary transformation of $\mathbf{x}$, and the canonical coordinates (3.30), which are found by a strictly linear transformation of $\mathbf{x}$. Both principal components and canonical coordinates are uncorrelated, i.e.,

$$E\,\xi_i\xi_j^* = E\,\xi_i\xi_j = 0 \quad \text{for } i \neq j, \tag{3.61}$$

but only the canonical coordinates have unit variance. Both principal components and canonical coordinates are generally improper with

$$E\,\xi_i^2 = \begin{cases} \frac{1}{2}(\lambda_{2i-1} - \lambda_{2i}), & \text{if } \boldsymbol{\xi} \text{ are principal components,} \\ k_i, & \text{if } \boldsymbol{\xi} \text{ are canonical coordinates.} \end{cases} \tag{3.62}$$

It is natural to ask whether there is a connection between the eigenvalues λ_i and the circularity coefficients k_i. There is indeed. The eigenvalue spectrum $\{\lambda_i\}_{i=1}^{2n}$ restricts the possibilities for the circularity spectrum $\{k_i\}_{i=1}^{n}$, albeit in a fairly intricate way. In the general setup – not restricted to the conjugate pair $\mathbf{x}$ and $\mathbf{x}^*$ – this has been explored by Drury (2002), who characterizes admissible k_is for given eigenvalues $\{\lambda_i\}$. The results are very involved, which is due to the fact that the singular values of the sum of two matrices are not easily characterized. It is much easier to develop bounds on certain functions of $\{k_i\}$ in terms of the eigenvalues $\{\lambda_i\}$. In particular, we are interested in bounds on the degree of impropriety ρ_1 if $\{\lambda_i\}$ are known. We first state the upper bound on ρ_1.

Result 3.5. *The degree of impropriety ρ_1 of a vector* **x** *with prescribed eigenvalues $\{\lambda_i\}$ of the augmented covariance matrix $\underline{\mathbf{R}}_{xx}$ is upper-bounded by*

$$\rho_1 = 1 - \prod_{i=1}^{r}(1 - k_i^2) \le 1 - \prod_{i=1}^{r} \frac{4\lambda_i \lambda_{2n+1-i}}{(\lambda_i + \lambda_{2n+1-i})^2}, \quad r = 1, \ldots, n. \tag{3.63}$$

This upper bound is attained when

$$\mathbf{R}_{xx} = \tfrac{1}{2}\,\mathbf{Diag}(\lambda_1 + \lambda_{2n}, \lambda_2 + \lambda_{2n-1}, \ldots, \lambda_n + \lambda_{n+1}), \tag{3.64}$$

$$\widetilde{\mathbf{R}}_{xx} = \tfrac{1}{2}\,\mathbf{Diag}(\lambda_1 - \lambda_{2n}, \lambda_2 - \lambda_{2n-1}, \ldots, \lambda_n - \lambda_{n+1}). \tag{3.65}$$

This bound has been derived by Bartmann and Bloomfield (1981) for the canonical correlations between arbitrary pairs of real vectors $(\mathbf{u}, \mathbf{v})$, and it holds *a forteriori* for the canonical correlations between $\mathbf{x}$ and $\mathbf{x}^*$. It is easy to see that $\underline{\mathbf{R}}_{xx}$, with $\mathbf{R}_{xx}$ and $\widetilde{\mathbf{R}}_{xx}$ as specified in (3.64) and (3.65), is a valid augmented covariance matrix with eigenvalues $\{\lambda_i\}$ and attains the bound.

There is no nontrivial lower bound on the canonical correlations between arbitrary pairs of random vectors $(\mathbf{x}, \mathbf{y})$. It is always possible to choose, for instance, $E\mathbf{x}\mathbf{x}^{\mathrm{H}} = \mathbf{Diag}(\lambda_1, \ldots, \lambda_n)$, $E\mathbf{y}\mathbf{y}^{\mathrm{H}} = \mathbf{Diag}(\lambda_{n+1}, \ldots, \lambda_{2n})$, and $E\mathbf{x}\mathbf{y}^{\mathrm{H}} = \mathbf{0}$, which has the required eigenvalues $\{\lambda_i\}$ and zero canonical correlation matrix $\mathbf{K} = \mathbf{0}$. That there is a lower bound on ρ_1 stems from the special structure of the augmented covariance matrix $\underline{\mathbf{R}}_{xx}$, where the northwest and southeast blocks must be complex conjugates.

We will now derive this lower bound for $r = n$. Let $\mathbf{x} = \mathbf{u} + \mathrm{j}\mathbf{v}$ and $\mathbf{z} = [\mathbf{u}^{\mathrm{T}}, \mathbf{v}^{\mathrm{T}}]^{\mathrm{T}}$. From (2.21) and (2.22), we know that

$$\mathbf{R}_{xx} = \mathbf{R}_{uu} + \mathbf{R}_{vv} + \mathrm{j}(\mathbf{R}_{uv}^{\mathrm{T}} - \mathbf{R}_{uv}), \tag{3.66}$$

$$\widetilde{\mathbf{R}}_{xx} = \mathbf{R}_{uu} - \mathbf{R}_{vv} + \mathrm{j}(\mathbf{R}_{uv}^{\mathrm{T}} + \mathbf{R}_{uv}). \tag{3.67}$$

Since the eigenvalues are given,

$$\det \underline{\mathbf{R}}_{xx} = \prod_{i=1}^{2n} \lambda_i \tag{3.68}$$

is fixed. Hence, it follows from (3.34) that the minimum ρ_1 is achieved when $\det \mathbf{R}_{xx}$ is minimized. We can assume without loss of generality that $\mathbf{R}_{xx}$ is diagonal. If it is not, it can be made diagonal with a strictly unitary transform that leaves $\det \mathbf{R}_{xx}$ and the eigenvalues $\{\lambda_i\}$ unchanged. Thus, we have

$$\min \det \mathbf{R}_{xx} = \min \prod_{i=1}^{n} (\mathbf{R}_{xx})_{ii} = \min \prod_{i=1}^{n} [(\mathbf{R}_{uu})_{ii} + (\mathbf{R}_{vv})_{ii}]. \tag{3.69}$$

Now let q_i be the ith largest diagonal element of $\mathbf{R}_{uu} + \mathbf{R}_{vv}$, and r_i the ith largest diagonal element of $\mathbf{R}_{zz}$. Then,

$$\sum_{i=1}^{r} q_i \le \sum_{i=1}^{r} r_{2i-1} + r_{2i} \le \sum_{i=1}^{r} \frac{\lambda_{2i-1} + \lambda_{2i}}{2}, \quad r = 1, \ldots, n, \tag{3.70}$$

with equality for $r = n$. We have the second inequality because the diagonal elements of $\mathbf{R}_{zz}$ are majorized by the eigenvalues of $\mathbf{R}_{zz}$ (cf. Result A3.5). Since $\prod q_i$ is Schur-concave, a consequence of (3.70) is the following variant of Hadamard's inequality:

$$\prod_{i=1}^{n} [(\mathbf{R}_{uu})_{ii} + (\mathbf{R}_{vv})_{ii}] \ge \prod_{i=1}^{n} \frac{\lambda_{2i-1} + \lambda_{2i}}{2}. \tag{3.71}$$

This implies

$$\min \det \mathbf{R}_{xx} = \prod_{i=1}^{n} \frac{\lambda_{2i-1} + \lambda_{2i}}{2}. \tag{3.72}$$

Using this result in (3.34), we get

$$\prod_{i=1}^{n} (1 - k_i^2) \le \frac{\prod_{i=1}^{2n} \lambda_i}{\prod_{i=1}^{n} \frac{(\lambda_{2i-1} + \lambda_{2i})^2}{4}}, \tag{3.73}$$

from which we obtain the following lower bound on ρ_1.

Result 3.6. *The degree of impropriety ρ_1 of a vector $\mathbf{x}$ with prescribed eigenvalues $\{\lambda_i\}$ of the augmented covariance matrix $\underline{\mathbf{R}}_{xx}$ is lower-bounded by*

$$\rho_1 = 1 - \prod_{i=1}^{n} (1 - k_i^2) \ge 1 - \prod_{i=1}^{n} \frac{4\lambda_{2i-1}\lambda_{2i}}{(\lambda_{2i-1} + \lambda_{2i})^2}. \tag{3.74}$$

The lower bound is attained if $\mathbf{R}_{uu} = \frac{1}{2}\mathbf{\Lambda}^{(1)}$, $\mathbf{R}_{vv} = \frac{1}{2}\mathbf{\Lambda}^{(2)}$, *and* $\mathbf{R}_{uv} = \mathbf{0}$, *or, equivalently,*

$$\mathbf{R}_{xx} = \tfrac{1}{2}(\mathbf{\Lambda}^{(1)} + \mathbf{\Lambda}^{(2)}) = \tfrac{1}{2}\,\mathbf{Diag}\,(\lambda_1 + \lambda_2, \lambda_3 + \lambda_4, \ldots, \lambda_{2n-1} + \lambda_{2n}), \tag{3.75}$$

$$\widetilde{\mathbf{R}}_{xx} = \tfrac{1}{2}(\mathbf{\Lambda}^{(1)} - \mathbf{\Lambda}^{(2)}) = \tfrac{1}{2}\,\mathbf{Diag}(\lambda_1 - \lambda_2, \lambda_3 - \lambda_4, \ldots, \lambda_{2n-1} - \lambda_{2n}), \tag{3.76}$$

where $\mathbf{\Lambda}^{(1)} = \mathbf{Diag}(\lambda_1, \lambda_3, \ldots, \lambda_{2n-1})$ *and* $\mathbf{\Lambda}^{(2)} = \mathbf{Diag}(\lambda_2, \lambda_4, \ldots, \lambda_{2n})$.

Example 3.2. In the scalar case $n = 1$, the upper bound equals the lower bound, which leads to the following expression for the degree of impropriety:

$$\rho_1 = k_1^2 = 1 - \frac{4\lambda_1\lambda_2}{(\lambda_1 + \lambda_2)^2} = \frac{(\lambda_1 - \lambda_2)^2}{(\lambda_1 + \lambda_2)^2}.$$

Thus we also have a simple expression for the circularity coefficient k_1 in terms of the eigenvalues λ_1 and λ_2:

$$k_1 = \frac{\lambda_1 - \lambda_2}{\lambda_1 + \lambda_2}.$$

While the upper bound (3.63) holds for $r = 1, \ldots, n$, we have been able to establish the lower bound (3.74) for $r = n$ only. A natural conjecture is to assume that the diagonal $\mathbf{R}_{xx}$ and $\widetilde{\mathbf{R}}_{xx}$ given by (3.75) and (3.76) also attain the lower bound for $r < n$. Let c_i be the ith largest of the factors

$$\frac{4\lambda_{2j-1}\lambda_{2j}}{(\lambda_{2j-1} + \lambda_{2j})^2}, \quad j = 1, \ldots, n.$$

Then the conjecture may be written as

$$\rho_1 = 1 - \prod_{i=1}^{r}(1 - k_i^2) \geq 1 - \prod_{i=n-r+1}^{n} c_i, \quad r = 1, \ldots, n. \tag{3.77}$$

We point out that the diagonal matrices $\mathbf{R}_{xx}$ and $\widetilde{\mathbf{R}}_{xx}$ that achieve the upper bound do not necessarily give upper bounds for other functions of $\{k_i\}$, such as ρ_2 and ρ_3. Drury *et al.* (2002) prove an upper bound for ρ_2 and conjecture an upper bound for ρ_3. Lower bounds for ρ_2 and ρ_3 are still unresolved problems.

Example 3.3. We have seen that the principal components minimize the degree of impropriety ρ_1 under widely unitary transformation. In this example, we show that they do not necessarily minimize other measures of impropriety such as ρ_2 and ρ_3.

Consider an augmented covariance matrix $\underline{\mathbf{R}}_{xx}$ with eigenvalues 100, 50, 50, and 2. The principal components $\boldsymbol{\xi}$ have covariance matrix $\mathbf{R}_{\xi\xi} = \mathbf{Diag}(75, 26)$, complementary covariance matrix $\widetilde{\mathbf{R}}_{\xi\xi} = \mathbf{Diag}(25, 24)$, and circularity coefficients $k_1 = 24/26$ and $k_2 = 25/75$. We compute $\rho_2 = 0.481$ and $\rho_3 = 0.095$.

On the other hand, there obviously exists a widely unitary transformation into coordinates $\mathbf{x}'$ with covariance matrix $\mathbf{R}_{x'x'} = \mathbf{Diag}(51, 50)$ and complementary covariance matrix $\widetilde{\mathbf{R}}_{x'x'} = \mathbf{Diag}(49, 0)$, and circularity coefficients $k_1 = 49/51$ and $k_2 = 0/50$. The description $\mathbf{x}'$ is less improper than the principal components $\boldsymbol{\xi}$ when measured in terms of $\rho_2 = 0.461$ and $\rho_3 = 0$.

Least improper analog

Given a random vector $\mathbf{x}$, we can produce a *least improper analog* $\underline{\boldsymbol{\xi}} = \underline{\mathbf{U}}^{\mathrm{H}}\underline{\mathbf{x}}$, using a widely unitary transformation $\underline{\mathbf{U}}$. It is clear from (3.75) and (3.76) that the principal components obtained from (3.14) are such an analog, with $\underline{\mathbf{U}}$ determined by the EVD $\underline{\mathbf{R}}_{xx} = \underline{\mathbf{U}}\,\underline{\boldsymbol{\Lambda}}\,\underline{\mathbf{U}}^{\mathrm{H}}$. The principal components $\boldsymbol{\xi}$ have the same eigenvalues and thus the same power as $\mathbf{x}$. They minimize ρ_1 and thus maximize entropy under widely unitary transformation. We note that a least improper analog is not unique, since any *strictly unitary* transform will leave both the eigenvalues $\{\lambda_i\}$ and the canonical correlations $\{k_i\}$ unchanged.

3.3.2 Eigenvalue spread of the augmented covariance matrix

Let us try to further illuminate the upper and lower bounds. Both the upper and the lower bounds are attained when $\mathbf{R}_{xx}$ and $\widetilde{\mathbf{R}}_{xx}$ are diagonal matrices. For $\mathbf{R}_{xx} = \mathbf{Diag}(R_{11}, \ldots, R_{nn})$ and $\widetilde{\mathbf{R}}_{xx} = \mathbf{Diag}(\widetilde{R}_{11}, \ldots, \widetilde{R}_{nn})$, we find that

$$\prod_{i=1}^{n}(1-k_i^2) = \prod_{i=1}^{n}\left(1 - \frac{|\widetilde{R}_{ii}|^2}{R_{ii}^2}\right). \tag{3.78}$$

If $\underline{\mathbf{R}}_{xx}$ has diagonal blocks $\mathbf{R}_{xx}$ and $\widetilde{\mathbf{R}}_{xx}$, it has eigenvalues $\{R_{ii} \pm |\widetilde{R}_{ii}|\}_{i=1}^{n}$. This gives

$$\prod_{i=1}^{n}(1-k_i^2) = \prod_{i=1}^{n}\left(1 - \frac{(a_i - b_i)^2}{(a_i + b_i)^2}\right) = \prod_{i=1}^{n}\frac{4a_i b_i}{(a_i + b_i)^2}, \tag{3.79}$$

where $\{a_i\}_{i=1}^{n}$ and $\{b_i\}_{i=1}^{n}$ are two disjoint subsets of $\{\lambda_i\}_{i=1}^{2n}$. Each factor

$$\frac{4a_i b_i}{(a_i + b_i)^2}$$

is the squared ratio of the geometric and arithmetic means of a_i and b_i. Hence, it is 1 if $a_i = b_i$, and 0 if a_i or b_i are 0, and thus measures the spread between a_i and b_i. Minimizing or maximizing (3.79) is a matter of choosing the subsets $\{a_i\}$ and $\{b_i\}$ from the eigenvalues $\{\lambda_i\}$ using a combinatorial argument presented by Bloomfield and Watson (1975). In order to minimize (3.79), we need maximum spread between the two sets $\{a_i\}$ and $\{b_i\}$, which is achieved by choosing $a_i = \lambda_i$ and $b_i = \lambda_{2n-i}$. In order to maximize (3.79), we need minimum spread between $\{a_i\}$ and $\{b_i\}$, which is achieved by $a_i = \lambda_{2i-1}$ and $b_i = \lambda_{2i}$. Hence, the degree of impropriety is related to the eigenvalue spread of $\underline{\mathbf{R}}_{xx}$.

3.3.3 Maximally improper vectors

Following this line of thought, one might expect a vector that is maximally improper – in the sense that $\mathbf{K} = \mathbf{I}$ – to correspond to an augmented covariance matrix with maximum possible eigenvalue spread. This was in fact claimed by Schreier *et al.* (2005) but, unfortunately, it is only partially true. Let $\mathbf{ev}(\mathbf{R}_{xx}) = [\mu_1, \mu_2, \ldots, \mu_n]^{\mathrm{T}}$ and $\mathbf{ev}(\underline{\mathbf{R}}_{xx}) = [\lambda_1, \lambda_2, \ldots, \lambda_{2n}]$. Let $\underline{\mathbf{R}}_{\#}$ be the augmented covariance matrix of a maximally improper vector with $\mathbf{K} = \mathbf{I}$. Using (3.46), we may write

$$\underline{\mathbf{R}}_{\#} = \begin{bmatrix} \mathbf{R}_{xx} & \mathbf{R}_{xx}^{1/2}\mathbf{F}\mathbf{F}^{\mathrm{T}}\mathbf{R}_{xx}^{\mathrm{T}/2} \\ \mathbf{R}_{xx}^{*/2}\mathbf{F}^{*}\mathbf{F}^{\mathrm{H}}\mathbf{R}_{xx}^{\mathrm{H}/2} & \mathbf{R}_{xx}^{*} \end{bmatrix} \tag{3.80}$$

for some unitary matrix $\mathbf{F}$. The matrix $\underline{\mathbf{R}}_{\#}$ has a vanishing Schur complement and thus $\widetilde{\mathbf{R}}_{\#} = \mathbf{R}_{xx}^{1/2}\mathbf{F}\mathbf{F}^{\mathrm{T}}\mathbf{R}_{xx}^{\mathrm{T}/2}$ is an extreme point in the set $\mathcal{Q}$. Schreier *et al.* (2005) incorrectly stated that $\mathbf{ev}(\underline{\mathbf{R}}_{\#}) = [2\mu_1, 2\mu_2, \ldots, 2\mu_n, \mathbf{0}_n^{\mathrm{T}}]^{\mathrm{T}}$ for *any* extreme point $\widetilde{\mathbf{R}}_{\#}$. While this is not true, there is indeed *at least one* extreme point $\widetilde{\mathbf{R}}_{\#\#}$ such that the augmented covariance matrix $\underline{\mathbf{R}}_{\#\#}$ has these eigenvalues. Let $\mathbf{R}_{xx} = \mathbf{UMU}^{\mathrm{H}}$ be the EVD of $\mathbf{R}_{xx}$. Choosing

$$\widetilde{\mathbf{R}}_{\#\#} = \mathbf{UMU}^{\mathrm{T}} = \mathbf{UM}^{1/2}\mathbf{U}^{\mathrm{H}}\mathbf{UU}^{\mathrm{T}}\mathbf{U}^{*}\mathbf{M}^{1/2}\mathbf{U}^{\mathrm{T}} = \mathbf{R}_{xx}^{1/2}\mathbf{FF}^{\mathrm{T}}\mathbf{R}_{xx}^{\mathrm{T}/2} \tag{3.81}$$

with $\mathbf{F} = \mathbf{U}$ means that

$$\underline{\mathbf{R}}_{\#\#} = \begin{bmatrix} \mathbf{R}_{xx} & \widetilde{\mathbf{R}}_{\#\#} \\ \widetilde{\mathbf{R}}_{\#\#}^{*} & \mathbf{R}_{xx}^{*} \end{bmatrix} \tag{3.82}$$

has eigenvalues $\mathbf{ev}(\underline{\mathbf{R}}_{\#\#}) = [2\mu_1, 2\mu_2, \ldots, 2\mu_n, \mathbf{0}_n^{\mathrm{T}}]^{\mathrm{T}}$. Let us now establish the following.

Result 3.7. *There is the majorization preordering*

$$\mathbf{ev}(\underline{\mathbf{R}}_0) \prec \mathbf{ev}(\underline{\mathbf{R}}_{xx}) \prec \mathbf{ev}(\underline{\mathbf{R}}_{\#\#}). \tag{3.83}$$

This says that, for given Hermitian covariance matrix $\mathbf{R}_{xx}$, *the vector whose augmented covariance matrix* $\underline{\mathbf{R}}_{xx}$ *has least eigenvalue spread must be proper:* $\widetilde{\mathbf{R}}_{xx} = \mathbf{0}$ *and* $\underline{\mathbf{R}}_0 = \mathbf{Diag}(\mathbf{R}_{xx}, \mathbf{R}_{xx}^{*})$. *The vector whose augmented covariance matrix* $\underline{\mathbf{R}}_{xx}$ *has maximum eigenvalue spread must be maximally improper, i.e.,* $\mathbf{K} = \mathbf{I}$.

In order to show this result we note that the left inequality is (3.23), and the right inequality is a consequence of Result A3.8. Applied to the matrix $\underline{\mathbf{R}}_{xx}$, Result A3.8 says that

$$\sum_{i=1}^{k} \lambda_i + \lambda_{2n-k+i} \leq \sum_{i=1}^{k} 2\mu_i, \quad k = 1, \ldots, n, \tag{3.84}$$

and, since $\lambda_{2n-k+i} \geq 0$,

$$\sum_{i=1}^{k} \lambda_i \leq \sum_{i=1}^{k} 2\mu_i, \quad k = 1, \ldots, n. \tag{3.85}$$

Moreover, the trace constraint $\operatorname{tr} \underline{\mathbf{R}}_{xx} = 2 \operatorname{tr} \mathbf{R}_{xx}$ and $\lambda_i \geq 0$ imply

$$\sum_{i=1}^{k} \lambda_i \leq \sum_{i=1}^{n} 2\mu_i, \quad k = n+1, \ldots, 2n. \tag{3.86}$$

Together, (3.85) and (3.86) prove the result.

Again, $\mathbf{K} = \mathbf{I}$ is a *necessary condition* only for maximum eigenvalue spread of $\underline{\mathbf{R}}_{xx}$. It is not sufficient, as Schreier *et al.* (2005) incorrectly claimed. In Section 5.4.2, we will use (3.83) to maximize/minimize Schur-convex/concave functions of $\mathbf{ev}(\underline{\mathbf{R}}_{xx})$ for fixed $\mathbf{R}_{xx}$. It follows from (3.83) that these maxima/minima will be achieved for the extreme point $\widetilde{\mathbf{R}}_{\#\#}$, but not necessarily for all extreme points.

3.4 Testing for impropriety

In practice, the complementary covariance matrix must often be estimated from the data available. Such an estimate will in general be nonzero even if the source is actually proper. So how do we classify a problem as proper or improper? In this section, we present a hypothesis test for impropriety that is based on a generalized likelihood-ratio test (GLRT), which is a special case of a more general class of tests presented in Section 4.5. A general introduction to likelihood-ratio tests is provided in Section 7.1.

In a GLR, the unknown parameters ($\mathbf{R}_{xx}$ and $\widetilde{\mathbf{R}}_{xx}$ in our case) are replaced by maximum-likelihood estimates. The GLR is always invariant with respect to transformations for which the hypothesis-testing problem itself is invariant. Since propriety is preserved by strictly linear, but not widely linear, transformations, the hypothesis test must be invariant with respect to strictly linear, but not widely linear, transformations. A maximal invariant statistic under linear transformation is given by the circularity coefficients. Since the GLR must be a function of a maximal invariant statistic the GLR is a function of the circularity coefficients.

Let $\mathbf{x}$ be a complex Gaussian random vector with probability density function

$$p(\mathbf{x}) = \pi^{-n}(\det \underline{\mathbf{R}}_{xx})^{-1/2} \exp\left\{-\tfrac{1}{2}(\underline{\mathbf{x}} - \underline{\boldsymbol{\mu}}_x)^{\mathrm{H}}\underline{\mathbf{R}}_{xx}^{-1}(\underline{\mathbf{x}} - \underline{\boldsymbol{\mu}}_x)\right\} \tag{3.87}$$

with augmented mean vector $\underline{\boldsymbol{\mu}}_x = E\underline{\mathbf{x}}$ and augmented covariance matrix $\underline{\mathbf{R}}_{xx} = E[(\underline{\mathbf{x}} - \underline{\boldsymbol{\mu}}_x)(\underline{\mathbf{x}} - \underline{\boldsymbol{\mu}}_x)^{\mathrm{H}}]$. Consider M independent and identically distributed (i.i.d.) random samples $\mathbf{X} = [\mathbf{x}_1, \mathbf{x}_2, \ldots, \mathbf{x}_M]$ drawn from this distribution, and let $\underline{\mathbf{X}} = [\underline{\mathbf{x}}_1, \underline{\mathbf{x}}_2, \ldots, \underline{\mathbf{x}}_M]$ denote the augmented sample matrix. As shown in Section 2.4, the joint probability density function of these samples is

$$p(\mathbf{X}) = \pi^{-Mn}(\det \underline{\mathbf{R}}_{xx})^{-M/2} \exp\left\{-\frac{M}{2}\operatorname{tr}(\underline{\mathbf{R}}_{xx}^{-1}\underline{\mathbf{S}}_{xx})\right\}, \tag{3.88}$$

where $\underline{\mathbf{S}}_{xx}$ is the augmented sample covariance matrix

$$\underline{\mathbf{S}}_{xx} = \begin{bmatrix} \mathbf{S}_{xx} & \widetilde{\mathbf{S}}_{xx} \\ \widetilde{\mathbf{S}}_{xx}^{*} & \mathbf{S}_{xx}^{*} \end{bmatrix} = \frac{1}{M}\sum_{m=1}^{M}(\underline{\mathbf{x}}_m - \underline{\mathbf{m}}_x)(\underline{\mathbf{x}}_m - \underline{\mathbf{m}}_x)^{\mathrm{H}} = \frac{1}{M}\underline{\mathbf{X}}\,\underline{\mathbf{X}}^{\mathrm{H}} - \underline{\mathbf{m}}_x\underline{\mathbf{m}}_x^{\mathrm{H}} \tag{3.89}$$

and $\underline{\mathbf{m}}_x$ is the augmented sample mean vector

$$\underline{\mathbf{m}}_x = \frac{1}{M}\sum_{m=1}^{M}\underline{\mathbf{x}}_m. \tag{3.90}$$

We will now develop the GLR test of the hypotheses

$$H_0\colon\ \mathbf{x} \text{ is proper } (\widetilde{\mathbf{R}}_{xx} = \mathbf{0}),$$

$$H_1\colon\ \mathbf{x} \text{ is improper } (\widetilde{\mathbf{R}}_{xx} \neq \mathbf{0}).$$

The GLRT statistic is

$$\lambda = \frac{\displaystyle\max_{\substack{\underline{\mathbf{R}}_{xx} \\ \widetilde{\mathbf{R}}_{xx}=\mathbf{0}}} p(\mathbf{X})}{\displaystyle\max_{\underline{\mathbf{R}}_{xx}} p(\mathbf{X})}. \tag{3.91}$$

This is the ratio of likelihood with $\underline{\mathbf{R}}_{xx}$ constrained to have zero off-diagonal blocks, $\widetilde{\mathbf{R}}_{xx} = \mathbf{0}$, to likelihood with $\underline{\mathbf{R}}_{xx}$ unconstrained. We are thus testing whether or not $\underline{\mathbf{R}}_{xx}$ is block-diagonal.

As discussed in Section 2.4, the unconstrained maximum-likelihood (ML) estimate of $\underline{\mathbf{R}}_{xx}$ is the augmented sample covariance matrix $\underline{\mathbf{S}}_{xx}$. The ML estimate of $\underline{\mathbf{R}}_{xx}$ under the constraint $\widetilde{\mathbf{R}}_{xx} = \mathbf{0}$ is

$$\underline{\mathbf{S}}_0 = \begin{bmatrix} \mathbf{S}_{xx} & \mathbf{0} \\ \mathbf{0} & \mathbf{S}_{xx}^* \end{bmatrix}. \tag{3.92}$$

Hence, the GLR (3.91) can be expressed as

$$\ell = \lambda^{2/M} = \det(\underline{\mathbf{S}}_0^{-1}\underline{\mathbf{S}}_{xx}) \left(\exp\left\{ -\frac{M}{2} \operatorname{tr}\left(\underline{\mathbf{S}}_0^{-1}\underline{\mathbf{S}}_{xx} - \mathbf{I} \right) \right\} \right)^{2/M} \tag{3.93}$$

$$= \det \begin{bmatrix} \mathbf{I} & \mathbf{S}_{xx}^{-1}\widetilde{\mathbf{S}}_{xx} \\ \mathbf{S}_{xx}^{-*}\widetilde{\mathbf{S}}_{xx}^* & \mathbf{I} \end{bmatrix} \tag{3.94}$$

$$= \det{}^{-2}\mathbf{S}_{xx} \det \underline{\mathbf{S}}_{xx} \tag{3.95}$$

$$= \prod_{i=1}^{n}(1 - \hat{k}_i^2) = 1 - \widehat{\rho}_1. \tag{3.96}$$

In the last line, $\{\hat{k}_i\}$ denotes the estimated circularity coefficients, which are computed from the augmented sample covariance matrix $\underline{\mathbf{S}}_{xx}$. The estimated circularity coefficients are then used to estimate the degree of impropriety ρ_1. Our main finding follows.

Result 3.8. *The estimated degree of impropriety $\widehat{\rho}_1$ is a test statistic for a GLRT for impropriety.*

Equations (3.95) and (3.96) are equivalent formulations of this GLR. A full-rank implementation of this test relies on (3.95) since it does not require computation of $\mathbf{S}_{xx}^{-1}$. However, a reduced-rank implementation considers only the r largest estimated circularity coefficients in the product (3.96).

This test was first proposed by Andersson and Perlman (1984), then in complex notation by Ollila and Koivunen (2004), and the connection with canonical correlations was established by Schreier *et al.* (2006). Andersson and Perlman (1984) also show that the estimated degree of impropriety

$$\widehat{\rho}_3 = \frac{1}{n}\sum_{i=1}^{n}\hat{k}_i^2 \tag{3.97}$$

rather than $\widehat{\rho}_1$ is the *locally most powerful* (LMP) test for impropriety. An LMP test has the highest possible power (i.e., probability of detection) for H_1 close to H_0, where all circularity coefficients are small.

Intuitively, it is clear that $\widehat{\rho}_1$ and $\widehat{\rho}_3$ behave quite differently. While $\widehat{\rho}_1$ is close to 1 if at least one circularity coefficient is close to 1, $\widehat{\rho}_3$ is close to 1 only if all circularity coefficients are close to 1. Walden and Rubin-Delanchy (2009) reexamined testing for impropriety, studying the null distributions of $\widehat{\rho}_1$ and $\widehat{\rho}_3$, and deriving a distributional approximation for $\widehat{\rho}_1$. They also point out that no uniformly powerful (UMP) test exists for this problem, simply because the GLR and the LMP tests are

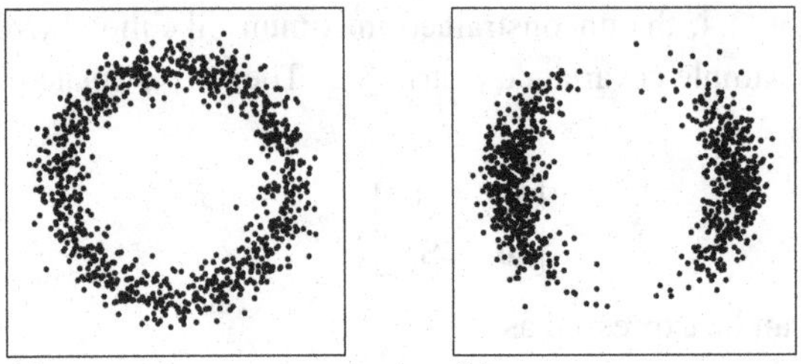

(a) (b)

Figure 3.4 BPSK symbols transmitted over (a) a noncoherent and (b) a partially coherent AWGN channel.

different for dimensions $n \geq 2$. (In the scalar case $n = 1$, the GLR and LMP tests are identical.)

Example 3.4. In the scalar case $n = 1$, the GLR becomes

$$\ell = 1 - \frac{|\widetilde{S}_{xx}|^2}{S_{xx}^2}, \tag{3.98}$$

with sample variance S_{xx} and sample complementary variance $\widetilde{S}_{xx}$. As a simple example, we consider the transmission of BPSK symbols – that is, equiprobable binary data $b_m \in \{\pm 1\}$ – over an AWGN channel that also rotates the phase of the transmitted bits by ϕ_m. The received statistic is

$$x_m = b_m \mathrm{e}^{\mathrm{j}\phi_m} + n_m, \tag{3.99}$$

where n_m are samples of white Gaussian noise and ϕ_m are samples of the channel phase. We are interested in classifying this channel as either noncoherent or partially coherent. We will see that, even though x_m is not Gaussian, the GLR (3.98) is well suited for this hypothesis test.

We evaluate the performance of the GLRT detector by Monte Carlo simulations. Under H_0, we assume that the phase samples ϕ_m are i.i.d. and uniformly distributed. Under H_1, we assume that the phase samples are i.i.d. and drawn from a Gaussian distribution. This means that, under H_0, no useful phase information can be extracted, whereas under H_1 a phase estimate is available, albeit with a tracking error. Figure 3.4 plots BPSK symbols that have been transmitted over a noncoherent additive white Gaussian noise channel with uniformly distributed phase in (a), and for a partially coherent channel in (b).

Figure 3.5 shows experimentally estimated receiver operator characteristics (ROC) for this detector for various signal-to-noise ratios (SNRs) and variances for the phase tracking error. In an ROC curve, the probability of detection P_{D}, which is the probability of correctly accepting H_1, is plotted versus the probability of false alarm P_{FA}, which is the probability of incorrectly rejecting H_0. The ROC curve will be taken up more thoroughly in Section 7.3.

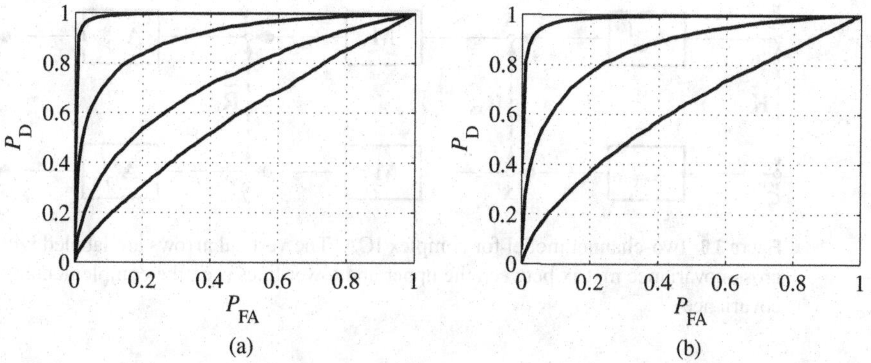

Figure 3.5 Receiver operating characteristics of the GLRT detector. In (a), the SNR is fixed at 0 dB. From northwest to southeast, the curves correspond to phase tracking-error variance of 0.7, 0.95, 1.2, and 1.5. In (b), the phase tracking-error variance is fixed at 1. From northwest to southeast, the curves correspond to SNR of 5 dB, 0 dB, and −5 dB. In all cases, the number of samples was $M = 1000$.

3.5 Independent component analysis

An interesting application of the invariance property of the circularity coefficients is independent component analysis (ICA). In ICA, we observe a *linear mixture* $\mathbf{y}$ of *independent complex components* (sources) $\mathbf{x}$, as described by

$$\mathbf{y} = \mathbf{M}\mathbf{x}. \tag{3.100}$$

We will make a few simplifying assumptions in this section. The dimensions of $\mathbf{y}$ and $\mathbf{x}$ are assumed to be equal, and the *mixing matrix* $\mathbf{M}$ is assumed to be nonsingular. The objective is to *blindly* recover the sources $\mathbf{x}$ from the observations $\mathbf{y}$, without knowledge of $\mathbf{M}$, using a linear transformation $\mathbf{M}^{\#}$. This transformation $\mathbf{M}^{\#}$ can be regarded as a *blind inverse* of $\mathbf{M}$, which is usually called a *separating matrix*. Note that, since the model (3.100) is linear, it is unnecessary to consider widely linear transformations.

ICA seeks to determine independent components. Arbitrary scaling of $\mathbf{x}$, i.e., multiplication by a diagonal matrix, and reordering the components of $\mathbf{x}$, i.e., multiplication by a permutation matrix, preserves the independence of its components. The product of a diagonal and a permutation matrix is a *monomial* matrix, which has exactly one nonzero entry in each column and row. Hence, we can determine $\mathbf{M}^{\#}$ up to multiplication with a monomial matrix.

Standard ICA requires the use of higher-order statistical information, and the blind recovery of $\mathbf{x}$ cannot work if more than one source x_i is Gaussian. If only second-order information is available, the best possible solution is to *decorrelate* the components, rather than to make them independent. This is done by determining the *principal* components $\mathbf{U}^{\mathrm{H}}\mathbf{y}$ using the EVD $\mathbf{R}_{yy} = E\mathbf{y}\mathbf{y}^{\mathrm{H}} = \mathbf{U}\boldsymbol{\Lambda}\mathbf{U}^{\mathrm{H}}$. However, the restriction to unitary

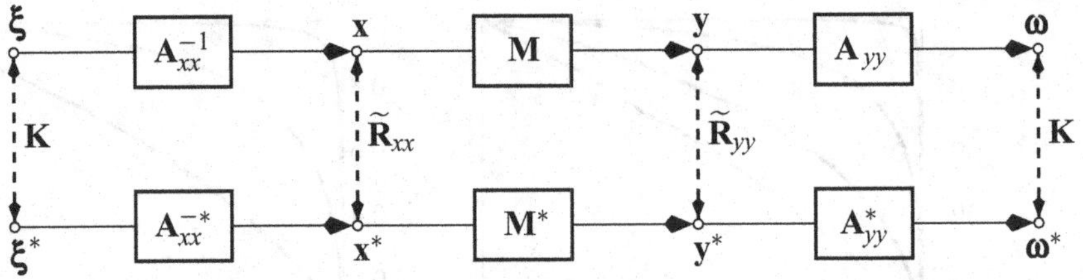

Figure 3.6 Two-channel model for complex ICA. The vertical arrows are labeled with the cross-covariance matrix between the upper and lower lines (i.e., the complementary covariance).

rather than general linear transformations wastes a considerable degree of freedom in designing the blind inverse $\mathbf{M}^{\#}$.

In this section, we demonstrate that, in the complex case, it can be possible to determine $\mathbf{M}^{\#}$ using *second-order* information only. This was first shown by DeLathauwer and DeMoor (2002) and independently discovered by Eriksson and Koivunen (2006). The key insight in our demonstration is that the independence of the components of $\mathbf{x}$ means that, up to simple scaling and permutation, $\mathbf{x}$ is already given in canonical coordinates. The idea is then to exploit the invariance of circularity coefficients of $\mathbf{x}$ under the *linear* mixing transformation $\mathbf{M}$.

The assumption of *independent* components $\mathbf{x}$ implies that the covariance matrix $\mathbf{R}_{xx}$ and the complementary covariance matrix $\widetilde{\mathbf{R}}_{xx}$ are both diagonal. It is therefore easy to compute canonical coordinates between $\mathbf{x}$ and $\mathbf{x}^*$, denoted by $\boldsymbol{\xi} = \mathbf{A}_{xx}\mathbf{x}$. In the strong uncorrelating transform $\mathbf{A}_{xx} = \mathbf{F}_{xx}^{\mathrm{H}}\mathbf{R}_{xx}^{-1/2}$, $\mathbf{R}_{xx}^{-1/2}$ is a *diagonal* scaling matrix, and $\mathbf{F}_{xx}^{\mathrm{H}}$ is a *permutation* matrix that rearranges the canonical coordinates $\boldsymbol{\xi}$ such that ξ_1 corresponds to the largest circularity coefficient k_1, ξ_2 to the second largest coefficient k_2, and so on. This makes the strong uncorrelating transform $\mathbf{A}_{xx}$ monomial. As a consequence, $\boldsymbol{\xi}$ also has *independent components*.

The mixture $\mathbf{y}$ has covariance matrix $\mathbf{R}_{yy} = \mathbf{M}\mathbf{R}_{xx}\mathbf{M}^{\mathrm{H}}$ and complementary covariance matrix $\widetilde{\mathbf{R}}_{yy} = \mathbf{M}\widetilde{\mathbf{R}}_{xx}\mathbf{M}^{\mathrm{T}}$. The canonical coordinates of $\mathbf{y}$ and $\mathbf{y}^*$ are computed as $\boldsymbol{\omega} = \mathbf{A}_{yy}\mathbf{y} = \mathbf{F}_{yy}^{\mathrm{H}}\mathbf{R}_{yy}^{-1/2}\mathbf{y}$, and $\boldsymbol{\omega}^* = \mathbf{A}_{yy}^*\mathbf{y}^*$. The strong uncorrelating transform $\mathbf{A}_{yy}$ is determined as explained in Section 3.2.2.

Figure 3.6 shows the connection between the different coordinate systems. The important observation is that $\boldsymbol{\xi}$ and $\boldsymbol{\omega}$ are both in canonical coordinates with the *same* circularity coefficients k_i. In the next paragraph, we will show that $\boldsymbol{\xi}$ and $\boldsymbol{\omega}$ are related as $\boldsymbol{\omega} = \mathbf{D}\boldsymbol{\xi}$ by a diagonal matrix $\mathbf{D}$ with diagonal entries ± 1, provided that all circularity coefficients are distinct. Since $\boldsymbol{\xi}$ has independent components, so does $\boldsymbol{\omega}$. Hence, we have a solution to the ICA problem.

Result 3.9. *The strong uncorrelating transform* $\mathbf{A}_{yy}$ *is a separating matrix for the complex linear ICA problem if all circularity coefficients are distinct.*

The only thing left to show is that $\mathbf{D} = \mathbf{A}_{yy}\mathbf{M}\mathbf{A}_{xx}^{-1}$ is indeed diagonal with diagonal elements ± 1. Since $\boldsymbol{\xi}$ and $\boldsymbol{\omega}$ are both in canonical coordinates with the same diagonal canonical correlation matrix **K**, we find

$$E\,\underline{\boldsymbol{\xi}}\,\underline{\boldsymbol{\xi}}^{\mathrm{H}} = \begin{bmatrix} \mathbf{I} & \mathbf{K} \\ \mathbf{K} & \mathbf{I} \end{bmatrix} = \begin{bmatrix} \mathbf{D} & \mathbf{0} \\ \mathbf{0} & \mathbf{D}^* \end{bmatrix} \begin{bmatrix} \mathbf{I} & \mathbf{K} \\ \mathbf{K} & \mathbf{I} \end{bmatrix} \begin{bmatrix} \mathbf{D}^{\mathrm{H}} & \mathbf{0} \\ \mathbf{0} & \mathbf{D}^{\mathrm{T}} \end{bmatrix} = E\,\underline{\boldsymbol{\omega}}\,\underline{\boldsymbol{\omega}}^{\mathrm{H}} \tag{3.101}$$

$$= \begin{bmatrix} \mathbf{D}\mathbf{D}^{\mathrm{H}} & \mathbf{D}\mathbf{K}\mathbf{D}^{\mathrm{T}} \\ \mathbf{D}^*\mathbf{K}\mathbf{D}^{\mathrm{H}} & \mathbf{D}^*\mathbf{D}^{\mathrm{T}} \end{bmatrix}. \tag{3.102}$$

This shows that **D** is unitary and $\mathbf{DKD}^{\mathrm{T}} = \mathbf{K}$. The latter can be true only if $D_{ij} = 0$ whenever $k_i \neq k_j$. Therefore, **D** is diagonal and unitary if all circularity coefficients are distinct. Since **K** is real, the corresponding diagonal entries of all nonzero circularity coefficients are actually ± 1. On the other hand, components with identical circularity coefficient cannot be separated.

Example 3.5. Consider a source $\mathbf{x} = [x_1, x_2]^{\mathrm{T}}$. The first component x_1 is the signal-space representation of a QPSK signal with amplitude 2 and phase offset $\pi/8$, i.e., $x_1 \in \{\pm 2\mathrm{e}^{\mathrm{j}\pi/8}, \pm 2\mathrm{j}\mathrm{e}^{\mathrm{j}\pi/8}\}$. The second component x_2, independent of x_1, is the signal-space representation of a BPSK signal with amplitude 1 and phase offset $\pi/4$, i.e., $x_1 \in \{\pm \mathrm{e}^{\mathrm{j}\pi/4}\}$. Hence,

$$\mathbf{R}_{xx} = \begin{bmatrix} 4 & 0 \\ 0 & 1 \end{bmatrix} \quad \text{and} \quad \widetilde{\mathbf{R}}_{xx} = \begin{bmatrix} 0 & 0 \\ 0 & \mathrm{j} \end{bmatrix}.$$

In order to take **x** into canonical coordinates, we use the strong uncorrelating transform

$$\mathbf{A}_{xx} = \mathbf{F}_{xx}^{\mathrm{H}} \mathbf{R}_{xx}^{-1/2} = \begin{bmatrix} 0 & 1 \\ 1 & 0 \end{bmatrix} \begin{bmatrix} \frac{1}{2} & 0 \\ 0 & 1 \end{bmatrix} = \begin{bmatrix} 0 & 1 \\ \frac{1}{2} & 0 \end{bmatrix}.$$

We see that $\mathbf{F}_{xx}^{\mathrm{H}}$ is a permutation matrix, $\mathbf{R}_{xx}^{-1/2}$ is diagonal, and the product of the two is monomial. The circularity coefficients are $k_1 = 1$ and $k_2 = 0$. Note that the circularity coefficients carry no information about the amplitude or phase of the two signals.

Now consider the linear mixture $\mathbf{y} = \mathbf{Mx}$ with

$$\mathbf{M} = \begin{bmatrix} -\mathrm{j} & 1 \\ 2-\mathrm{j} & 1+\mathrm{j} \end{bmatrix}.$$

With $\mathbf{R}_{yy} = \mathbf{M}\mathbf{R}_{xx}\mathbf{M}^{\mathrm{H}}$ and $\widetilde{\mathbf{R}}_{yy} = \mathbf{M}\widetilde{\mathbf{R}}_{xx}\mathbf{M}^{\mathrm{T}}$, we compute the SVD of the coherence matrix $\mathbf{C} = \mathbf{R}_{yy}^{-1/2}\widetilde{\mathbf{R}}_{yy}\mathbf{R}_{yy}^{-\mathrm{T}/2} = \mathbf{UKV}^{\mathrm{H}}$. The unitary Takagi factor $\mathbf{F}_{yy}$ is then obtained as $\mathbf{F}_{yy} = \mathbf{U}(\mathbf{V}^{\mathrm{H}}\mathbf{U}^*)^{1/2}$. The circularity coefficients of **y** are the same as those of **x**.

In order to take **y** into canonical coordinates, we use the strong uncorrelating transform (rounded to four decimals)

$$\mathbf{A}_{yy} = \mathbf{F}_{yy}^{\mathrm{H}} \mathbf{R}_{yy}^{-1/2} = \begin{bmatrix} 0.7071 - 2.1213\mathrm{j} & 0.7071 + 0.7071\mathrm{j} \\ -0.7045 - 0.0605\mathrm{j} & 0.3825 - 0.3220\mathrm{j} \end{bmatrix}.$$

We see that

$$\mathbf{A}_{yy}\mathbf{M} = \begin{bmatrix} 0 & 0.7071 - 0.7071\mathrm{j} \\ 0.3825 - 0.3220\mathrm{j} & 0 \end{bmatrix}$$

is monomial, and

$$\mathbf{A}_{yy}\mathbf{M}\mathbf{A}_{xx}^{-1} = \begin{bmatrix} 0.7071 - 0.7071\mathrm{j} & 0 \\ 0 & 0.7650 - 0.6440\mathrm{j} \end{bmatrix}$$

is diagonal and unitary.

One final comment is in order. The technique presented in this section enables the blind separation of mixtures using second-order statistics only, under two crucial assumptions. First, the sources must be complex and uncorrelated with distinct circularity coefficients. Second, $\mathbf{y}$ must be a *linear* mixture. If $\mathbf{y}$ does not satisfy the linear model (3.100), the objective of ICA is to find components that are as independent as possible. The degree of independence is measured by a *contrast function* such as mutual information or negentropy. It is important to realize that the strong uncorrelating transform $\mathbf{A}_{yy}$ is not guaranteed to optimize any contrast function. Finding maximally independent components in the *nonlinear* case requires the use of higher-order statistics.[4]

Notes

1 Much of the material presented in this chapter has been drawn from Schreier and Scharf (2003a) and Schreier *et al.* (2005).

2 The circularity coefficients and the strong uncorrelating transform were introduced by Eriksson and Koivunen (2006). The facts that the circularity coefficients are canonical correlations between $\mathbf{x}$ and $\mathbf{x}^*$, and that the strong uncorrelating transform takes $\mathbf{x}$ into canonical coordinates were shown by Schreier *et al.* (2006) and Schreier (2008a). These two papers are ©IEEE, and portions of them are reused with permission. More mathematical background on the Takagi factorization can be found in Horn and Johnson (1985).

3 The degree of impropriety and bounds in terms of eigenvalues were developed by Schreier (2008a).

4 Our discussion of independent component analysis barely scratches the surface of this rich topic. A readable introductory paper to ICA is Comon (1994). The proof of complex second-order ICA in Section 3.5 was first presented by Schreier *et al.* (2009). This paper is ©IEEE, and portions are reused with permission.

4 Correlation analysis

Assessing multivariate association between two random vectors **x** and **y** is an important problem in many research areas, ranging from the natural sciences (e.g., oceanography and geophysics) to the social sciences (in particular psychometrics and behaviormetrics) and to engineering. While "multivariate association" is often simply visualized as "similarity" between two random vectors, there are many different ways of measuring it. In this chapter, we provide a unifying treatment of three popular correlation analysis techniques: canonical correlation analysis (CCA), multivariate linear regression (MLR), and partial least squares (PLS).[1] Each of these techniques transforms **x** and **y** into its respective internal representation $\boldsymbol{\xi}$ and $\boldsymbol{\omega}$. Different correlation coefficients may then be defined as functions of the diagonal cross-correlations $\{k_i\}$ between the internal representations ξ_i and ω_i.

The key differences among CCA, MLR, and PLS are revealed in their invariance properties. CCA is invariant under nonsingular linear transformation of **x** and **y**, MLR is invariant under nonsingular linear transformation of **y** but only unitary transformation of **x**, and PLS is invariant under unitary transformation of **x** and **y**. Correlation coefficients then share the invariance properties of the correlation analysis technique on which they are based.

Analyzing multivariate association of complex data is further complicated by the fact that there are different types of correlation.[2] Two scalar complex random variables x and y are called *rotationally dependent* if $x = ky$ for some complex constant k. This term is motivated by the observation that, in the complex plane, sample pairs of x and y rotate in the same direction (counterclockwise or clockwise), by the same angle. They are called *reflectionally dependent* if $x = \tilde{k}y^*$ for some complex constant $\tilde{k}$. This means that sample pairs of x and y rotate by the same angle, but in opposite directions – one rotating clockwise and the other counterclockwise. *Rotational* and *reflectional correlations* measure the degree of rotational and reflectional dependence. The combined effect of rotational and reflectional correlation is assessed by a *total correlation*.

Thus, there are two fundamental choices that must be made when analyzing multivariate association between two complex random vectors: the desired invariance properties of the analysis (linear/linear, unitary/linear, or unitary/unitary) and the type of correlation (rotational, reflectional, or total). These choices will have to be motivated by the problem under consideration.

The techniques in this chapter are developed for random vectors using *ensemble* averages. This assumes that the necessary second-order information is available, namely, the correlation matrices of $\mathbf{x}$ and $\mathbf{y}$ and their cross-correlation matrix are known. However, it is straightforward to apply these techniques to sample data, using *sample* correlation matrices. Then M independent snapshots of $(\mathbf{x}, \mathbf{y})$ would be assembled into matrices $\mathbf{X} = [\mathbf{x}_1, \ldots, \mathbf{x}_M]$ and $\mathbf{Y} = [\mathbf{y}_1, \ldots, \mathbf{y}_M]$, and the correlation matrices would be estimated as $\mathbf{S}_{xx} = M^{-1}\mathbf{X}\mathbf{X}^{\mathrm{H}}$, $\mathbf{S}_{xy} = M^{-1}\mathbf{X}\mathbf{Y}^{\mathrm{H}}$, and $\mathbf{S}_{yy} = M^{-1}\mathbf{Y}\mathbf{Y}^{\mathrm{H}}$.

The structure of this chapter is as follows. In Section 4.1, we look at the foundations for measuring multivariate association between a pair of complex vectors, which will lead to the introduction of the three correlation analysis techniques CCA, MLR, and PLS. In Section 4.2, we discuss their invariance properties. In particular, we show that the diagonal cross-correlations $\{k_i\}$ produced by CCA, MLR, and PLS are *maximal invariants* under linear/linear, unitary/linear, and unitary/unitary transformation, respectively, of $\mathbf{x}$ and $\mathbf{y}$. In Section 4.3, we introduce a few scalar-valued correlation coefficients as different functions of the diagonal cross-correlations $\{k_i\}$, and show how these coefficients can be interpreted.

An important feature of CCA, MLR, and PLS is that they all produce diagonal cross-correlations that have maximum spread in the sense of majorization (see Appendix 3 for background on majorization). Therefore, any correlation coefficient that is an increasing and Schur-convex function of $\{k_i\}$ is maximized, for arbitrary rank r. This allows assessment of correlation in a lower-dimensional subspace of dimension r. In Section 4.4, we introduce the *correlation spread* as a measure that indicates how much of the overall correlation can be compressed into a lower-dimensional subspace.

Finally, in Section 4.5, we present several generalized likelihood-ratio tests for the correlation structure of complex Gaussian data, such as sphericity, independence within one data set, and independence between two data sets. All these tests have natural invariance properties, and the generalized likelihood ratio is a function of an appropriate maximal invariant.

4.1 Foundations for measuring multivariate association between two complex random vectors

The correlation coefficient between two *scalar real* zero-mean random variables u and v is defined as

$$\rho_{uv} = \frac{Euv}{\sqrt{Eu^2}\sqrt{Ev^2}} = \frac{R_{uv}}{\sqrt{R_{uu}}\sqrt{R_{vv}}}. \tag{4.1}$$

The correlation coefficient is a convenient measure for how closely u and v are related. It satisfies $-1 \leq \rho_{uv} \leq 1$. If $\rho_{uv} = 0$, then u and v are uncorrelated. If $|\rho_{uv}| = 1$, then u is a linear function of v, or vice versa, with probability 1:

$$u = \frac{R_{uv}}{R_{vv}} v. \tag{4.2}$$

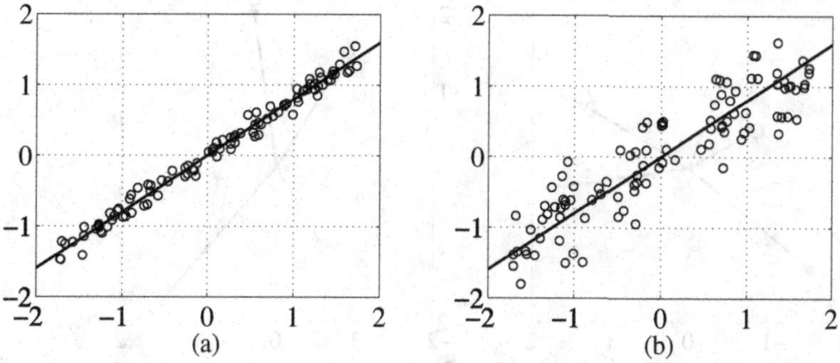

Figure 4.1 Scatter plots of 100 sample pairs of u and v with different ρ_{uv}.

In general, for $|\rho_{uv}| \leq 1$,

$$\hat{u} = \frac{R_{uv}}{R_{vv}} v \tag{4.3}$$

is a linear minimum mean-squared error (LMMSE) estimate of u from v, and the sign of ρ_{uv} is the sign of the slope of the LMMSE estimate. An expression for the MSE is

$$E|\hat{u} - u|^2 = R_{uu} - \frac{R_{uv}^2}{R_{vv}} = R_{uu}(1 - \rho_{uv}^2). \tag{4.4}$$

Example 4.1. Consider a zero-mean, unit-variance, real random variable u and a zero-mean, unit-variance, real random variable n, uncorrelated with u. Now let $v = au + bn$ for given real a and b. We find

$$\rho_{uv} = \frac{R_{uv}}{\sqrt{R_{uu}}\sqrt{R_{vv}}} = \frac{a}{\sqrt{a^2 + b^2}}.$$

Figure 4.1 depicts 100 sample pairs of u and v, for uniform u and Gaussian n. Plot (a) shows the case $a = 0.8, b = 0.1$, which results in $\rho_{uv} = 0.9923$. Plot (b) shows $a = 0.8, b = 0.4$, which results in $\rho_{uv} = 0.8944$. The line is the LMMSE estimate $\hat{v} = au$ (or, equivalently, $\hat{u} = a^{-1}v$).

How may we define a correlation coefficient between a pair of *complex* random *vectors* to measure their multivariate association? We shall consider the extension from real to complex quantities and the extension from scalars to vectors separately, and then combine our findings.

4.1.1 Rotational, reflectional, and total correlations for complex scalars

Consider a pair of *scalar complex* zero-mean random variables x and y. As a straightforward extension of the real case, let us define the complex correlation coefficient

$$\rho_{xy} = \frac{Exy^*}{\sqrt{E|x|^2}\sqrt{E|y|^2}} = \frac{R_{xy}}{\sqrt{R_{xx}}\sqrt{R_{yy}}}, \tag{4.5}$$

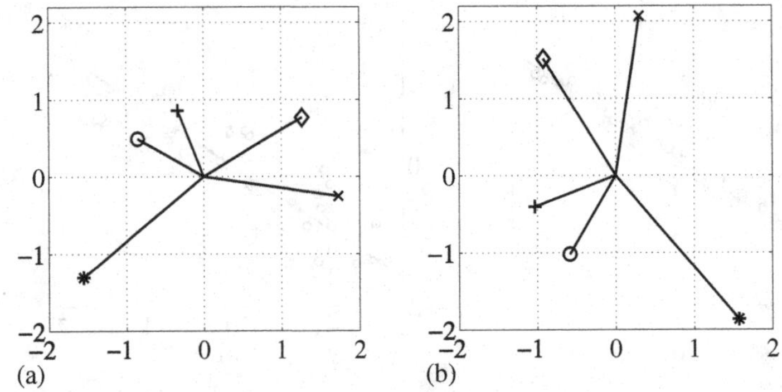

Figure 4.2 Sample pairs of two complex random variables x and y with $\rho_{xy} = \exp(\mathrm{j}\pi/2)$ and $|R_{xy}|/R_{yy} = 1.2$. Plot (a) depicts samples of x and (b) samples of y, in the complex plane. For corresponding samples we use the same symbol.

which satisfies $0 \leq |\rho_{xy}| \leq 1$. The LMMSE estimate of x from y is

$$\hat{x}(y) = \frac{R_{xy}}{R_{yy}} y = \frac{|R_{xy}|}{R_{yy}} \mathrm{e}^{\mathrm{j}\measuredangle R_{xy}} y, \tag{4.6}$$

which achieves the minimum error

$$E|\hat{x}(y) - x|^2 = R_{xx} - \frac{|R_{xy}|^2}{R_{yy}} = R_{xx}(1 - |\rho_{xy}|^2). \tag{4.7}$$

Hence, if $|\rho_{xy}| = 1$, $\hat{x}(y)$ is a perfect estimate of x from y. Figure 4.2 depicts five sample pairs of two complex random variables x and y with $\rho_{xy} = \exp(\mathrm{j}\pi/2)$, in the complex plane. Plot (a) shows samples of x and (b) the corresponding samples of y. We observe that (b) is simply a *scaled and rotated* version of (a). The amplitude is scaled by the factor $|R_{xy}|/R_{yy}$, preserving the aspect ratio, and the rotation angle is $\measuredangle R_{xy} = \measuredangle \rho_{xy}$.

Now what about complementary correlations? Instead of estimating x as a linear function of y, we may employ the *conjugate linear* minimum mean-squared error (CLMMSE) estimator

$$\hat{x}(y^*) = \frac{E\,xy}{E|y|^2} y^* = \frac{\widetilde{R}_{xy}}{R_{yy}} y^* = \frac{|\widetilde{R}_{xy}|}{R_{yy}} \mathrm{e}^{\mathrm{j}\measuredangle \widetilde{R}_{xy}} y^*. \tag{4.8}$$

The corresponding correlation coefficient is

$$\widetilde{\rho}_{xy} = \frac{\widetilde{R}_{xy}}{\sqrt{R_{xx}}\sqrt{R_{yy}}}, \tag{4.9}$$

with $0 \leq |\widetilde{\rho}_{xy}| \leq 1$, and the CLMMSE is

$$E|\hat{x}(y^*) - x|^2 = R_{xx} - \frac{|\widetilde{R}_{xy}|^2}{R_{yy}} = R_{xx}(1 - |\widetilde{\rho}_{xy}|^2). \tag{4.10}$$

Hence, if $|\widetilde{\rho}_{xy}| = 1$, $\hat{x}(y^*)$ is a perfect estimate of x from y^*. Figure 4.3 depicts five sample pairs of two complex random variables x and y with $\widetilde{\rho}_{xy} = \exp(\mathrm{j}\pi/2)$, in the complex plane. Plot (a) shows samples of x and (b) the corresponding samples of y.

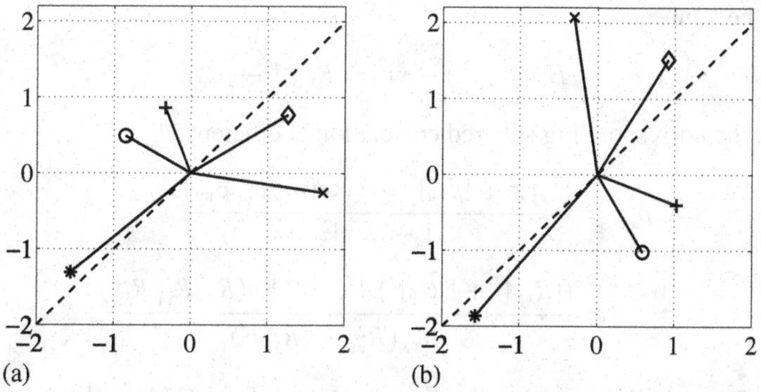

Figure 4.3 Sample pairs of two complex random variables x and y with $\widetilde{\rho}_{xy} = \exp(j\pi/2)$ and $|\widetilde{R}_{xy}|/R_{yy} = 1.2$. Plot (a) depicts samples of x and (b) samples of y, in the complex plane. Samples of y correspond to amplified and reflected samples of x. The reflection axis is the dashed line, which is given by $\measuredangle\widetilde{R}_{xy}/2 = \measuredangle\widetilde{\rho}_{xy}/2 = \pi/4$.

We observe that (b) is a *scaled and reflected* version of (a). The amplitude is scaled by the factor $|\widetilde{R}_{xy}|/R_{yy}$, preserving the aspect ratio. Since $\measuredangle x = \measuredangle\widetilde{R}_{xy} - \measuredangle y$, we have, with probability 1:

$$\measuredangle x - \tfrac{1}{2}\measuredangle\widetilde{R}_{xy} = -\left(\measuredangle y - \tfrac{1}{2}\measuredangle\widetilde{R}_{xy}\right). \tag{4.11}$$

Thus, the reflection axis is $\measuredangle\widetilde{R}_{xy}/2 = \measuredangle\widetilde{\rho}_{xy}/2$, which is the dashed line in Fig. 4.3.

Depending on whether *rotation* or *reflection* better models the relationship between x and y, $|\rho_{xy}|$ or $|\widetilde{\rho}_{xy}|$ will be greater. We note the ease with which the best possible reflection axis is determined as half the angle of the complementary correlation $\widetilde{R}_{xy}$ (or half the angle of the correlation coefficient $\widetilde{\rho}_{xy}$). This would be significantly more cumbersome in real-valued notation.

Of course, data might exhibit a combination of rotational and reflectional correlation, motivating use of a *widely linear* minimum mean-squared error (WLMMSE) estimator

$$\hat{x}(y, y^*) = \alpha y + \beta y^*, \tag{4.12}$$

where α and β are chosen to minimize $E|\hat{x}(y, y^*) - x|^2$. We will be discussing WLMMSE estimation in much detail in Section 5.4. At this point, we content ourselves with stating that the solution is

$$\begin{aligned}\hat{x}(y, y^*) &= \frac{\hat{x}(y) - \hat{x}[\hat{y}^*(y)] + \hat{x}(y^*) - \hat{x}[\hat{y}(y^*)]}{1 - |\widetilde{\rho}_{yy}|^2} \\ &= \frac{(R_{xy}R_{yy} - \widetilde{R}_{xy}\widetilde{R}^*_{yy})y + (\widetilde{R}_{xy}R_{yy} - R_{xy}\widetilde{R}_{yy})y^*}{R^2_{yy} - |\widetilde{R}_{yy}|^2},\end{aligned} \tag{4.13}$$

where $\widetilde{R}_{yy} = Ey^2$ is the complementary variance of y, $|\widetilde{\rho}_{yy}|^2 = |\widetilde{R}_{yy}|^2/R^2_{yy}$ is the degree of impropriety of y, $\hat{y}^*(y) = [\widetilde{R}^*_{yy}/R_{yy}]y$ is the CLMMSE estimate of y^* from y, and $\hat{y}(y^*) = [\widetilde{R}_{yy}/R_{yy}]y^*$ is the CLMMSE estimate of y from y^*.

Through the connection

$$E|\hat{x}(y, y^*) - x|^2 = R_{xx}(1 - \bar{\rho}_{xy}^2), \tag{4.14}$$

we obtain the corresponding squared correlation coefficient

$$\bar{\rho}_{xy}^2 = \frac{|\rho_{xy}|^2 + |\widetilde{\rho}_{xy}|^2 - 2\,\mathrm{Re}[\rho_{xy}\widetilde{\rho}_{xy}^*\widetilde{\rho}_{yy}]}{1 - |\widetilde{\rho}_{yy}|^2}$$

$$= \frac{(|R_{xy}|^2 + |\widetilde{R}_{xy}|^2)R_{yy} - 2\,\mathrm{Re}(R_{xy}\widetilde{R}_{xy}^*\widetilde{R}_{yy})}{R_{xx}(R_{yy}^2 - |\widetilde{R}_{yy}|^2)}, \tag{4.15}$$

with $0 \le \bar{\rho}_{xy}^2 \le 1$. We note that the correlation coefficient $\bar{\rho}_{xy}$, unlike $\rho_{xy} = \rho_{yx}^*$ and $\widetilde{\rho}_{xy} = \widetilde{\rho}_{yx}$, is not symmetric in x and y: in general, $\bar{\rho}_{xy}^2 \ne \bar{\rho}_{yx}^2$.

The correlation coefficient $\bar{\rho}_{xy}$ is bounded in terms of the coefficients ρ_{xy} and $\widetilde{\rho}_{xy}$ as

$$\max(|\rho_{xy}|^2, |\widetilde{\rho}_{xy}|^2) \le \bar{\rho}_{xy}^2 \le \min(|\rho_{xy}|^2 + |\widetilde{\rho}_{xy}|^2, 1). \tag{4.16}$$

The lower bound holds because a WLMMSE estimator subsumes both the LMMSE and CLMMSE estimators, so we must have WLMMSE $\le$ LMMSE and WLMMSE $\le$ CLMMSE. However, there is no general ordering of LMMSE and CLMMSE, which we write as LMMSE $\gtreqless$ CLMMSE.

A common scenario in which the lower bound is attained is when y is maximally improper, i.e., $R_{yy} = |\widetilde{R}_{yy}| \Leftrightarrow |\widetilde{\rho}_{yy}|^2 = 1$, which yields a zero denominator in (4.15). This means that, with probability 1, $y^* = \mathrm{e}^{\mathrm{j}\alpha}y$ for some constant α, and $R_{xy} = \mathrm{e}^{\mathrm{j}\alpha}\widetilde{R}_{xy}$. In this case, y and y^* carry exactly the same information about x. Therefore, WLMMSE estimation is unnecessary, and can be replaced with either LMMSE or CLMMSE estimation. In the maximally improper case, $\bar{\rho}_{xy}^2 = |\rho_{xy}|^2 = |\widetilde{\rho}_{xy}|^2$. Two other examples of attaining the lower bound in (4.16) are either $\widetilde{R}_{xy} = 0$ and $\widetilde{R}_{yy} = 0$ (i.e., $\widetilde{\rho}_{xy} = 0$ and $\widetilde{\rho}_{yy} = 0$), which leads to $\bar{\rho}_{xy}^2 = |\rho_{xy}|^2$, or $R_{xy} = 0$ and $\widetilde{R}_{yy} = 0$ (i.e., $\rho_{xy} = 0$ and $\widetilde{\rho}_{yy} = 0$), which yields $\bar{\rho}_{xy}^2 = |\widetilde{\rho}_{xy}|^2$; cf. (4.15).

The upper bound $\bar{\rho}_{xy}^2 = |\rho_{xy}|^2 + |\widetilde{\rho}_{xy}|^2$ is attained when the WLMMSE estimator is the sum of the LMMSE and CLMMSE estimators: $\hat{x}(y, y^*) = \hat{x}(y) + \hat{x}(y^*)$. In this case, y and y^* carry completely complementary information about x. This is possible only for uncorrelated y and y^*, that is, a *proper* y. It is easy to see that $\widetilde{R}_{yy} = 0 \Leftrightarrow |\widetilde{\rho}_{yy}|^2 = 0$ in (4.15) leads to $\bar{\rho}_{xy}^2 = |\rho_{xy}|^2 + |\widetilde{\rho}_{xy}|^2$. The following example gives two scenarios in which the lower and upper bounds are attained.

Example 4.2. *Attaining the lower bound.* Consider a complex random variable

$$y = \mathrm{e}^{\mathrm{j}\alpha}(u + n),$$

where u is a *real* random variable and α is a fixed constant. Further, assume that n is a *real* random variable, uncorrelated with u, and $R_{nn} = R_{uu}$. Let $x = \mathrm{Re}\,(\mathrm{e}^{\mathrm{j}\alpha}u) = \cos(\alpha)u$. The LMMSE estimator $\hat{x}(y) = \frac{1}{2}\cos(\alpha)e^{-\mathrm{j}\alpha}y$ and the CLMMSE estimator $\hat{x}(y^*) = \frac{1}{2}\cos(\alpha)\mathrm{e}^{\mathrm{j}\alpha}y^*$ both perform equally well. However, they both extract the *same* information from y because y is maximally improper. Hence, a WLMMSE estimator has no performance advantage over an LMMSE or CLMMSE estimator: $|\rho_{xy}|^2 = |\widetilde{\rho}_{xy}|^2 = \bar{\rho}_{xy}^2 = \frac{1}{2}$.

Attaining the upper bound. Consider a *proper* complex random variable y and let x be its real part. The LMMSE estimator of x from y is $\hat{x}(y) = \frac{1}{2}y$, the CLMMSE estimator is $\hat{x}(y^*) = \frac{1}{2}y^*$, and the WLMMSE estimator $\hat{x}(y, y^*) = \frac{1}{2}y + \frac{1}{2}y^*$ produces a *perfect* estimate of x. Here, rotational and reflectional models are equally appropriate. Each tells only half the story, but they *complement* each other perfectly. It is easy to see that $|\rho_{xy}|^2 = |\widetilde{\rho}_{xy}|^2 = \frac{1}{2}$ and $\bar{\rho}_{xy}^2 = |\rho_{xy}|^2 + |\widetilde{\rho}_{xy}|^2 = 1$. (This also shows that $\bar{\rho}_{xy}^2 \neq \bar{\rho}_{yx}^2$ since it is obviously impossible to perfectly reconstruct y as a widely linear function of x.)

The following definition sums up the main findings of this section, and will be used to classify correlation coefficients throughout this chapter.

Definition 4.1. *A correlation coefficient that measures how well a*

(1) linear function
(2) conjugate linear function
(3) widely linear function

models the relationship between two complex random variables is respectively called a

(1) rotational correlation coefficient
(2) reflectional correlation coefficient
(3) total correlation coefficient.

4.1.2 Principle of multivariate correlation analysis

We would now like to define a scalar-valued correlation coefficient that gives an overall measure of the association between two zero-mean random *vectors*. The following definition sets out the minimum requirements for such a correlation coefficient.

Definition 4.2. *A correlation coefficient ρ_{xy} between two random vectors* **x** *and* **y** *must satisfy the following conditions for all nonzero scalars α and β, provided that* **x** *and* **y** *are not both zero.*

$$0 \leq \rho_{xy} \leq 1, \tag{4.17}$$

$$\rho_{xy} = \rho_{x'y} = \rho_{xy'} \quad \textit{for } \mathbf{x}' = \alpha\mathbf{x}, \mathbf{y}' = \beta\mathbf{y}, \tag{4.18}$$

$$\rho_{xy} = 1 \quad \textit{if } \mathbf{y} = \beta\mathbf{x}, \tag{4.19}$$

$$\rho_{xy} = 0 \quad \textit{if } \mathbf{x} \textit{ and } \mathbf{y} \textit{ are uncorrelated.} \tag{4.20}$$

Note that we do not require the symmetry $\rho_{xy} = \rho_{yx}$.[3] If a correlation coefficient is allowed to be negative or complex-valued (as we have seen in the previous section), these conditions apply to its absolute value. However, the correlation coefficients considered hereafter are all real and nonnegative. For simplicity, *we consider only rotational*

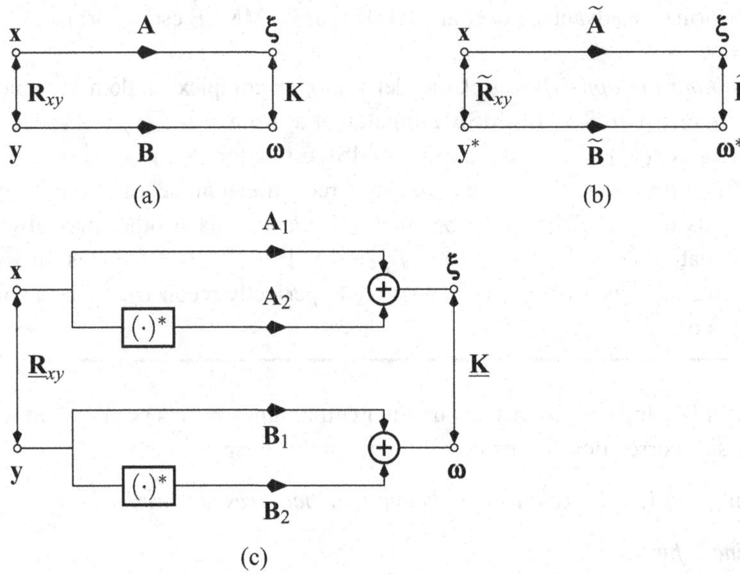

Figure 4.4 The principles of multivariate correlation analysis: (a) rotational correlations, (b) reflectional correlations, and (c) total correlations.

correlations and strictly linear transforms in this section. Reflectional and total correlations will be discussed in the following section.

A correlation coefficient that only satisfies (4.17)–(4.20) will probably not be very useful. It is usually required to have further cases that result in a unit correlation coefficient $\rho_{xy} = 1$, such as when $\mathbf{y} = \mathbf{Mx}$, where $\mathbf{M}$ is any nonsingular matrix, or $\mathbf{y} = \mathbf{Ux}$, where $\mathbf{U}$ is any unitary matrix. There are further desirable properties. Chief among them are invariance under specified classes of transformations on $\mathbf{x}$ and $\mathbf{y}$, and the ability to assess correlation in a lower-dimensional subspace. What exactly this means will become clearer as we move along in our development.

The cross-correlation properties between $\mathbf{x}$ and $\mathbf{y}$ are described by the cross-correlation matrix $\mathbf{R}_{xy} = E\mathbf{x}\mathbf{y}^{\mathrm{H}}$, but this matrix is generally difficult to interpret. In order to illuminate the underlying cross-correlation structure, we shall transform n-dimensional $\mathbf{x}$ and m-dimensional $\mathbf{y}$ into p-dimensional internal (latent) representations $\boldsymbol{\xi} = \mathbf{Ax}$ and $\boldsymbol{\omega} = \mathbf{By}$, with $p = \min(m, n)$, as shown in Fig. 4.4(a). The way in which the full-rank matrices $\mathbf{A} \in \mathbb{C}^{p\times n}$ and $\mathbf{B} \in \mathbb{C}^{p\times m}$ are chosen will determine the type of correlation analysis. In the statistical literature, the latent vectors $\boldsymbol{\xi}$ and $\boldsymbol{\omega}$ are usually called *score vectors*, and the matrices $\mathbf{A}$ and $\mathbf{B}$ are called the *matrices of loadings*.

Our goal is to define different correlation coefficients as different functions of the correlations $k_i = E\,\xi_i\omega_i^*$, $i = 1, \ldots, p$, which are the diagonal elements of the cross-correlation matrix $\mathbf{K} = E\,\boldsymbol{\xi}\boldsymbol{\omega}^{\mathrm{H}}$ in the internal coordinate system of $(\boldsymbol{\xi}, \boldsymbol{\omega})$. We would like as much correlation as possible concentrated in the first r coefficients $\{k_1, k_2, \ldots, k_r\}$, for any $r \leq p$, because this will allow us to assess correlation in a lower-dimensional subspace of dimension r. Hence, our aim is to choose $\mathbf{A}$ and $\mathbf{B}$ such that all partial sums over the absolute values of the diagonal cross-correlations k_i are maximized:

$$\max_{\mathbf{A},\mathbf{B}} \sum_{i=1}^{r} |k_i|, \quad r = 1, \ldots, p \tag{4.21}$$

In order to make this a well-defined maximization problem, we need to impose some constraints on **A** and **B**. The following three choices are most compelling.

- Require that the internal representations $\boldsymbol{\xi}$ and $\boldsymbol{\omega}$ each have identity correlation matrix (we avoid using the term "white" because $\boldsymbol{\xi}$ and $\boldsymbol{\omega}$ may have non-identity *complementary* correlation matrices): $\mathbf{R}_{\xi\xi} = \mathbf{A}\mathbf{R}_{xx}\mathbf{A}^{\mathrm{H}} = \mathbf{I}$ and $\mathbf{R}_{\omega\omega} = \mathbf{B}\mathbf{R}_{yy}\mathbf{B}^{\mathrm{H}} = \mathbf{I}$. This choice leads to *canonical correlation analysis* (CCA). The corresponding diagonal cross-correlations k_i are called the *canonical correlations*, and the latent vectors $\boldsymbol{\xi}$ and $\boldsymbol{\omega}$ are given in *canonical coordinates*.
- Require that **A** have unitary rows (which we will simply call *row-unitary*) and $\boldsymbol{\omega}$ have identity correlation matrix: $\mathbf{A}\mathbf{A}^{\mathrm{H}} = \mathbf{I}$ and $\mathbf{R}_{\omega\omega} = \mathbf{B}\mathbf{R}_{yy}\mathbf{B}^{\mathrm{H}} = \mathbf{I}$. This choice leads to *multivariate linear regression* (MLR), also known as *half-canonical correlation analysis*. The corresponding diagonal cross-correlations k_i are called the *half-canonical correlations*, and the latent vectors $\boldsymbol{\xi}$ and $\boldsymbol{\omega}$ are given in *half-canonical coordinates*.
- Require that **A** and **B** be row-unitary: $\mathbf{A}\mathbf{A}^{\mathrm{H}} = \mathbf{I}$ and $\mathbf{B}\mathbf{B}^{\mathrm{H}} = \mathbf{I}$. This choice leads to *partial least-squares* (PLS) analysis. The corresponding diagonal cross-correlations k_i are called the *PLS correlations*, and the latent vectors $\boldsymbol{\xi}$ and $\boldsymbol{\omega}$ are given in *PLS coordinates*.

Sometimes, when there is a risk of confusion, we will use the subscript C, M, or P to emphasize that quantities were derived using CCA, MLR, or PLS.

Example 4.3. If x and y are scalars, then the CCA constraints are $|A|^2 = R_{xx}^{-1}$ and $|B|^2 = R_{yy}^{-1}$. Thus, the latent variables are $\xi = R_{xx}^{-1/2}e^{j\phi_1}x$ and $\omega = R_{yy}^{-1/2}e^{j\phi_2}y$ for arbitrary ϕ_1 and ϕ_2. We then obtain

$$|k| = |E\xi\omega^*| = \frac{|R_{xy}|}{\sqrt{R_{xx}}\sqrt{R_{yy}}} = |\rho_{xy}|,$$

with ρ_{xy} defined in (4.5).

We will find in Section 4.1.4 that the solution to the maximization problem (4.21) for CCA, MLR, or PLS results in a diagonal cross-correlation matrix

$$\mathbf{K} = E\,\boldsymbol{\xi}\boldsymbol{\omega}^{\mathrm{H}} = \mathbf{Diag}(k_1, \ldots, k_p), \tag{4.22}$$

with $k_1 \geq k_2 \geq \cdots \geq k_p \geq 0$. Of course, the k_is depend on which of the principles CCA, MLR, and PLS is employed, so they are canonical correlations, half-canonical correlations, or PLS correlations. Furthermore, we will show that CCA, MLR, and PLS each produce a set $\{k_i\}$ that has maximum spread in the sense of majorization (see Appendix 3). Therefore, any correlation coefficient that is an increasing, Schur-convex function of $\{k_i\}$ is maximized, for arbitrary rank r.

The key difference among CCA, MLR, and PLS lies in their invariance properties. We will see that CCA is invariant under nonsingular linear transformation of both $\mathbf{x}$ and $\mathbf{y}$, MLR is invariant under nonsingular linear transformation of $\mathbf{y}$ but only unitary transformation of $\mathbf{x}$, and PLS is invariant under unitary transformation of both $\mathbf{x}$ and $\mathbf{y}$. Therefore, CCA and PLS provide a symmetric assessment of correlation since the roles of $\mathbf{x}$ and $\mathbf{y}$ are interchangeable. MLR, on the other hand, distinguishes between the *message* (or predictor/explanatory variables) $\mathbf{x}$ and the *measurement* (or criterion/response variables) $\mathbf{y}$. The correlation analysis technique must be chosen to match the invariance properties of the problem at hand.

4.1.3 Rotational, reflectional, and total correlations for complex vectors

It is easy to see how to apply these principles to reflectional and total correlations, as shown in Figs. 4.4(b) and (c). For *reflectional* correlations, the internal representation is $\boldsymbol{\xi} = \widetilde{\mathbf{A}}\mathbf{x}$ and $\boldsymbol{\omega}^* = \widetilde{\mathbf{B}}\mathbf{y}^*$, whose complementary cross-correlation matrix is $\widetilde{\mathbf{K}} = E\boldsymbol{\xi}\boldsymbol{\omega}^{\mathrm{T}}$, and $\tilde{k}_i = E\,\xi_i\omega_i$. The maximization problem is then to maximize all partial sums over the absolute value of the *complementary* cross-correlations

$$\max_{\widetilde{\mathbf{A}},\widetilde{\mathbf{B}}} \sum_{i=1}^{r} |\tilde{k}_i|, \quad r = 1, \ldots, p, \tag{4.23}$$

with the following constraints on $\widetilde{\mathbf{A}}$ and $\widetilde{\mathbf{B}}$.

- For CCA, $\widetilde{\mathbf{A}}\mathbf{R}_{xx}\widetilde{\mathbf{A}}^{\mathrm{H}} = \mathbf{I}$ and $\widetilde{\mathbf{B}}\mathbf{R}_{yy}^*\widetilde{\mathbf{B}}^{\mathrm{H}} = \mathbf{I}$.
- For MLR, $\widetilde{\mathbf{A}}\widetilde{\mathbf{A}}^{\mathrm{H}} = \mathbf{I}$ and $\widetilde{\mathbf{B}}\mathbf{R}_{yy}^*\widetilde{\mathbf{B}}^{\mathrm{H}} = \mathbf{I}$.
- For PLS, $\widetilde{\mathbf{A}}\widetilde{\mathbf{A}}^{\mathrm{H}} = \mathbf{I}$ and $\widetilde{\mathbf{B}}\widetilde{\mathbf{B}}^{\mathrm{H}} = \mathbf{I}$.

For *total* correlations, the internal representation is computed as a *widely linear* function: $\boldsymbol{\xi} = \underline{\mathbf{A}}\,\underline{\mathbf{x}}$ (i.e., $\boldsymbol{\xi} = \mathbf{A}_1\mathbf{x} + \mathbf{A}_2\mathbf{x}^*$) and $\boldsymbol{\omega} = \underline{\mathbf{B}}\,\underline{\mathbf{y}}$ (i.e., $\boldsymbol{\omega} = \mathbf{B}_1\mathbf{y} + \mathbf{B}_2\mathbf{y}^*$). Here, our goal is to maximize the diagonal cross-correlations between the vectors of real and imaginary parts of $\boldsymbol{\xi}$ and $\boldsymbol{\omega}$. That is, for $i = 1, \ldots, p$, we let

$$\bar{k}_{2i-1} = 2E(\operatorname{Re}\xi_i \operatorname{Re}\omega_i) = \operatorname{Re}(E\,\xi_i\omega_i^* + E\,\xi_i\omega_i), \tag{4.24}$$

$$\bar{k}_{2i} = 2E(\operatorname{Im}\xi_i \operatorname{Im}\omega_i) = \operatorname{Re}(E\,\xi_i\omega_i^* - E\,\xi_i\omega_i), \tag{4.25}$$

and maximize

$$\max_{\underline{\mathbf{A}},\underline{\mathbf{B}}} \sum_{i=1}^{r} |\bar{k}_i|, \quad r = 1, \ldots, 2p, \tag{4.26}$$

with the following constraints placed on $\underline{\mathbf{A}}$ and $\underline{\mathbf{B}}$.

- For CCA, $\underline{\mathbf{A}}\,\underline{\mathbf{R}}_{xx}\underline{\mathbf{A}}^{\mathrm{H}} = \mathbf{I}$ and $\underline{\mathbf{B}}\,\underline{\mathbf{R}}_{yy}\underline{\mathbf{B}}^{\mathrm{H}} = \mathbf{I}$. Hence, $\boldsymbol{\xi}$ and $\boldsymbol{\omega}$ are each white and proper.
- For MLR, $\underline{\mathbf{A}}\,\underline{\mathbf{A}}^{\mathrm{H}} = \mathbf{I}$ and $\underline{\mathbf{B}}\,\underline{\mathbf{R}}_{yy}\underline{\mathbf{B}}^{\mathrm{H}} = \mathbf{I}$. While $\boldsymbol{\omega}$ is white and proper, $\boldsymbol{\xi}$ is generally improper.
- For PLS, $\underline{\mathbf{A}}\,\underline{\mathbf{A}}^{\mathrm{H}} = \mathbf{I}$ and $\underline{\mathbf{B}}\,\underline{\mathbf{B}}^{\mathrm{H}} = \mathbf{I}$. Both $\boldsymbol{\xi}$ and $\boldsymbol{\omega}$ are generally improper.

Hence, we have a total of *nine* possible combinations between the three *correlation analysis techniques* (CCA, MLR, PLS) and the three different *correlation types* (rotational, reflectional, total). Each of these nine cases leads to different latent vectors $(\boldsymbol{\xi}, \boldsymbol{\omega})$, and different diagonal cross-correlations k_i, $\tilde{k}_i$, or $\bar{k}_i$. We thus speak of rotational canonical correlations, reflectional half-canonical correlations, total canonical correlations, and so on.

4.1.4 Transformations into latent variables

We will now derive the transformations that solve the maximization problems for rotational, reflectional, and total correlations using CCA, MLR, or PLS. In doing so, we determine the internal (latent) coordinate system for $(\boldsymbol{\xi}, \boldsymbol{\omega})$. The approach is the same in all cases, and is based on majorization theory. The background on majorization necessary to understand this material can be found in Appendix 3.

Result 4.1. *The solutions to the maximization problems for rotational correlations (4.21), reflectional correlations (4.23), and total correlations (4.26), using CCA, MLR, or PLS, all yield an internal (latent) coordinate system with mutually uncorrelated components ξ_i and ω_j for all $i \neq j$.*

Consider first the maximization problem (4.21) subject to the CCA constraints $\mathbf{A}\mathbf{R}_{xx}\mathbf{A}^{\mathrm{H}} = \mathbf{I}$ and $\mathbf{B}\mathbf{R}_{yy}\mathbf{B}^{\mathrm{H}} = \mathbf{I}$. These two constraints determine $\mathbf{A}$ and $\mathbf{B}$ up to multiplication from the left by $\mathbf{F}^{\mathrm{H}}$ and $\mathbf{G}^{\mathrm{H}}$, respectively, where $\mathbf{F} \in \mathbb{C}^{n\times p}$ and $\mathbf{G} \in \mathbb{C}^{m\times p}$ are both column-unitary:

$$\mathbf{A} = \mathbf{F}^{\mathrm{H}}\mathbf{R}_{xx}^{-1/2}, \tag{4.27}$$

$$\mathbf{B} = \mathbf{G}^{\mathrm{H}}\mathbf{R}_{yy}^{-1/2}. \tag{4.28}$$

The cross-correlation matrix between $\boldsymbol{\xi} = \mathbf{A}\mathbf{x}$ and $\boldsymbol{\omega} = \mathbf{B}\mathbf{y}$ is therefore

$$\mathbf{K} = \mathbf{F}^{\mathrm{H}}\mathbf{R}_{xx}^{-1/2}\mathbf{R}_{xy}\mathbf{R}_{yy}^{-\mathrm{H}/2}\mathbf{G}, \tag{4.29}$$

and the singular values of $\mathbf{K}$ are invariant to the choice of $\mathbf{F}$ and $\mathbf{G}$. Maximizing all partial sums (4.21) requires maximum spread (in the sense of majorization) among the diagonal elements of $\mathbf{K}$. As discussed in Appendix 3 in Result A3.6, the absolute values of the diagonal elements of an $n \times m$ matrix are weakly majorized by its singular values:

$$|\mathbf{diag}(\mathbf{K})| \prec_{\mathrm{w}} \mathbf{sv}(\mathbf{K}). \tag{4.30}$$

Maximum spread is therefore achieved by making $\mathbf{K}$ *diagonal*, which means that $\mathbf{F}$ and $\mathbf{G}$ are determined by the singular value decomposition (SVD) of

$$\mathbf{C} = \mathbf{R}_{xx}^{-1/2}\mathbf{R}_{xy}\mathbf{R}_{yy}^{-\mathrm{H}/2} = \mathbf{F}\mathbf{K}\mathbf{G}^{\mathrm{H}}. \tag{4.31}$$

We call the matrix $\mathbf{C}$ the *rotational coherence matrix*. The diagonal elements of the diagonal matrix $\mathbf{K} = \mathbf{Diag}(k_1, k_2, \ldots, k_p)$ are nonnegative, and arranged in decreasing order $k_1 \geq k_2 \geq \cdots \geq k_p \geq 0$.

Table 4.1 SVDs and optimum transformations for various correlation types and correlation analysis techniques. This table has been adapted from Schreier (2008c) ©IEEE, and is used with permission.

		Correlation analysis technique		
Correlation type	SVD	CCA	MLR	PLS
Rotational $(\mathbf{x}, \mathbf{y})$	$\mathbf{C} = \mathbf{F}\mathbf{K}\mathbf{G}^{\mathrm{H}}$	$\mathbf{C} = \mathbf{R}_{xx}^{-1/2}\mathbf{R}_{xy}\mathbf{R}_{yy}^{-\mathrm{H}/2}$ $\mathbf{A} = \mathbf{F}^{\mathrm{H}}\mathbf{R}_{xx}^{-1/2}$ $\mathbf{B} = \mathbf{G}^{\mathrm{H}}\mathbf{R}_{yy}^{-1/2}$	$\mathbf{C} = \mathbf{R}_{xy}\mathbf{R}_{yy}^{-\mathrm{H}/2}$ $\mathbf{A} = \mathbf{F}^{\mathrm{H}}$ $\mathbf{B} = \mathbf{G}^{\mathrm{H}}\mathbf{R}_{yy}^{-1/2}$	$\mathbf{C} = \mathbf{R}_{xy}$ $\mathbf{A} = \mathbf{F}^{\mathrm{H}}$ $\mathbf{B} = \mathbf{G}^{\mathrm{H}}$
Reflectional $(\mathbf{x}, \mathbf{y}^*)$	$\tilde{\mathbf{C}} = \tilde{\mathbf{F}}\tilde{\mathbf{K}}\tilde{\mathbf{G}}^{\mathrm{T}}$	$\tilde{\mathbf{C}} = \mathbf{R}_{xx}^{-1/2}\tilde{\mathbf{R}}_{xy}\mathbf{R}_{yy}^{-\mathrm{T}/2}$ $\tilde{\mathbf{A}} = \tilde{\mathbf{F}}^{\mathrm{H}}\mathbf{R}_{xx}^{-1/2}$ $\tilde{\mathbf{B}} = \tilde{\mathbf{G}}^{\mathrm{T}}\mathbf{R}_{yy}^{-*/2}$	$\tilde{\mathbf{C}} = \tilde{\mathbf{R}}_{xy}\mathbf{R}_{yy}^{-\mathrm{T}/2}$ $\tilde{\mathbf{A}} = \tilde{\mathbf{F}}^{\mathrm{H}}$ $\tilde{\mathbf{B}} = \tilde{\mathbf{G}}^{\mathrm{T}}\mathbf{R}_{yy}^{-*/2}$	$\tilde{\mathbf{C}} = \tilde{\mathbf{R}}_{xy}$ $\tilde{\mathbf{A}} = \tilde{\mathbf{F}}^{\mathrm{H}}$ $\tilde{\mathbf{B}} = \tilde{\mathbf{G}}^{\mathrm{T}}$
Total $(\underline{\mathbf{x}}, \underline{\mathbf{y}})$	$\underline{\mathbf{C}} = \underline{\mathbf{F}}\underline{\mathbf{K}}\underline{\mathbf{G}}^{\mathrm{H}}$	$\underline{\mathbf{C}} = \underline{\mathbf{R}}_{xx}^{-1/2}\underline{\mathbf{R}}_{xy}\underline{\mathbf{R}}_{yy}^{-\mathrm{H}/2}$ $\underline{\mathbf{A}} = \underline{\mathbf{F}}^{\mathrm{H}}\underline{\mathbf{R}}_{xx}^{-1/2}$ $\underline{\mathbf{B}} = \underline{\mathbf{G}}^{\mathrm{H}}\underline{\mathbf{R}}_{yy}^{-1/2}$	$\underline{\mathbf{C}} = \underline{\mathbf{R}}_{xy}\underline{\mathbf{R}}_{yy}^{-\mathrm{H}/2}$ $\underline{\mathbf{A}} = \underline{\mathbf{F}}^{\mathrm{H}}$ $\underline{\mathbf{B}} = \underline{\mathbf{G}}^{\mathrm{H}}\underline{\mathbf{R}}_{yy}^{-1/2}$	$\underline{\mathbf{C}} = \underline{\mathbf{R}}_{xy}$ $\underline{\mathbf{A}} = \underline{\mathbf{F}}^{\mathrm{H}}$ $\underline{\mathbf{B}} = \underline{\mathbf{G}}^{\mathrm{H}}$

It is straightforward to extend this result to the remaining eight combinations of correlation analysis technique (CCA, MLR, or PLS) and correlation type (rotational, reflectional, or total). This is shown in Table 4.1. In each of these cases, the solution involves the SVD of a rotational matrix $\mathbf{C}$, a reflectional matrix $\tilde{\mathbf{C}}$, or a total matrix $\underline{\mathbf{C}}$, whose definition depends on the correlation analysis technique.

For *reflectional* correlations, the expression $\tilde{\mathbf{C}} = \tilde{\mathbf{F}}\tilde{\mathbf{K}}\tilde{\mathbf{G}}^{\mathrm{T}}$ in Table 4.1 is the usual SVD but with column-unitary $\tilde{\mathbf{G}}^*$. Using the regular transpose rather than the Hermitian transpose matches the structure of $\tilde{\mathbf{C}}$.

For *total* correlations, the augmented SVD of the augmented matrix $\underline{\mathbf{C}} = \underline{\mathbf{F}}\underline{\mathbf{K}}\underline{\mathbf{G}}^{\mathrm{H}}$ is obtained completely analogously to the augmented EVD of an augmented covariance matrix, which was given in Result 3.1. The singular values of $\underline{\mathbf{C}}$ are $\bar{k}_1 \geq \bar{k}_2 \geq \cdots \geq \bar{k}_{2p} \geq 0$ and the matrix

$$\underline{\mathbf{K}} = \begin{bmatrix} \mathbf{K}_1 & \mathbf{K}_2 \\ \mathbf{K}_2 & \mathbf{K}_1 \end{bmatrix} \tag{4.32}$$

consists of a diagonal block $\mathbf{K}_1$ with diagonal elements $\frac{1}{2}(\bar{k}_{2i-1} + \bar{k}_{2i})$ and a diagonal block $\mathbf{K}_2$ with diagonal elements $\frac{1}{2}(\bar{k}_{2i-1} - \bar{k}_{2i})$. Therefore, the internal description $(\boldsymbol{\xi}, \boldsymbol{\omega})$ is mutually uncorrelated,

$$E\,\xi_i\omega_j^* = 0 \quad \text{and} \quad E\,\xi_i\omega_j = 0 \quad \text{for } i \neq j. \tag{4.33}$$

However, besides the *real* Hermitian cross-correlation

$$E\,\xi_i\omega_i^* = \tfrac{1}{2}(\bar{k}_{2i-1} + \bar{k}_{2i}), \tag{4.34}$$

there is also a generally nonzero *real* complementary cross-correlation

$$E\,\xi_i\omega_i = \tfrac{1}{2}(\bar{k}_{2i-1} - \bar{k}_{2i}), \tag{4.35}$$

unless all singular values of $\underline{\mathbf{C}}$ have even multiplicity. Thus, unlike in the rotational and reflectional case, the solution to the maximization problem (4.26) does not lead to a *diagonal* matrix, but a matrix $\underline{\mathbf{K}} = E\,\underline{\boldsymbol{\xi}}\,\underline{\boldsymbol{\omega}}^{\mathrm{H}}$ with *diagonal blocks* instead. This means that generally the internal description $(\boldsymbol{\xi}, \boldsymbol{\omega})$ cannot be made cross-proper because this would require a transformation *jointly* operating on $\mathbf{x}$ and $\mathbf{y}$ – yet all we have at our disposal are widely linear transformations *separately* operating on $\mathbf{x}$ and $\mathbf{y}$.

4.2 Invariance properties

In this section, we take a closer look at the properties of CCA, MLR, and PLS. Our focus will be on the invariance properties that characterize these correlation analysis techniques.

4.2.1 Canonical correlations

Canonical correlation analysis (CCA), which was introduced by Hotelling (1936), is an extremely popular classical tool for assessing multivariate association. Canonical correlations have a number of important properties. The following is an immediate consequence of the fact that the canonical vectors have identity correlation matrix, i.e., $\mathbf{R}_{\xi\xi} = \mathbf{I}$ and $\mathbf{R}_{\omega\omega} = \mathbf{I}$ in the rotational and reflectional case, and $\underline{\mathbf{R}}_{\xi\xi} = \mathbf{I}$ and $\underline{\mathbf{R}}_{\omega\omega} = \mathbf{I}$ in the total case.

Result 4.2. *Canonical correlations (rotational, reflectional, and total) satisfy* $0 \le k_i \le 1$.

A key property of canonical correlations is their invariance under nonsingular linear transformation, which can be more precisely stated as follows.

Result 4.3. *Rotational and reflectional canonical correlations are invariant under nonsingular linear transformation of* $\mathbf{x}$ *and* $\mathbf{y}$, *i.e.,* $(\mathbf{x}, \mathbf{y})$ *and* $(\mathbf{N}\mathbf{x}, \mathbf{M}\mathbf{y})$ *have the same rotational and reflectional canonical correlations for all nonsingular* $\mathbf{N} \in \mathbb{C}^{n\times n}$ *and* $\mathbf{M} \in \mathbb{C}^{m\times m}$. *Total canonical correlations are invariant under nonsingular widely linear transformation of* $\mathbf{x}$ *and* $\mathbf{y}$, *i.e.,* $(\underline{\mathbf{x}}, \underline{\mathbf{y}})$ *and* $(\underline{\mathbf{N}}\,\underline{\mathbf{x}}, \underline{\mathbf{M}}\,\underline{\mathbf{y}})$ *have the same total canonical correlations for all nonsingular* $\underline{\mathbf{N}} \in \mathcal{W}^{n\times n}$ *and* $\underline{\mathbf{M}} \in \mathcal{W}^{m\times m}$. *Moreover, the canonical correlations of* $(\mathbf{x}, \mathbf{y})$ *and* $(\mathbf{y}, \mathbf{x})$ *are identical.*

We will show this for rotational correlations. The rotational canonical correlations k_i are the singular values of $\mathbf{C}$, or, equivalently, the nonnegative roots of the eigenvalues of $\mathbf{C}\mathbf{C}^{\mathrm{H}}$. Keeping in mind that the nonzero eigenvalues of $\mathbf{XY}$ are the nonzero eigenvalues of $\mathbf{YX}$, we obtain

$$
\begin{aligned}
\mathbf{ev}(\mathbf{C}\mathbf{C}^{\mathrm{H}}) &= \mathbf{ev}(\mathbf{R}_{xx}^{-1/2}\mathbf{R}_{xy}\mathbf{R}_{yy}^{-1}\mathbf{R}_{xy}^{\mathrm{H}}\mathbf{R}_{xx}^{-\mathrm{H}/2}) \\
&= \mathbf{ev}(\mathbf{R}_{xx}^{-1}\mathbf{R}_{xy}\mathbf{R}_{yy}^{-1}\mathbf{R}_{xy}^{\mathrm{H}}). \qquad (4.36)
\end{aligned}
$$

Now consider the coherence matrix $\mathbf{C}'$ between $\mathbf{x}' = \mathbf{N}\mathbf{x}$ and $\mathbf{y}' = \mathbf{M}\mathbf{y}$. We find

$$\begin{aligned}\mathbf{ev}(\mathbf{C}'\mathbf{C}'^{\mathrm{H}}) &= \mathbf{ev}(\mathbf{R}_{x'x'}^{-1}\mathbf{R}_{x'y'}\mathbf{R}_{y'y'}^{-1}\mathbf{R}_{x'y'}^{\mathrm{H}}) \\ &= \mathbf{ev}(\mathbf{N}^{-\mathrm{H}}\mathbf{R}_{xx}^{-1}\mathbf{N}^{-1}\mathbf{N}\mathbf{R}_{xy}\mathbf{M}^{\mathrm{H}}\mathbf{M}^{-\mathrm{H}}\mathbf{R}_{yy}^{-1}\mathbf{M}^{-1}\mathbf{M}\mathbf{R}_{xy}^{\mathrm{H}}\mathbf{N}^{\mathrm{H}}) \\ &= \mathbf{ev}(\mathbf{N}^{-\mathrm{H}}\mathbf{R}_{xx}^{-1}\mathbf{R}_{xy}\mathbf{R}_{yy}^{-1}\mathbf{R}_{xy}^{\mathrm{H}}\mathbf{N}^{\mathrm{H}}) \\ &= \mathbf{ev}(\mathbf{C}\mathbf{C}^{\mathrm{H}}). \end{aligned} \tag{4.37}$$

Finally, the canonical correlations of $(\mathbf{x}, \mathbf{y})$ and $(\mathbf{y}, \mathbf{x})$ are identical because of

$$\mathbf{ev}(\mathbf{R}_{xx}^{-1}\mathbf{R}_{xy}\mathbf{R}_{yy}^{-1}\mathbf{R}_{xy}^{\mathrm{H}}) = \mathbf{ev}\,(\mathbf{R}_{yy}^{-1}\mathbf{R}_{xy}^{\mathrm{H}}\mathbf{R}_{xx}^{-1}\mathbf{R}_{xy}). \tag{4.38}$$

A similar proof works for reflectional and total correlations.

Canonical correlations are actually not just invariant under nonsingular linear transformation, they are *maximal invariant*. The following is the formal definition of a maximal invariant function (see Eaton (1983)).

Definition 4.3. *Let G be a group acting on a set $\mathcal{P}$. A function f defined on $\mathcal{P}$ is* invariant *if $f(\mathbf{R}) = f(g(\mathbf{R}))$ for all $\mathbf{R} \in \mathcal{P}$ and $g \in G$. A function f is* maximal invariant *if f is invariant and $f(\mathbf{R}) = f(\mathbf{R}')$ implies that $\mathbf{R} = g(\mathbf{R}')$ for some $g \in G$.*

Let us first discuss how this definition applies to *rotational* canonical correlations. Here, the set $\mathcal{P}$ is the set of all positive definite composite correlation matrices

$$\mathbb{R}_{xy} = \begin{bmatrix}\mathbf{R}_{xx} & \mathbf{R}_{xy} \\ \mathbf{R}_{xy}^{\mathrm{H}} & \mathbf{R}_{yy}\end{bmatrix}. \tag{4.39}$$

The group G consists of all nonsingular linear transformations applied to $\mathbf{x}$ and to $\mathbf{y}$. We write $g = (\mathbf{N}, \mathbf{M}) \in G$ to describe the nonsingular linear transformation

$$\mathbf{x} \longrightarrow \mathbf{N}\mathbf{x}, \tag{4.40}$$

$$\mathbf{y} \longrightarrow \mathbf{M}\mathbf{y}, \tag{4.41}$$

which acts on $\mathbb{R}_{xy}$ as

$$g(\mathbb{R}_{xy}) = \begin{bmatrix}\mathbf{N}\mathbf{R}_{xx}\mathbf{N}^{\mathrm{H}} & \mathbf{N}\mathbf{R}_{xy}\mathbf{M}^{\mathrm{H}} \\ \mathbf{M}\mathbf{R}_{xy}^{\mathrm{H}}\mathbf{N}^{\mathrm{H}} & \mathbf{M}\mathbf{R}_{yy}\mathbf{M}^{\mathrm{H}}\end{bmatrix}. \tag{4.42}$$

The function f produces the set of rotational canonical correlations $\{k_i\}$ from $\mathbb{R}_{xy}$. We have already proved above that f is invariant because $f(\mathbb{R}_{xy}) = f(g(\mathbb{R}_{xy}))$.

What is left to show is that if two composite correlation matrices $\mathbb{R}_{xy}$ and $\mathbb{R}'_{xy}$ have the same rotational canonical correlation matrix $\mathbf{K}$, then there exists $g \in G$ that relates the two matrices as $\mathbb{R}_{xy} = g(\mathbb{R}'_{xy})$. The fact that $\mathbb{R}_{xy}$ and $\mathbb{R}'_{xy}$ have the same $\mathbf{K}$ means that $g = (\mathbf{A}, \mathbf{B}) \in G$, with $\mathbf{A}$ and $\mathbf{B}$ determined by (4.27) and (4.28), and $g' = (\mathbf{A}', \mathbf{B}') \in G$ act on $\mathbb{R}_{xy}$ and $\mathbb{R}'_{xy}$ as

$$g(\mathbb{R}_{xy}) = \begin{bmatrix}\mathbf{I} & \mathbf{K} \\ \mathbf{K} & \mathbf{I}\end{bmatrix} = g'(\mathbb{R}'_{xy}). \tag{4.43}$$

Therefore, $\mathbb{R}_{xy} = (g^{-1} \circ g')(\mathbb{R}'_{xy})$, which makes f maximal invariant. In other words, the rotational canonical correlations are a *complete*, or *maximal*, *set of invariants* for $\mathbb{R}_{xy}$ under nonsingular linear transformation of $\mathbf{x}$ and $\mathbf{y}$. This maximal invariance property accounts for the name "canonical correlations."

For *reflectional* canonical correlations, the set to be considered contains all positive definite composite correlation matrices

$$\widetilde{\mathbb{R}}_{xy} = \begin{bmatrix} \mathbf{R}_{xx} & \widetilde{\mathbf{R}}_{xy} \\ \widetilde{\mathbf{R}}_{xy}^{\mathrm{H}} & \mathbf{R}_{yy}^{*} \end{bmatrix}, \tag{4.44}$$

on which g acts as

$$g(\widetilde{\mathbb{R}}_{xy}) = \begin{bmatrix} \mathbf{N}\mathbf{R}_{xx}\mathbf{N}^{\mathrm{H}} & \mathbf{N}\widetilde{\mathbf{R}}_{xy}\mathbf{M}^{\mathrm{T}} \\ \mathbf{M}^{*}\widetilde{\mathbf{R}}_{xy}^{\mathrm{H}}\mathbf{N}^{\mathrm{H}} & \mathbf{M}^{*}\mathbf{R}_{yy}^{*}\mathbf{M}^{\mathrm{T}} \end{bmatrix}. \tag{4.45}$$

It is obvious that the reflectional canonical correlations are a maximal set of invariants for $\widetilde{\mathbb{R}}_{xy}$ under nonsingular linear transformation of $\mathbf{x}$ and $\mathbf{y}$.

For *total* canonical correlations, we deal with the set of all positive definite composite augmented correlation matrices

$$\underline{\mathbb{R}}_{xy} = \begin{bmatrix} \underline{\mathbf{R}}_{xx} & \underline{\mathbf{R}}_{xy} \\ \underline{\mathbf{R}}_{xy}^{\mathrm{H}} & \underline{\mathbf{R}}_{yy} \end{bmatrix}. \tag{4.46}$$

Now let the group G consist of all nonsingular widely linear transformations applied to $\mathbf{x}$ and $\mathbf{y}$. That is, $g = (\underline{\mathbf{N}}, \underline{\mathbf{M}}) \in G$ describes

$$\underline{\mathbf{x}} \longrightarrow \underline{\mathbf{N}}\,\underline{\mathbf{x}}, \tag{4.47}$$

$$\underline{\mathbf{y}} \longrightarrow \underline{\mathbf{M}}\,\underline{\mathbf{y}}, \tag{4.48}$$

and acts on $\underline{\mathbb{R}}_{xy}$ as

$$g(\underline{\mathbb{R}}_{xy}) = \begin{bmatrix} \underline{\mathbf{N}}\,\underline{\mathbf{R}}_{xx}\underline{\mathbf{N}}^{\mathrm{H}} & \underline{\mathbf{N}}\,\underline{\mathbf{R}}_{xy}\underline{\mathbf{M}}^{\mathrm{H}} \\ \underline{\mathbf{M}}\,\underline{\mathbf{R}}_{xy}^{\mathrm{H}}\underline{\mathbf{N}}^{\mathrm{H}} & \underline{\mathbf{M}}\,\underline{\mathbf{R}}_{yy}\underline{\mathbf{M}}^{\mathrm{H}} \end{bmatrix}. \tag{4.49}$$

A maximal set of invariants for $\underline{\mathbb{R}}_{xy}$ under widely linear transformation of $\mathbf{x}$ and $\mathbf{y}$ is given by the total canonical correlations $\{\bar{k}_i\}_{i=1}^{2p}$ or, alternatively, by the Hermitian cross-correlations

$$\left\{E\,\xi_i\omega_i^* = \tfrac{1}{2}(\bar{k}_{2i-1} + \bar{k}_{2i})\right\}_{i=1}^{p}$$

together with the complementary cross-correlations

$$\left\{E\,\xi_i\omega_i = \tfrac{1}{2}(\bar{k}_{2i-1} - \bar{k}_{2i})\right\}_{i=1}^{p}.$$

The importance of a maximal invariant function is that a function is invariant *if and only if* it is (only) a function of a maximal invariant. For our purposes, this yields the following key result whose importance is difficult to overstate.

Result 4.4. *Any function of* $\mathbf{R}_{xx}$, $\mathbf{R}_{xy}$, *and* $\mathbf{R}_{yy}$ *that is invariant under nonsingular linear transformation of* $\mathbf{x}$ *and* $\mathbf{y}$ *is a function of the rotational canonical correlations only.*

Any function of $\mathbf{R}_{xx}$, $\widetilde{\mathbf{R}}_{xy}$, *and* $\mathbf{R}_{yy}^*$ *that is invariant under nonsingular linear transformation of* $\mathbf{x}$ *and* $\mathbf{y}$ *is a function of the reflectional canonical correlations only.*

Any function of $\underline{\mathbf{R}}_{xx}$, $\underline{\mathbf{R}}_{xy}$, *and* $\underline{\mathbf{R}}_{yy}$ *that is invariant under nonsingular widely linear transformation of* $\mathbf{x}$ *and* $\mathbf{y}$ *is a function of the total canonical correlations only.*

This result is the basis for testing for independence between two Gaussian random vectors $\mathbf{x}$ and $\mathbf{y}$ (see Section 4.5), and testing for propriety, i.e., uncorrelatedness of $\mathbf{x}$ and $\mathbf{x}^*$ (see Section 3.4).

4.2.2 Multivariate linear regression (half-canonical correlations)

We now consider multivariate linear regression (MLR), which is also called half-canonical correlation analysis. A key difference between MLR and CCA is that MLR is not symmetric because it distinguishes between the message (or predictor/explanatory variables) $\mathbf{x}$ and the measurement (or criterion/response variables) $\mathbf{y}$. The invariance properties of MLR are as follows.

Result 4.5. *Rotational and reflectional half-canonical correlations are invariant under unitary transformation of* $\mathbf{x}$ *and nonsingular linear transformation of* $\mathbf{y}$, *i.e.,* $(\mathbf{x}, \mathbf{y})$ *and* $(\mathbf{Ux}, \mathbf{My})$ *have the same rotational and reflectional half-canonical correlations for all unitary* $\mathbf{U} \in \mathbb{C}^{n\times n}$ *and nonsingular* $\mathbf{M} \in \mathbb{C}^{m\times m}$. *Total half-canonical correlations are invariant under widely unitary transformation of* $\mathbf{x}$ *and nonsingular widely linear transformation of* $\mathbf{y}$, *i.e.,* $(\underline{\mathbf{x}}, \underline{\mathbf{y}})$ *and* $(\underline{\mathbf{U}}\,\underline{\mathbf{x}}, \underline{\mathbf{M}}\,\underline{\mathbf{y}})$ *have the same total half-canonical correlations for all unitary* $\underline{\mathbf{U}} \in \mathcal{W}^{n\times n}$, $\underline{\mathbf{U}}\,\underline{\mathbf{U}}^{\mathrm{H}} = \underline{\mathbf{U}}^{\mathrm{H}}\underline{\mathbf{U}} = \mathbf{I}$, *and nonsingular* $\underline{\mathbf{M}} \in \mathcal{W}^{m\times m}$.

This can be shown similarly to the invariance properties of CCA. The rotational half-canonical correlations k_i are the singular values of $\mathbf{C}$, or equivalently, the nonnegative roots of the eigenvalues of $\mathbf{CC}^{\mathrm{H}}$. Let $\mathbf{C}'$ be the half-coherence matrix between $\mathbf{x}' = \mathbf{Ux}$ and $\mathbf{y}' = \mathbf{My}$. We obtain

$$\begin{aligned}\mathbf{ev}(\mathbf{C}'\mathbf{C}'^{\mathrm{H}}) &= \mathbf{ev}(\mathbf{R}_{x'y'}\mathbf{R}_{y'y'}^{-1}\mathbf{R}_{x'y'}^{\mathrm{H}}) \\ &= \mathbf{ev}(\mathbf{U}\mathbf{R}_{xy}\mathbf{M}^{\mathrm{H}}\mathbf{M}^{-\mathrm{H}}\mathbf{R}_{yy}^{-1}\mathbf{M}^{-1}\mathbf{M}\mathbf{R}_{xy}^{\mathrm{H}}\mathbf{U}^{\mathrm{H}}) \\ &= \mathbf{ev}(\mathbf{CC}^{\mathrm{H}}). \end{aligned} \tag{4.50}$$

Similar proofs work for reflectional and total half-canonical correlations.

Rotational half-canonical correlations are a maximal set of invariants for $\mathbf{R}_{xy}$ and $\mathbf{R}_{yy}$, and reflectional half-canonical correlations are a maximal set of invariants for $\widetilde{\mathbf{R}}_{xy}$ and $\mathbf{R}_{yy}^*$, under the transformation

$$\mathbf{x} \longrightarrow \mathbf{Ux}, \tag{4.51}$$

$$\mathbf{y} \longrightarrow \mathbf{My} \tag{4.52}$$

for unitary $\mathbf{U} \in \mathbb{C}^{n\times n}$ and nonsingular $\mathbf{M} \in \mathbb{C}^{m\times m}$.

Total half-canonical correlations are a maximal set of invariants for $\underline{\mathbf{R}}_{xy}$ and $\underline{\mathbf{R}}_{yy}$ under the transformation

$$\underline{\mathbf{x}} \longrightarrow \underline{\mathbf{U}}\,\underline{\mathbf{x}}, \tag{4.53}$$

$$\underline{\mathbf{y}} \longrightarrow \underline{\mathbf{M}}\,\underline{\mathbf{y}} \tag{4.54}$$

for unitary $\underline{\mathbf{U}} \in \mathcal{W}^{n\times n}$ and nonsingular $\underline{\mathbf{M}} \in \mathcal{W}^{m\times m}$. This maximal invariance property is proved along the same lines as the maximal invariance property of canonical correlations in the previous section. Thus, we have the following important result.

Result 4.6. *Any function of $\mathbf{R}_{xy}$ and $\mathbf{R}_{yy}$ that is invariant under unitary transformation of $\mathbf{x}$ and nonsingular linear transformation of $\mathbf{y}$ is a function of the rotational half-canonical correlations only.*

Any function of $\widetilde{\mathbf{R}}_{xy}$ and $\mathbf{R}^*_{yy}$ that is invariant under unitary transformation of $\mathbf{x}$ and nonsingular linear transformation of $\mathbf{y}$ is a function of the reflectional half-canonical correlations only.

Any function of $\underline{\mathbf{R}}_{xy}$ and $\underline{\mathbf{R}}_{yy}$ that is invariant under widely unitary transformation of $\mathbf{x}$ and nonsingular widely linear transformations of $\mathbf{y}$ is a function of the total half-canonical correlations only.

One may have been tempted to assume that half-canonical correlations are maximal invariants for the composite correlation matrices $\mathbb{R}_{xy}$ in (4.39), $\widetilde{\mathbb{R}}_{xy}$ in (4.44), or $\underline{\mathbb{R}}_{xy}$ in (4.46). But it is clear that half-canonical correlations do not characterize $\mathbf{R}_{xx}$ or $\underline{\mathbf{R}}_{xx}$. A maximal set of invariants for $\mathbf{R}_{xx}$ under unitary transformation of $\mathbf{x}$ is the set of eigenvalues of $\mathbf{R}_{xx}$, and a maximal set of invariants for $\underline{\mathbf{R}}_{xx}$ under widely unitary transformation of $\mathbf{x}$ is the set of eigenvalues of $\underline{\mathbf{R}}_{xx}$. For instance, $\operatorname{tr}\mathbf{R}_{xx}$, which is invariant under unitary transformation, is the sum of the eigenvalues of $\mathbf{R}_{xx}$.

Weighted MLR

A generalization of MLR replaces the condition that $\mathbf{A}$ have unitary rows, $\mathbf{A}\mathbf{A}^{\mathrm{H}} = \mathbf{I}$, with $\mathbf{A}\mathbf{W}\mathbf{A}^{\mathrm{H}} = \mathbf{I}$, where $\mathbf{W} \in \mathbb{C}^{n\times n}$ is a Hermitian, positive definite weighting matrix. In the rotational case, the optimum transformations and corresponding SVD are then given by

$$\mathbf{A} = \mathbf{F}^{\mathrm{H}}\mathbf{W}^{-1/2}, \tag{4.55}$$

$$\mathbf{B} = \mathbf{G}^{\mathrm{H}}\mathbf{R}_{yy}^{-1/2}, \tag{4.56}$$

$$\mathbf{C} = \mathbf{W}^{-1/2}\mathbf{R}_{xy}\mathbf{R}_{yy}^{-\mathrm{H}/2} = \mathbf{F}\mathbf{K}\mathbf{G}^{\mathrm{H}}. \tag{4.57}$$

Extensions to the reflectional and total case work analogously. The invariance properties of weighted half-canonical correlations are different than those of half-canonical correlations and depend on the choice of $\mathbf{W}$. It is interesting to note that canonical correlations are actually weighted half-canonical correlations with weighting matrix $\mathbf{W} = \mathbf{R}_{xx}$.

4.2.3 Partial least squares

Finally, we turn to partial least squares (PLS). We emphasize that there are many variants of PLS. Our description of PLS follows Sampson *et al.* (1989), which differs from the original PLS algorithm presented by Wold (1975, 1985). In the original PLS algorithm, both $\mathbf{R}_{\xi\xi}$ and $\mathbf{R}_{\omega\omega}$ are diagonal matrices – albeit not identity matrices – and $\mathbf{K}$ is not generally diagonal. Moreover, $\mathbf{A}$ and $\mathbf{B}$ are not unitary. In the version of Sampson *et al.* (1989), which we follow, $\mathbf{K}$ is diagonal but $\mathbf{R}_{\xi\xi}$ and $\mathbf{R}_{\omega\omega}$ are generally not, and $\mathbf{A}$ and $\mathbf{B}$ are both unitary.

The maximal invariance properties of PLS should be obvious and straightforward to prove because PLS is obtained directly by the SVD of the cross-correlation matrix between $\mathbf{x}$ and $\mathbf{y}$. The following two results are stated for completeness.

Result 4.7. *Rotational and reflectional PLS correlations are invariant under unitary transformation of* $\mathbf{x}$ *and* $\mathbf{y}$*, i.e.,* $(\mathbf{x}, \mathbf{y})$ *and* $(\mathbf{Ux}, \mathbf{Vy})$ *have the same rotational and reflectional PLS correlations for all unitary* $\mathbf{U} \in \mathbb{C}^{n\times n}$ *and* $\mathbf{V} \in \mathbb{C}^{m\times m}$*. Total PLS correlations are invariant under widely unitary transformation of* $\mathbf{x}$ *and* $\mathbf{y}$*, i.e.,* $(\underline{\mathbf{x}}, \underline{\mathbf{y}})$ *and* $(\underline{\mathbf{U}}\,\underline{\mathbf{x}}, \underline{\mathbf{V}}\,\underline{\mathbf{y}})$ *have the same total PLS correlations for all unitary* $\underline{\mathbf{U}} \in \mathcal{W}^{n\times n}$ *and* $\underline{\mathbf{V}} \in \mathcal{W}^{m\times m}$*. Moreover, the PLS correlations of* $(\mathbf{x}, \mathbf{y})$ *and* $(\mathbf{y}, \mathbf{x})$ *are identical.*

Result 4.8. *Any function of* $\mathbf{R}_{xy}$ *that is invariant under unitary transformation of* $\mathbf{x}$ *and* $\mathbf{y}$ *is a function of the rotational PLS correlations only.*

Any function of $\widetilde{\mathbf{R}}_{xy}$ *that is invariant under unitary transformation of* $\mathbf{x}$ *and* $\mathbf{y}$ *is a function of the reflectional PLS correlations only.*

Any function of $\underline{\mathbf{R}}_{xy}$ *that is invariant under widely unitary transformation of* $\mathbf{x}$ *and* $\mathbf{y}$ *is a function of the total PLS correlations only.*

Both PLS and CCA provide a symmetric measure of multivariate association since the roles of $\mathbf{x}$ and $\mathbf{y}$ are interchangeable. PLS has an advantage over CCA if it is applied to sample correlation matrices. Since the computation of the sample coherence matrix $\mathbf{S}_{xx}^{-1/2}\mathbf{S}_{xy}\mathbf{S}_{yy}^{-H/2}$ requires the computation of inverses $\mathbf{S}_{xx}^{-1}$ and $\mathbf{S}_{yy}^{-1}$, we run into stability problems if $\mathbf{S}_{xx}$ or $\mathbf{S}_{yy}$ are close to singular. That is, the SVD of the coherence matrix can change significantly after recomputing sample correlation matrices with added samples that are nearly collinear with previous samples. These problems do not arise with PLS, which is why PLS has been called "robust canonical analysis" by Tishler and Lipovetsky (2000). However, it is possible to alleviate some of the numerical problems of CCA by applying appropriate penalties to the sample correlation matrices $\mathbf{S}_{xx}$ and $\mathbf{S}_{yy}$.

4.3 Correlation coefficients for complex vectors

In order to summarize the degree of multivariate association between $\mathbf{x}$ and $\mathbf{y}$, we now define different scalar-valued overall correlation coefficients ρ as functions of the diagonal cross-correlations $\{k_i\}$, which may be computed as rotational, reflectional, or total CCA, MLR, or PLS coefficients. Therefore, these correlation coefficients inherit their invariance properties from $\{k_i\}$. It is easy to show that all correlation coefficients presented in this chapter are increasing, Schur-convex functions. (Refer to Appendix 3,

Section A3.1.2, for background on increasing, Schur-convex functions.) This means they are maximized by CCA, MLR, or PLS, for all ranks r.

4.3.1 Canonical correlations

There is a number of commonly used correlation coefficients that are defined on the basis of the first r canonical correlations $\{k_i\}_{i=1}^r$. Three particularly compelling coefficients are[4]

$$\rho_{C_1}^2 = \frac{1}{p}\sum_{i=1}^{r} k_i^2, \tag{4.58}$$

$$\rho_{C_2}^2 = 1 - \prod_{i=1}^{r}(1-k_i^2), \tag{4.59}$$

$$\rho_{C_3}^2 = \frac{\displaystyle\sum_{i=1}^{r}\frac{k_i^2}{1-k_i^2}}{\displaystyle\sum_{i=1}^{r}\frac{1}{1-k_i^2}+(p-r)}. \tag{4.60}$$

These are *rotational* coefficients, and their *reflectional* versions $\tilde{\rho}$ are obtained simply by replacing k_i with $\tilde{k}_i$. Their *total* versions use $\bar{k}_i$ but require slightly different normalizations because there are $2r$ rather than r coefficients:

$$\bar{\rho}_{C_1}^2 = \frac{1}{2p}\sum_{i=1}^{2r} \bar{k}_i^2, \tag{4.61}$$

$$\bar{\rho}_{C_2}^2 = 1 - \prod_{i=1}^{2r}(1-\bar{k}_i^2)^{1/2}, \tag{4.62}$$

$$\bar{\rho}_{C_3}^2 = \frac{\displaystyle\sum_{i=1}^{2r}\frac{\bar{k}_i^2}{1-\bar{k}_i^2}}{\displaystyle\sum_{i=1}^{2r}\frac{1}{1-\bar{k}_i^2}+2(p-r)}. \tag{4.63}$$

In the full-rank case $r = p = \min(m, n)$, these coefficients can also be expressed directly in terms of the correlation and cross-correlation matrices. For the rotational coefficients, we find the following expressions:

$$\rho_{C_1}^2 = \frac{1}{p}\operatorname{tr}\mathbf{K}^2 = \frac{1}{p}\operatorname{tr}(\mathbf{R}_{xx}^{-1}\mathbf{R}_{xy}\mathbf{R}_{yy}^{-1}\mathbf{R}_{xy}^{\mathrm{H}}), \tag{4.64}$$

$$\rho_{C_2}^2 = 1 - \det(\mathbf{I}-\mathbf{K}^2) = 1 - \det(\mathbf{I}-\mathbf{R}_{xx}^{-1}\mathbf{R}_{xy}\mathbf{R}_{yy}^{-1}\mathbf{R}_{xy}^{\mathrm{H}}), \tag{4.65}$$

$$\rho_{C_3}^2 = \frac{\operatorname{tr}\big(\mathbf{K}^2(\mathbf{I}-\mathbf{K}^2)^{-1}\big)}{\operatorname{tr}\big((\mathbf{I}-\mathbf{K}^2)^{-1}\big)} = \frac{\operatorname{tr}\big(\mathbf{R}_{xy}\mathbf{R}_{yy}^{-1}\mathbf{R}_{xy}^{\mathrm{H}}(\mathbf{R}_{xx}-\mathbf{R}_{xy}\mathbf{R}_{yy}^{-1}\mathbf{R}_{xy}^{\mathrm{H}})^{-1}\big)}{\operatorname{tr}\big(\mathbf{R}_{xx}(\mathbf{R}_{xx}-\mathbf{R}_{xy}\mathbf{R}_{yy}^{-1}\mathbf{R}_{xy}^{\mathrm{H}})^{-1}\big)}. \tag{4.66}$$

The formulae for the reflectional versions are obtained by replacing $\mathbf{R}_{xy}$ with $\widetilde{\mathbf{R}}_{xy}$, and $\mathbf{R}_{yy}^{-1}$ with $\mathbf{R}_{yy}^{-*}$. The expressions for the total full-rank coefficients are

$$\bar{\rho}_{\mathrm{C}_1}^2 = \frac{1}{2p}\,\mathrm{tr}\,\underline{\mathbf{K}}^2 = \frac{1}{2p}\,\mathrm{tr}(\underline{\mathbf{R}}_{xx}^{-1}\underline{\mathbf{R}}_{xy}\underline{\mathbf{R}}_{yy}^{-1}\underline{\mathbf{R}}_{xy}^{\mathrm{H}}), \tag{4.67}$$

$$\bar{\rho}_{\mathrm{C}_2}^2 = 1 - \det{}^{1/2}(\mathbf{I} - \underline{\mathbf{K}}^2) = 1 - \det{}^{1/2}(\mathbf{I} - \underline{\mathbf{R}}_{xx}^{-1}\underline{\mathbf{R}}_{xy}\underline{\mathbf{R}}_{yy}^{-1}\underline{\mathbf{R}}_{xy}^{\mathrm{H}}) \tag{4.68}$$

$$\bar{\rho}_{\mathrm{C}_3}^2 = \frac{\mathrm{tr}\big(\underline{\mathbf{K}}^2(\mathbf{I} - \underline{\mathbf{K}}^2)^{-1}\big)}{\mathrm{tr}\big((\mathbf{I} - \underline{\mathbf{K}}^2)^{-1}\big)} = \frac{\mathrm{tr}\big(\underline{\mathbf{R}}_{xy}\underline{\mathbf{R}}_{yy}^{-1}\underline{\mathbf{R}}_{xy}^{\mathrm{H}}(\underline{\mathbf{R}}_{xx} - \underline{\mathbf{R}}_{xy}\underline{\mathbf{R}}_{yy}^{-1}\underline{\mathbf{R}}_{xy}^{\mathrm{H}})^{-1}\big)}{\mathrm{tr}\big(\underline{\mathbf{R}}_{xx}(\underline{\mathbf{R}}_{xx} - \underline{\mathbf{R}}_{xy}\underline{\mathbf{R}}_{yy}^{-1}\underline{\mathbf{R}}_{xy}^{\mathrm{H}})^{-1}\big)}. \tag{4.69}$$

These coefficients inherit the invariance properties of the canonical correlations. That is, the rotational and reflectional coefficients are invariant under nonsingular linear transformation of $\mathbf{x}$ and $\mathbf{y}$, and the total coefficients are invariant under nonsingular widely linear transformation.

How should we interpret these coefficients? The rotational version of ρ_{C_1} characterizes the MMSE when constructing a *linear* estimate of the canonical vector $\boldsymbol{\xi}$ from $\mathbf{y}$. This estimate is

$$\hat{\boldsymbol{\xi}}(\mathbf{y}) = \mathbf{R}_{\xi y}\mathbf{R}_{yy}^{-1}\mathbf{y} = \mathbf{F}^{\mathrm{H}}\mathbf{R}_{xx}^{-1/2}\mathbf{R}_{xy}\mathbf{R}_{yy}^{-1}\mathbf{y} = \mathbf{F}^{\mathrm{H}}\mathbf{C}\mathbf{R}_{yy}^{-1/2}\mathbf{y} = \mathbf{K}\mathbf{B}\mathbf{y} = \mathbf{K}\boldsymbol{\omega}, \tag{4.70}$$

and the resulting MMSE is

$$E\|\hat{\boldsymbol{\xi}}(\mathbf{y}) - \boldsymbol{\xi}\|^2 = \mathrm{tr}(\mathbf{R}_{\xi\xi} - \mathbf{R}_{\xi\omega}\mathbf{R}_{\omega\omega}^{-1}\mathbf{R}_{\xi\omega}^{\mathrm{H}}) = \mathrm{tr}(\mathbf{I} - \mathbf{K}^2) = p(1 - \rho_{\mathrm{C}_1}^2). \tag{4.71}$$

Since CCA is symmetric in $\mathbf{x}$ and $\mathbf{y}$, the same MMSE is obtained when estimating $\boldsymbol{\omega}$ from $\mathbf{x}$. In a similar fashion, the reflectional version $\widetilde{\rho}_{\mathrm{C}_1}$ is related to the MMSE of the *conjugate linear* estimator

$$\hat{\boldsymbol{\xi}}(\mathbf{y}^*) = \widetilde{\mathbf{K}}\boldsymbol{\omega}^*, \tag{4.72}$$

and the total version $\bar{\rho}_{\mathrm{C}_1}$ is related to the MMSE of the *widely linear* estimator

$$\hat{\boldsymbol{\xi}}(\mathbf{y}, \mathbf{y}^*) = \mathbf{K}_1\boldsymbol{\omega} + \mathbf{K}_2\boldsymbol{\omega}^*. \tag{4.73}$$

For jointly Gaussian $\mathbf{x}$ and $\mathbf{y}$, the second coefficient, ρ_{C_2}, determines the mutual information between $\mathbf{x}$ and $\mathbf{y}$,

$$I(\mathbf{x};\mathbf{y}) = -\log\left(\frac{\det(\mathbf{R}_{xx} - \mathbf{R}_{xy}\mathbf{R}_{yy}^{-1}\mathbf{R}_{xy}^{\mathrm{H}})}{\det\mathbf{R}_{xx}}\right) = -\log\det(\mathbf{I} - \mathbf{K}^2) = -\log(1 - \rho_{\mathrm{C}_2}^2). \tag{4.74}$$

The rotational and reflectional versions only take rotational and reflectional dependences into account, respectively, whereas the total version characterizes the total mutual information between $\mathbf{x}$ and $\mathbf{y}$.

Finally, ρ_{C_3} has an interesting interpretation in the signal-plus-uncorrelated-noise case $\mathbf{y} = \mathbf{x} + \mathbf{n}$. Let $\mathbf{R}_{xx} = \mathbf{R}_{xy} = \mathbf{S}$ and $\mathbf{R}_{yy} = \mathbf{S} + \mathbf{N}$. It is easy to show that the eigenvalues of the signal-to-noise-ratio (SNR) matrix $\mathbf{S}\mathbf{N}^{-1}$ are $\{k_i^2/(1 - k_i^2)\}$. Hence, they are invariant under nonsingular linear transformation of the signal $\mathbf{x}$, and the numerator of $\rho_{\mathrm{C}_3}^2$ in (4.66) is $\mathrm{tr}(\mathbf{S}\mathbf{N}^{-1})$. Correlation coefficient ρ_{C_3} can thus be interpreted as a normalized SNR.

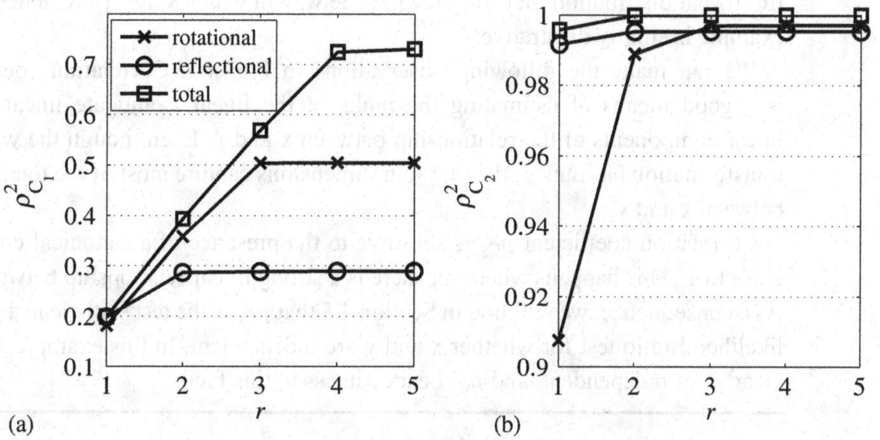

Figure 4.5 Correlation coefficients $\rho^2_{C_1}$ (a) and $\rho^2_{C_2}$ (b) for various ranks r.

Example 4.4. Consider the widely linear model

$$\mathbf{y} = \mathbf{H}_1\mathbf{x} + \mathbf{H}_2\mathbf{x}^* + \mathbf{n}, \tag{4.75}$$

where $\mathbf{x}$ has dimension $n = 5$ and $\mathbf{y}$ and $\mathbf{n}$ have dimension $m = 6$. Furthermore, $\mathbf{x}$ and $\mathbf{n}$ are independent Gaussians, both proper, with zero mean and correlation matrices $\mathbf{R}_{xx} = \mathbf{I}$ and $\mathbf{R}_{nn} = \mathbf{I}$. The matrix describing the linear part of the transformation

$$\mathbf{H}_1 = \begin{bmatrix} 1 & 1+\mathrm{j} & 1-\mathrm{j} & 2 & 1-2\mathrm{j} \\ 1 & 0 & 0 & \mathrm{j} & -\mathrm{j} \\ \mathrm{j} & 1-\mathrm{j} & -2\mathrm{j} & 1+\mathrm{j} & 0 \\ -1 & 1+\mathrm{j} & 1-\mathrm{j} & 2-2\mathrm{j} & 1 \\ 0 & 1+\mathrm{j} & 1-\mathrm{j} & 2-\mathrm{j} & 1-\mathrm{j} \\ \mathrm{j} & 2 & 1-3\mathrm{j} & 3 & 1-\mathrm{j} \end{bmatrix}$$

has rank 3, the matrix describing the conjugate linear part

$$\mathbf{H}_2 = \begin{bmatrix} 1 & -\mathrm{j} & 0 & 1-\mathrm{j} & 0 \\ 1+\mathrm{j} & 2-\mathrm{j} & 2-\mathrm{j} & 1 & 1 \\ \mathrm{j} & 1+\mathrm{j} & 1+2\mathrm{j} & 1 & \mathrm{j} \\ 3-2\mathrm{j} & -2-5\mathrm{j} & -2-4\mathrm{j} & 1-3\mathrm{j} & -2\mathrm{j} \\ 3+3\mathrm{j} & 5-2\mathrm{j} & 5 & 4-\mathrm{j} & 2+\mathrm{j} \\ 0 & -1+\mathrm{j} & -1+3\mathrm{j} & 1-\mathrm{j} & -1+\mathrm{j} \end{bmatrix}$$

has rank 2, and $[\mathbf{H}_1, \mathbf{H}_2]$, which models the widely linear transformation, has rank 5.

From 1000 sample pairs of $(\mathbf{x}, \mathbf{y})$ we computed estimates of the correlation and cross-correlation matrices of $\mathbf{x}$ and $\mathbf{y}$. From these, we estimated the correlation coefficients $\rho^2_{C_1}$ and $\rho^2_{C_2}$ for $r = 1, \ldots, 5$, as depicted in Fig. 4.5. We emphasize that this computation does not assume any knowledge whatsoever about how $\mathbf{x}$ and $\mathbf{y}$ are generated (i.e., from what distribution they are drawn). The way in which $\mathbf{x}$ and $\mathbf{y}$ are generated in this example is simply illustrative.

We can make the following observations in Fig. 4.5. Correlation coefficient ρ_{C_1} is a good means of estimating the ranks of the linear, conjugate linear, or widely linear components of the relationship between $\mathbf{x}$ and $\mathbf{y}$. Even though the widely linear transformation has rank 5, the first four dimensions capture most of the total correlation between $\mathbf{x}$ and $\mathbf{y}$.

Correlation coefficient ρ_{C_2} is sensitive to the presence of a canonical correlation k_i close to 1. This happens whenever there is a strong linear relationship between $\mathbf{x}$ and $\mathbf{y}$. As a consequence, we will find in Section 4.5 that ρ_{C_2} is the test statistic in a generalized likelihood-ratio test for whether $\mathbf{x}$ and $\mathbf{y}$ are independent. In this example, $\mathbf{x}$ and $\mathbf{y}$ are clearly not independent, and ρ_{C_2} bears witness to this fact.

4.3.2 Multivariate linear regression (half-canonical correlations)

There are many problems in which the roles of $\mathbf{x}$ and $\mathbf{y}$ are not interchangeable. For these, the symmetric assessment of correlation by CCA is unsatisfactory. The most obvious example is multivariate linear regression, where a message $\mathbf{x}$ is estimated from a measurement $\mathbf{y}$. The resulting MMSE is invariant under nonsingular transformation of $\mathbf{y}$, but only under *unitary* transformation of $\mathbf{x}$. Half-canonical correlations have exactly these invariance properties. While there are many conceivable correlation coefficients based on the first r half-canonical correlations $\{k_i\}_{i=1}^r$, the most common correlation coefficient is

$$\rho_{\mathrm{M}}^2 = \frac{1}{\operatorname{tr} \mathbf{R}_{xx}} \sum_{i=1}^{r} k_i^2. \tag{4.76}$$

For full rank $r = p$, it can also be written as

$$\rho_{\mathrm{M}}^2 = \frac{\operatorname{tr} \mathbf{K}^2}{\operatorname{tr} \mathbf{R}_{xx}} = \frac{\operatorname{tr}(\mathbf{R}_{xy}\mathbf{R}_{yy}^{-1}\mathbf{R}_{xy}^{\mathrm{H}})}{\operatorname{tr} \mathbf{R}_{xx}}. \tag{4.77}$$

The reflectional version $\widetilde{\rho}_{\mathrm{M}}$ is obtained by replacing k_i with $\tilde{k}_i$, $\mathbf{R}_{xy}$ with $\widetilde{\mathbf{R}}_{xy}$, and $\mathbf{R}_{yy}^{-1}$ with $\mathbf{R}_{yy}^{-*}$. The total version is obtained by including a normalizing factor of $1/2$,

$$\bar{\rho}_{\mathrm{M}}^2 = \frac{1}{2 \operatorname{tr} \mathbf{R}_{xx}} \sum_{i=1}^{2r} \bar{k}_i^2, \tag{4.78}$$

which, for $r = p$, yields

$$\bar{\rho}_{\mathrm{M}}^2 = \frac{\operatorname{tr} \underline{\mathbf{K}}^2}{2 \operatorname{tr} \mathbf{R}_{xx}} = \frac{\operatorname{tr}(\underline{\mathbf{R}}_{xy}\underline{\mathbf{R}}_{yy}^{-1}\underline{\mathbf{R}}_{xy}^{\mathrm{H}})}{\operatorname{tr} \underline{\mathbf{R}}_{xx}}. \tag{4.79}$$

The coefficient ρ_{M}^2 has been called the *redundancy index*[5] because it is the fraction of the total variance of $\mathbf{x}$ that is explained by a linear estimate from the measurement $\mathbf{y}$.

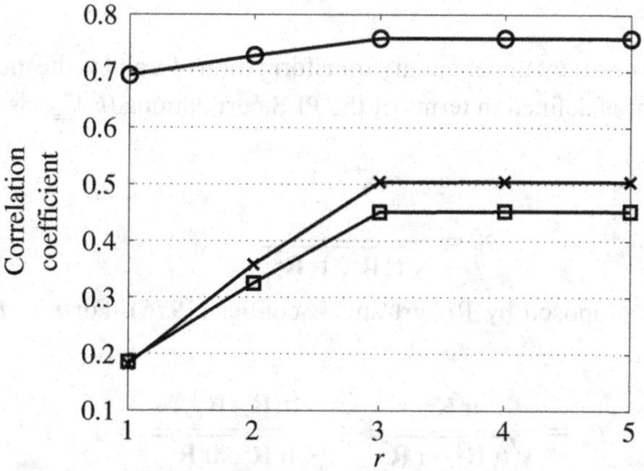

Figure 4.6 Correlation coefficients $\rho^2_{C_1} = \rho^2_M$ for $\mathbf{R}_{xx} = \mathbf{I}$ (shown as $\times$), $\rho^2_{C_1}$ for $\mathbf{R}_{xx} = \mathbf{Diag}(4, \frac{1}{4}, \frac{1}{4}, \frac{1}{4}, \frac{1}{4})$ (shown as $\square$), and ρ^2_M for $\mathbf{R}_{xx} = \mathbf{Diag}(4, \frac{1}{4}, \frac{1}{4}, \frac{1}{4}, \frac{1}{4})$ (shown as $\circ$).

This estimate is

$$\hat{\mathbf{x}}(\mathbf{y}) = \mathbf{R}_{xy}\mathbf{R}_{yy}^{-1}\mathbf{y} = \mathbf{C}\mathbf{R}_{yy}^{-1/2}\mathbf{y} = \mathbf{F}\mathbf{K}\mathbf{G}^{\mathrm{H}}\mathbf{R}_{yy}^{-1/2}\mathbf{y} = \mathbf{A}\mathbf{K}\mathbf{B}\mathbf{y}, \tag{4.80}$$

and the MMSE is

$$E\|\hat{\mathbf{x}}(\mathbf{y}) - \mathbf{x}\|^2 = \mathrm{tr}(\mathbf{R}_{xx} - \mathbf{R}_{xy}\mathbf{R}_{yy}^{-1}\mathbf{R}_{xy}^{\mathrm{H}}) = \mathrm{tr}\,\mathbf{R}_{xx} - \mathrm{tr}\,\mathbf{K}^2 = \mathrm{tr}\,\mathbf{R}_{xx}(1 - \rho^2_M). \tag{4.81}$$

If $\rho^2_M = 1$, then $\mathbf{x}$ is perfectly linearly estimable from $\mathbf{y}$. Similarly, the reflectional version $\tilde{\rho}_M$ is related to the MMSE of the conjugate linear estimate

$$\hat{\mathbf{x}}(\mathbf{y}^*) = \tilde{\mathbf{A}}\tilde{\mathbf{K}}\tilde{\mathbf{B}}\mathbf{y}^*, \tag{4.82}$$

and the total version $\bar{\rho}_M$ characterizes the MMSE of the widely linear estimate

$$\hat{\underline{\mathbf{x}}}(\mathbf{y}, \mathbf{y}^*) = \underline{\mathbf{A}}\,\underline{\mathbf{K}}\,\underline{\mathbf{B}}\,\underline{\mathbf{y}}. \tag{4.83}$$

If there is a perfect linear relationship between $\mathbf{x}$ and $\mathbf{y}$, then both the redundancy index $\rho^2_M = 1$ and the corresponding CCA correlation coefficient $\rho^2_{C_1} = 1$. In general, however, ρ^2_M and $\rho^2_{C_1}$ can behave quite differently, as the following example shows.

Example 4.5. Consider again the setup in the previous example, but we will only look at rotational correlations here. Since $\mathbf{R}_{xx} = \mathbf{I}$, $\rho^2_{C_1} = \rho^2_M$ for all ranks r.

Now consider $\mathbf{R}_{xx} = \mathbf{Diag}(4, \frac{1}{4}, \frac{1}{4}, \frac{1}{4}, \frac{1}{4})$, which has the same trace as before but now most of its variance is concentrated in one dimension. This change affects the coefficients ρ^2_M and $\rho^2_{C_1}$ quite differently, as shown in Fig. 4.6. While $\rho^2_{C_1}$ decreases for each r, ρ^2_M increases. This indicates that it is easier to linearly estimate $\mathbf{x}$ from $\mathbf{y}$ if most of the variance of $\mathbf{x}$ is concentrated in a one-dimensional subspace. Moreover, while CCA still spreads the correlation over the first three dimensions, MLR concentrates most of the correlation in the first half-canonical correlation k_1^2.

4.3.3 Partial least squares

For problems that are invariant under unitary transformation of **x** and **y**, the most common correlation coefficient is defined in terms of the PLS correlations $\{k_i\}_{i=1}^r$ as

$$\rho_{\mathrm{P}}^2 = \frac{\sum_{i=1}^{r} k_i^2}{\sqrt{\operatorname{tr} \mathbf{R}_{xx}^2 \operatorname{tr} \mathbf{R}_{yy}^2}}. \tag{4.84}$$

This coefficient was proposed by Robert and Escoufier (1976). For $r = p$, it can be expressed as

$$\rho_{\mathrm{P}}^2 = \frac{\operatorname{tr} \mathbf{K}^2}{\sqrt{\operatorname{tr} \mathbf{R}_{xx}^2 \operatorname{tr} \mathbf{R}_{yy}^2}} = \frac{\operatorname{tr}(\mathbf{R}_{xy}\mathbf{R}_{xy}^{\mathrm{H}})}{\sqrt{\operatorname{tr} \mathbf{R}_{xx}^2 \operatorname{tr} \mathbf{R}_{yy}^2}}. \tag{4.85}$$

This correlation coefficient measures how closely **x** and **y** are related under unitary transformation. For $\mathbf{y} = \mathbf{U}\mathbf{x}$, with $\mathbf{U}^{\mathrm{H}}\mathbf{U} = \mathbf{I}$ but not necessarily $\mathbf{U}\mathbf{U}^{\mathrm{H}} = \mathbf{I}$, we have perfect correlation $\rho_{\mathrm{P}}^2 = 1$.

The reflectional version $\tilde{\rho}_{\mathrm{P}}$ is obtained by replacing k_i with $\tilde{k}_i$ and $\mathbf{R}_{xy}$ with $\tilde{\mathbf{R}}_{xy}$. The total version, which is invariant under widely unitary transformation of **x** and **y**, is

$$\bar{\rho}_{\mathrm{P}}^2 = \frac{\sum_{i=1}^{2r} \bar{k}_i^2}{\sqrt{\operatorname{tr} \underline{\mathbf{R}}_{xx}^2 \operatorname{tr} \underline{\mathbf{R}}_{yy}^2}}. \tag{4.86}$$

In the full-rank case $r = p$, this yields

$$\bar{\rho}_{\mathrm{P}}^2 = \frac{\operatorname{tr} \underline{\mathbf{K}}^2}{\sqrt{\operatorname{tr} \underline{\mathbf{R}}_{xx}^2 \operatorname{tr} \underline{\mathbf{R}}_{yy}^2}} = \frac{\operatorname{tr}(\underline{\mathbf{R}}_{xy}\underline{\mathbf{R}}_{xy}^{\mathrm{H}})}{\sqrt{\operatorname{tr} \underline{\mathbf{R}}_{xx}^2 \operatorname{tr} \underline{\mathbf{R}}_{yy}^2}}. \tag{4.87}$$

4.4 Correlation spread

Using the results from Appendix 3, Section A3.2, it is not difficult to show that all correlation coefficients presented in this chapter are Schur-convex and increasing functions of the diagonal cross-correlations $\{k_i\}$. By maximizing all partial sums over their absolute values, subject to the constraints imposed by the correlation analysis technique, the correlation coefficients are then also maximized for all ranks r.

The importance of this result is that we may assess correlation in a low(er)-dimensional subspace of dimension $r < p$. In fact, the development in this chapter anticipates some closely related results on reduced-rank estimation, which we will study in Section 5.5. There we will find that canonical coordinates are the *optimum* coordinate system for reduced-rank estimation and transform coding if the aim is to maximize information rate, whereas half-canonical coordinates are the *optimum* coordinate system for reduced-rank estimation and transform coding if we want to minimize mean-squared error.

An interesting question in this context is how much of the overall correlation is captured by r coefficients $\{k_i\}_{i=1}^r$, which can be rotational, reflectional, or total coefficients defined either through CCA, MLR, or PLS. One could, of course, compute the fraction $\rho^2(r)/\rho^2(p)$ for all $1 \le r < p$. The following definition, however, provides a more convenient approach.

Definition 4.4. *The* rotational correlation spread *is defined as*

$$\sigma^2 = \frac{1}{p}\frac{\operatorname{var}(\{k_i^2\})}{\mu^2(\{k_i^2\})} = \frac{p}{p-1}\left(\frac{\sum_{i=1}^{p} k_i^4}{\left(\sum_{i=1}^{p} k_i^2\right)^2} - \frac{1}{p}\right), \tag{4.88}$$

where $\operatorname{var}(\{k_i^2\})$ *denotes the variance of the correlations* $\{k_i^2\}$ *and* $\mu^2(\{k_i^2\})$ *denotes their squared mean.*

The correlation spread provides a *single, normalized measure* of how concentrated the overall correlation is. If there is only one nonzero coefficient k_1, then $\sigma^2 = 1$. If all coefficients are equal, $k_1 = k_2 = \cdots = k_p$, then $\sigma^2 = 0$. In essence, the correlation spread gives an indication of how *compressible* the cross-correlation between $\mathbf{x}$ and $\mathbf{y}$ is.

The definition (4.88) is inspired by the definition of the *degree of polarization* of a random vector $\mathbf{x}$. The degree of polarization measures the spread among the eigenvalues of $\mathbf{R}_{xx}$. A random vector $\mathbf{x}$ is said to be completely polarized if all of its energy is concentrated in one direction, i.e., if there is only one nonzero eigenvalue. On the other hand, $\mathbf{x}$ is unpolarized if its energy is equally distributed among all dimensions, i.e., if all eigenvalues are equal. The correlation spread σ^2 generalizes this idea to the correlation between two random vectors $\mathbf{x}$ and $\mathbf{y}$.

Example 4.6. Continuing Example 4.5 for $\mathbf{R}_{xx} = \mathbf{I}$, we find $\sigma_{\mathrm{C}}^2 = \sigma_{\mathrm{M}}^2 = 0.167$ for both CCA and MLR. For $\mathbf{R}_{xx} = \mathbf{Diag}(4, \frac{1}{4}, \frac{1}{4}, \frac{1}{4}, \frac{1}{4})$, we find $\sigma_{\mathrm{C}}^2 = 0.178$ for CCA and $\sigma_{\mathrm{M}}^2 = 0.797$ for MLR. A σ^2-value close to 1 indicates that the correlation is highly concentrated. Indeed, as we have found in Example 4.5, most of the MLR correlation is concentrated in a one-dimensional subspace.

The reflectional correlation spread is defined as a straightforward extension by replacing k_i^2 with $\tilde{k}_i^2$ in Definition 4.4, but the total correlation spread is defined in a slightly different manner.

Definition 4.5. *The* total correlation spread *is defined as*

$$\overline{\sigma}^2 = \frac{1}{p}\frac{\operatorname{var}\left(\{\frac{1}{2}(\bar{k}_{2i-1}^2 + \bar{k}_{2i}^2)\}\right)}{\mu^2\left(\{\frac{1}{2}(\bar{k}_{2i-1}^2 + \bar{k}_{2i}^2)\}\right)} \tag{4.89}$$

$$= \frac{p}{p-1}\left(\frac{\sum_{i=1}^{p}(\bar{k}_{2i-1}^2 + \bar{k}_{2i}^2)^2}{\left(\sum_{i=1}^{p}\bar{k}_{2i-1}^2 + \bar{k}_{2i}^2\right)^2} - \frac{1}{p}\right). \tag{4.90}$$

The motivation for defining $\overline{\sigma}^2$ in this way is twofold. First, since $\bar{k}_{2i-1}$ is the cross-correlation between the real parts and $\bar{k}_{2i}$ is the cross-correlation between the imaginary parts of ξ_i and ω_i, they belong to the same complex dimension. Secondly, the expression (4.90) becomes (4.88) if $\bar{k}_{2i-1} = \bar{k}_{2i} = k_i$ for all i.

4.5 Testing for correlation structure

In practice, correlation matrices will be estimated from measurements. Given these estimates, how can we make informed decisions such as whether **x** is white, or whether **x** and **y** are independent? Questions such as these must be answered by employing statistical tests for correlation structure. In this section, we develop generalized likelihood-ratio tests (GLRTs) for complex Gaussian data.[6] That is, we assume that the data can be modeled as a complex Gaussian random vector $\mathbf{x}\colon \Omega \longrightarrow \mathbb{C}^n$ with probability density function

$$p(\mathbf{x}) = \frac{1}{\pi^n \det \mathbf{R}_{xx}} \exp\left\{-(\mathbf{x}-\boldsymbol{\mu}_x)^{\mathrm{H}}\mathbf{R}_{xx}^{-1}(\mathbf{x}-\boldsymbol{\mu}_x)\right\}, \tag{4.91}$$

mean $\boldsymbol{\mu}_x$, and covariance matrix $\mathbf{R}_{xx}$. Later we will also replace **x** with other vectors such as $[\mathbf{x}^{\mathrm{T}}, \mathbf{y}^{\mathrm{T}}]^{\mathrm{T}}$. We note that (4.91) is a proper Gaussian pdf, and the story we will be telling here is strictly linear, for simplicity. Appropriate extensions exist for augmented vectors, where the pdf (4.91) is adapted by replacing det with $\det^{1/2}$ and including a factor of $1/2$ in the quadratic form of the exponential (cf. Result 2.4). These two changes to the pdf have no significant consequences for the following development.

Consider M independent and identically distributed (i.i.d.) random samples $\mathbf{X} = [\mathbf{x}_1, \mathbf{x}_2, \ldots, \mathbf{x}_M]$ drawn from this distribution. The joint probability density function of these samples is given by

$$p(\mathbf{X}) = \pi^{-Mn}(\det \mathbf{R}_{xx})^{-M} \exp\left\{-\sum_{m=1}^{M}(\mathbf{x}_m-\boldsymbol{\mu}_x)^{\mathrm{H}}\mathbf{R}_{xx}^{-1}(\mathbf{x}_m-\boldsymbol{\mu}_x)\right\} \tag{4.92}$$

$$= \pi^{-Mn}(\det \mathbf{R}_{xx})^{-M} \exp\left\{-M \operatorname{tr}(\mathbf{R}_{xx}^{-1}\mathbf{S}_{xx})\right\}, \tag{4.93}$$

where $\mathbf{S}_{xx}$ is the sample covariance matrix

$$\mathbf{S}_{xx} = \frac{1}{M}\sum_{m=1}^{M}(\mathbf{x}_m - \mathbf{m}_x)(\mathbf{x}_m - \mathbf{m}_x)^{\mathrm{H}} = \frac{1}{M}\mathbf{X}\mathbf{X}^{\mathrm{H}} - \mathbf{m}_x\mathbf{m}_x^{\mathrm{H}} \tag{4.94}$$

and $\mathbf{m}_x$ is the sample mean vector

$$\mathbf{m}_x = \frac{1}{M}\sum_{m=1}^{M}\mathbf{x}_m. \tag{4.95}$$

We would now like to test whether $\mathbf{R}_{xx}$ has structure 0 or the alternative structure 1. We write this hypothesis-testing problem as

$$H_0\colon \mathbf{R}_{xx} \in \mathcal{R}_0,$$
$$H_1\colon \mathbf{R}_{xx} \in \mathcal{R}_1.$$

The GLRT statistic is

$$\lambda = \frac{\max\limits_{\mathbf{R}_{xx}\in\mathcal{R}_0} p(\mathbf{X})}{\max\limits_{\mathbf{R}_{xx}\in\mathcal{R}_1} p(\mathbf{X})}. \tag{4.96}$$

This ratio compares the likelihood of drawing the samples $\mathbf{x}_1, \ldots, \mathbf{x}_M$ from a distribution whose covariance matrix has structure 0 with the likelihood of drawing them from a distribution whose covariance matrix has structure 1. Since the actual covariance matrices are not known, they are replaced with their maximum-likelihood (ML) estimates computed from the samples. If we denote by $\widehat{\mathbf{R}}_0$ the ML estimate of $\mathbf{R}_{xx}$ under H_0 and by $\widehat{\mathbf{R}}_1$ the ML estimate of $\mathbf{R}_{xx}$ under H_1, we find

$$\lambda = \det{}^M\!\left(\widehat{\mathbf{R}}_0^{-1}\widehat{\mathbf{R}}_1\right)\exp\!\left[-M\,\mathrm{tr}\!\left(\widehat{\mathbf{R}}_0^{-1}\mathbf{S}_{xx} - \widehat{\mathbf{R}}_1^{-1}\mathbf{S}_{xx}\right)\right]. \tag{4.97}$$

If we further assume that $\mathcal{R}_1$ is the set of positive semidefinite matrices (i.e., no special constraints are imposed), then $\widehat{\mathbf{R}}_1 = \mathbf{S}_{xx}$, and

$$\lambda = \det{}^M\!\left(\widehat{\mathbf{R}}_0^{-1}\mathbf{S}_{xx}\right)\exp\!\left[Mn - M\,\mathrm{tr}\!\left(\widehat{\mathbf{R}}_0^{-1}\mathbf{S}_{xx}\right)\right]. \tag{4.98}$$

This may be expressed in the following result, which is Theorem 5.3.2 in Mardia *et al.* (1979).

Result 4.9. *The generalized likelihood ratio for testing whether* $\mathbf{R}_{xx}$ *has structure* $\mathcal{R}_0$ *is*

$$\ell = \lambda^{1/(Mn)} = g e^{1-a} \tag{4.99}$$

where a and g are the arithmetic and geometric means of the eigenvalues of $\widehat{\mathbf{R}}_0^{-1}\mathbf{S}_{xx}$*:*

$$a = \frac{1}{n}\mathrm{tr}\!\left(\widehat{\mathbf{R}}_0^{-1}\mathbf{S}_{xx}\right), \tag{4.100}$$

$$g = \left[\det\!\left(\widehat{\mathbf{R}}_0^{-1}\mathbf{S}_{xx}\right)\right]^{1/n}. \tag{4.101}$$

In the hypothesis-testing problem, a threshold ℓ_0 is chosen to achieve a desired probability of false alarm or probability of detection. Then, if $\ell \geq \ell_0$, we accept hypothesis H_0, and if $\ell < \ell_0$, we reject it.

The GLRT is always invariant with respect to transformations for which the hypothesis-testing problem itself is invariant. The GLR ℓ will therefore turn out to be a function of a *maximal invariant*, which is determined by the hypothesis-testing problem. Let's consider a few interesting special cases.

4.5.1 Sphericity

We would like to test whether $\mathbf{x}$ has a *spherical distribution*, i.e.,

$$\mathcal{R}_0 = \{\mathbf{R}_{xx} = \sigma_x^2 \mathbf{I}\}, \tag{4.102}$$

where σ_x^2 is the variance of each component of $\mathbf{x}$. The ML estimate of $\mathbf{R}_{xx}$ under H_0 is $\widehat{\mathbf{R}}_0 = \widehat{\sigma}_x^2 \mathbf{I}$, where the variance is estimated as $\widehat{\sigma}_x^2 = n^{-1} \operatorname{tr} \mathbf{S}_{xx}$. The GLR in Result 4.9 is therefore

$$\ell = \frac{[\det \mathbf{S}_{xx}]^{1/n}}{\frac{1}{n} \operatorname{tr} \mathbf{S}_{xx}}. \tag{4.103}$$

This test is invariant with respect to scale and unitary transformation. That is, $\mathbf{x} \longrightarrow \alpha \mathbf{U} \mathbf{x}$ with $\alpha \in \mathbb{C}$ and unitary $\mathbf{U} \in \mathbb{C}^{n \times n}$ leaves the GLR unchanged. The GLR is therefore a function of a maximal invariant under scale and unitary transformation. Such a maximal invariant is the set of estimated eigenvalues of $\mathbf{R}_{xx}$ (i.e., the eigenvalues of $\mathbf{S}_{xx}$), denoted by $\{\hat{\lambda}_i\}_{i=1}^n$, normalized by $\operatorname{tr} \mathbf{S}_{xx}$:

$$\left\{ \frac{\hat{\lambda}_i}{\operatorname{tr} \mathbf{S}_{xx}} \right\}_{i=1}^n .$$

Indeed, we can express the GLR as

$$(n\ell)^n = \prod_{i=1}^n \frac{\hat{\lambda}_i}{\operatorname{tr} \mathbf{S}_{xx}}, \tag{4.104}$$

which also allows a reduced-rank implementation of the GLRT by considering only the r largest estimated eigenvalues.

4.5.2 Independence within one data set

Now we would like to test whether $\mathbf{x}$ has *independent components*, i.e., whether $\mathbf{R}_{xx}$ is diagonal:

$$\mathcal{R}_0 = \{\mathbf{R}_{xx} = \mathbf{Diag}(R_{11}, R_{22}, \ldots, R_{nn})\}. \tag{4.105}$$

The ML estimate of $\mathbf{R}_{xx}$ under H_0 is $\widehat{\mathbf{R}}_0 = \mathbf{Diag}(S_{11}, S_{22}, \ldots, S_{nn})$, where S_{ii} is the ith diagonal element of $\mathbf{S}_{xx}$. The GLR in Result 4.9 is therefore the Hadamard ratio

$$\ell^n = \frac{\det \mathbf{S}_{xx}}{\prod_{i=1}^{n} S_{ii}}. \tag{4.106}$$

This test is invariant with respect to multiplication with a nonsingular diagonal matrix. That is, $\mathbf{x} \longrightarrow \mathbf{D}\mathbf{x}$ with nonsingular diagonal $\mathbf{D} \in \mathbb{C}^{n\times n}$ leaves the GLR unchanged. The GLR is therefore a function of a maximal invariant under multiplication with a diagonal matrix. Such a maximal invariant is obtained as follows. Let $\mathbf{Z}_{ii}$ denote the submatrix obtained from rows $i+1$ through n and columns $i+1$ through n of the sample covariance matrix $\mathbf{S}_{xx}$. Let $\mathbf{z}_i$ be the column-vector obtained from rows $i+1$ through n in column i of the sample covariance matrix $\mathbf{S}_{xx}$. For $i = 1, \ldots, n-1$, they can be combined in

$$\mathbf{Z}_{i-1,i-1} = \begin{bmatrix} S_{ii} & \mathbf{z}_i^{\mathrm{H}} \\ \mathbf{z}_i & \mathbf{Z}_{ii} \end{bmatrix} \tag{4.107}$$

and $\mathbf{Z}_{00} = \mathbf{S}_{xx}$. Now define

$$\hat{\beta}_i^2 = \frac{\mathbf{z}_i^{\mathrm{H}}\mathbf{Z}_{ii}^{-1}\mathbf{z}_i}{S_{ii}}, \quad i = 1, \ldots, n-1, \tag{4.108}$$

$$\hat{\beta}_n^2 = 0. \tag{4.109}$$

We notice that $\hat{\beta}_i^2$ is the estimated squared canonical correlation between the ith component of $\mathbf{x}$ and components $i+1$ through n of $\mathbf{x}$, and thus $0 \le \hat{\beta}_i^2 \le 1$. It can be shown that the set $\{\hat{\beta}_i^2\}_{i=1}^{n-1}$ is the maximal invariant we are looking for. Indeed, we can express the GLR as

$$\ell^n = \frac{\prod_{i=1}^{n} S_{ii}(1-\hat{\beta}_i^2)}{\prod_{i=1}^{n} S_{ii}} = \prod_{i=1}^{n-1}(1-\hat{\beta}_i^2), \tag{4.110}$$

which also allows a reduced-rank implementation of the GLRT by considering only the r largest estimated canonical correlations $\hat{\beta}_i^2$.

4.5.3 Independence between two data sets

Now replace $\mathbf{x}$ with the composite vector $[\mathbf{x}^{\mathrm{T}}, \mathbf{y}^{\mathrm{T}}]^{\mathrm{T}}$, where $\mathbf{x}\colon \Omega \longrightarrow \mathbb{C}^n$ and $\mathbf{y}\colon \Omega \longrightarrow \mathbb{C}^m$, and replace $\mathbf{R}_{xx}$ with the composite covariance matrix of $\mathbf{x}$ and $\mathbf{y}$:

$$\mathbb{R}_{xy} = \begin{bmatrix} \mathbf{R}_{xx} & \mathbf{R}_{xy} \\ \mathbf{R}_{xy}^{\mathrm{H}} & \mathbf{R}_{yy} \end{bmatrix}. \tag{4.111}$$

We would like to test whether $\mathbf{x}$ and $\mathbf{y}$ are *independent*, i.e., whether $\mathbf{x}$ and $\mathbf{y}$ have a block-diagonal composite covariance matrix $\mathbb{R}_{xy}$ with $\mathbf{R}_{xy} = \mathbf{0}$:

$$\mathcal{R}_0 = \left\{ \mathbb{R}_{xy} = \begin{bmatrix} \mathbf{R}_{xx} & \mathbf{0} \\ \mathbf{0} & \mathbf{R}_{yy} \end{bmatrix} \right\}. \tag{4.112}$$

The ML estimate of $\mathbb{R}_{xy}$ under H_0 is

$$\widehat{\mathbb{R}}_0 = \begin{bmatrix} \mathbf{S}_{xx} & \mathbf{0} \\ \mathbf{0} & \mathbf{S}_{yy} \end{bmatrix}. \tag{4.113}$$

The GLR in Result 4.9 is therefore

$$\ell^n = \det \begin{bmatrix} \mathbf{I} & \mathbf{S}_{xx}^{-1}\mathbf{S}_{xy} \\ \mathbf{S}_{yy}^{-1}\mathbf{S}_{xy}^{\mathrm{H}} & \mathbf{I} \end{bmatrix} = \det(\mathbf{I} - \mathbf{S}_{xx}^{-1}\mathbf{S}_{xy}\mathbf{S}_{yy}^{-1}\mathbf{S}_{xy}^{\mathrm{H}}). \tag{4.114}$$

This test is invariant with respect to multiplication of $\mathbf{x}$ and $\mathbf{y}$, each with a nonsingular matrix. That is, $\mathbf{x} \longrightarrow \mathbf{N}\mathbf{x}$ and $\mathbf{y} \longrightarrow \mathbf{M}\mathbf{y}$ for nonsingular $\mathbf{N} \in \mathbb{C}^{n\times n}$ and $\mathbf{M} \in \mathbb{C}^{m\times m}$ leaves the GLR unchanged. A maximal invariant for these invariances is, of course, the set of canonical correlations. Hence, the GLR can also be written as

$$\ell^n = \prod_{i=1}^{p}(1 - \hat{k}_i^2) = 1 - \widehat{\rho}_{C_2}^2, \tag{4.115}$$

where $p = \min(m, n)$, $\{\hat{k}_i\}_{i=1}^p$ are the estimated canonical correlations between $\mathbf{x}$ and $\mathbf{y}$, computed from the sample covariances, and $\widehat{\rho}_{C_2}^2$ is the estimated correlation coefficient defined in (4.59). This expression allows a reduced-rank implementation of the GLRT by considering only the r largest estimated canonical correlations. The test for impropriety in Section 3.4 is a special case of the test presented here, where $\mathbf{y} = \mathbf{x}^*$. However, the test for impropriety tests for *uncorrelatedness* between $\mathbf{x}$ and $\mathbf{x}^*$ rather than independence: $\mathbf{x}$ and $\mathbf{x}^*$ cannot be independent because $\mathbf{x}^*$ is perfectly determined by $\mathbf{x}$ through complex conjugation.[7]

Notes

1 Since the assessment of multivariate association between random vectors is of importance in many research areas, there is a rich literature on this topic. Canonical correlation analysis is the oldest technique, which was invented by Hotelling (1936). Partial least squares was introduced by Wold (1975) in the field of chemometrics but our description of it follows a different variant suggested by Sampson *et al.* (1989).

Much of this chapter, in particular the discussion of majorization and correlation spread, closely follows Schreier (2008c). This paper is ©IEEE, and portions are used with permission. An excellent unifying discussion of correlation analysis techniques with emphasis on their invariance properties is given by Ramsay *et al.* (1984). Different research areas often use quite different jargon, and we have attempted to use terms that are as generic as possible.

2 To the best of our knowledge, there are not many papers on correlation analysis of complex data. Usually, the analysis of rotational and reflectional correlations proceeds in terms of real data of double dimension rather than complex data, see, for instance, Jupp and Mardia (1980). Hanson *et al.* (1992) proposed working directly with complex data when analyzing rotational and reflectional dependences.

3 Many definitions of correlation coefficients (see, e.g., Renyi (1959)) require the symmetry $\rho_{xy} = \rho_{yx}$. However, this does not seem like a fundamental requirement. In fact, if the correlation analysis technique itself is not symmetric (as for instance is the case for MLR) it would not be reasonable to ask for symmetry in correlation coefficients based on it.

4 Correlation coefficients based on CCA were discussed by Rozeboom (1965), Coxhead (1974), Yanai (1974), and Cramer and Nicewander (1979).
5 The redundancy index, which is based on MLR, was analyzed by Stewart and Love (1968) and Gleason (1976).
6 Section 4.5 draws on material from Mardia *et al.* (1979). A general introduction to likelihood-ratio tests is provided in Section 7.1.
7 Much more can be done in Section 4.5. There are the obvious extensions to augmented vectors and widely linear transformations. The results of Sections 4.5.2 and 4.5.3 can be combined to test for block-diagonal structure with more than two diagonal blocks. Then hypothesis tests that have the invariance properties of half-canonical correlations and partial least squares can be developed, for which half-canonical correlations and partial least-squares correlations will be the maximal invariants. Finally, we have completely ignored the distributional properties of the GLRT statistics. The moments of the statistic ℓ in (4.114) are well known, cf. Mardia *et al.* (1979) and Lehmann and Romano (2005). The latter reference provides more background on the rich topic of hypothesis testing.

5 Estimation

One of the most important applications of probability in science and engineering is to the theory of *statistical inference*, wherein the problem is to draw defensible conclusions from experimental evidence. The three main branches of statistical inference are *parameter estimation*, *hypothesis testing*, and *time-series analysis*. Or, as we say in the engineering sciences, the three main branches of *statistical signal processing* are *estimation*, *detection*, and *signal analysis*.

A common problem is to estimate the value of a parameter, or vector of parameters, from a sequence of measurements. The underlying probability law that governs the generation of the measurements depends on the parameter. Engineering language would say that a source of information, loosely speaking, generates a signal $\mathbf{x}$ and a channel carries this information in a measurement $\mathbf{y}$, whose probability law $p(\mathbf{y}|\mathbf{x})$ depends on the signal. There is usually little controversy over this aspect of the problem because the measurement scheme generally determines the probability law. There is, however, a philosophical divide about the modeling of the signal $\mathbf{x}$. *Frequentists* adopt the point of view that to assign a probability law to the signal assumes too much. They argue that the signal should be treated as an *unknown constant* and the data should be allowed to speak for itself. *Bayesians* argue that the signal should be treated as a *random variable* whose *prior* probability distribution is to be updated to a *posterior distribution* as measurements are made.

In the realm of exploratory data analysis and in the absence of a physical model for the experiment or measurement scheme, the difference between these points of view is profound. But, for most problems of parameter estimation and hypothesis testing in the engineering and applied sciences, the question resolves itself on the basis of physical reasoning, either theoretical or empirical. For example, in radar, sonar, and geophysical imaging, the question of estimating a signal, or resolving a hypothesis about the signal, usually proceeds without the assignment of a probability law or the assignment of a prior probability to the hypothesis that a nonzero signal will be returned. The point of view is decidedly frequentist. On the other hand, in data communication, where the problem is to determine which signal from a set of signals was transmitted, it is quite appropriate to assign prior probabilities to elements of the set. The appropriate view is Bayesian. Finally, in most problems of engineering and applied science, measurements are plentiful by statistical standards, and the weight of experimental evidence overwhelms the prior model, making the practical distinction between the frequentist and the Bayesian a philosophical one of marginal practical effect.

In this chapter on estimation, we shall be interested, by and large, in *linear and widely linear least-squares* problems, wherein parameter estimators are constrained to be linear or widely linear in the measurement and the performance criterion is mean-squared error or squared error under a constraint. The estimators we compute are then linear or widely linear minimum mean-squared error (LMMSE or WLMMSE) estimators. These estimators strike a balance between frequentism and Bayesianism by using only second-order statistical models for signal and noise. When the underlying joint distribution between the measurement $\mathbf{y}$ and the signal $\mathbf{x}$ is multivariate Gaussian, these estimators are also conditional mean and maximum a-posteriori likelihood estimators that would appeal to a Bayesian. No matter what your point of view is, the resulting estimators exploit all of the Hermitian and complementary correlation that can be exploited, within the measurements, within the signals, and between the signals and measurements.

As a prelude to our development of estimators, Section 5.1 establishes a Hilbert-space geometry for augmented random variables that will be central to our reasoning about WLMMSE estimators. We then review a few fundamental results for MMSE estimation in Section 5.2, without the constraint that the estimator be linear or widely linear. From there we proceed to the development of linear channel and filtering models, since these form the foundation for LMMSE estimation, which is discussed in Section 5.3. Most of these results translate to widely linear estimation in a straightforward fashion, but it is still worthwhile to point out a few peculiarities of WLMMSE estimation and compare it with LMMSE estimation, which is done in Section 5.4. Section 5.5 considers reduced-rank widely linear estimators, which either minimize mean-squared error or maximize information rate. We derive linear and widely linear minimum-variance distortionless response (MVDR) receivers in Section 5.6. These estimators assign no statistical model to the signal, and therefore appeal to the frequentist. Finally, Section 5.7 presents widely linear-quadratic estimators.

5.1 Hilbert-space geometry of second-order random variables

The Hilbert space geometry of random variables is based on second-order moment properties only. Why should such a geometry be important in applications? The obvious answer is that the dimension of signals to be estimated is generally small and the dimension of measurements large, so that quality estimates can be obtained from sample means and generalizations of these that are based on correlation structure. The subtle answer is that the geometry of second-order random variables enables us to exploit a comfortable imagery and intuition based on our familiarity with Euclidean geometry.

An algebraic approach to Euclidean geometry, often called analytic geometry, is founded on the concept of an inner product between two vectors. To exploit this point of view, we will have to think of a random variable as a vector. Then norm, distance, and orthogonality can be defined in terms of inner products.

A random experiment is modeled by the sample space Ω, where each point $\omega \in \Omega$ corresponds to a possible outcome, and a probability measure P on Ω. The probability measure associates with each event A (a measurable subset of Ω) a probability $P(A)$. A random variable x is a measurable function $x\colon \Omega \longrightarrow \mathbb{C}$, which assigns each outcome $\omega \in \Omega$ a complex number x. The set of second-order random variables (for which $E|x|^2 < \infty$) forms a Hilbert space with respect to the inner product defined as

$$\langle x, y\rangle = E(x^* y) = \int_\Omega x^*(\omega) y(\omega) \mathrm{d}P(\omega), \tag{5.1}$$

which is the correlation between x and y. We may thus identify the space of second-order random variables with the space of square-integrable functions $L^2(\Omega, P)$. This space is closed under addition and multiplication by complex scalars. From this inner product, we obtain the following geometrical quantities:

- the norm (length) of a random variable x: $\|x\| = \sqrt{\langle x, x\rangle} = \sqrt{E|x|^2}$,
- the distance between two random variables x and y: $\|x - y\| = \sqrt{\langle x - y, x - y\rangle} = \sqrt{E|x - y|^2}$,
- the angle between two random variables x and y:

$$\cos^2\alpha = \frac{|\langle x, y\rangle|^2}{\|x\|^2\|y\|^2} = \frac{|E(x^* y)|^2}{E|x|^2 E|y|^2},$$

- orthogonality: x and y are orthogonal if $\langle x, y\rangle = E(x^* y) = 0$,
- the Cauchy–Schwarz inequality: $|\langle x, y\rangle| \le \|x\|\|y\|$, i.e., $|E(x^* y)| \le \sqrt{E|x|^2 E|y|^2}$.

The random variable x may be replaced by the random variable x^*, in which case the inner product is the *complementary* correlation

$$\langle x^*, y\rangle = E(xy) = \int_\Omega x(\omega) y(\omega) \mathrm{d}P(\omega). \tag{5.2}$$

This inner product is nonzero if and only if x and y are cross-improper. For $y = x$ the complementary correlation is $\langle x^*, x\rangle = E\,x^2$, which is nonzero if and only if x is improper. It follows from the Cauchy–Schwarz inequality that $|E\,x^2| \le E|x|^2$.

In the preceding chapters we have worked with augmented random variables $\underline{\mathbf{x}} = [x, x^*]^{\mathrm{T}}$, which are measurable functions $\underline{\mathbf{x}}\colon \Omega \longrightarrow \mathbb{C}^2_*$. The vector space $\mathbb{C}^2_*$ contains all complex pairs of the form $[x, x^*]^{\mathrm{T}}$. It is a *real linear* (or *complex widely linear*) subspace of $\mathbb{C}^2$ because it is closed under addition and multiplication with a real scalar, but not with a complex scalar. The inner product for augmented random variables is

$$\langle \underline{\mathbf{x}}, \underline{\mathbf{y}}\rangle = E(\underline{\mathbf{x}}^{\mathrm{H}} \underline{\mathbf{y}}) = 2\,\mathrm{Re}\,E(x^* y) = 2\,\mathrm{Re}\,\langle x, y\rangle, \tag{5.3}$$

which amounts to taking (twice) the real part of the inner product between x and y. In this book, we use all three inner products, $\langle x, y\rangle$, $\langle x^*, y\rangle$, and $\langle \underline{\mathbf{x}}, \underline{\mathbf{y}}\rangle$.

Example 5.1. Let x be a real random variable and y a purely imaginary random variable. The inner product $\langle x, y\rangle$ is generally nonzero, but it must be purely imaginary. Therefore, $\langle \underline{\mathbf{x}}, \underline{\mathbf{y}}\rangle = 2\,\mathrm{Re}\,\langle x, y\rangle = 0$. This means that $\underline{\mathbf{x}}$ and $\underline{\mathbf{y}}$ are orthogonal even though x and y might not be.

If $z = x + y$, we may perfectly reconstruct x from z. This, however, requires the *widely linear* operation $x = \operatorname{Re} z$, which assumes that we know the phase of z. It is not generally possible to perfectly reconstruct x from z using strictly linear operations.

The preceding generalizes to random vectors $\mathbf{x} = [x_1, x_2, \ldots, x_n]^{\mathrm{T}}: \Omega \longrightarrow \mathbb{C}^n$ and $\underline{\mathbf{x}} = [\mathbf{x}^{\mathrm{T}}, \mathbf{x}^{\mathrm{H}}]^{\mathrm{T}}: \Omega \longrightarrow \mathbb{C}_*^{2n}$. Let $H(\underline{\mathbf{x}})$ denote the Hilbert space of second-order augmented random vectors $\underline{\mathbf{x}}$. The Euclidean view characterizes the angle between spanning vectors in terms of a Grammian matrix, consisting of all inner products between vectors. In the Hilbert space $H(\underline{\mathbf{x}})$, this Grammian matrix consists of all inner products of the form $E(x_i^* x_j)$ and $E(x_i x_j)$. We have come to know this Grammian matrix as the augmented correlation matrix of $\mathbf{x}$:

$$\underline{\mathbf{R}}_{xx} = E(\underline{\mathbf{x}}\,\underline{\mathbf{x}}^{\mathrm{H}}) = E\begin{bmatrix} \mathbf{x} \\ \mathbf{x}^* \end{bmatrix}\begin{bmatrix} \mathbf{x}^{\mathrm{H}} & \mathbf{x}^{\mathrm{T}} \end{bmatrix} = \begin{bmatrix} \mathbf{R}_{xx} & \widetilde{\mathbf{R}}_{xx} \\ \widetilde{\mathbf{R}}_{xx}^* & \mathbf{R}_{xx}^* \end{bmatrix}. \tag{5.4}$$

It is commonplace in adaptive systems to use a vector of N realizations $\mathbf{s}_i \in \mathbb{C}^N$ of a random variable x_i to estimate the Hilbert-space inner products $E(x_i^* x_j)$ and $E(x_i x_j)$ by the Euclidean inner products $N^{-1}\mathbf{s}_i^{\mathrm{H}}\mathbf{s}_j$ and $N^{-1}\mathbf{s}_i^{\mathrm{T}}\mathbf{s}_j$.

There is one more connection to be made with Euclidean spaces. Given two sets of (deterministic) spanning vectors $\mathbf{S} = [\mathbf{s}_1, \mathbf{s}_2, \ldots, \mathbf{s}_n]$ and $\mathbf{R} = [\mathbf{r}_1, \mathbf{r}_2, \ldots, \mathbf{r}_n]$, we may define the cosines of the principal angles between their linear subspaces $\langle \mathbf{S} \rangle$ and $\langle \mathbf{R} \rangle$ as the singular values of the matrix $(\mathbf{S}^{\mathrm{H}}\mathbf{S})^{-1/2}\mathbf{S}^{\mathrm{H}}\mathbf{R}(\mathbf{R}^{\mathrm{H}}\mathbf{R})^{-\mathrm{H}/2}$. In a similar way we define the cosines of the principal angles between the subspaces $H(\mathbf{x})$ and $H(\mathbf{y})$ as the singular values of the matrix $\mathbf{R}_{xx}^{-1/2}\mathbf{R}_{xy}\mathbf{R}_{yy}^{-\mathrm{H}/2}$, which are called the canonical correlations between $\mathbf{x}$ and $\mathbf{y}$. They were introduced in Section 4.1. We will use them again in Section 5.5 to develop reduced-rank linear estimators.

5.2 Minimum mean-squared error estimation

The general problem we address here is the estimation of a complex random vector $\mathbf{x}$ from a complex random vector $\mathbf{y}$. As usual we think of $\mathbf{x}$ as the signal or source, and $\mathbf{y}$ as the measurement or observation. Presumably, $\mathbf{y}$ carries information about $\mathbf{x}$ that can be extracted with an estimation algorithm.

We begin with the consideration of an arbitrary estimator $\hat{\mathbf{x}}$, which is a function of $\mathbf{y}$, and its mean-squared error matrix $\mathbf{Q} = E[(\hat{\mathbf{x}} - \mathbf{x})(\hat{\mathbf{x}} - \mathbf{x})^{\mathrm{H}}]$. We first expand the expression for $\mathbf{Q}$ using the conditional mean $E[\mathbf{x}|\mathbf{y}]$ – which is itself a random vector since it depends on $\mathbf{y}$ – as

$$\mathbf{Q} = E[(\hat{\mathbf{x}} - E[\mathbf{x}|\mathbf{y}] + E[\mathbf{x}|\mathbf{y}] - \mathbf{x})(\hat{\mathbf{x}} - E[\mathbf{x}|\mathbf{y}] + E[\mathbf{x}|\mathbf{y}] - \mathbf{x})^{\mathrm{H}}], \tag{5.5}$$

where the outer E stands for expectation over the joint distribution of $\mathbf{x}$ and $\mathbf{y}$. We may then write

$$\begin{aligned}\mathbf{Q} = {} & E[(E[\mathbf{x}|\mathbf{y}] - \mathbf{x})(E[\mathbf{x}|\mathbf{y}] - \mathbf{x})^{\mathrm{H}}] + E[(\hat{\mathbf{x}} - E[\mathbf{x}|\mathbf{y}])(E[\mathbf{x}|\mathbf{y}] - \mathbf{x})^{\mathrm{H}}] \\ & + E[(E[\mathbf{x}|\mathbf{y}] - \mathbf{x})(\hat{\mathbf{x}} - E[\mathbf{x}|\mathbf{y}])^{\mathrm{H}}] + E[(\hat{\mathbf{x}} - E[\mathbf{x}|\mathbf{y}])(\hat{\mathbf{x}} - E[\mathbf{x}|\mathbf{y}])^{\mathrm{H}}].\end{aligned} \tag{5.6}$$

The third term is of the form $E[(E[\mathbf{x}|\mathbf{y}] - \mathbf{x})\mathbf{g}^{\mathrm{H}}(\mathbf{y})]$ with $\mathbf{g}(\mathbf{y}) = \hat{\mathbf{x}} - E[\mathbf{x}|\mathbf{y}]$. Using the law of total expectation, we see that this term vanishes:

$$E\{E[(E[\mathbf{x}|\mathbf{y}] - \mathbf{x})\mathbf{g}^{\mathrm{H}}(\mathbf{y})|\mathbf{y}]\} = \mathbf{0}. \tag{5.7}$$

The same reasoning can be applied to conclude that the second term in (5.6) is also zero. Therefore, the optimum estimator, obtained by making the fourth term in (5.6) equal to zero, turns out to be the *conditional mean estimator* $\hat{\mathbf{x}} = E[\mathbf{x}|\mathbf{y}]$. Let

$$\mathbf{e} = \hat{\mathbf{x}} - \mathbf{x} = E[\mathbf{x}|\mathbf{y}] - \mathbf{x} \tag{5.8}$$

be the error vector. Its mean $E[\mathbf{e}] = \mathbf{0}$, and thus $E[\hat{\mathbf{x}}] = E[\mathbf{x}]$. This says that $\hat{\mathbf{x}}$ is an *unbiased estimator* of $\mathbf{x}$. The covariance matrix of the error vector is

$$\mathbf{Q} = E[\mathbf{e}\mathbf{e}^{\mathrm{H}}] = E[(E[\mathbf{x}|\mathbf{y}] - \mathbf{x})(E[\mathbf{x}|\mathbf{y}] - \mathbf{x})^{\mathrm{H}}]. \tag{5.9}$$

Any competing estimator $\hat{\mathbf{x}}'$ with mean-squared error matrix $\mathbf{Q}' = E[\mathbf{e}'\mathbf{e}'^{\mathrm{H}}] = E[(\hat{\mathbf{x}}' - \mathbf{x})(\hat{\mathbf{x}}' - \mathbf{x})^{\mathrm{H}}]$ will be suboptimum in the sense that $\mathbf{Q}' \geq \mathbf{Q}$, meaning that $\mathbf{Q}' - \mathbf{Q}$ is positive semidefinite. As a consequence, the conditional mean estimator is a minimum mean-squared error (MMSE) estimator:

$$E\,\|\mathbf{e}\|^2 = \operatorname{tr}\mathbf{Q} \leq \operatorname{tr}\mathbf{Q}' = E\,\|\mathbf{e}'\|^2. \tag{5.10}$$

But we can say more. For the class of real-valued *increasing* functions of matrices, $\mathbf{Q} \leq \mathbf{Q}'$ implies $f(\mathbf{Q}) \leq f(\mathbf{Q}')$. Thus we have the following result.

Result 5.1. *The conditional mean estimator* $E[\mathbf{x}|\mathbf{y}]$ *minimizes (maximizes) any increasing (decreasing) function of the mean-squared error matrix* $\mathbf{Q}$.

Besides the trace, the determinant is another example of an increasing function. Thus, the volume of the error covariance ellipsoid is also minimized:

$$\det\mathbf{Q} \leq \det\mathbf{Q}'. \tag{5.11}$$

A very important property of the MMSE estimator is the so-called *orthogonality principle*, a special case of which we have already encountered in the derivation (5.7). However, it holds much more generally.

Result 5.2. *The error vector* $\mathbf{e}$ *is orthogonal to every measurable function of* $\mathbf{y}$, $\mathbf{g}(\mathbf{y})$. *That is,* $E[\mathbf{e}\mathbf{g}^{\mathrm{H}}(\mathbf{y})] = E\{(E[\mathbf{x}|\mathbf{y}] - \mathbf{x})\mathbf{g}^{\mathrm{H}}(\mathbf{y})\} = \mathbf{0}$.

For $\mathbf{g}(\mathbf{y}) = E[\mathbf{x}|\mathbf{y}]$, this orthogonality condition says that the estimator error $\mathbf{e}$ is orthogonal to the estimator $E[\mathbf{x}|\mathbf{y}]$. Moreover, since the conditional mean estimator is an idempotent operator $\mathbf{P}\mathbf{x} = E[\mathbf{x}|\mathbf{y}]$ – which is to say that $\mathbf{P}^2\mathbf{x} = E\{E[\mathbf{x}|\mathbf{y}]|\mathbf{y}\} = E[\mathbf{x}|\mathbf{y}] = \mathbf{P}\mathbf{x}$ – we may think of the conditional mean estimator as a projection operator that orthogonally resolves $\mathbf{x}$ into its estimator $E[\mathbf{x}|\mathbf{y}]$ minus its estimator error $\mathbf{e}$:

$$\mathbf{x} = E[\mathbf{x}|\mathbf{y}] - (E[\mathbf{x}|\mathbf{y}] - \mathbf{x}). \tag{5.12}$$

The conditional mean estimator summarizes everything of use for MMSE estimation of $\mathbf{x}$ from $\mathbf{y}$. This means conditioning $\mathbf{x}$ on $\mathbf{y}$ and $\mathbf{y}^*$ would change nothing, since this has already been done, so to speak, in the conditional mean $E[\mathbf{x}|\mathbf{y}]$. Similarly the conditional mean estimator of $\mathbf{x}^*$ is just the complex conjugate of the conditional mean estimator of $\mathbf{x}$. Thus it makes no sense to consider the estimation of $\mathbf{x}$ from $\mathbf{y}$ and $\mathbf{y}^*$, since this brings no refinement to the conditional mean estimator. As we shall see, this statement is decidedly false when the estimator is constrained to be a *(widely) linear* or *(widely) linear-quadratic* estimator.

5.3 Linear MMSE estimation

We shall treat the problem of linearly or widely linearly estimating a signal from a measurement as a *virtual two-channel* estimation problem, wherein the underlying experiment consists of generating a *composite* vector of signal and measurement, only one of which is observed.[1] Our point of view is that the *augmented covariance matrix* for this composite vector then encapsulates all of the second-order information that can be extracted. In order to introduce our methods, we shall begin here with the Hermitian case, wherein complementary covariances are ignored, and then extend our methods to include Hermitian *and* complementary covariances in the next section.

Let us begin with a signal (or source) $\mathbf{x}: \Omega \longrightarrow \mathbb{C}^n$ and a measurement vector $\mathbf{y}: \Omega \longrightarrow \mathbb{C}^m$. There is no requirement that the signal dimension n be smaller than the measurement dimension m. We first assume that the signal and measurement have zero mean, but remove this restriction later on. Their composite covariance matrix is the matrix

$$\mathbb{R}_{xy} = E\left\{\begin{bmatrix}\mathbf{x}\\ \mathbf{y}\end{bmatrix}\begin{bmatrix}\mathbf{x}^{\mathrm{H}} & \mathbf{y}^{\mathrm{H}}\end{bmatrix}\right\} = \begin{bmatrix}\mathbf{R}_{xx} & \mathbf{R}_{xy}\\ \mathbf{R}_{xy}^{\mathrm{H}} & \mathbf{R}_{yy}\end{bmatrix}. \tag{5.13}$$

The error between the signal $\mathbf{x}$ and the linear estimator $\hat{\mathbf{x}} = \mathbf{W}\mathbf{y}$ is $\mathbf{e} = \hat{\mathbf{x}} - \mathbf{x}$ and the error covariance matrix is $\mathbf{Q} = E[(\hat{\mathbf{x}} - \mathbf{x})(\hat{\mathbf{x}} - \mathbf{x})^{\mathrm{H}}]$. This error covariance is

$$\mathbf{Q} = E[(\mathbf{W}\mathbf{y} - \mathbf{x})(\mathbf{W}\mathbf{y} - \mathbf{x})^{\mathrm{H}}] = \mathbf{W}\mathbf{R}_{yy}\mathbf{W}^{\mathrm{H}} - \mathbf{R}_{xy}\mathbf{W}^{\mathrm{H}} - \mathbf{W}\mathbf{R}_{xy}^{\mathrm{H}} + \mathbf{R}_{xx}. \tag{5.14}$$

After completing the square, this may be written

$$\mathbf{Q} = \mathbf{R}_{xx} - \mathbf{R}_{xy}\mathbf{R}_{yy}^{-1}\mathbf{R}_{xy}^{\mathrm{H}} + (\mathbf{W} - \mathbf{R}_{xy}\mathbf{R}_{yy}^{-1})\mathbf{R}_{yy}(\mathbf{W} - \mathbf{R}_{xy}\mathbf{R}_{yy}^{-1})^{\mathrm{H}}. \tag{5.15}$$

This quadratic form in $\mathbf{W}$ is positive semidefinite, so $\mathbf{Q} \geq \mathbf{R}_{xx} - \mathbf{R}_{xy}\mathbf{R}_{yy}^{-1}\mathbf{R}_{xy}^{\mathrm{H}}$ with equality for

$$\mathbf{W} = \mathbf{R}_{xy}\mathbf{R}_{yy}^{-1} \quad \text{and} \quad \mathbf{Q} = \mathbf{R}_{xx} - \mathbf{R}_{xy}\mathbf{R}_{yy}^{-1}\mathbf{R}_{xy}^{\mathrm{H}}. \tag{5.16}$$

The solution for $\mathbf{W}$ may be written as the solution to the *normal* equations $\mathbf{W}\mathbf{R}_{yy} - \mathbf{R}_{xy} = \mathbf{0}$, or more insightfully as[2]

$$E[(\mathbf{W}\mathbf{y} - \mathbf{x})\mathbf{y}^{\mathrm{H}}] = \mathbf{0}. \tag{5.17}$$

Thus the orthogonality principle is at work: the estimator error $\mathbf{e} = \mathbf{W}\mathbf{y} - \mathbf{x}$ is orthogonal to the measurement $\mathbf{y}$, as illustrated in Fig. 5.1. There $H(\mathbf{y})$ denotes the Hilbert space of measurement vectors $\mathbf{y}$ and orthogonality means $E[(\mathbf{W}\mathbf{y} - \mathbf{x})_i y_j^*] = 0$ for all pairs (i, j). Moreover, the linear operator $\mathbf{W}$ is a projection operator, since the LMMSE estimator of $\mathbf{x}$ from $\mathbf{W}\mathbf{y}$ remains $\mathbf{W}\mathbf{y}$. The LMMSE estimator is sometimes referred to as the

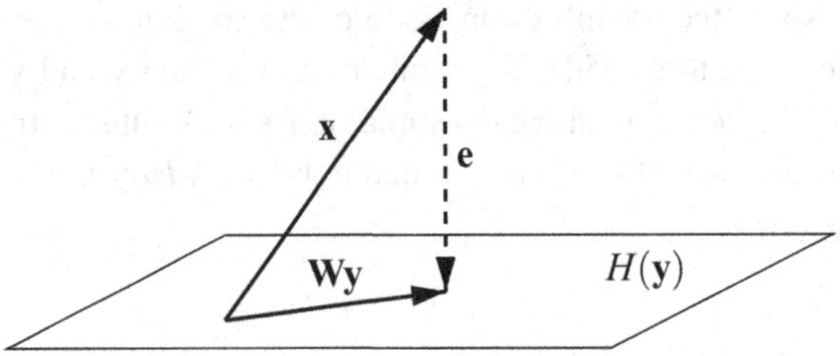

Figure 5.1 Orthogonality between the error $\mathbf{e} = \mathbf{W}\mathbf{y} - \mathbf{x}$ and $H(\mathbf{y})$ in LMMSE estimation.

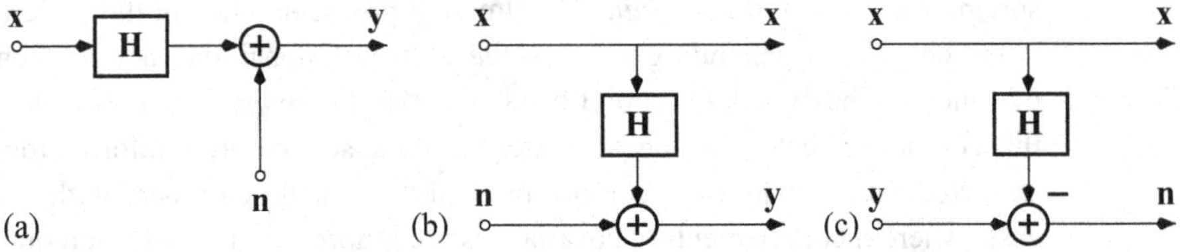

Figure 5.2 The signal-plus-noise channel model: (a) channel, (b) synthesis, and (c) analysis.

(discrete) Wiener filter, which explains our choice of notation $\mathbf{W}$. However, the LMMSE estimator predates Wiener's work on *causal* LMMSE prediction and smoothing of time series.

5.3.1 The signal-plus-noise channel model

We intend to argue that it is *as if* the source and measurement vectors were drawn from a linear channel model $\mathbf{y} = \mathbf{H}\mathbf{x} + \mathbf{n}$ that draws a *source vector* $\mathbf{x}$, linearly filters the source vector with a *channel filter* $\mathbf{H}$, and adds an *uncorrelated noise vector* $\mathbf{n}$ to produce the measurement vector. This scheme is illustrated with the *signal-plus-noise channel model* of Fig. 5.2(a). According to this channel model, the source and measurement have the *synthesis* and *analysis* representations of Figs. 5.2(b) and (c):

$$\begin{bmatrix} \mathbf{x} \\ \mathbf{y} \end{bmatrix} = \begin{bmatrix} \mathbf{I} & \mathbf{0} \\ \mathbf{H} & \mathbf{I} \end{bmatrix} \begin{bmatrix} \mathbf{x} \\ \mathbf{n} \end{bmatrix}, \tag{5.18}$$

$$\begin{bmatrix} \mathbf{x} \\ \mathbf{n} \end{bmatrix} = \begin{bmatrix} \mathbf{I} & \mathbf{0} \\ -\mathbf{H} & \mathbf{I} \end{bmatrix} \begin{bmatrix} \mathbf{x} \\ \mathbf{y} \end{bmatrix}. \tag{5.19}$$

The synthesis model of Fig. 5.2(b) and the analysis model of Fig. 5.2(c) produce these block LDU (Lower triangular–Diagonal–Upper triangular) Cholesky factorizations:

$$\underline{\mathbf{R}}_{xy} = \begin{bmatrix} \mathbf{R}_{xx} & \mathbf{R}_{xy} \\ \mathbf{R}_{xy}^{\mathrm{H}} & \mathbf{R}_{yy} \end{bmatrix} = \begin{bmatrix} \mathbf{I} & \mathbf{0} \\ \mathbf{H} & \mathbf{I} \end{bmatrix} \begin{bmatrix} \mathbf{R}_{xx} & \mathbf{0} \\ \mathbf{0} & \mathbf{R}_{nn} \end{bmatrix} \begin{bmatrix} \mathbf{I} & \mathbf{H}^{\mathrm{H}} \\ \mathbf{0} & \mathbf{I} \end{bmatrix}, \tag{5.20}$$

$$\begin{bmatrix} \mathbf{R}_{xx} & \mathbf{0} \\ \mathbf{0} & \mathbf{R}_{nn} \end{bmatrix} = \begin{bmatrix} \mathbf{I} & \mathbf{0} \\ -\mathbf{H} & \mathbf{I} \end{bmatrix} \begin{bmatrix} \mathbf{R}_{xx} & \mathbf{R}_{xy} \\ \mathbf{R}_{xy}^{\mathrm{H}} & \mathbf{R}_{yy} \end{bmatrix} \begin{bmatrix} \mathbf{I} & -\mathbf{H}^{\mathrm{H}} \\ \mathbf{0} & \mathbf{I} \end{bmatrix}. \tag{5.21}$$

For this channel model and its corresponding analysis and synthesis models to work, we must choose

$$\mathbf{H} = \mathbf{R}_{xy}^{\mathrm{H}}\mathbf{R}_{xx}^{-1} \quad \text{and} \quad \mathbf{R}_{nn} = \mathbf{R}_{yy} - \mathbf{R}_{xy}^{\mathrm{H}}\mathbf{R}_{xx}^{-1}\mathbf{R}_{xy}. \tag{5.22}$$

The noise covariance matrix $\mathbf{R}_{nn}$ is the Schur complement of $\mathbf{R}_{xx}$ within the composite covariance matrix $\mathbb{R}_{xy}$. Thus, up to second order, *every* virtual two-channel estimation problem is a problem of representing the measurement as a noisy and linearly filtered version of the signal, which is to say that the LDU Cholesky factorization of the composite covariance matrix $\mathbb{R}_{xy}$ has actually produced a channel model, a synthesis model, and an analysis model for the source and measurement.

From these block Cholesky factorizations we can also extract block UDL (Upper triangular–Diagonal–Lower triangular) Cholesky factorizations of inverses:

$$\mathbb{R}_{xy}^{-1} = \begin{bmatrix} \mathbf{R}_{xx} & \mathbf{R}_{xy} \\ \mathbf{R}_{xy}^{\mathrm{H}} & \mathbf{R}_{yy} \end{bmatrix}^{-1} = \begin{bmatrix} \mathbf{I} & -\mathbf{H}^{\mathrm{H}} \\ \mathbf{0} & \mathbf{I} \end{bmatrix} \begin{bmatrix} \mathbf{R}_{xx}^{-1} & \mathbf{0} \\ \mathbf{0} & \mathbf{R}_{nn}^{-1} \end{bmatrix} \begin{bmatrix} \mathbf{I} & \mathbf{0} \\ -\mathbf{H} & \mathbf{I} \end{bmatrix}, \tag{5.23}$$

$$\begin{bmatrix} \mathbf{R}_{xx}^{-1} & \mathbf{0} \\ \mathbf{0} & \mathbf{R}_{nn}^{-1} \end{bmatrix} = \begin{bmatrix} \mathbf{I} & \mathbf{H}^{\mathrm{H}} \\ \mathbf{0} & \mathbf{I} \end{bmatrix} \begin{bmatrix} \mathbf{R}_{xx} & \mathbf{R}_{xy} \\ \mathbf{R}_{xy}^{\mathrm{H}} & \mathbf{R}_{yy} \end{bmatrix}^{-1} \begin{bmatrix} \mathbf{I} & \mathbf{0} \\ \mathbf{H} & \mathbf{I} \end{bmatrix}. \tag{5.24}$$

It follows that $\mathbf{R}_{nn}^{-1}$ is the southeast block of $\mathbb{R}_{xy}^{-1}$. The block Cholesky factors, in turn, produce these factorizations of determinants:

$$\det \mathbb{R}_{xy} = \det \mathbf{R}_{xx} \det \mathbf{R}_{nn}, \tag{5.25}$$

$$\det \mathbb{R}_{xy}^{-1} = \det \mathbf{R}_{xx}^{-1} \det \mathbf{R}_{nn}^{-1}. \tag{5.26}$$

Let us summarize. From the synthesis model we say that every composite source and measurement vector $[\mathbf{x}^{\mathrm{T}}, \mathbf{y}^{\mathrm{T}}]^{\mathrm{T}}$ may be modeled as if it were synthesized in a virtual two-channel experiment, wherein the source is the unobserved channel and the measurement is the observed channel $\mathbf{y} = \mathbf{H}\mathbf{x} + \mathbf{n}$. Of course, the filter and the noise covariance must be chosen just right. This model then decomposes the composite covariance matrix for the source and measurement, and its inverse, into block Cholesky factors.

5.3.2 The measurement-plus-error channel model

We now interchange the roles of the signal and measurement in order to obtain the *measurement-plus-error channel model*, wherein the source $\mathbf{x}$ is produced as a noisy measurement of a linearly filtered version of the measurement $\mathbf{y}$. That is, $\mathbf{x} = \mathbf{W}\mathbf{y} - \mathbf{e}$, as illustrated in Fig. 5.3(a). It will soon become clear that our choice of $\mathbf{W}$ to denote this filter is not accidental. It will turn out to be the LMMSE filter.

According to this channel model, the source and measurement have the representations of Figs. 5.3(b) and (c),

$$\begin{bmatrix} \mathbf{x} \\ \mathbf{y} \end{bmatrix} = \begin{bmatrix} -\mathbf{I} & \mathbf{W} \\ \mathbf{0} & \mathbf{I} \end{bmatrix} \begin{bmatrix} \mathbf{e} \\ \mathbf{y} \end{bmatrix}, \tag{5.27}$$

$$\begin{bmatrix} \mathbf{e} \\ \mathbf{y} \end{bmatrix} = \begin{bmatrix} -\mathbf{I} & \mathbf{W} \\ \mathbf{0} & \mathbf{I} \end{bmatrix} \begin{bmatrix} \mathbf{x} \\ \mathbf{y} \end{bmatrix}, \tag{5.28}$$

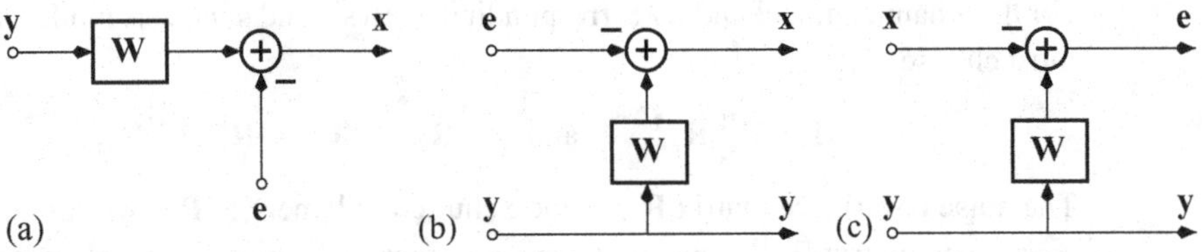

Figure 5.3 The measurement-plus-error channel model: (a) channel, (b) synthesis, and (c) analysis

where the error $\mathbf{e}$ and the measurement $\mathbf{y}$ are uncorrelated. The synthesis model of Fig. 5.3 (b) and the analysis model of Fig. 5.3(c) produce these block UDL Cholesky factorizations:

$$\mathbb{R}_{xy} = \begin{bmatrix} \mathbf{R}_{xx} & \mathbf{R}_{xy} \\ \mathbf{R}_{xy}^{\mathrm{H}} & \mathbf{R}_{yy} \end{bmatrix} = \begin{bmatrix} -\mathbf{I} & \mathbf{W} \\ \mathbf{0} & \mathbf{I} \end{bmatrix} \begin{bmatrix} \mathbf{Q} & \mathbf{0} \\ \mathbf{0} & \mathbf{R}_{yy} \end{bmatrix} \begin{bmatrix} -\mathbf{I} & \mathbf{0} \\ \mathbf{W}^{\mathrm{H}} & \mathbf{I} \end{bmatrix}, \tag{5.29}$$

$$\begin{bmatrix} \mathbf{Q} & \mathbf{0} \\ \mathbf{0} & \mathbf{R}_{yy} \end{bmatrix} = \begin{bmatrix} -\mathbf{I} & \mathbf{W} \\ \mathbf{0} & \mathbf{I} \end{bmatrix} \begin{bmatrix} \mathbf{R}_{xx} & \mathbf{R}_{xy} \\ \mathbf{R}_{xy}^{\mathrm{H}} & \mathbf{R}_{yy} \end{bmatrix} \begin{bmatrix} -\mathbf{I} & \mathbf{0} \\ \mathbf{W}^{\mathrm{H}} & \mathbf{I} \end{bmatrix}. \tag{5.30}$$

For this channel model and its corresponding analysis and synthesis models to work, we must choose $\mathbf{W}$ as the LMMSE filter and $\mathbf{Q} = E[\mathbf{e}\mathbf{e}^{\mathrm{H}}]$ as the corresponding error covariance matrix:

$$\mathbf{W} = \mathbf{R}_{xy}\mathbf{R}_{yy}^{-1} \quad \text{and} \quad \mathbf{Q} = \mathbf{R}_{xx} - \mathbf{R}_{xy}\mathbf{R}_{yy}^{-1}\mathbf{R}_{xy}^{\mathrm{H}}. \tag{5.31}$$

The error covariance matrix $\mathbf{Q}$ is the Schur complement of $\mathbf{R}_{yy}$ within $\mathbb{R}_{xy}$. Thus, up to second order, every virtual two-channel estimation problem is a problem of representing the signal as a noisy and linearly filtered version of the measurement, which is to say that the block UDL Cholesky factorization of the composite covariance matrix $\mathbb{R}_{xy}$ has actually produced a channel model, a synthesis model, and an analysis model for the source and measurement. The orthogonality between the estimator error and the measurement is expressed by the northeast and southwest zeros of the composite covariance matrix for the error $\mathbf{e}$ and the measurement $\mathbf{y}$.

From these block Cholesky factorizations of the composite covariance matrix $\mathbb{R}_{xy}$ we can also extract block LDU Cholesky factorizations of inverses:

$$\mathbb{R}_{xy}^{-1} = \begin{bmatrix} \mathbf{R}_{xx} & \mathbf{R}_{xy} \\ \mathbf{R}_{xy}^{\mathrm{H}} & \mathbf{R}_{yy} \end{bmatrix}^{-1} = \begin{bmatrix} -\mathbf{I} & \mathbf{0} \\ \mathbf{W}^{\mathrm{H}} & \mathbf{I} \end{bmatrix} \begin{bmatrix} \mathbf{Q}^{-1} & \mathbf{0} \\ \mathbf{0} & \mathbf{R}_{yy}^{-1} \end{bmatrix} \begin{bmatrix} -\mathbf{I} & \mathbf{W} \\ \mathbf{0} & \mathbf{I} \end{bmatrix}, \tag{5.32}$$

$$\begin{bmatrix} \mathbf{Q}^{-1} & \mathbf{0} \\ \mathbf{0} & \mathbf{R}_{yy}^{-1} \end{bmatrix} = \begin{bmatrix} -\mathbf{I} & \mathbf{0} \\ \mathbf{W}^{\mathrm{H}} & \mathbf{I} \end{bmatrix} \begin{bmatrix} \mathbf{R}_{xx} & \mathbf{R}_{xy} \\ \mathbf{R}_{xy}^{\mathrm{H}} & \mathbf{R}_{yy} \end{bmatrix}^{-1} \begin{bmatrix} -\mathbf{I} & \mathbf{W} \\ \mathbf{0} & \mathbf{I} \end{bmatrix}. \tag{5.33}$$

It follows that $\mathbf{Q}^{-1}$ is the northwest block of $\mathbb{R}_{xy}^{-1}$. The block Cholesky factors produce the following factorizations of determinants:

$$\det \mathbb{R}_{xy} = \det \mathbf{Q} \det \mathbf{R}_{yy}, \tag{5.34}$$

$$\det \mathbb{R}_{xy}^{-1} = \det \mathbf{Q}^{-1} \det \mathbf{R}_{yy}^{-1}. \tag{5.35}$$

Let us summarize. From the analysis model we say that every composite signal and measurement vector $[\mathbf{x}^T, \mathbf{y}^T]^T$ may be modeled *as if* it were a virtual two-channel experiment, wherein the signal is subtracted from a linearly filtered measurement to produce an error that is orthogonal to the measurement. Of course, the filter and the error covariance must be chosen just right. This model then decomposes the composite covariance matrix for the signal and measurement, and its inverse, into block Cholesky factors.

5.3.3 Filtering models

We might say that the models we have developed give us two alternative parameterizations:

- the *signal-plus-noise channel model* $(\mathbf{R}_{xx}, \mathbf{H}, \mathbf{R}_{nn})$ with $\mathbf{H} = \mathbf{R}_{xy}^H \mathbf{R}_{xx}^{-1}$ and $\mathbf{R}_{nn} = \mathbf{R}_{yy} - \mathbf{R}_{xy}^H \mathbf{R}_{xx}^{-1} \mathbf{R}_{xy}$; and
- the *measurement-plus-error channel model* $(\mathbf{R}_{yy}, \mathbf{W}, \mathbf{Q})$ with $\mathbf{W} = \mathbf{R}_{xy}\mathbf{R}_{yy}^{-1}$ and $\mathbf{Q} = \mathbf{R}_{xx} - \mathbf{R}_{xy}\mathbf{R}_{yy}^{-1}\mathbf{R}_{xy}^H$.

These correspond to the two factorizations (A1.2) and (A1.1) of $\mathbb{R}_{xy}$, and $\mathbf{R}_{nn}$ and $\mathbf{Q}$ are the Schur complements of $\mathbf{R}_{xx}$ and $\mathbf{R}_{yy}$, respectively, within $\mathbb{R}_{xy}$. Let's mix the synthesis equation for the signal-plus-noise channel model with the analysis equation of the measurement-plus-error channel model to solve for the filter $\mathbf{W}$ and error covariance matrix $\mathbf{Q}$ in terms of the channel parameters $\mathbf{H}$ and $\mathbf{R}_{nn}$:

$$\begin{bmatrix} \mathbf{e} \\ \mathbf{y} \end{bmatrix} = \begin{bmatrix} -\mathbf{I} & \mathbf{W} \\ \mathbf{0} & \mathbf{I} \end{bmatrix} \begin{bmatrix} \mathbf{I} & \mathbf{0} \\ \mathbf{H} & \mathbf{I} \end{bmatrix} \begin{bmatrix} \mathbf{x} \\ \mathbf{n} \end{bmatrix}. \tag{5.36}$$

This composition of maps produces these factorizations:

$$\begin{bmatrix} \mathbf{Q} & \mathbf{0} \\ \mathbf{0} & \mathbf{R}_{yy} \end{bmatrix} = \begin{bmatrix} -\mathbf{I} & \mathbf{W} \\ \mathbf{0} & \mathbf{I} \end{bmatrix} \begin{bmatrix} \mathbf{I} & \mathbf{0} \\ \mathbf{H} & \mathbf{I} \end{bmatrix} \begin{bmatrix} \mathbf{R}_{xx} & \mathbf{0} \\ \mathbf{0} & \mathbf{R}_{nn} \end{bmatrix} \begin{bmatrix} \mathbf{I} & \mathbf{H}^H \\ \mathbf{0} & \mathbf{I} \end{bmatrix} \begin{bmatrix} -\mathbf{I} & \mathbf{0} \\ \mathbf{W}^H & \mathbf{I} \end{bmatrix}, \tag{5.37}$$

$$\begin{bmatrix} \mathbf{Q}^{-1} & \mathbf{0} \\ \mathbf{0} & \mathbf{R}_{yy}^{-1} \end{bmatrix} = \begin{bmatrix} -\mathbf{I} & \mathbf{0} \\ \mathbf{W}^H & \mathbf{I} \end{bmatrix} \begin{bmatrix} \mathbf{I} & -\mathbf{H}^H \\ \mathbf{0} & \mathbf{I} \end{bmatrix} \begin{bmatrix} \mathbf{R}_{xx}^{-1} & \mathbf{0} \\ \mathbf{0} & \mathbf{R}_{nn}^{-1} \end{bmatrix} \begin{bmatrix} \mathbf{I} & \mathbf{0} \\ -\mathbf{H} & \mathbf{I} \end{bmatrix} \begin{bmatrix} -\mathbf{I} & \mathbf{W} \\ \mathbf{0} & \mathbf{I} \end{bmatrix}. \tag{5.38}$$

We now evaluate the northeast block of (5.37) and the southwest block of (5.38) to obtain two formulae for the filter $\mathbf{W}$:

$$\mathbf{W} = \mathbf{R}_{xx}\mathbf{H}^H(\mathbf{H}\mathbf{R}_{xx}\mathbf{H}^H + \mathbf{R}_{nn})^{-1} = (\mathbf{R}_{xx}^{-1} + \mathbf{H}^H\mathbf{R}_{nn}^{-1}\mathbf{H})^{-1}\mathbf{H}^H\mathbf{R}_{nn}^{-1}. \tag{5.39}$$

In a similar fashion, we evaluate the northwest blocks of both (5.37) and (5.38) to get two formulae for the error covariance $\mathbf{Q}$:

$$\mathbf{Q} = \mathbf{R}_{xx} - \mathbf{R}_{xx}\mathbf{H}^H(\mathbf{H}\mathbf{R}_{xx}\mathbf{H}^H + \mathbf{R}_{nn})^{-1}\mathbf{H}\mathbf{R}_{xx} = (\mathbf{R}_{xx}^{-1} + \mathbf{H}^H\mathbf{R}_{nn}^{-1}\mathbf{H})^{-1}. \tag{5.40}$$

These equations are Woodbury identities, (A1.43) and (A1.44). They determine the LMMSE inversion of the measurement $\mathbf{y}$ for the signal $\mathbf{x}$. In the absence of noise, if the signal and measurement were of the same dimension, then $\mathbf{W}$ would be $\mathbf{H}^{-1}$, assuming that this inverse existed. However, generally $\mathbf{WH} \neq \mathbf{I}$, but is approximately so if $\mathbf{R}_{nn}$ is small compared with $\mathbf{HR}_{xx}\mathbf{H}^{\mathrm{H}}$. The LMMSE estimator may be written as

$$\mathbf{Wy} = \mathbf{W}(\mathbf{Hx} + \mathbf{n}) = \mathbf{WHx} + \mathbf{Wn}. \tag{5.41}$$

So, the LMMSE estimator $\mathbf{W}$, sometimes called a *deconvolution* filter, decidedly does not equalize $\mathbf{H}$ to produce $\mathbf{WH} = \mathbf{I}$. Rather, it approximates $\mathbf{I}$ so that the error $\mathbf{e} = \mathbf{Wy} - \mathbf{x} = (\mathbf{WH} - \mathbf{I})\mathbf{x} + \mathbf{Wn}$, with covariance matrix $\mathbf{Q} = (\mathbf{WH} - \mathbf{I})\mathbf{R}_{xx}(\mathbf{WH} - \mathbf{I})^{\mathrm{H}} + \mathbf{WR}_{nn}\mathbf{W}^{\mathrm{H}}$ provides the best tradeoff between model-bias-squared $(\mathbf{WH} - \mathbf{I})\mathbf{R}_{xx}(\mathbf{WH} - \mathbf{I})^{\mathrm{H}}$ and filtered noise variance $\mathbf{WR}_{nn}\mathbf{W}^{\mathrm{H}}$ to minimize the error covariance $\mathbf{Q}$. The importance of these results cannot be overstated.

Let us summarize. Every problem of LMMSE estimation requiring only first- and second-order moments can be phrased as a problem of estimating one unobserved channel from another observed channel in a virtual two-channel experiment. In this virtual two-channel experiment, there are two different channel models.

The first channel model says it is *as if* the measurement were a linear combination of filtered source and uncorrelated additive noise. The second channel model says it is *as if* the source vector were a linear combination of the filtered measurement and the estimator error.

The first channel model produces a block LDU Cholesky factorization for the composite covariance matrix $\mathbb{R}_{xy}$ and a block UDL Cholesky factorization for its inverse. The second channel model produces a block UDL Cholesky factorization for the composite covariance matrix $\mathbb{R}_{xy}$ and a block LDU Cholesky factorization for its inverse. By mixing these two factorizations, two different solutions are found for the LMMSE estimator and its error covariance.

Example 5.2. In many problems of signal estimation in communication, radar, and sonar, or imaging in geophysics and radio astronomy, the measurement model may be taken to be the *rank-one linear model*

$$\mathbf{y} = \boldsymbol{\psi} x + \mathbf{n}, \tag{5.42}$$

where $\boldsymbol{\psi} \in \mathbb{C}^m$ is the channel vector that carries the unknown complex signal amplitude x to the measurement, x is a zero-mean random variable with variance σ_x^2, and $\mathbf{R}_{nn}$ is the covariance matrix of the noise vector $\mathbf{n}$. Using the results of the previous subsections, we may write down the following formulae for the LMMSE estimator of x and its mean-squared error:

$$\hat{x} = \sigma_x^2\boldsymbol{\psi}^{\mathrm{H}}(\sigma_x^2\boldsymbol{\psi}\boldsymbol{\psi}^{\mathrm{H}} + \mathbf{R}_{nn})^{-1}\mathbf{y} = (\sigma_x^{-2} + \boldsymbol{\psi}^{\mathrm{H}}\mathbf{R}_{nn}^{-1}\boldsymbol{\psi})^{-1}\boldsymbol{\psi}^{\mathrm{H}}\mathbf{R}_{nn}^{-1}\mathbf{y}, \tag{5.43}$$

$$Q = \sigma_x^2 - \sigma_x^4\boldsymbol{\psi}^{\mathrm{H}}(\sigma_x^2\boldsymbol{\psi}\boldsymbol{\psi}^{\mathrm{H}} + \mathbf{R}_{nn})^{-1}\boldsymbol{\psi} = \frac{1}{\sigma_x^{-2} + \boldsymbol{\psi}^{\mathrm{H}}\mathbf{R}_{nn}^{-1}\boldsymbol{\psi}}. \tag{5.44}$$

In both forms the estimator consists of a matrix inverse operator, followed by a correlator, followed by a scalar multiplication. When the channel filter is a vector's worth of a geometric sequence, $\boldsymbol{\psi} = [1, e^{j\theta}, \ldots, e^{jm\theta}]^{\mathrm{T}}$, as in a uniformly sampled complex exponential time series or a uniformly sampled complex exponential plane wave, then the correlation step is a correlation with a discrete-time Fourier-transform (DTFT) vector.

5.3.4 Nonzero means

We have assumed until now that the signal and measurement have zero means. What if the signal has known mean $\boldsymbol{\mu}_x$ and the measurement has known mean $\boldsymbol{\mu}_y$? How do these filtering formulae change? The centered signal and measurement $\mathbf{x} - \boldsymbol{\mu}_x$ and $\mathbf{y} - \boldsymbol{\mu}_y$ then share the composite covariance matrix $\mathbb{R}_{xy}$. So the LMMSE estimator of $\mathbf{x} - \boldsymbol{\mu}_x$ from $\mathbf{y} - \boldsymbol{\mu}_y$ should obey all of the equations already derived. That is,

$$\hat{\mathbf{x}} - \boldsymbol{\mu}_x = \mathbf{W}(\mathbf{y} - \boldsymbol{\mu}_y) \Leftrightarrow \hat{\mathbf{x}} = \mathbf{W}(\mathbf{y} - \boldsymbol{\mu}_y) + \boldsymbol{\mu}_x. \tag{5.45}$$

But what about the orthogonality principle which says that the error between the estimator and the signal is orthogonal to the measurement? This is still so due to

$$E[(\hat{\mathbf{x}} - \mathbf{x})\mathbf{y}^{\mathrm{H}}] = E\{[\hat{\mathbf{x}} - \boldsymbol{\mu}_x - (\mathbf{x} - \boldsymbol{\mu}_x)](\mathbf{y} - \boldsymbol{\mu}_y)^{\mathrm{H}}\} + E\{[\hat{\mathbf{x}} - \boldsymbol{\mu}_x - (\mathbf{x} - \boldsymbol{\mu}_x)]\boldsymbol{\mu}_y^{\mathrm{H}}\} = \mathbf{0}. \tag{5.46}$$

The first term on the right is zero due to the orthogonality principle already established for zero-mean LMMSE estimators. The second term is zero because $\hat{\mathbf{x}}$ is an unbiased estimator of $\mathbf{x}$, i.e., $E[\hat{\mathbf{x}}] = \boldsymbol{\mu}_x$.

5.3.5 Concentration ellipsoids

Let's call $B_{xx} = \{\mathbf{x}\colon \mathbf{x}^{\mathrm{H}}\mathbf{R}_{xx}^{-1}\mathbf{x} = 1\}$ the concentration ellipsoid for the signal vector $\mathbf{x}$. For scalar x, this is the circle $B_{xx} = \{x\colon |x|^2 = R_{xx}\}$. The concentration ellipsoid for the error vector $\mathbf{e}$ is $B_{ee} = \{\mathbf{e}\colon \mathbf{e}^{\mathrm{H}}\mathbf{Q}^{-1}\mathbf{e} = 1\}$. From the equation $\mathbf{Q} = \mathbf{R}_{xx} - \mathbf{R}_{xy}\mathbf{R}_{yy}^{-1}\mathbf{R}_{xy}^{\mathrm{H}}$ we know that $\mathbf{R}_{xx} \geq \mathbf{Q}$ and hence $\mathbf{R}_{xx}^{-1} \leq \mathbf{Q}^{-1}$. The posterior concentration ellipsoid B_{ee} is therefore smaller than, and completely embedded within, the prior concentration ellipsoid B_{xx}. When the signal and measurement are jointly Gaussian, these ellipsoids are probability-density contour lines (level curves).

Among measures of effectiveness for LMMSE are relative values for the trace and determinant of $\mathbf{Q}$ and $\mathbf{R}_{xx}$. Various forms of these may be derived from the matrix identities obtained from the channel models. They may be given insightful forms in canonical coordinate systems, as we will see in Section 5.5. One particularly illuminating formula is the *gain* of LMMSE filtering, which is closely related to mutual information (cf. Section 5.5.2):

$$\frac{\det \mathbf{R}_{xx}}{\det \mathbf{Q}} = \frac{\det \mathbf{R}_{yy}}{\det \mathbf{R}_{nn}}. \tag{5.47}$$

So the gain of LMMSE estimation is the ratio of the volume of the measurement concentration ellipsoid to the volume of the noise concentration ellipsoid.

Example 5.3. Let's consider the problem of linearly estimating the zero-mean signal $\mathbf{x}$ from the zero-mean measurement $\mathbf{y}$ when the measurement is in fact the complex conjugate of the signal, namely $\mathbf{y} = \mathbf{x}^*$. The MMSE estimator is obviously the *widely linear* estimator $\hat{\mathbf{x}} = \mathbf{y}^*$ and its error covariance matrix is $\mathbf{Q} = \mathbf{0}$.

The composite covariance matrix for $\mathbf{x}$ and $\mathbf{x}^*$ is the augmented covariance matrix $\underline{\mathbf{R}}_{xx}$. From the structure of this matrix, namely

$$\underline{\mathbf{R}}_{xx} = \begin{bmatrix} \mathbf{R}_{xx} & \widetilde{\mathbf{R}}_{xx} \\ \widetilde{\mathbf{R}}^*_{xx} & \mathbf{R}^*_{xx} \end{bmatrix},$$

we know that the LMMSE estimator of $\mathbf{x}$ from $\mathbf{x}^*$ is $\hat{\mathbf{x}} = \widetilde{\mathbf{R}}_{xx}\mathbf{R}^{-*}_{xx}\mathbf{x}^*$, with error covariance matrix $\mathbf{Q} = \mathbf{R}_{xx} - \widetilde{\mathbf{R}}_{xx}\mathbf{R}^{-*}_{xx}\widetilde{\mathbf{R}}^*_{xx}$. This error covariance matrix satisfies the inequality $\mathbf{0} \leq \mathbf{Q} \leq \mathbf{R}_{xx}$. Of course, if $\mathbf{x}$ is proper, $\widetilde{\mathbf{R}}_{xx} = \mathbf{0}$, then linear MMSE estimation is no good at all. But, for improper signals, the error concentration ellipsoid for the error $\mathbf{e} = \hat{\mathbf{x}} - \mathbf{x}$ lies inside the concentration ellipsoid for $\mathbf{x}$.

5.3.6 Special cases

Two special cases of LMMSE estimation warrant particular attention: the pure signal-plus-noise case and the Gaussian case.

Signal plus noise

The pure signal-plus-noise problem is $\mathbf{y} = \mathbf{x} + \mathbf{n}$, with $\mathbf{H} = \mathbf{I}$ and $\mathbf{x}$ and $\mathbf{n}$ uncorrelated. As a consequence, $\mathbf{R}_{xy} = \mathbf{R}_{xx}$, $\mathbf{R}_{yy} = \mathbf{R}_{xx} + \mathbf{R}_{nn}$, and the composite covariance matrix is

$$\mathbb{R}_{xy} = \begin{bmatrix} \mathbf{R}_{xx} & \mathbf{R}_{xx} \\ \mathbf{R}_{xx} & \mathbf{R}_{xx} + \mathbf{R}_{nn} \end{bmatrix}. \tag{5.48}$$

The prior signal-plus-noise covariance is the series formula $\mathbf{R}_{yy} = \mathbf{R}_{xx} + \mathbf{R}_{nn}$, but the posterior error covariance is the parallel formula $\mathbf{Q} = (\mathbf{R}^{-1}_{xx} + \mathbf{R}^{-1}_{nn})^{-1}$. The gain of LMMSE filtering is

$$\frac{\det \mathbf{R}_{xx}}{\det \mathbf{Q}} = \frac{\det(\mathbf{R}_{xx} + \mathbf{R}_{nn})}{\det \mathbf{R}_{nn}} = \det(\mathbf{I} + \mathbf{R}^{-1/2}_{nn}\mathbf{R}_{xx}\mathbf{R}^{-\mathrm{H}/2}_{nn}). \tag{5.49}$$

The matrix $\mathbf{R}^{-1/2}_{nn}\mathbf{R}_{xx}\mathbf{R}^{-\mathrm{H}/2}_{nn}$ can reasonably be called a *signal-to-noise-ratio matrix*. These formulae are characteristic of LMMSE problems.

The Gaussian case

Suppose the signal and measurement are multivariate Gaussian, meaning that the composite vector $[\mathbf{x}^{\mathrm{T}}, \mathbf{y}^{\mathrm{T}}]^{\mathrm{T}}$ is multivariate normal with zero mean and composite covariance matrix $\mathbb{R}_{xy}$. Then the conditional pdf for $\mathbf{x}$, given $\mathbf{y}$, is

$$p(\mathbf{x}|\mathbf{y}) = \frac{p(\mathbf{x}, \mathbf{y})}{p(\mathbf{y})} = \frac{\det \mathbf{R}_{yy}}{\pi^n \det \mathbb{R}_{xy}} \exp\left\{ \begin{bmatrix} \mathbf{x}^{\mathrm{H}} & \mathbf{y}^{\mathrm{H}} \end{bmatrix} \mathbb{R}^{-1}_{xy} \begin{bmatrix} \mathbf{x} \\ \mathbf{y} \end{bmatrix} - \mathbf{y}^{\mathrm{H}}\mathbf{R}^{-1}_{yy}\mathbf{y} \right\}. \tag{5.50}$$

Using the identity $\det \mathbb{R}_{xy} = \det \mathbf{R}_{yy} \det \mathbf{Q}$ and one of the Cholesky factors for $\mathbb{R}_{xy}^{-1}$, this pdf may be written

$$p(\mathbf{x}|\mathbf{y}) = \frac{1}{\pi^n \det \mathbf{Q}} \exp\left\{(\mathbf{x} - \mathbf{W}\mathbf{y})^{\mathrm{H}}\mathbf{Q}^{-1}(\mathbf{x} - \mathbf{W}\mathbf{y})\right\}. \tag{5.51}$$

Thus the posterior pdf for $\mathbf{x}$, given $\mathbf{y}$, is Gaussian with conditional mean $\mathbf{W}\mathbf{y}$ and conditional covariance $\mathbf{Q}$. This means that the MMSE estimator – without the qualifier that it be linear – is this conditional mean $\mathbf{W}\mathbf{y}$, and its error covariance is $\mathbf{Q}$.

5.4 Widely linear MMSE estimation

We have already established that widely linear transformations are required in order to access the information contained in the complementary covariance matrix. In this section, we consider widely linear minimum mean-squared error (WLMMSE) estimation of the zero-mean signal $\mathbf{x}\colon \Omega \longrightarrow \mathbb{C}^n$ from the zero-mean measurement $\mathbf{y}\colon \Omega \longrightarrow \mathbb{C}^m$. To extend our results for LMMSE estimation to WLMMSE estimation we need only replace the signal $\mathbf{x}$ by the augmented signal $\underline{\mathbf{x}}$ and the measurement $\mathbf{y}$ by the augmented measurement $\underline{\mathbf{y}}$, specify the composite augmented covariance matrix for these two vectors, and proceed as before. Nothing else changes. Thus, most of the results from Section 5.3 apply straightforwardly to WLMMSE estimation. It is, however, still worthwhile to summarize some of these results and to compare LMMSE with WLMMSE estimation.[3] The widely linear (or linear–conjugate linear) estimator is

$$\underline{\hat{\mathbf{x}}} = \underline{\mathbf{W}}\,\underline{\mathbf{y}} \Longleftrightarrow \hat{\mathbf{x}} = \mathbf{W}_1\mathbf{y} + \mathbf{W}_2\mathbf{y}^*, \tag{5.52}$$

where

$$\underline{\mathbf{W}} = \begin{bmatrix} \mathbf{W}_1 & \mathbf{W}_2 \\ \mathbf{W}_2^* & \mathbf{W}_1^* \end{bmatrix} \tag{5.53}$$

is determined such that the mean-squared error

$$E\|\hat{\mathbf{x}} - \mathbf{x}\|^2 = \tfrac{1}{2}E\|\underline{\hat{\mathbf{x}}} - \underline{\mathbf{x}}\|^2 \tag{5.54}$$

is minimized. This can be done by applying the *orthogonality principle*,

$$(\hat{\mathbf{x}} - \mathbf{x}) \perp \mathbf{y} \qquad \text{and} \qquad (\hat{\mathbf{x}} - \mathbf{x}) \perp \mathbf{y}^*, \tag{5.55}$$

or $(\underline{\hat{\mathbf{x}}} - \underline{\mathbf{x}}) \perp \underline{\mathbf{y}}$. This says that the error between the augmented estimator and the augmented signal must be orthogonal to the augmented measurement. This leads to

$$E(\underline{\hat{\mathbf{x}}}\,\underline{\mathbf{y}}^{\mathrm{H}}) - E(\underline{\mathbf{x}}\,\underline{\mathbf{y}}^{\mathrm{H}}) = \mathbf{0},$$
$$\underline{\mathbf{W}}\,\underline{\mathbf{R}}_{yy} - \underline{\mathbf{R}}_{xy} = \mathbf{0} \Leftrightarrow \underline{\mathbf{W}} = \underline{\mathbf{R}}_{xy}\underline{\mathbf{R}}_{yy}^{-1}. \tag{5.56}$$

Using the matrix-inversion lemma (A1.42) for $\underline{\mathbf{R}}_{yy}^{-1}$, we obtain the following result.

Result 5.3. *Given an observation* $\mathbf{y}\colon \Omega \longrightarrow \mathbb{C}^m$, *the WLMMSE estimator of the signal* $\mathbf{x}\colon \Omega \longrightarrow \mathbb{C}^n$ *in augmented notation is*

$$\underline{\hat{\mathbf{x}}} = \underline{\mathbf{R}}_{xy}\underline{\mathbf{R}}_{yy}^{-1}\underline{\mathbf{y}}. \tag{5.57}$$

Equivalently,

$$\hat{\mathbf{x}} = (\mathbf{R}_{xy} - \widetilde{\mathbf{R}}_{xy}\mathbf{R}_{yy}^{-*}\widetilde{\mathbf{R}}_{yy}^{*})\mathbf{P}_{yy}^{-1}\mathbf{y} + (\widetilde{\mathbf{R}}_{xy} - \mathbf{R}_{xy}\mathbf{R}_{yy}^{-1}\widetilde{\mathbf{R}}_{yy})\mathbf{P}_{yy}^{-*}\mathbf{y}^{*}. \tag{5.58}$$

In this equation, the Schur complement $\mathbf{P}_{yy} = \mathbf{R}_{yy} - \widetilde{\mathbf{R}}_{yy}\mathbf{R}_{yy}^{-*}\widetilde{\mathbf{R}}_{yy}^{*}$ *is the error covariance matrix for linearly estimating* $\mathbf{y}$ *from* $\mathbf{y}^*$. *The augmented error covariance matrix* $\underline{\mathbf{Q}}$ *of the error vector* $\mathbf{e} = \hat{\mathbf{x}} - \mathbf{x}$ *is*

$$\underline{\mathbf{Q}} = E[\underline{\mathbf{e}}\,\underline{\mathbf{e}}^{\mathrm{H}}] = \underline{\mathbf{R}}_{xx} - \underline{\mathbf{R}}_{xy}\underline{\mathbf{R}}_{yy}^{-1}\underline{\mathbf{R}}_{xy}^{\mathrm{H}}. \tag{5.59}$$

A competing estimator $\underline{\hat{\mathbf{x}}}' = \underline{\mathbf{W}}'\underline{\mathbf{y}}$ will produce an augmented error $\underline{\mathbf{e}}' = \underline{\hat{\mathbf{x}}}' - \underline{\mathbf{x}}$ with covariance matrix

$$\underline{\mathbf{Q}}' = E\left[\underline{\mathbf{e}}'\underline{\mathbf{e}}'^{\mathrm{H}}\right] = \underline{\mathbf{Q}} + (\underline{\mathbf{W}} - \underline{\mathbf{W}}')\underline{\mathbf{R}}_{yy}(\underline{\mathbf{W}} - \underline{\mathbf{W}}')^{\mathrm{H}}, \tag{5.60}$$

which shows that $\underline{\mathbf{Q}} \leq \underline{\mathbf{Q}}'$. As a consequence, all real-valued increasing functions of $\underline{\mathbf{Q}}$ are minimized, in particular,

$$E\,\|\underline{\mathbf{e}}\|^2 = \operatorname{tr}\underline{\mathbf{Q}} \leq \operatorname{tr}\underline{\mathbf{Q}}' = E\,\|\underline{\mathbf{e}}'\|^2, \tag{5.61}$$

$$\det\underline{\mathbf{Q}} \leq \det\underline{\mathbf{Q}}'. \tag{5.62}$$

These statements hold for the error vector $\mathbf{e}$ as well as the augmented error vector $\underline{\mathbf{e}}$ because

$$\underline{\mathbf{Q}} \leq \underline{\mathbf{Q}}' \Rightarrow \mathbf{Q} \leq \mathbf{Q}'. \tag{5.63}$$

The error covariance matrix $\mathbf{Q}$ of the error vector $\mathbf{e} = \hat{\mathbf{x}} - \mathbf{x}$ is the northwest block of the augmented error covariance matrix $\underline{\mathbf{Q}}$, which can be evaluated as

$$\mathbf{Q} = E[\mathbf{e}\mathbf{e}^{\mathrm{H}}] = \mathbf{R}_{xx} - (\mathbf{R}_{xy} - \widetilde{\mathbf{R}}_{xy}\mathbf{R}_{yy}^{-*}\widetilde{\mathbf{R}}_{yy}^{*})\mathbf{P}_{yy}^{-1}\mathbf{R}_{xy}^{\mathrm{H}} - (\widetilde{\mathbf{R}}_{xy} - \mathbf{R}_{xy}\mathbf{R}_{yy}^{-1}\widetilde{\mathbf{R}}_{yy})\mathbf{P}_{yy}^{-*}\widetilde{\mathbf{R}}_{xy}^{\mathrm{H}}. \tag{5.64}$$

A particular choice for a generally suboptimum filter is the LMMSE filter

$$\underline{\mathbf{W}}' = \begin{bmatrix} \mathbf{R}_{xy}\mathbf{R}_{yy}^{-1} & \mathbf{0} \\ \mathbf{0} & \mathbf{R}_{xy}^{*}\mathbf{R}_{yy}^{-*} \end{bmatrix} \Longleftrightarrow \mathbf{W}' = \mathbf{R}_{xy}\mathbf{R}_{yy}^{-1}, \tag{5.65}$$

which ignores complementary covariance matrices. We will examine the relation between LMMSE and WLMMSE filters in the following subsections.

5.4.1 Special cases

If the signal $\mathbf{x}$ is *real*, we have $\widetilde{\mathbf{R}}_{xy} = \mathbf{R}_{xy}^{*}$. This leads to the simplified expression

$$\hat{\mathbf{x}} = 2\,\mathrm{Re}\left\{(\mathbf{R}_{xy} - \widetilde{\mathbf{R}}_{xy}\mathbf{R}_{yy}^{-*}\widetilde{\mathbf{R}}_{yy}^{*})\mathbf{P}_{yy}^{-1}\mathbf{y}\right\}. \tag{5.66}$$

While the WLMMSE estimate of a real signal is always real, the LMMSE estimate is generally complex.

Result 5.4. *The WLMMSE and LMMSE estimates are identical if and only if the error of the LMMSE estimate is orthogonal to* $\mathbf{y}^*$, *i.e.,*

$$(\mathbf{W}'\mathbf{y} - \mathbf{x}) \perp \mathbf{y}^* \Longleftrightarrow \mathbf{R}_{xy}\mathbf{R}_{yy}^{-1}\widetilde{\mathbf{R}}_{yy} - \widetilde{\mathbf{R}}_{xy} = \mathbf{0}. \tag{5.67}$$

There are two important special cases in which (5.67) holds.

- The signal and measurement are cross-proper, $\widetilde{\mathbf{R}}_{xy} = \mathbf{0}$, and the measurement is proper, $\widetilde{\mathbf{R}}_{yy} = \mathbf{0}$. Joint propriety of **x** and **y** will suffice but it is not necessary that **x** be proper.
- The measurement is *maximally improper*, i.e., $\mathbf{y} = \alpha\mathbf{y}^*$ with probability 1 for constant α with $|\alpha| = 1$. In this case, $\widetilde{\mathbf{R}}_{xy} = \alpha\mathbf{R}_{xy}$ and $\widetilde{\mathbf{R}}_{yy} = \alpha\mathbf{R}_{yy}$ and $\mathbf{R}_{xy}\mathbf{R}_{yy}^{-1}\mathbf{R}_{yy}\alpha - \mathbf{R}_{xy}\alpha = \mathbf{0}$. WL estimation is unnecessary since **y** and $\mathbf{y}^*$ both carry exactly the same information about **x**. This is irrespective of whether or not **x** is proper.

It may be surprising that (5.67) puts no restrictions on whether or not **x** be proper. This is true for the performance criterion of MMSE. Other criteria such as weighted MMSE or mutual information do depend on the complementary covariance matrix of **x**.

Finally, the following result for Gaussian random vectors is now obvious by comparing $\hat{\mathbf{x}}$ with $\boldsymbol{\mu}_{x|y}$ in (2.60) and $\mathbf{Q}$ with $\underline{\mathbf{R}}_{xx|y}$ in (2.61).

Result 5.5. *Let* **x** *and* **y** *be two jointly Gaussian random vectors. The mean of the conditional distribution for* **x** *given* **y** *is the WLMMSE estimator of* **x** *from* **y**, *and the covariance matrix of the conditional distribution is the error covariance matrix of the WLMMSE estimator.*

5.4.2 Performance comparison between LMMSE and WLMMSE estimation

It is interesting to compare the performance of LMMSE and WLMMSE estimation. We will consider the signal-plus-uncorrelated-noise case $\mathbf{y} = \mathbf{x} + \mathbf{n}$ with white and proper noise **n**, i.e., $\underline{\mathbf{R}}_{nn} = N_0\mathbf{I}$, $\underline{\mathbf{R}}_{xy} = \underline{\mathbf{R}}_{xx}$, and $\underline{\mathbf{R}}_{yy} = \underline{\mathbf{R}}_{xx} + N_0\mathbf{I}$.

The LMMSE is

$$\begin{aligned}\text{LMMSE} &= E\|\mathbf{e}'\|^2 = \operatorname{tr}(\mathbf{R}_{xx} - \mathbf{R}_{xx}(\mathbf{R}_{xx} + N_0\mathbf{I})^{-1}\mathbf{R}_{xx}^{\mathrm{H}}) \\ &= \sum_{i=1}^{n}\left(\mu_i - \frac{\mu_i^2}{\mu_i + N_0}\right) \\ &= N_0\sum_{i=1}^{n}\frac{\mu_i}{\mu_i + N_0},\end{aligned} \tag{5.68}$$

where $\{\mu_i\}_{i=1}^{n}$ are the eigenvalues of $\mathbf{R}_{xx}$. On the other hand, the WLMMSE is

$$\begin{aligned}\text{WLMMSE} &= E\|\mathbf{e}\|^2 = \tfrac{1}{2}\operatorname{tr}(\underline{\mathbf{R}}_{xx} - \underline{\mathbf{R}}_{xx}(\underline{\mathbf{R}}_{xx} + N_0\mathbf{I})^{-1}\underline{\mathbf{R}}_{xx}^{\mathrm{H}}) \\ &= \frac{N_0}{2}\sum_{i=1}^{2n}\frac{\lambda_i}{\lambda_i + N_0},\end{aligned} \tag{5.69}$$

where $\{\lambda_i\}_{i=1}^{2n}$ are the eigenvalues of $\underline{\mathbf{R}}_{xx}$. In order to evaluate the maximum performance advantage of WLMMSE over LMMSE processing, we need to minimize the WLMMSE for fixed $\mathbf{R}_{xx}$ and varying $\widetilde{\mathbf{R}}_{xx}$. Using (A3.13) from Appendix 3, we see that the WLMMSE is a Schur-concave function of the eigenvalues $\{\lambda_i\}$. Therefore, minimizing the WLMMSE requires maximum spread of the $\{\lambda_i\}$ in the sense of majorization.

According to Result 3.7, this is achieved for

$$\lambda_i = 2\mu_i, \quad i = 1, \ldots, n, \quad \text{and} \quad \lambda_i = 0, \quad i = n+1, \ldots, 2n. \tag{5.70}$$

On plugging this into (5.69), we obtain

$$\min_{\widetilde{\mathbf{R}}_{xx}} \text{WLMMSE} = \frac{N_0}{2} \sum_{i=1}^{n} \frac{2\mu_i}{2\mu_i + N_0}, \tag{5.71}$$

and thus

$$\min_{N_0} \frac{\min_{\widetilde{\mathbf{R}}_{xx}} \text{WLMMSE}}{\text{LMMSE}} = \frac{1}{2}, \tag{5.72}$$

which is attained for $N_0 = 0$. This gives the following result.

Result 5.6. *When estimating an improper complex signal in uncorrelated additive white (i.e., proper) noise, the maximum performance advantage of WLMMSE estimation over LMMSE estimation is a factor of 2. This is achieved as the noise level N_0 approaches zero.*

The advantage diminishes for larger N_0 and disappears completely for $N_0 \to \infty$. We note that, if (5.70) is not satisfied, the maximum performance advantage, which is then less than a factor of 2, occurs for some noise level $N_0 > 0$.

The factor of 2 is a very conservative performance-advantage bound because it assumes the worst-case scenario of *white* additive noise. If the noise is improper or colored, the performance difference between WLMMSE and LMMSE processing can be much larger. Consider for instance the scenario in which the signal $\mathbf{x}$ is real and the noise $\mathbf{n}$ is purely imaginary. It is clear that the WL operation $\text{Re}\,\mathbf{y}$ will yield a perfect estimate of $\mathbf{x}$. The LMMSE estimator, on the other hand, is not real-valued and therefore incurs a nonzero estimation error. In this special case, the performance advantage of WLMMSE over LMMSE estimation would be infinite.

5.5 Reduced-rank widely linear estimation

We now consider reduced-rank widely linear estimators. There are two different aims for rank reduction that we pursue: either keep the mean-squared error as small as possible, or keep the concentration ellipsoid for the error as small as possible. In the Gaussian case, the latter keeps the mutual information between the reduced-rank estimator and the signal as large as possible. The first goal requires the minimization of the *trace* of the error covariance matrix and will be referred to as the *min-trace problem*. The second goal requires the minimization of the *determinant* of the error covariance matrix and correspondingly will be called the *min-det problem*.

The mean-squared estimation error is invariant under nonsingular widely linear transformation of the measurement and widely unitary transformation of the signal, and mutual information is invariant under nonsingular widely linear transformation of both measurement and signal. In Chapter 4, we have already found that half-canonical correlations and canonical correlations are maximal invariants for these transformation groups. Therefore, it should come as no surprise that the min-trace problem is solved by performing rank reduction in half-canonical coordinates, and the min-det problem by performing rank reduction in canonical coordinates.[4]

5.5.1 Minimize mean-squared error (min-trace problem)

We first consider the min-trace problem, in which we minimize the reduced-rank MSE. The full-rank WLMMSE is

$$E\|\hat{\mathbf{x}} - \mathbf{x}\|^2 = \operatorname{tr}\mathbf{Q} = \tfrac{1}{2}E\|\underline{\hat{\mathbf{x}}} - \underline{\mathbf{x}}\|^2 = \tfrac{1}{2}\operatorname{tr}\underline{\mathbf{Q}} = \tfrac{1}{2}\operatorname{tr}(\underline{\mathbf{R}}_{xx} - \underline{\mathbf{R}}_{xy}\underline{\mathbf{R}}_{yy}^{-1}\underline{\mathbf{R}}_{xy}^{\mathrm{H}}), \tag{5.73}$$

and it is easy to see that it is invariant under widely linear transformation of the measurement $\mathbf{y}$ and widely unitary transformation of the signal $\mathbf{x}$. Therefore, the full-rank WLMMSE must be a function of the augmented covariance matrix $\underline{\mathbf{R}}_{xx}$ and the *total half-canonical correlations* $k_1 \geq k_2 \geq \cdots \geq k_{2p} \geq 0$ between $\mathbf{x}$ and $\mathbf{y}$. As discussed in Section 4.1.4, the total half-canonical correlations are obtained by the SVD

$$\underline{\mathbf{C}} = \underline{\mathbf{R}}_{xy}\underline{\mathbf{R}}_{yy}^{-\mathrm{H}/2} = \underline{\mathbf{F}}\,\underline{\mathbf{K}}\,\underline{\mathbf{G}}^{\mathrm{H}}, \tag{5.74}$$

where $\underline{\mathbf{F}}^{\mathrm{H}}\underline{\mathbf{F}} = \mathbf{I}$, $\underline{\mathbf{F}} \in \mathcal{W}^{n\times p}$, and $\underline{\mathbf{G}}^{\mathrm{H}}\underline{\mathbf{G}} = \mathbf{I}$, $\underline{\mathbf{G}} \in \mathcal{W}^{m\times p}$. The matrix

$$\underline{\mathbf{K}} = \begin{bmatrix} \mathbf{K}_1 & \mathbf{K}_2 \\ \mathbf{K}_2 & \mathbf{K}_1 \end{bmatrix}$$

consists of a diagonal block $\mathbf{K}_1$ with diagonal elements $\frac{1}{2}(k_{2i-1} + k_{2i})$ and a diagonal block $\mathbf{K}_2$ with diagonal elements $\frac{1}{2}(k_{2i-1} - k_{2i})$. (In Chapter 4 we used a bar over $\bar{k}_i$ to differentiate total correlations from rotational correlations, but this is not necessary here.)

The full-rank WLMMSE is

$$E\|\hat{\mathbf{x}} - \mathbf{x}\|^2 = \tfrac{1}{2}\operatorname{tr}(\underline{\mathbf{R}}_{xx} - \underline{\mathbf{K}}^2) = \operatorname{tr}\mathbf{R}_{xx} - \tfrac{1}{2}\sum_{i=1}^{2p} k_i^2. \tag{5.75}$$

We may also rewrite the widely linear estimator as

$$\begin{aligned} \underline{\hat{\mathbf{x}}} &= \underline{\mathbf{W}}\,\underline{\mathbf{y}} = \underline{\mathbf{R}}_{xy}\underline{\mathbf{R}}_{yy}^{-1}\underline{\mathbf{y}} = \underline{\mathbf{R}}_{xy}\underline{\mathbf{R}}_{yy}^{-\mathrm{H}/2}\underline{\mathbf{R}}_{yy}^{-1/2}\underline{\mathbf{y}} \\ &= \underline{\mathbf{F}}\,\underline{\mathbf{K}}\,\underline{\mathbf{G}}^{\mathrm{H}}\underline{\mathbf{R}}_{yy}^{-1/2}\underline{\mathbf{y}}. \end{aligned} \tag{5.76}$$

We thus break up the estimator into three blocks, as shown in Fig. 5.4. We first transform the measurement $\underline{\mathbf{y}}$ into *white* and *proper* half-canonical measurement coordinates $\underline{\boldsymbol{\omega}} = \underline{\mathbf{B}}\,\underline{\mathbf{y}}$ using the *coder* $\underline{\mathbf{B}} = \underline{\mathbf{G}}^{\mathrm{H}}\underline{\mathbf{R}}_{yy}^{-1/2}$. We then estimate the signal in half-canonical coordinates as $\underline{\hat{\boldsymbol{\xi}}} = \underline{\mathbf{K}}\,\underline{\boldsymbol{\omega}}$. A key observation is that $\underline{\mathbf{K}}$ has *diagonal blocks*. Thus, for a given i, every component $\hat{\xi}_i$ is a widely linear function of ω_i only and does not depend on ω_j, $j \neq i$. The estimator $\underline{\hat{\boldsymbol{\xi}}}$ still has mutually uncorrelated components but is generally *no longer white or proper*. Finally, the estimate $\underline{\hat{\mathbf{x}}}$ is produced by passing $\underline{\hat{\boldsymbol{\xi}}}$ through the *decoder* $\underline{\mathbf{A}}^{-1} = \underline{\mathbf{F}}$ (where the inverse is a *right* inverse).

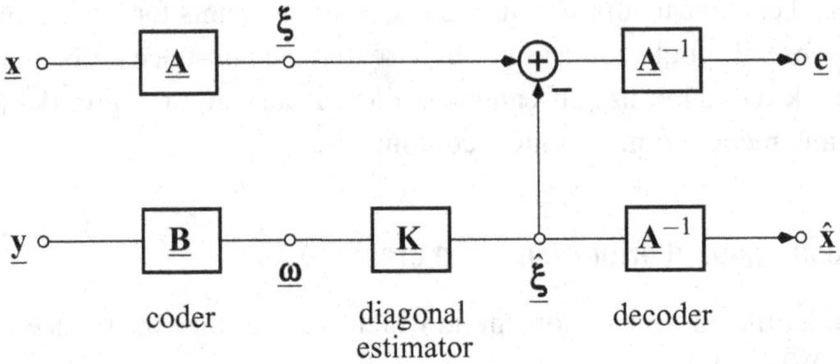

Figure 5.4 WLMMSE estimation in half-canonical coordinates.

So how do we obtain the optimum rank-r approximation of this estimator? Looking at (5.75), it is tempting to assume that we should discard the $2(p-r)$ smallest half-canonical correlations. That is, we would build the rank-r estimator

$$\hat{\underline{\mathbf{x}}}_r = \underline{\mathbf{W}}_r\underline{\mathbf{y}} = \underline{\mathbf{F}}\,\underline{\mathbf{K}}_r\underline{\mathbf{G}}^{\mathrm{H}}\underline{\mathbf{R}}_{yy}^{-1/2}\underline{\mathbf{y}}, \tag{5.77}$$

where $\underline{\mathbf{K}}_r$ is obtained from $\underline{\mathbf{K}}$ by replacing the $2(p-r)$ smallest half-canonical correlations with zeros. This is indeed the optimum solution. In order to turn this into a rigorous proof, we need to establish that the reduced-rank estimator uses the same half-canonical coordinate system as the full-rank estimator. This can be done by proceeding from (5.60). In order to determine the rank-$2r$ matrix $\underline{\mathbf{W}}_r$ that minimizes the trace of the extra covariance

$$\mathrm{tr}(\underline{\mathbf{W}}-\underline{\mathbf{W}}_r)\underline{\mathbf{R}}_{yy}(\underline{\mathbf{W}}-\underline{\mathbf{W}}_r)^{\mathrm{H}} = \mathrm{tr}(\underline{\mathbf{W}}\,\underline{\mathbf{R}}_{yy}^{1/2}-\underline{\mathbf{W}}_r\underline{\mathbf{R}}_{yy}^{1/2})(\underline{\mathbf{W}}\,\underline{\mathbf{R}}_{yy}^{1/2}-\underline{\mathbf{W}}_r\underline{\mathbf{R}}_{yy}^{1/2})^{\mathrm{H}}, \tag{5.78}$$

we need to find the rank-$2r$ matrix $\underline{\mathbf{W}}_r\underline{\mathbf{R}}_{yy}^{1/2}$ that is closest to $\underline{\mathbf{W}}\,\underline{\mathbf{R}}_{yy}^{1/2} = \underline{\mathbf{R}}_{xy}\underline{\mathbf{R}}_{yy}^{-\mathrm{H}/2}$, where closeness is measured by the trace norm $\|\mathbf{X}\|^2 = \mathrm{tr}(\mathbf{X}\mathbf{X}^{\mathrm{H}})$. It is a classical result that the reduced-rank matrix closest to a given matrix, as measured by any unitarily invariant norm including the trace norm, is obtained via the SVD (see, e.g., Marshall and Olkin (1979)). That is, with the SVD $\underline{\mathbf{C}} = \underline{\mathbf{R}}_{xy}\underline{\mathbf{R}}_{yy}^{-\mathrm{H}/2} = \underline{\mathbf{F}}\,\underline{\mathbf{K}}\,\underline{\mathbf{G}}^{\mathrm{H}}$ the best rank-$2r$ approximation is

$$\underline{\mathbf{W}}_r\underline{\mathbf{R}}_{yy}^{1/2} = \underline{\mathbf{F}}\,\underline{\mathbf{K}}_r\underline{\mathbf{G}}^{\mathrm{H}} \tag{5.79}$$

and therefore

$$\underline{\mathbf{W}}_r = \underline{\mathbf{F}}\,\underline{\mathbf{K}}_r\underline{\mathbf{G}}^{\mathrm{H}}\underline{\mathbf{R}}_{yy}^{-1/2}. \tag{5.80}$$

This confirms our proposition (5.77).

Result 5.7. *Optimum rank reduction for WLMMSE estimation is performed in total half-canonical coordinates.*

The reduced-rank WLMMSE can be expressed as the sum of the full-rank WLMMSE and the WLMMSE due to rank reduction:

$$E\|\hat{\mathbf{x}}_r-\mathbf{x}\|^2 = \tfrac{1}{2}\,\mathrm{tr}\,\underline{\mathbf{Q}}_r = \left(\mathrm{tr}\,\mathbf{R}_{xx} - \tfrac{1}{2}\sum_{i=1}^{2p} k_i^2\right) + \tfrac{1}{2}\sum_{i=2(r+1)}^{2p} k_i^2. \tag{5.81}$$

Rank reduction, or data compaction, is therefore most effective if much of the total correlation is concentrated in a few half-canonical correlations $\{k_i\}_{i=1}^{2r}$. In Section 4.4, we introduced the *correlation spread* as a single, normalized measure of the effectiveness of rank reduction.

There are analogous suboptimum stories for strictly linear estimation and conjugate linear estimation, which utilize rotational and reflectional half-canonical correlations, respectively, in place of total half-canonical correlations. For any given rank r, strictly linear and conjugate linear estimators can never outperform widely linear estimators.

5.5.2 Maximize mutual information (min-det problem)

Mutual information has different invariance properties than the mean-squared error. The full-rank mutual information between jointly Gaussian $\mathbf{x}$ and $\mathbf{y}$ is $I(\mathbf{x};\mathbf{y}) = H(\mathbf{x}) - H(\mathbf{x}|\mathbf{y})$. The differential entropy of the signal is

$$H(\mathbf{x}) = \tfrac{1}{2}\log[(\pi e)^{2n}\det \underline{\mathbf{R}}_{xx}] \tag{5.82}$$

and the conditional differential entropy of the signal given the measurement is

$$H(\mathbf{x}|\mathbf{y}) = H(\mathbf{e}) = \tfrac{1}{2}\log[(\pi e)^{2n}\det \underline{\mathbf{Q}}]. \tag{5.83}$$

Thus, the mutual information is

$$I(\mathbf{x};\mathbf{y}) = H(\mathbf{x}) - H(\mathbf{x}|\mathbf{y}) = -\tfrac{1}{2}\log\left(\frac{\det \underline{\mathbf{Q}}}{\det \underline{\mathbf{R}}_{xx}}\right). \tag{5.84}$$

It is easy to see that $I(\mathbf{x};\mathbf{y})$ is invariant under widely linear transformation of the signal $\mathbf{x}$ and the measurement $\mathbf{y}$. Hence, it must be a function of the *total canonical correlations* $\{k_i\}$ between $\mathbf{x}$ and $\mathbf{y}$. As discussed in Section 4.1.4, the total canonical correlations are obtained by the SVD of the augmented coherence matrix $\underline{\mathbf{C}}$,

$$\underline{\mathbf{C}} = \underline{\mathbf{R}}_{xx}^{-1/2}\underline{\mathbf{R}}_{xy}\underline{\mathbf{R}}_{yy}^{-\mathrm{H}/2} = \underline{\mathbf{F}}\underline{\mathbf{K}}\underline{\mathbf{G}}^{\mathrm{H}}, \tag{5.85}$$

with $\underline{\mathbf{F}}^{\mathrm{H}}\underline{\mathbf{F}} = \mathbf{I}$, $\underline{\mathbf{F}} \in \mathcal{W}^{n\times p}$, and $\underline{\mathbf{G}}^{\mathrm{H}}\underline{\mathbf{G}} = \mathbf{I}$, $\underline{\mathbf{G}} \in \mathcal{W}^{m\times p}$, and $\underline{\mathbf{K}}$ defined as in the min-trace problem above. We assume the ordering $1 \geq k_1 \geq k_2 \geq \cdots \geq k_{2p} \geq 0$. The mutual information can thus be written in terms of canonical correlations as

$$\begin{aligned} I(\mathbf{x};\mathbf{y}) &= -\tfrac{1}{2}\log\left(\frac{\det \underline{\mathbf{R}}_{xx}\det(\mathbf{I} - \underline{\mathbf{R}}_{xx}^{-1/2}\underline{\mathbf{R}}_{xy}\underline{\mathbf{R}}_{yy}^{-1}\underline{\mathbf{R}}_{xy}^{\mathrm{H}}\underline{\mathbf{R}}_{xx}^{-\mathrm{H}/2})}{\det \underline{\mathbf{R}}_{xx}}\right) \\ &= -\tfrac{1}{2}\log\det(\mathbf{I} - \underline{\mathbf{K}}^2) = -\tfrac{1}{2}\log\left(\prod_{i=1}^{2p}(1-k_i^2)\right) = -\tfrac{1}{2}\sum_{i=1}^{2p}\log(1-k_i^2). \end{aligned} \tag{5.86}$$

We may also express the WLMMSE estimator in canonical coordinates as

$$\begin{aligned} \underline{\hat{\mathbf{x}}} &= \underline{\mathbf{W}}\,\underline{\mathbf{y}} = \underline{\mathbf{R}}_{xy}\underline{\mathbf{R}}_{yy}^{-1}\underline{\mathbf{y}} \\ &= \underline{\mathbf{R}}_{xx}^{1/2}\underline{\mathbf{R}}_{xx}^{-1/2}\underline{\mathbf{R}}_{xy}\underline{\mathbf{R}}_{yy}^{-\mathrm{H}/2}\underline{\mathbf{R}}_{yy}^{-1/2}\underline{\mathbf{y}} \\ &= \underline{\mathbf{R}}_{xx}^{1/2}\underline{\mathbf{F}}\underline{\mathbf{K}}\underline{\mathbf{G}}^{\mathrm{H}}\underline{\mathbf{R}}_{yy}^{-1/2}\underline{\mathbf{y}}. \end{aligned} \tag{5.87}$$

As in the previous section, this breaks up the estimator into three blocks. However, this time we use a *canonical coordinate system* rather than a *half-canonical coordinate system*. We first transform the measurement $\underline{\mathbf{y}}$ into white and proper canonical measurement coordinates $\underline{\boldsymbol{\omega}} = \underline{\mathbf{B}}\,\underline{\mathbf{y}}$ using the *coder* $\underline{\mathbf{B}} = \underline{\mathbf{G}}^{\mathrm{H}}\underline{\mathbf{R}}_{yy}^{-1/2}$. We then estimate the signal in canonical coordinates as $\hat{\underline{\boldsymbol{\xi}}} = \underline{\mathbf{K}}\,\underline{\boldsymbol{\omega}}$. While the estimator $\hat{\underline{\boldsymbol{\xi}}}$ has mutually uncorrelated components but is generally *not white or proper*, the canonical signal $\underline{\boldsymbol{\xi}}$ is white and proper. Finally, the estimate $\hat{\underline{\mathbf{x}}}$ is produced by passing $\hat{\underline{\boldsymbol{\xi}}}$ through the *decoder* $\underline{\mathbf{A}}^{-1} = \underline{\mathbf{F}}\,\underline{\mathbf{R}}_{xx}^{1/2}$ (where the inverse is a *right* inverse).

We now ask how to find a rank-r widely linear estimator that provides as much mutual information about the signal as possible. Looking at (5.86), we are inclined to discard the $2(p-r)$ smallest canonical correlations. We would thus construct the rank-r estimator

$$\hat{\underline{\mathbf{x}}}_r = \underline{\mathbf{W}}_r\underline{\mathbf{y}} = \underline{\mathbf{R}}_{xx}^{1/2}\underline{\mathbf{F}}\,\underline{\mathbf{K}}_r\underline{\mathbf{G}}^{\mathrm{H}}\underline{\mathbf{R}}_{yy}^{-1/2}\underline{\mathbf{y}}, \tag{5.88}$$

which is similar to the solution of the min-trace problem (5.77) except that it uses a canonical coordinate system in place of a half-canonical coordinate system. The rank-r matrix $\underline{\mathbf{K}}_r$ is obtained from $\underline{\mathbf{K}}$ by replacing the $2(p-r)$ smallest canonical correlations with zeros.

As in the min-trace case, our intuition is correct but it still requires a proof that the reduced-rank maximum mutual information estimator does indeed use the same canonical coordinate system as the full-rank estimator. The proof given below follows Hua *et al.* (2001).

Starting from (5.60), maximizing mutual information means minimizing

$$\begin{aligned}\det \underline{\mathbf{Q}}_r &= \det(\underline{\mathbf{Q}} + (\underline{\mathbf{W}} - \underline{\mathbf{W}}_r)\underline{\mathbf{R}}_{yy}(\underline{\mathbf{W}} - \underline{\mathbf{W}}_r)^{\mathrm{H}}) \\ &= \det \underline{\mathbf{R}}_{xx} \det\left(\mathbf{I} - \underline{\mathbf{C}}\,\underline{\mathbf{C}}^{\mathrm{H}} + (\underline{\mathbf{C}} - \underline{\mathbf{R}}_{xx}^{-1/2}\underline{\mathbf{W}}_r\underline{\mathbf{R}}_{yy}^{1/2})(\underline{\mathbf{C}} - \underline{\mathbf{R}}_{xx}^{-1/2}\underline{\mathbf{W}}_r\underline{\mathbf{R}}_{yy}^{1/2})^{\mathrm{H}}\right).\end{aligned} \tag{5.89}$$

For notational convenience, let

$$\underline{\mathbf{X}}_r = \underline{\mathbf{C}} - \underline{\mathbf{R}}_{xx}^{-1/2}\underline{\mathbf{W}}_r\underline{\mathbf{R}}_{yy}^{1/2}. \tag{5.90}$$

The minimum $\det \underline{\mathbf{Q}}_r$ is zero if there is at least one canonical correlation k_i equal to 1. We may thus assume that all canonical correlations are strictly less than 1, which allows us to write

$$\begin{aligned}&\det \underline{\mathbf{R}}_{xx} \det\left(\mathbf{I} - \underline{\mathbf{C}}\,\underline{\mathbf{C}}^{\mathrm{H}} + \underline{\mathbf{X}}_r\underline{\mathbf{X}}_r^{\mathrm{H}}\right) \\ &= \det \underline{\mathbf{R}}_{xx} \det(\mathbf{I} - \underline{\mathbf{C}}\,\underline{\mathbf{C}}^{\mathrm{H}})\det\left(\mathbf{I} + (\mathbf{I} - \underline{\mathbf{C}}\,\underline{\mathbf{C}}^{\mathrm{H}})^{-1/2}\underline{\mathbf{X}}_r\underline{\mathbf{X}}_r^{\mathrm{H}}(\mathbf{I} - \underline{\mathbf{C}}\,\underline{\mathbf{C}}^{\mathrm{H}})^{-\mathrm{H}/2}\right).\end{aligned} \tag{5.91}$$

Since $\det \underline{\mathbf{R}}_{xx}$ and $\det(\mathbf{I} - \underline{\mathbf{C}}\,\underline{\mathbf{C}}^{\mathrm{H}})$ are independent of $\underline{\mathbf{W}}_r$, they can be disregarded in the minimization. The third determinant in (5.91) is of the form

$$\det(\mathbf{I} + \underline{\mathbf{Y}}\,\underline{\mathbf{Y}}^{\mathrm{H}}) = \prod_{i=1}^{2p}(1 + \sigma_i^2), \tag{5.92}$$

where $\{\sigma_i\}_{i=1}^{2p}$ are the singular values of $\underline{\mathbf{Y}} = (\mathbf{I} - \underline{\mathbf{C}}\,\underline{\mathbf{C}}^{\mathrm{H}})^{-1/2}\underline{\mathbf{X}}_r$. This amounts to minimizing $\|\underline{\mathbf{Y}}\|$ with respect to an arbitrary unitarily invariant norm. In other words, we need to find the rank-$2r$ matrix $(\mathbf{I} - \underline{\mathbf{C}}\,\underline{\mathbf{C}}^{\mathrm{H}})^{-1/2}\underline{\mathbf{R}}_{xx}^{-1/2}\underline{\mathbf{W}}_r\underline{\mathbf{R}}_{yy}^{1/2}$ closest to $(\mathbf{I} - \underline{\mathbf{C}}\,\underline{\mathbf{C}}^{\mathrm{H}})^{-1/2}\underline{\mathbf{C}}$. This is achieved by a rank-$2r$ SVD of $(\mathbf{I} - \underline{\mathbf{C}}\,\underline{\mathbf{C}}^{\mathrm{H}})^{-1/2}\underline{\mathbf{C}}$. However, because the eigenvectors of $(\mathbf{I} - \underline{\mathbf{C}}\,\underline{\mathbf{C}}^{\mathrm{H}})^{-1/2}$ are the left singular vectors of $\underline{\mathbf{C}}$, a simpler solution is to use directly the SVD of the augmented coherence matrix $\underline{\mathbf{C}} = \underline{\mathbf{R}}_{xx}^{-1/2}\underline{\mathbf{R}}_{xy}\underline{\mathbf{R}}_{yy}^{-\mathrm{H}/2} = \underline{\mathbf{F}}\,\underline{\mathbf{K}}\,\underline{\mathbf{G}}^{\mathrm{H}}$. The best rank-$2r$ approximation is then

$$\underline{\mathbf{R}}_{xx}^{-1/2}\underline{\mathbf{W}}_r\underline{\mathbf{R}}_{yy}^{1/2} = \underline{\mathbf{F}}\,\underline{\mathbf{K}}_r\underline{\mathbf{G}}^{\mathrm{H}} \tag{5.93}$$

and thus

$$\underline{\mathbf{W}}_r = \underline{\mathbf{R}}_{xx}^{1/2}\underline{\mathbf{F}}\,\underline{\mathbf{K}}_r\underline{\mathbf{G}}^{\mathrm{H}}\underline{\mathbf{R}}_{yy}^{-1/2}. \tag{5.94}$$

This confirms our initial claim.

Result 5.8. *For Gaussian data, optimum rank reduction for widely linear maximum information rate estimation is performed in total canonical coordinates.*

The reduced-rank information rate is

$$I_r(\mathbf{x};\mathbf{y}) = -\tfrac{1}{2}\sum_{i=1}^{2r}\log(1-k_i^2), \tag{5.95}$$

and the loss of information rate due to rank reduction is

$$I(\mathbf{x};\mathbf{y}) - I_r(\mathbf{x};\mathbf{y}) = -\tfrac{1}{2}\sum_{i=2(r+1)}^{2p}\log(1-k_i^2). \tag{5.96}$$

Two remarks made in the min-trace case hold for the min-det case as well. Firstly, rank reduction is most effective if much of the total correlation is concentrated in a few canonical correlations $\{k_i\}_{i=1}^{2r}$. A good measure for the degree of concentration is the correlation spread introduced in Section 4.4. Secondly, there are completely analogous, suboptimum, developments for strictly linear min-det estimation and conjugate linear min-det estimation. These utilize rotational and reflectional canonical correlations, respectively, in place of total canonical correlations. For any given rank, widely linear min-det estimation is at least as good as strictly linear or conjugate linear min-det estimation, but superior if the signals are improper.

5.6 Linear and widely linear minimum-variance distortionless response estimators

In the engineering literature, especially in the fields of radar, sonar, and wireless communication, it is commonplace to design a *linear minimum-variance unbiased* or *linear minimum variance-distortionless response* (LMVDR) estimator (or receiver). Actually, it would be more accurate to call these receivers *best linear unbiased estimators* (BLUEs), which conforms to standard language in the statistics literature. In this section, we will first review the design and performance of the LMVDR receiver using Hermitian covariances only, and then we shall extend our results to *widely linear minimum-variance distortionless response* (WLMVDR) receivers by accounting for complementary covariances.

5.6.1 Rank-one LMVDR receiver

We shall begin by assuming the signal component of the received signal to lie in a known one-dimensional subspace of $\mathbb{C}^m$, denoted $\langle \boldsymbol{\psi} \rangle$. We will later generalize to p-dimensional subspaces $\langle \boldsymbol{\Psi} \rangle$. Without loss of generality we may assume $\boldsymbol{\psi}^{\mathrm{H}}\boldsymbol{\psi} = 1$, which is to say that $\boldsymbol{\psi}$ is a unit vector. Thus the measurement model for the received signal $\mathbf{y}: \Omega \longrightarrow \mathbb{C}^m$ is

$$\mathbf{y} = \boldsymbol{\psi} x + \mathbf{n}, \tag{5.97}$$

where the complex scalar x is unknown and to be estimated, but modeled as deterministic, i.e., x is not assigned a probability distribution. The additive noise $\mathbf{n}$ is zero-mean with Hermitian covariance matrix $\mathbf{R}_{nn} = E[\mathbf{n}\mathbf{n}^{\mathrm{H}}]$. The problem is to design a linear estimator of x that is unbiased with variance smaller than any unbiased competitor. A first guess would be the *matched filter*

$$\hat{x} = \boldsymbol{\psi}^{\mathrm{H}}\mathbf{y}, \tag{5.98}$$

which has mean $E[\hat{x}] = x$ and variance

$$E[(\hat{x} - x)(\hat{x} - x)^*] = \boldsymbol{\psi}^{\mathrm{H}}\mathbf{R}_{nn}\boldsymbol{\psi}. \tag{5.99}$$

This solution is linear and unbiased, but, as we now show, not minimum variance. To design the LMVDR receiver $\hat{x} = \mathbf{w}^{\mathrm{H}}\mathbf{y}$, let's minimize its variance, under the constraint that its mean be x:

$$\min \mathbf{w}^{\mathrm{H}}\mathbf{R}_{nn}\mathbf{w} \quad \text{under constraint } \mathbf{w}^{\mathrm{H}}\boldsymbol{\psi} = 1. \tag{5.100}$$

It is a simple matter to set up a Lagrangian, solve for the set of candidate solutions, and enforce the constraint. This leads to the LMVDR receiver

$$\hat{x} = \frac{\boldsymbol{\psi}^{\mathrm{H}}\mathbf{R}_{nn}^{-1}}{\boldsymbol{\psi}^{\mathrm{H}}\mathbf{R}_{nn}^{-1}\boldsymbol{\psi}}\mathbf{y}, \tag{5.101}$$

which, for obvious reasons, is sometimes also called a *whitened matched filter*. This solution too has mean $E[\hat{x}] = x$, but its variance is smaller:

$$Q = E[(\hat{x} - x)(\hat{x} - x)^*] = \frac{1}{\boldsymbol{\psi}^{\mathrm{H}}\mathbf{R}_{nn}^{-1}\boldsymbol{\psi}}. \tag{5.102}$$

The inequality $(\boldsymbol{\psi}^{\mathrm{H}}\mathbf{R}_{nn}^{-1}\boldsymbol{\psi})^{-1} \leq \boldsymbol{\psi}^{\mathrm{H}}\mathbf{R}_{nn}\boldsymbol{\psi}$ is a special case of the Kantorovich inequality. The LMVDR or BLUE estimator is also maximum likelihood when the measurement noise $\mathbf{n}$ is zero-mean Gaussian with covariance $\mathbf{R}_{nn}$, in which case the measurement $\mathbf{y}$ is Gaussian with mean $\boldsymbol{\psi}x$ and covariance $\mathbf{R}_{nn}$.

Relation to LMMSE estimator

If the unknown variable x is modeled as in Example 5.2 as a random variable that is uncorrelated with $\mathbf{n}$ and has zero mean and variance σ_x^2, then its LMMSE estimator is

$$\hat{x} = \frac{\boldsymbol{\psi}^{\mathrm{H}}\mathbf{R}_{nn}^{-1}}{\sigma_x^{-2} + \boldsymbol{\psi}^{\mathrm{H}}\mathbf{R}_{nn}^{-1}\boldsymbol{\psi}}\mathbf{y}. \tag{5.103}$$

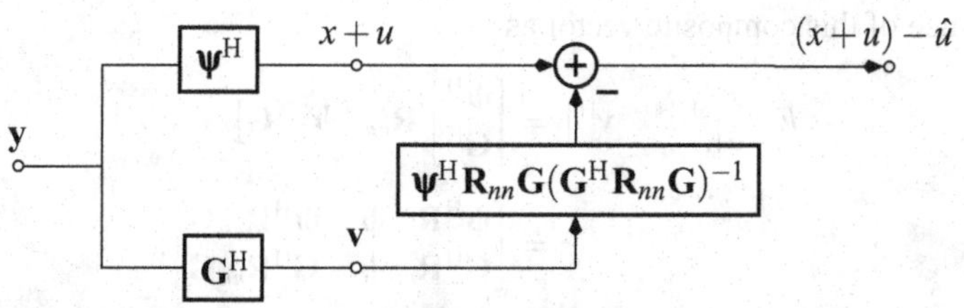

Figure 5.5 Rank-one generalized sidelobe canceler.

This estimator is conditionally biased, but unconditionally unbiased. Its mean-squared error is $(\sigma_x^{-2} + \boldsymbol{\psi}^{\mathrm{H}}\mathbf{R}_{nn}^{-1}\boldsymbol{\psi})^{-1}$. If nothing is known a priori about the scale of x (i.e., its variance σ_x^2), then the LMVDR estimator

$$\hat{x} = \frac{\boldsymbol{\psi}^{\mathrm{H}}\mathbf{R}_{nn}^{-1}}{\boldsymbol{\psi}^{\mathrm{H}}\mathbf{R}_{nn}^{-1}\boldsymbol{\psi}}\mathbf{y} \tag{5.104}$$

might be used as an approximation to the LMMSE estimator. It is conditionally unbiased and unconditionally unbiased, so its mean-squared error is the average of its conditional variance, which, when averaged over the distribution of x, remains unchanged at $(\boldsymbol{\psi}^{\mathrm{H}}\mathbf{R}_{nn}^{-1}\boldsymbol{\psi})^{-1}$. Obviously the LMVDR approximation to the LMMSE estimator has worse mean-squared error than the LMMSE estimator. So there is a tradeoff between minimum mean-squared error and conditional unbiasedness. As a practical matter, in many problems of communication, radar, sonar, geophysics, and radio astronomy, the scale of x is unknown and the LMVDR approximation is routinely used, with the virtue that it is conditionally and unconditionally unbiased and its mean-squared error is near the minimum achievable at high SNRs.

5.6.2 Generalized sidelobe canceler

The LMVDR estimator can also be derived by considering the generalized sidelobe canceler (GSC) structure in Fig. 5.5. The top branch of the figure is matched to the signal subspace $\langle\boldsymbol{\psi}\rangle$ and the bottom branch is mismatched, which is to say that $\mathbf{G}^{\mathrm{H}}\boldsymbol{\psi} = \mathbf{0}$ for $\mathbf{G} \in \mathbb{C}^{m\times(m-1)}$. Moreover, the matrix $\mathbf{U} = [\boldsymbol{\psi}, \mathbf{G}]$ is a unitary matrix, meaning that $\boldsymbol{\psi}^{\mathrm{H}}\boldsymbol{\psi} = 1$, $\mathbf{G}^{\mathrm{H}}\boldsymbol{\psi} = \mathbf{0}$, and $\mathbf{G}^{\mathrm{H}}\mathbf{G} = \mathbf{I}$. When the Hermitian transpose of this matrix is applied to the measurement $\mathbf{y}$, the result is

$$\mathbf{U}^{\mathrm{H}}\mathbf{y} = \begin{bmatrix}\boldsymbol{\psi}^{\mathrm{H}}\\ \mathbf{G}^{\mathrm{H}}\end{bmatrix}\mathbf{y} = \begin{bmatrix}x + \boldsymbol{\psi}^{\mathrm{H}}\mathbf{n}\\ \mathbf{G}^{\mathrm{H}}\mathbf{n}\end{bmatrix} = \begin{bmatrix}x + u\\ \mathbf{v}\end{bmatrix}. \tag{5.105}$$

The output of the top branch is $x + u$, the unbiased matched filter output, with variance $\boldsymbol{\psi}^{\mathrm{H}}\mathbf{R}_{nn}\boldsymbol{\psi}$, and the output of the GSC in the bottom branch is $\mathbf{v} = \mathbf{G}^{\mathrm{H}}\mathbf{y}$. It contains no signal, but is correlated with the noise variable $u = \boldsymbol{\psi}^{\mathrm{H}}\mathbf{n}$ in the output of the top branch. Therefore it may be used to estimate u, and this estimate may be subtracted from the top branch to reduce variance. This is the essence of the LMVDR receiver.

Let us implement this program, using the formalism we have developed in our discussion of LMMSE estimators. We organize u and $\mathbf{v}$ into a composite vector and compute the covariance of this composite vector as

$$E\begin{bmatrix} u \\ \mathbf{v} \end{bmatrix}\begin{bmatrix} u^* & \mathbf{v}^{\mathrm{H}} \end{bmatrix} = \begin{bmatrix} \boldsymbol{\psi}^{\mathrm{H}} \\ \mathbf{G}^{\mathrm{H}} \end{bmatrix} \mathbf{R}_{nn} \begin{bmatrix} \boldsymbol{\psi} & \mathbf{G} \end{bmatrix}$$
$$= \begin{bmatrix} \boldsymbol{\psi}^{\mathrm{H}}\mathbf{R}_{nn}\boldsymbol{\psi} & \boldsymbol{\psi}^{\mathrm{H}}\mathbf{R}_{nn}\mathbf{G} \\ \mathbf{G}^{\mathrm{H}}\mathbf{R}_{nn}\boldsymbol{\psi} & \mathbf{G}^{\mathrm{H}}\mathbf{R}_{nn}\mathbf{G} \end{bmatrix}. \tag{5.106}$$

From the unitarity of the matrix $[\boldsymbol{\psi}, \mathbf{G}]$, it follows that the inverse of this composite covariance matrix is

$$\begin{bmatrix} \boldsymbol{\psi}^{\mathrm{H}} \\ \mathbf{G}^{\mathrm{H}} \end{bmatrix} \mathbf{R}_{nn}^{-1} \begin{bmatrix} \boldsymbol{\psi} & \mathbf{G} \end{bmatrix} = \begin{bmatrix} \boldsymbol{\psi}^{\mathrm{H}}\mathbf{R}_{nn}^{-1}\boldsymbol{\psi} & \boldsymbol{\psi}^{\mathrm{H}}\mathbf{R}_{nn}^{-1}\mathbf{G} \\ \mathbf{G}^{\mathrm{H}}\mathbf{R}_{nn}^{-1}\boldsymbol{\psi} & \mathbf{G}^{\mathrm{H}}\mathbf{R}_{nn}^{-1}\mathbf{G} \end{bmatrix}. \tag{5.107}$$

Recall that the northwest entry of this matrix is the inverse of the error variance for estimating u from $\mathbf{v}$:

$$E[(\hat{u} - u)(\hat{u} - u)^*] = \frac{1}{\boldsymbol{\psi}^{\mathrm{H}}\mathbf{R}_{nn}^{-1}\boldsymbol{\psi}}. \tag{5.108}$$

This result establishes that the variance in estimating u from $\mathbf{v}$ is the same as the variance in estimating x from $\mathbf{y}$. Let us now show that it is also the variance in estimating x from $\mathbf{v}$, thus establishing that the GSC is indeed an implementation of the LMVDR or BLUE estimator. The LMMSE estimator of u from $\mathbf{v}$ may be read out of the composite covariance matrix for u and $\mathbf{v}$:

$$\hat{u} = \boldsymbol{\psi}^{\mathrm{H}}\mathbf{R}_{nn}\mathbf{G}(\mathbf{G}^{\mathrm{H}}\mathbf{R}_{nn}\mathbf{G})^{-1}\mathbf{v}. \tag{5.109}$$

The resulting LMVDR estimator of x from $\mathbf{y}$, illustrated in Fig. 5.5, is the difference between $x + u$ and $\hat{u}$, namely $\hat{x} = x + (u - \hat{u}) = (x + u) - \hat{u}$:

$$\hat{x} = \boldsymbol{\psi}^{\mathrm{H}}[\mathbf{I} - \mathbf{R}_{nn}\mathbf{G}(\mathbf{G}^{\mathrm{H}}\mathbf{R}_{nn}\mathbf{G})^{-1}\mathbf{G}^{\mathrm{H}}]\mathbf{y}. \tag{5.110}$$

From the orthogonality of $\boldsymbol{\psi}$ and $\mathbf{G}$, it follows that this estimator is unbiased. Moreover, its variance is the mean-squared error between $\hat{u}$ and u:

$$Q = E[(\hat{x} - x)(\hat{x} - x)^*] = E[(\hat{u} - u)(\hat{u} - u)^*]$$
$$= \boldsymbol{\psi}^{\mathrm{H}}\mathbf{R}_{nn}\boldsymbol{\psi} - \boldsymbol{\psi}^{\mathrm{H}}\mathbf{R}_{nn}\mathbf{G}(\mathbf{G}^{\mathrm{H}}\mathbf{R}_{nn}\mathbf{G})^{-1}\mathbf{G}^{\mathrm{H}}\mathbf{R}_{nn}\boldsymbol{\psi} = \mathbf{h}^{\mathrm{H}}(\mathbf{I} - \mathbf{P}_{\mathbf{F}})\mathbf{h}. \tag{5.111}$$

Here, $\mathbf{h} = \mathbf{R}_{nn}^{\mathrm{H}/2}\boldsymbol{\psi}$, $\mathbf{F} = \mathbf{R}_{nn}^{\mathrm{H}/2}\mathbf{G}$, and $\mathbf{P}_{\mathbf{F}} = \mathbf{F}(\mathbf{F}^{\mathrm{H}}\mathbf{F})^{-1}\mathbf{F}^{\mathrm{H}}$ is the orthogonal projection onto the subspace $\langle\mathbf{F}\rangle$. This result produces the following identity for the variance of the LMVDR receiver:

$$Q = \frac{1}{\boldsymbol{\psi}^{\mathrm{H}}\mathbf{R}_{nn}^{-1}\boldsymbol{\psi}} = \mathbf{h}^{\mathrm{H}}(\mathbf{I} - \mathbf{P}_{\mathbf{F}})\mathbf{h} = \mathbf{h}^{\mathrm{H}}\mathbf{h} - \mathbf{h}^{\mathrm{H}}\mathbf{P}_{\mathbf{F}}\mathbf{h}. \tag{5.112}$$

Actually, this equality is a decomposition of the variance of u, namely $\mathbf{h}^{\mathrm{H}}\mathbf{h} = \boldsymbol{\psi}^{\mathrm{H}}\mathbf{R}_{nn}\boldsymbol{\psi}$, into the sum of the variance for $(\hat{u} - u)$, namely $(\boldsymbol{\psi}^{\mathrm{H}}\mathbf{R}_{nn}^{-1}\boldsymbol{\psi})^{-1}$, plus the variance of $\hat{u}$, namely $\mathbf{h}^{\mathrm{H}}\mathbf{P}_{\mathbf{F}}\mathbf{h} = \boldsymbol{\psi}^{\mathrm{H}}\mathbf{R}_{nn}\mathbf{G}(\mathbf{G}^{\mathrm{H}}\mathbf{R}_{nn}\mathbf{G})^{-1}\mathbf{G}^{\mathrm{H}}\mathbf{R}_{nn}\boldsymbol{\psi}$. This decomposition comes from the orthogonal decomposition of u into $\hat{u} - (\hat{u} - u)$.

The virtue of the identity (5.112) for the variance of the LMVDR receiver is that it shows the variance to depend on the angle between the subspaces $\langle\mathbf{h}\rangle$ and $\langle\mathbf{F}\rangle$, as

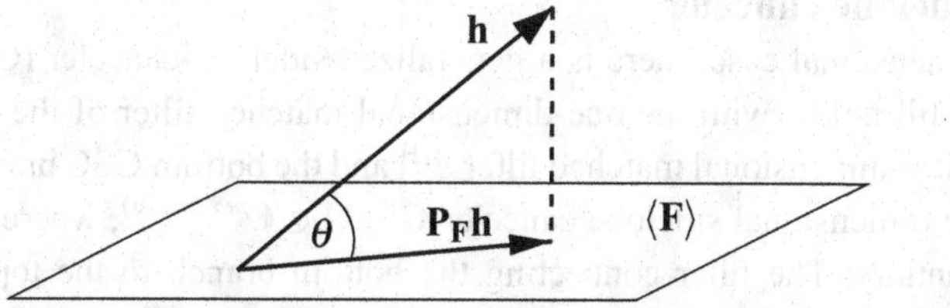

Figure 5.6 The performance gain of the LMVDR estimator over a matched filter depends on the angle between $\langle \mathbf{h} \rangle$ and $\langle \mathbf{F} \rangle$.

illustrated in Fig. 5.6. In fact, the gain of the LMVDR receiver over the matched filter receiver is

$$\frac{\boldsymbol{\psi}^{\mathrm{H}}\mathbf{R}_{nn}\boldsymbol{\psi}}{1/(\boldsymbol{\psi}^{\mathrm{H}}\mathbf{R}_{nn}^{-1}\boldsymbol{\psi})} = \frac{\mathbf{h}^{\mathrm{H}}\mathbf{h}}{\mathbf{h}^{\mathrm{H}}(\mathbf{I}-\mathbf{P}_{\mathrm{F}})\mathbf{h}} = \frac{1}{\sin^2\theta}, \tag{5.113}$$

where θ is the angle between the colored subspaces $\langle \mathbf{h} \rangle$ and $\langle \mathbf{F} \rangle$. When these two subspaces are close in angle, then $\sin^2\theta$ is small and the gain is large, meaning that the bottom branch of the GSC can estimate and cancel out the noise component u in the top branch, thus reducing the variance of the error $\hat{u} - u = \hat{x} - x$.

5.6.3 Multi-rank LMVDR receiver

These results are easily extended to the multi-rank LMVDR receiver, wherein the measurement model is

$$\mathbf{y} = \boldsymbol{\Psi}\mathbf{x} + \mathbf{n}. \tag{5.114}$$

In this model, the matrix $\boldsymbol{\Psi} = [\boldsymbol{\psi}_1, \boldsymbol{\psi}_2, \ldots, \boldsymbol{\psi}_p] \in \mathbb{C}^{m\times p}$ consists of p modes and the vector $\mathbf{x} = [x_1, x_2, \ldots, x_p]$ consists of p complex amplitudes. Without loss of generality we assume $\boldsymbol{\Psi}^{\mathrm{H}}\boldsymbol{\Psi} = \mathbf{I}$. The multi-rank matched filter estimator of $\mathbf{x}$ is

$$\hat{\mathbf{x}} = \boldsymbol{\Psi}^{\mathrm{H}}\mathbf{y} \tag{5.115}$$

with mean $\mathbf{x}$ and error covariance matrix $E[(\hat{\mathbf{x}} - \mathbf{x})(\hat{\mathbf{x}} - \mathbf{x})^{\mathrm{H}}] = \boldsymbol{\Psi}^{\mathrm{H}}\mathbf{R}_{nn}\boldsymbol{\Psi}$. The LMVDR estimator $\hat{\mathbf{x}} = \mathbf{W}^{\mathrm{H}}\mathbf{y}$, with $\mathbf{W} \in \mathbb{C}^{m\times p}$, is derived by minimizing the trace of the error covariance under an unbiasedness constraint:

$$\min \operatorname{tr}[\mathbf{W}^{\mathrm{H}}\mathbf{R}_{nn}\mathbf{W}] \quad \text{under constraint } \mathbf{W}^{\mathrm{H}}\boldsymbol{\Psi} = \mathbf{I}. \tag{5.116}$$

The solution is

$$\hat{\mathbf{x}} = (\boldsymbol{\Psi}^{\mathrm{H}}\mathbf{R}_{nn}^{-1}\boldsymbol{\Psi})^{-1}\boldsymbol{\Psi}^{\mathrm{H}}\mathbf{R}_{nn}^{-1}\mathbf{y} \tag{5.117}$$

with mean $\mathbf{x}$ and error covariance matrix

$$\mathbf{Q} = E[(\hat{\mathbf{x}} - \mathbf{x})(\hat{\mathbf{x}} - \mathbf{x})^{\mathrm{H}}] = (\boldsymbol{\Psi}^{\mathrm{H}}\mathbf{R}_{nn}^{-1}\boldsymbol{\Psi})^{-1}. \tag{5.118}$$

The inequality $(\boldsymbol{\Psi}^{\mathrm{H}}\mathbf{R}_{nn}^{-1}\boldsymbol{\Psi})^{-1} \le \boldsymbol{\Psi}^{\mathrm{H}}\mathbf{R}_{nn}\boldsymbol{\Psi}$ is again a special case of the Kantorovich inequality.

Generalized sidelobe canceler

As in the one-dimensional case, there is a generalized sidelobe-canceler (GSC) structure. Figure 5.5 still holds, with the one-dimensional matched filter of the top branch relabeled with the p-dimensional matched filter $\boldsymbol{\Psi}^{\mathrm{H}}$ and the bottom GSC branch labeled with the $(m-p)$-dimensional sidelobe canceler $\mathbf{G}^{\mathrm{H}}$, $\mathbf{G} \in \mathbb{C}^{m\times(m-p)}$, where the matrix $\mathbf{U} = [\boldsymbol{\Psi}, \mathbf{G}]$ is unitary. The filter connecting the bottom branch to the top branch is $\boldsymbol{\Psi}^{\mathrm{H}}\mathbf{R}_{nn}\mathbf{G}(\mathbf{G}^{\mathrm{H}}\mathbf{R}_{nn}\mathbf{G})^{-1}\mathbf{v}$ and the resulting LMVDR estimator of $\mathbf{x}$ is the difference between $\mathbf{x}+\mathbf{u}$ and $\hat{\mathbf{u}}$, namely $\hat{\mathbf{x}} = \mathbf{x} + (\mathbf{u} - \hat{\mathbf{u}}) = (\mathbf{x}+\mathbf{u}) - \hat{\mathbf{u}}$:

$$\hat{\mathbf{x}} = \boldsymbol{\Psi}^{\mathrm{H}}[\mathbf{I} - \mathbf{R}_{nn}\mathbf{G}(\mathbf{G}^{\mathrm{H}}\mathbf{R}_{nn}\mathbf{G})^{-1}\mathbf{G}^{\mathrm{H}}]\mathbf{y}. \tag{5.119}$$

From the orthogonality of $\boldsymbol{\Psi}$ and $\mathbf{G}$, it follows that this estimator is unbiased. Moreover, its variance is the mean-squared error between $\mathbf{u}$ and $\hat{\mathbf{u}}$:

$$\begin{aligned}\mathbf{Q} &= E[(\hat{\mathbf{x}}-\mathbf{x})(\hat{\mathbf{x}}-\mathbf{x})^{\mathrm{H}}] = E[(\hat{\mathbf{u}}-\mathbf{u})(\hat{\mathbf{u}}-\mathbf{u})^{\mathrm{H}}] \\ &= \boldsymbol{\Psi}^{\mathrm{H}}\mathbf{R}_{nn}\boldsymbol{\Psi} - \boldsymbol{\Psi}^{\mathrm{H}}\mathbf{R}_{nn}\mathbf{G}(\mathbf{G}^{\mathrm{H}}\mathbf{R}_{nn}\mathbf{G})^{-1}\mathbf{G}^{\mathrm{H}}\mathbf{R}_{nn}\boldsymbol{\Psi} = \mathbf{H}^{\mathrm{H}}(\mathbf{I} - \mathbf{P}_{\mathbf{F}})\mathbf{H}.\end{aligned} \tag{5.120}$$

Here, $\mathbf{H} = \mathbf{R}_{nn}^{\mathrm{H}/2}\boldsymbol{\Psi}$ and $\mathbf{F} = \mathbf{R}_{nn}^{\mathrm{H}/2}\mathbf{G}$. As in the one-dimensional case, this result produces the following identity for the error covariance matrix of the LMVDR receiver:

$$\mathbf{Q} = (\boldsymbol{\Psi}^{\mathrm{H}}\mathbf{R}_{nn}^{-1}\boldsymbol{\Psi})^{-1} = \mathbf{H}^{\mathrm{H}}(\mathbf{I} - \mathbf{P}_{\mathbf{F}})\mathbf{H}. \tag{5.121}$$

The virtue of this identity is that it shows the error covariance to depend on the principal angles between the subspaces $\langle\mathbf{H}\rangle$ and $\langle\mathbf{F}\rangle$. Figure 5.6 applies correspondingly, if one accounts for the fact that $\langle\mathbf{H}\rangle$ is now multidimensional. The gain of the multi-rank LMVDR receiver over the multi-rank matched filter receiver may be written as

$$\frac{\det(\mathbf{H}^{\mathrm{H}}\mathbf{H})}{\det[\mathbf{H}^{\mathrm{H}}(\mathbf{I} - \mathbf{P}_{\mathbf{F}})\mathbf{H}]} = \frac{1}{\prod_{i=1}^{p}\sin^2\theta_i}, \tag{5.122}$$

where the θ_i are the principal angles between the two subspaces $\langle\mathbf{H}\rangle$ and $\langle\mathbf{F}\rangle$. When these angles are small, the gain is large, meaning that the bottom branch of the GSC can estimate and cancel out the noise component $\mathbf{u}$ in the top branch, thus reducing the variance of the error $\hat{\mathbf{u}} - \mathbf{u}$.

5.6.4 Subspace identification for beamforming and spectrum analysis

Often the measurement model $\mathbf{y} = \boldsymbol{\psi}x + \mathbf{n}$ is hypothesized, but the channel vector $\boldsymbol{\psi}$ that carries the unknown complex parameter x to the measurement is unknown. This kind of model is called a *separable linear model*. Call the true unknown vector $\boldsymbol{\psi}_0$. A standard ploy in subspace identification, spectrum analysis, and beamforming is to *scan* the LMVDR estimator $\hat{x}$ through candidates for $\boldsymbol{\psi}_0$ and choose the candidate that best models the measurement $\mathbf{y}$. That is to say, the weighted squared error between the measurement $\mathbf{y}$ and its prediction $\boldsymbol{\psi}\hat{x}$ is evaluated for each candidate $\boldsymbol{\psi}$ and the minimizing candidate is taken to be the best estimate of $\boldsymbol{\psi}_0$. Such scanners are called *goniometers*. This minimizing candidate is actually the maximum-likelihood estimate in a multivariate Gaussian model for the noise and it is also the weighted least-squared error estimate, as we now demonstrate.

Recall that the LMVDR estimate of the complex amplitude x in the measurement model $\mathbf{y} = \boldsymbol{\psi}x + \mathbf{n}$ is $\hat{x} = (\boldsymbol{\psi}^{\mathrm{H}}\mathbf{R}_{nn}^{-1}\boldsymbol{\psi})^{-1}\boldsymbol{\psi}^{\mathrm{H}}\mathbf{R}_{nn}^{-1}\mathbf{y}$. The corresponding LMVDR estimator of the whitened noise $\mathbf{R}_{nn}^{-1/2}\mathbf{n}$ is $\mathbf{R}_{nn}^{-1/2}\hat{\mathbf{n}} = \mathbf{R}_{nn}^{-1/2}(\mathbf{y} - \boldsymbol{\psi}\hat{x})$. It is just a few steps of algebra to write this estimator as

$$\mathbf{R}_{nn}^{-1/2}\hat{\mathbf{n}} = (\mathbf{I} - \mathbf{P}_{\omega})\mathbf{w}, \tag{5.123}$$

where $\mathbf{w} = \mathbf{R}_{nn}^{-1/2}\mathbf{y}$ is a whitened version of the measurement, $\boldsymbol{\omega} = \mathbf{R}_{nn}^{-1/2}\boldsymbol{\psi}$ is a whitened version of $\boldsymbol{\psi}$, and $\mathbf{P}_{\omega} = (\boldsymbol{\omega}^{\mathrm{H}}\boldsymbol{\omega})^{-1}\boldsymbol{\omega}\boldsymbol{\omega}^{\mathrm{H}}$ is the orthogonal projector onto the subspace $\langle\boldsymbol{\omega}\rangle$. The mean value of $\mathbf{R}_{nn}^{-1/2}\hat{\mathbf{n}}$ is the bias

$$\mathbf{b} = E[\mathbf{R}_{nn}^{-1/2}\hat{\mathbf{n}}] = (\mathbf{I} - \mathbf{P}_{\omega})\boldsymbol{\omega}_0 x, \tag{5.124}$$

with $\boldsymbol{\omega}_0 = \mathbf{R}_{nn}^{-1/2}\boldsymbol{\psi}_0$, and its covariance is $E[(\mathbf{R}_{nn}^{-1/2}\hat{\mathbf{n}} - \mathbf{b})(\mathbf{R}_{nn}^{-1/2}\hat{\mathbf{n}} - \mathbf{b})^{\mathrm{H}}] = (\mathbf{I} - \mathbf{P}_{\omega})$. Thus the bias is zero when $\boldsymbol{\omega} = \boldsymbol{\omega}_0$, as expected.

The output power of the goniometer is the quadratic form

$$\hat{\mathbf{n}}^{\mathrm{H}}\mathbf{R}_{nn}^{-1}\hat{\mathbf{n}} = \mathbf{w}^{\mathrm{H}}(\mathbf{I} - \mathbf{P}_{\omega})\mathbf{w} = \mathbf{y}^{\mathrm{H}}\mathbf{R}_{nn}^{-1}\mathbf{y} - \boldsymbol{\omega}^{\mathrm{H}}\boldsymbol{\omega}|\hat{x}|^2 = \mathbf{y}^{\mathrm{H}}\mathbf{R}_{nn}^{-1}\mathbf{y} - \frac{|\boldsymbol{\psi}^{\mathrm{H}}\mathbf{R}_{nn}^{-1}\mathbf{y}|^2}{\boldsymbol{\psi}^{\mathrm{H}}\mathbf{R}_{nn}^{-1}\boldsymbol{\psi}}. \tag{5.125}$$

This is a biased estimator of the noise-free squared bias $\mathbf{b}^{\mathrm{H}}\mathbf{b}$:

$$\begin{aligned} E(\hat{\mathbf{n}}^{\mathrm{H}}\mathbf{R}_{nn}^{-1}\hat{\mathbf{n}}) &= \operatorname{tr}(\mathbf{I} - \mathbf{P}_{\omega}) + \mathbf{b}^{\mathrm{H}}\mathbf{b} = (m-1) + \mathbf{b}^{\mathrm{H}}\mathbf{b} \\ &= (m-1) + |x|^2\boldsymbol{\omega}_0^{\mathrm{H}}(\mathbf{I} - \mathbf{P}_{\omega})\boldsymbol{\omega}_0. \end{aligned} \tag{5.126}$$

At the true value of the parameter $\boldsymbol{\psi}_0$ the mean value of the goniometer, or maximum-likelihood function, is $m - 1$. At other values of $\boldsymbol{\psi}$ the mean value is $(m-1) + \mathbf{b}^{\mathrm{H}}\mathbf{b}$. So the output power of the goniometer is scanned through candidate values of $\boldsymbol{\psi}$ to search for a minimum.

On the right-hand side of (5.125), only the second term depends on $\boldsymbol{\psi}$. So a reasonable strategy for approximating the goniometer is to compute this second term, which is sometimes called the *noncoherent adaptive matched filter*. When $\boldsymbol{\psi}$ is a DFT vector, the adaptive matched filter is a *spectrum analyzer* or *beamformer*.[5]

5.6.5 Extension to WLMVDR receiver

Let us now extend the results for multi-rank LMVDR estimators to multi-rank WLMVDR estimators that account for complementary covariance.[6] The multi-rank measurement model (5.114) in augmented form is

$$\underline{\mathbf{y}} = \underline{\boldsymbol{\Psi}}\,\underline{\mathbf{x}} + \underline{\mathbf{n}}, \tag{5.127}$$

where

$$\underline{\boldsymbol{\Psi}} = \begin{bmatrix} \boldsymbol{\Psi} & \mathbf{0} \\ \mathbf{0} & \boldsymbol{\Psi}^* \end{bmatrix} \tag{5.128}$$

and the noise $\mathbf{n}$ is generally improper with augmented covariance matrix $\underline{\mathbf{R}}_{nn}$. The matched filter estimator of $\mathbf{x}$, however, does not take into account noise, and thus the widely linear matched filter solution is still the linear solution (5.115)

$$\underline{\hat{\mathbf{x}}} = \underline{\mathbf{\Psi}}^{\mathrm{H}}\underline{\mathbf{y}} \Longleftrightarrow \hat{\mathbf{x}} = \mathbf{\Psi}^{\mathrm{H}}\mathbf{y}. \tag{5.129}$$

The whitened matched filter or WLMVDR estimator, on the other hand, is obtained as the solution to

$$\min \operatorname{tr}[\underline{\mathbf{W}}^{\mathrm{H}}\underline{\mathbf{R}}_{nn}\underline{\mathbf{W}}] \quad \text{under constraint } \underline{\mathbf{W}}^{\mathrm{H}}\underline{\mathbf{\Psi}} = \mathbf{I} \tag{5.130}$$

with

$$\underline{\mathbf{W}} = \begin{bmatrix} \mathbf{W}_1 & \mathbf{W}_2 \\ \mathbf{W}_2^* & \mathbf{W}_1^* \end{bmatrix}. \tag{5.131}$$

This solution is widely linear,

$$\underline{\hat{\mathbf{x}}} = (\underline{\mathbf{\Psi}}^{\mathrm{H}}\underline{\mathbf{R}}_{nn}^{-1}\underline{\mathbf{\Psi}})^{-1}\underline{\mathbf{\Psi}}^{\mathrm{H}}\underline{\mathbf{R}}_{nn}^{-1}\underline{\mathbf{y}}, \tag{5.132}$$

with mean $\mathbf{x}$ and augmented error covariance matrix

$$\underline{\mathbf{Q}} = E[(\underline{\hat{\mathbf{x}}} - \underline{\mathbf{x}})(\underline{\hat{\mathbf{x}}} - \underline{\mathbf{x}})^{\mathrm{H}}] = (\underline{\mathbf{\Psi}}^{\mathrm{H}}\underline{\mathbf{R}}_{nn}^{-1}\underline{\mathbf{\Psi}})^{-1}. \tag{5.133}$$

It requires a few steps of algebra to show that the variance of the WLMVDR estimator is indeed less than or equal to the variance of the LMVDR estimator. Alternatively, we can argue as follows. The optimization (5.130) is performed under the constraint $\underline{\mathbf{W}}^{\mathrm{H}}\underline{\mathbf{\Psi}} = \mathbf{I}$, or, equivalently, $\mathbf{W}_1^{\mathrm{H}}\mathbf{\Psi} = \mathbf{I}$ and $\mathbf{W}_2^{\mathrm{H}}\mathbf{\Psi} = \mathbf{0}$. Thus, the WLMMSE optimization problem contains the LMMSE optimization problem as a special case in which $\mathbf{W}_2 = \mathbf{0}$ is enforced. Hence, WLMVDR estimation cannot be worse than LMVDR estimation. However, the additional degree of freedom of being able to choose $\mathbf{W}_2 \neq \mathbf{0}$ can reduce the variance if the noise $\mathbf{n}$ is improper. The results from the preceding subsections – for instance, the generalized sidelobe canceler – apply in a straightforward manner to WLMVDR estimation.

The reduction in variance of the WLMVDR compared with the LMVDR estimator is entirely due to exploiting the complementary correlation of the noise $\mathbf{n}$. Indeed, it is easy to see that, for proper noise $\mathbf{n}$, the WLMVDR solution (5.132) simplifies to the LMVDR solution (5.117). Since $\mathbf{x}$ is not assigned statistical properties, the solution is independent of whether or not $\mathbf{x}$ is improper. This stands in marked contrast to the case of WLMMSE estimation discussed in Section 5.4.

5.7 Widely linear-quadratic estimation

While widely linear estimation is optimum in the Gaussian case, it may be improved upon for non-Gaussian data if we have access to higher-order statistics. The next logical extension of widely linear processing is to widely linear-quadratic (WLQ) processing, which requires statistical information up to fourth order.[7] Because we would like to keep the notation simple, we shall restrict our attention to systems with vector-valued input but *scalar* output.

5.7.1 Connection between real and complex quadratic forms

What does the most general widely quadratic transformation look like? To answer this question, we begin with the real-valued quadratic form

$$u = \mathbf{w}^{\mathrm{T}}\mathbf{M}\mathbf{w}, \tag{5.134}$$

$$\mathbf{w}^{\mathrm{T}} = \begin{bmatrix}\mathbf{a}^{\mathrm{T}} & \mathbf{b}^{\mathrm{T}}\end{bmatrix}, \tag{5.135}$$

$$\mathbf{M} = \begin{bmatrix}\mathbf{M}_{11} & \mathbf{M}_{12} \\ \mathbf{M}_{21} & \mathbf{M}_{22}\end{bmatrix} = \begin{bmatrix}\mathbf{M}_{11} & \mathbf{M}_{12} \\ \mathbf{M}_{12}^{\mathrm{T}} & \mathbf{M}_{22}\end{bmatrix}, \tag{5.136}$$

with $u \in \mathbb{R}$, $\mathbf{a}, \mathbf{b} \in \mathbb{R}^m$, and $\mathbf{M}_{11}, \mathbf{M}_{12}, \mathbf{M}_{21}, \mathbf{M}_{22} \in \mathbb{R}^{m\times m}$. Without loss of generality, we may assume that $\mathbf{M}$ is symmetric, and therefore $\mathbf{M}_{11} = \mathbf{M}_{11}^{\mathrm{T}}$, $\mathbf{M}_{21} = \mathbf{M}_{12}^{\mathrm{T}}$, and $\mathbf{M}_{22} = \mathbf{M}_{22}^{\mathrm{T}}$. Using

$$\mathbf{T} = \begin{bmatrix}\mathbf{I} & \mathrm{j}\mathbf{I} \\ \mathbf{I} & -\mathrm{j}\mathbf{I}\end{bmatrix}, \qquad \mathbf{T}\mathbf{T}^{\mathrm{H}} = \mathbf{T}^{\mathrm{H}}\mathbf{T} = 2\mathbf{I}, \tag{5.137}$$

the complex version of this real-valued quadratic form is

$$u = \tfrac{1}{2}(\mathbf{w}^{\mathrm{T}}\mathbf{T}^{\mathrm{H}})(\tfrac{1}{2}\mathbf{T}\mathbf{M}\mathbf{T}^{\mathrm{H}})(\mathbf{T}\mathbf{w}) = \tfrac{1}{2}\underline{\mathbf{y}}^{\mathrm{H}}\underline{\mathbf{N}}\,\underline{\mathbf{y}} \tag{5.138}$$

with $\mathbf{y} = \mathbf{a} + \mathrm{j}\mathbf{b}$ and $\underline{\mathbf{y}} = \mathbf{T}\mathbf{w}$. There is the familiar connection

$$\underline{\mathbf{N}} = \tfrac{1}{2}\mathbf{T}\mathbf{M}\mathbf{T}^{\mathrm{H}} = \begin{bmatrix}\mathbf{N}_1 & \mathbf{N}_2 \\ \mathbf{N}_2^* & \mathbf{N}_1^*\end{bmatrix}, \tag{5.139}$$

$$\mathbf{N}_1 = \tfrac{1}{2}[\mathbf{M}_{11} + \mathbf{M}_{22} + \mathrm{j}(\mathbf{M}_{12}^{\mathrm{T}} - \mathbf{M}_{12})], \tag{5.140}$$

$$\mathbf{N}_2 = \tfrac{1}{2}[\mathbf{M}_{11} - \mathbf{M}_{22} + \mathrm{j}(\mathbf{M}_{12}^{\mathrm{T}} + \mathbf{M}_{12})], \tag{5.141}$$

and $\underline{\mathbf{N}} \in \mathcal{W}^{m\times m}$ is an augmented Hermitian matrix, i.e., $\mathbf{N}_1^{\mathrm{H}} = \mathbf{N}_1$ and $\mathbf{N}_2^{\mathrm{T}} = \mathbf{N}_2$. In general, $\underline{\mathbf{N}}$ does not have to be positive or negative (semi)definite. However, if it is, $\underline{\mathbf{N}}$ will have the same definiteness property as $\mathbf{M}$. Because of the special structure of $\underline{\mathbf{N}}$, the quadratic form can be expressed as

$$\begin{aligned} u &= \tfrac{1}{2}[\mathbf{y}^{\mathrm{H}}\mathbf{N}_1\mathbf{y} + \mathbf{y}^{\mathrm{T}}\mathbf{N}_1^*\mathbf{y}^* + \mathbf{y}^{\mathrm{T}}\mathbf{N}_2^*\mathbf{y} + \mathbf{y}^{\mathrm{H}}\mathbf{N}_2\mathbf{y}^*] \\ &= \mathbf{y}^{\mathrm{H}}\mathbf{N}_1\mathbf{y} + \mathrm{Re}\,(\mathbf{y}^{\mathrm{H}}\mathbf{N}_2\mathbf{y}^*). \end{aligned} \tag{5.142}$$

So far, we have considered only a quadratic system with *real* output u. We may combine two quadratic systems $\underline{\mathbf{N}}_{\mathrm{r}}$ and $\underline{\mathbf{N}}_{\mathrm{i}}$ to produce the real and imaginary parts, respectively, of a complex-valued quadratic form:

$$x = u + \mathrm{j}v = \tfrac{1}{2}[\underline{\mathbf{y}}^{\mathrm{H}}\underline{\mathbf{N}}_{\mathrm{r}}\underline{\mathbf{y}} + \mathrm{j}\underline{\mathbf{y}}^{\mathrm{H}}\underline{\mathbf{N}}_{\mathrm{i}}\underline{\mathbf{y}}] = \tfrac{1}{2}\underline{\mathbf{y}}^{\mathrm{H}}(\underline{\mathbf{N}}_{\mathrm{r}} + \mathrm{j}\underline{\mathbf{N}}_{\mathrm{i}})\underline{\mathbf{y}}. \tag{5.143}$$

The resulting

$$\mathbf{H} = \begin{bmatrix}\mathbf{H}_{11} & \mathbf{H}_{12} \\ \mathbf{H}_{21} & \mathbf{H}_{22}\end{bmatrix} = \underline{\mathbf{N}}_{\mathrm{r}} + \mathrm{j}\underline{\mathbf{N}}_{\mathrm{i}} = \begin{bmatrix}\mathbf{N}_{\mathrm{r},1} + \mathrm{j}\mathbf{N}_{\mathrm{i},1} & \mathbf{N}_{\mathrm{r},2} + \mathrm{j}\mathbf{N}_{\mathrm{i},2} \\ \mathbf{N}_{\mathrm{r},2}^* + \mathrm{j}\mathbf{N}_{\mathrm{i},2}^* & \mathbf{N}_{\mathrm{r},1}^* + \mathrm{j}\mathbf{N}_{\mathrm{i},1}^*\end{bmatrix} \tag{5.144}$$

no longer has the block pattern of an augmented matrix, i.e., $\mathbf{H} \notin \mathcal{W}^{m\times m}$*, nor is it generally Hermitian*. It does, however, satisfy $\mathbf{H}_{22} = \mathbf{H}_{11}^{\mathrm{T}}$, $\mathbf{H}_{12} = \mathbf{H}_{12}^{\mathrm{T}}$, and $\mathbf{H}_{21} = \mathbf{H}_{21}^{\mathrm{T}}$.

Thus, a general complex-valued quadratic form can be written as

$$x = \frac{1}{2}\underline{\mathbf{y}}^{\mathrm{H}}\underline{\mathbf{H}}\underline{\mathbf{y}} = \frac{1}{2}[\mathbf{y}^{\mathrm{H}}\mathbf{H}_{11}\mathbf{y} + \mathbf{y}^{\mathrm{T}}\mathbf{H}_{11}^{\mathrm{T}}\mathbf{y}^* + \mathbf{y}^{\mathrm{T}}\mathbf{H}_{21}\mathbf{y} + \mathbf{y}^{\mathrm{H}}\mathbf{H}_{12}\mathbf{y}^*]$$
$$= \mathbf{y}^{\mathrm{H}}\mathbf{H}_{11}\mathbf{y} + \frac{1}{2}[\mathbf{y}^{\mathrm{T}}\mathbf{H}_{21}\mathbf{y} + \mathbf{y}^{\mathrm{H}}\mathbf{H}_{12}\mathbf{y}^*]. \tag{5.145}$$

Combining our results so far, the complex-valued output of a WLQ system can be expressed as

$$x = c + \frac{1}{2}\mathbf{g}^{\mathrm{H}}\underline{\mathbf{y}} + \frac{1}{2}\underline{\mathbf{y}}^{\mathrm{H}}\mathbf{H}\underline{\mathbf{y}}$$
$$= c + \frac{1}{2}[\mathbf{g}_1^{\mathrm{H}}\mathbf{y} + \mathbf{g}_2^{\mathrm{H}}\mathbf{y}^*] + \mathbf{y}^{\mathrm{H}}\mathbf{H}_{11}\mathbf{y} + \frac{1}{2}[\mathbf{y}^{\mathrm{T}}\mathbf{H}_{21}\mathbf{y} + \mathbf{y}^{\mathrm{H}}\mathbf{H}_{12}\mathbf{y}^*], \tag{5.146}$$

where c is a complex constant, and $\mathbf{g}^{\mathrm{H}}\underline{\mathbf{y}}$ with $\mathbf{g}^{\mathrm{H}} = [\mathbf{g}_1^{\mathrm{H}}, \mathbf{g}_2^{\mathrm{H}}] \in \mathbb{C}^{2m}$ constitutes the widely linear part and $\underline{\mathbf{y}}^{\mathrm{H}}\mathbf{H}\,\underline{\mathbf{y}}$ the widely quadratic part.

If x is real, then $\mathbf{g}_1 = \mathbf{g}_2^*$, i.e., $\mathbf{g}^{\mathrm{H}} = \underline{\mathbf{g}}_1^{\mathrm{H}} = [\mathbf{g}_1^{\mathrm{H}}, \mathbf{g}_1^{\mathrm{T}}] \in \mathbb{C}_*^{2m}$ has the structure of an augmented vector, and $\mathbf{H}_{11} = \mathbf{H}_{11}^{\mathrm{H}}$, $\mathbf{H}_{21} = \mathbf{H}_{12}^*$, i.e., $\mathbf{H}$ has the structure of an augmented matrix (where $\mathbf{N}_1 = \mathbf{H}_{11}$ and $\mathbf{N}_2 = \mathbf{H}_{12}$). Thus, for real x,

$$x = u = c + \frac{1}{2}\underline{\mathbf{g}}_1^{\mathrm{H}}\underline{\mathbf{y}} + \frac{1}{2}\underline{\mathbf{y}}^{\mathrm{H}}\underline{\mathbf{N}}\,\underline{\mathbf{y}}$$
$$= c + \mathrm{Re}(\mathbf{g}_1^{\mathrm{H}}\mathbf{y}) + \mathbf{y}^{\mathrm{H}}\mathbf{N}_1\mathbf{y} + \mathrm{Re}(\mathbf{y}^{\mathrm{H}}\mathbf{N}_2\mathbf{y}^*). \tag{5.147}$$

We may consider a pair $[\mathbf{g}, \mathbf{H}]$ to be a vector in a linear space,[8] with addition defined by

$$[\mathbf{g}, \mathbf{H}] + [\mathbf{g}', \mathbf{H}'] = [\mathbf{g} + \mathbf{g}', \mathbf{H} + \mathbf{H}'] \tag{5.148}$$

and multiplication by a complex scalar a by

$$a[\mathbf{g}, \mathbf{H}] = [a\mathbf{g}, a\mathbf{H}]. \tag{5.149}$$

The inner product in this space is defined by

$$\langle[\mathbf{g}, \mathbf{H}], [\mathbf{g}', \mathbf{H}']\rangle = \mathbf{g}^{\mathrm{H}}\mathbf{g}' + \mathrm{tr}(\mathbf{H}^{\mathrm{H}}\mathbf{H}'). \tag{5.150}$$

5.7.2 WLQMMSE estimation

We would now like to estimate a scalar complex signal $x\colon \Omega \longrightarrow \mathbb{C}$ from a measurement $\mathbf{y}\colon \Omega \longrightarrow \mathbb{C}^m$, using the WLQ estimator

$$\hat{x} = c + \mathbf{g}^{\mathrm{H}}\underline{\mathbf{y}} + \underline{\mathbf{y}}^{\mathrm{H}}\mathbf{H}\underline{\mathbf{y}}. \tag{5.151}$$

For convenience, we have absorbed the factors of $1/2$ in (5.146) into $\mathbf{g} \in \mathbb{C}^{2m}$ and $\mathbf{H} \in \mathbb{C}^{2m\times 2m}$. We derive the WLQ estimator as a fairly straightforward extension of Picinbono and Duvaut (1988) to the complex noncircular case, which avoids the use of tensor notation. We assume that both x and $\mathbf{y}$ have zero mean. In order to ensure that $\hat{x}$ has zero mean as well, we need to choose $c = -\mathrm{tr}\,(\underline{\mathbf{R}}_{yy}\mathbf{H})$ so that

$$\hat{x} = \mathbf{g}^{\mathrm{H}}\underline{\mathbf{y}} + \underline{\mathbf{y}}^{\mathrm{H}}\mathbf{H}\underline{\mathbf{y}} - \mathrm{tr}\,(\underline{\mathbf{R}}_{yy}\mathbf{H}). \tag{5.152}$$

Using the definition of the inner product (5.150), the estimator is

$$\hat{x} = \left\langle[\mathbf{g}, \mathbf{H}^{\mathrm{H}}], [\underline{\mathbf{y}}, \underline{\mathbf{y}}\,\underline{\mathbf{y}}^{\mathrm{H}} - \underline{\mathbf{R}}_{yy}]\right\rangle. \tag{5.153}$$

In order to make $\hat{x}$ a WLQ minimum mean-squared error (WLQMMSE) estimator that minimizes $E|\hat{x} - x|^2$, we apply the *orthogonality principle* of Result 5.2. That is,

$$(\hat{x} - x) \perp \hat{x}', \tag{5.154}$$

where

$$\hat{x}' = \mathbf{g}'^{\mathrm{H}}\underline{\mathbf{y}} + \underline{\mathbf{y}}^{\mathrm{H}}\mathbf{H}'\underline{\mathbf{y}} - \operatorname{tr}(\underline{\mathbf{R}}_{yy}\mathbf{H}') \tag{5.155}$$

is any WLQ estimator with arbitrary $[\mathbf{g}', \mathbf{H}']$. The orthogonality in (5.154) can be expressed as

$$E\{(\hat{x} - x)^*\hat{x}'\} = 0, \quad \text{for all } \mathbf{g}' \text{ and } \mathbf{H}'. \tag{5.156}$$

We now obtain

$$E(x^*\hat{x}') = E(x^*\mathbf{g}'^{\mathrm{H}}\underline{\mathbf{y}}) + E(x^*\underline{\mathbf{y}}^{\mathrm{H}}\mathbf{H}'\underline{\mathbf{y}}) - E(x^*)\operatorname{tr}(\underline{\mathbf{R}}_{yy}\mathbf{H}'), \tag{5.157}$$

and, because x has zero mean,

$$\begin{aligned} E(x^*\hat{x}') &= \mathbf{g}'^{\mathrm{H}}E(x^*\underline{\mathbf{y}}) + \operatorname{tr}[\mathbf{H}'E(x^*\underline{\mathbf{y}}\,\underline{\mathbf{y}}^{\mathrm{H}})] \\ &= \left\langle [\mathbf{g}', \mathbf{H}'^{\mathrm{H}}], [E(x^*\underline{\mathbf{y}}), E(x^*\underline{\mathbf{y}}\,\underline{\mathbf{y}}^{\mathrm{H}})] \right\rangle. \end{aligned} \tag{5.158}$$

Because $\mathbf{y}$ has zero mean, we find

$$\begin{aligned} E(\hat{x}'\hat{x}^*) &= \mathbf{g}'^{\mathrm{H}}E\left[\underline{\mathbf{y}}\mathbf{g}^{\mathrm{T}}\underline{\mathbf{y}}^* + \underline{\mathbf{y}}\,\underline{\mathbf{y}}^{\mathrm{T}}\mathbf{H}^*\underline{\mathbf{y}}^*\right] \\ &\quad + E\left[\underline{\mathbf{y}}^{\mathrm{H}}\mathbf{H}'\underline{\mathbf{y}}\mathbf{g}^{\mathrm{T}}\underline{\mathbf{y}}^* + \underline{\mathbf{y}}^{\mathrm{H}}\mathbf{H}'\underline{\mathbf{y}}\,\underline{\mathbf{y}}^{\mathrm{T}}\mathbf{H}^*\underline{\mathbf{y}}^* - \underline{\mathbf{y}}^{\mathrm{H}}\mathbf{H}'\underline{\mathbf{y}}\operatorname{tr}(\underline{\mathbf{R}}_{yy}^*\mathbf{H}^*)\right]. \end{aligned} \tag{5.159}$$

By repeatedly using the permutation property of the trace, $\operatorname{tr}(\mathbf{AB}) = \operatorname{tr}(\mathbf{BA})$, we obtain

$$\begin{aligned} E(\hat{x}'\hat{x}^*) &= \mathbf{g}'^{\mathrm{H}}E\left[\underline{\mathbf{y}}\mathbf{g}^{\mathrm{T}}\underline{\mathbf{y}}^* + \underline{\mathbf{y}}\,\underline{\mathbf{y}}^{\mathrm{T}}\mathbf{H}^*\underline{\mathbf{y}}^*\right] \\ &\quad + \operatorname{tr}\left[\mathbf{H}'E(\underline{\mathbf{y}}\mathbf{g}^{\mathrm{T}}\underline{\mathbf{y}}^*\underline{\mathbf{y}}^{\mathrm{H}} + \underline{\mathbf{y}}\,\underline{\mathbf{y}}^{\mathrm{T}}\mathbf{H}^*\underline{\mathbf{y}}^*\underline{\mathbf{y}}^{\mathrm{H}} - \operatorname{tr}(\underline{\mathbf{R}}_{yy}^*\mathbf{H}^*)\underline{\mathbf{y}}\,\underline{\mathbf{y}}^{\mathrm{H}})\right] \\ &= \left\langle [\mathbf{g}', \mathbf{H}'^{\mathrm{H}}], [\underline{\mathbf{R}}_{yy}\mathbf{g} + E(\underline{\mathbf{y}}\,\underline{\mathbf{y}}^{\mathrm{T}}\mathbf{H}^*\underline{\mathbf{y}}^*), \right. \\ &\qquad \left. E(\underline{\mathbf{y}}\,\underline{\mathbf{y}}^{\mathrm{H}}\underline{\mathbf{y}}^{\mathrm{H}}\mathbf{g}) + E(\underline{\mathbf{y}}\,\underline{\mathbf{y}}^{\mathrm{H}}\,\underline{\mathbf{y}}^{\mathrm{T}}\mathbf{H}^*\underline{\mathbf{y}}^*) - \operatorname{tr}(\underline{\mathbf{R}}_{yy}^*\mathbf{H}^*)\underline{\mathbf{R}}_{yy}]\right\rangle \end{aligned} \tag{5.160}$$

$$= \left\langle [\mathbf{g}', \mathbf{H}'^{\mathrm{H}}], \mathbb{K}[\mathbf{g}, \mathbf{H}] \right\rangle. \tag{5.161}$$

In the last equation, we have introduced the positive definite operator $\mathbb{K}$, which is defined in terms of the second-, third-, and fourth-order moments of $\mathbf{y}$. It acts on $[\mathbf{g}, \mathbf{H}]$ as in (5.160). We can find an explicit expression for this. Let $\underline{y}_i$, $i = 1, \ldots, 2m$, denote the ith element of $\underline{\mathbf{y}}$, i.e., $\underline{y}_i = y_i$, $i = 1, \ldots, m$, and $\underline{y}_i = y_{i-m}^*$, $i = m+1, \ldots, 2m$. Also let $(\underline{\mathbf{R}}_{yy})_{ij}$ denote the (i, j)th element of $\underline{\mathbf{R}}_{yy}$ and $(\underline{\mathbf{R}}_{yy})_i$ the ith row of $\underline{\mathbf{R}}_{yy}$. Then

$$\mathbb{K}[\mathbf{g}, \mathbf{H}] = [\mathbf{p}, \mathbf{N}] \tag{5.162}$$

with

$$p_i = (\underline{\mathbf{R}}_{yy})_i \mathbf{g} + \sum_{k=1}^{2m}\sum_{l=1}^{2m} E(\underline{y}_i \underline{y}_k \underline{y}_l^*) H_{kl}^*, \quad i = 1, \ldots, 2m,$$

$$\begin{aligned} N_{ij} &= \sum_{k=1}^{2m} E(\underline{y}_i \underline{y}_j^* \underline{y}_k^*) g_k + \sum_{k=1}^{2m}\sum_{l=1}^{2m} E(\underline{y}_i \underline{y}_j^* \underline{y}_k \underline{y}_l^*) H_{kl}^* \\ &\quad - \operatorname{tr}(\underline{\mathbf{R}}_{yy}^* \mathbf{H}^*)(\underline{\mathbf{R}}_{yy})_{ij}, \quad i, j = 1, \ldots, 2m \\ &= \sum_{k=1}^{2m} E(\underline{y}_i \underline{y}_j^* \underline{y}_k^*) g_k + \sum_{k=1}^{2m}\sum_{l=1}^{2m} [E(\underline{y}_i \underline{y}_j^* \underline{y}_k \underline{y}_l^*) - E(\underline{y}_i \underline{y}_j^*) E(\underline{y}_k \underline{y}_l^*)] H_{kl}^*. \end{aligned} \tag{5.163}$$

Putting all the pieces together, (5.156) becomes

$$\left\langle [\mathbf{g}', \mathbf{H}'^{\mathrm{H}}], \mathbb{K}[\mathbf{g}, \mathbf{H}] - [E(x^* \underline{\mathbf{y}}), E(x^* \underline{\mathbf{y}}\,\underline{\mathbf{y}}^{\mathrm{H}})] \right\rangle = 0. \tag{5.164}$$

Since this must hold for all $\mathbf{g}'$ and $\mathbf{H}'$, this implies

$$\mathbb{K}[\mathbf{g}, \mathbf{H}] - [E(x^* \underline{\mathbf{y}}), E(x^* \underline{\mathbf{y}}\,\underline{\mathbf{y}}^{\mathrm{H}})] = 0, \tag{5.165}$$

$$[\mathbf{g}, \mathbf{H}] = \mathbb{K}^{-1}[E(x^* \underline{\mathbf{y}}), E(x^* \underline{\mathbf{y}}\,\underline{\mathbf{y}}^{\mathrm{H}})]. \tag{5.166}$$

The MSE achieved by this WLQMMSE estimator is

$$\begin{aligned} E|\hat{x} - x|^2 &= E|x|^2 - \langle [E(x^* \underline{\mathbf{y}}), E(x^* \underline{\mathbf{y}}\,\underline{\mathbf{y}}^{\mathrm{H}})], [\mathbf{g}, \mathbf{H}] \rangle \\ &= E|x|^2 - \langle [E(x^* \underline{\mathbf{y}}), E(x^* \underline{\mathbf{y}}\,\underline{\mathbf{y}}^{\mathrm{H}})], \mathbb{K}^{-1}[E(x^* \underline{\mathbf{y}}), E(x^* \underline{\mathbf{y}}\,\underline{\mathbf{y}}^{\mathrm{H}})] \rangle. \end{aligned} \tag{5.167}$$

Two remarks are in order. First, the widely linear part $\mathbf{g}$ of a WLQMMSE filter is *not* the WLMMSE filter determined in Result 5.3. Second, it is worth pointing out that the WLQMMSE filter depends on the complete statistical information up to fourth order. That is, *all* conjugation patterns of the second-, third-, and fourth-order moments are required.

Example 5.4. Consider, for instance, the third-order moments. The formula (5.163) implicitly utilizes all eight conjugation patterns $E(y_i y_j y_k)$, $E(y_i^* y_j y_k)$, $E(y_i y_j^* y_k)$, $E(y_i y_j y_k^*)$, $E(y_i^* y_j^* y_k)$, $E(y_i^* y_j y_k^*)$, $E(y_i y_j^* y_k^*)$, and $E(y_i^* y_j^* y_k^*)$. Of course, only two of these eight conjugation patterns are distinct, due to permutation, e.g., $E(y_i^* y_j y_k) = E(y_j y_k y_i^*)$, or complex conjugation, e.g., $E(y_i y_j^* y_k) = [E(y_i^* y_j y_k^*)]^*$.

We may consider $E(\underline{y}_i \underline{y}_j \underline{y}_k)$ with $i, j, k = 1, \ldots, 2m$ as a $2m \times 2m \times 2m$ array, more precisely called a *tensor* of order 3. This tensor is depicted in Fig. 5.7, and it has many symmetry relationships. Each $m \times m \times m$ cube is the complex conjugate of the cube with which it does not share a side or edge.

A common linear estimation scenario is the estimation of a signal in second-order white (and proper) noise $\mathbf{n}$ whose augmented covariance matrix is $\underline{\mathbf{R}}_{nn} = \mathbf{I}$. For WLQMMSE estimation, we need to know statistical properties of the noise up to fourth

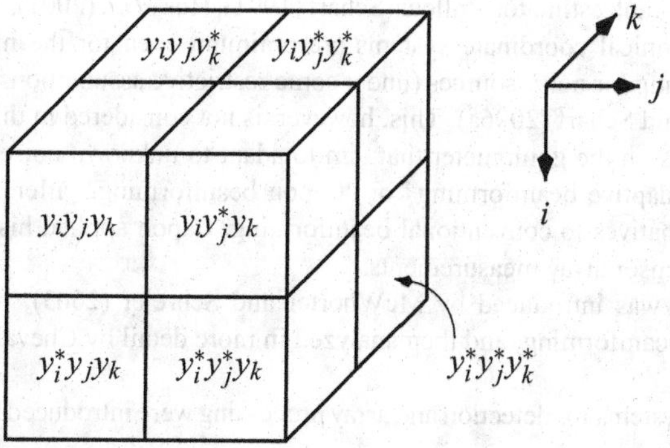

Figure 5.7 Tensor of third-order moments.

order. There exists a wide variety of higher-order noise models but the simplest model is *fourth-order white (and circular) noise*, which is defined as follows.

Definition 5.1. *A random vector* **n** *is called* fourth-order white *(and circular) if the only nonzero moments up to fourth order are*

$$E(n_i n_j^*) = \delta_{ij} \tag{5.168}$$

$$E(n_i n_j n_k^* n_l^*) = \delta_{ik}\delta_{jl} + \delta_{il}\delta_{jk} + K\delta_{ijkl}, \tag{5.169}$$

where $\delta_{ijkl} = 1$ *if* $i = j = k = l$ *and zero otherwise, and* K *is the kurtosis of the noise.*

This means that all moments must have the same number of conjugated and non-conjugated terms. This type of noise is "Gaussian-like" because it shares some of the properties of proper Gaussian random vectors (compare this with Result 2.10). However, Gaussian noise always has zero kurtosis, $K = 0$, whereas fourth-order white noise is allowed to have $K \neq 0$.

Notes

1 The introduction of channel, analysis, and synthesis models for the representation of composite covariance matrices of signal and measurement is fairly standard, but we have guided by Mullis and Scharf (1996).

2 The solution of LMMSE problems requires the solution of normal equations $\mathbf{W}\mathbf{R}_{yy} = \mathbf{R}_{xy}$. There has been a flurry of interest in "multistage" approximations to these equations, beginning with the work of Goldstein *et al.* (1998). The connection between multistage and conjugate gradient algorithms has been clarified by Dietl *et al.* (2001), Weippert *et al.* (2002), and Scharf *et al.* (2008). The attraction of these algorithms is that they converge rapidly (in N steps) if the matrix $\mathbf{R}_{yy}$ has only a small number N of distinct eigenvalues.

3 WLMMSE estimation was introduced by Picinbono and Chevalier (1995), but widely linear (or linear–linear-conjugate) filtering had been considered already by Brown and Crane (1969). The performance comparison between LMMSE and WLMMSE estimation is due to Schreier *et al.* (2005).

4 Our discussion of reduced-rank estimators follows Scharf (1991), Hua *et al.* (2001), and Schreier and Scharf (2003a). Canonical coordinate systems are optimum even for the more general problem of transform coding for noisy sources (under some restrictive assumptions), which has been shown by Schreier and Scharf (2006a). This, however, is not considered in this chapter.
5 There are many variations on the goniometer that aim to adapt to unknown noise covariance. They go by the name "adaptive beamforming" or "Capon beamforming," after Jack Capon, who first advocated alternatives to conventional beamforming. Capon applied his methods to the processing of multi-sensor array measurements.
6 The WLMVDR receiver was introduced by McWhorter and Schreier (2003), where it was applied to widely linear beamforming, and then analyzed in more detail by Chevalier and Blin (2007).
7 Widely linear-quadratic systems for detection and array processing were introduced by Chevalier and Picinbono (1996).
8 The linear space consisting of vectors $[\mathbf{g}, \mathbf{H}]$ is the *direct sum* of the Hilbert spaces $\mathbb{C}^{2m}$ and $\mathbb{C}^{2m\times 2m}$, usually denoted as $\mathbb{C}^{2m} \oplus \mathbb{C}^{2m\times 2m}$.

6 Performance bounds for parameter estimation

All parameter estimation begins with a *measurement* and an *algorithm* for extracting a parameter *estimate* from the measurement. The algorithm is the estimator.

There are two ways to think about performance analysis. One way is to begin with a particular estimator and then to compute its performance. Typically this would amount to computing the *bias* of the estimator and its *error covariance matrix*. The practitioner then draws or analyzes concentration ellipsoids to decide whether or not the estimator meets specifications. But the other, more general, way is to establish a limit on the accuracy of *any* estimator of the parameter. We might call this a uniform limit, uniform over an entire class of estimators. Such a limit would speak to the *information* that the measurement carries about the underlying parameter, independently of how the information is extracted.

Performance bounds are fundamental to signal processing because they tell us when the number and quality of spatial, temporal, or spatial–temporal measurements is sufficient to meet performance specifications. That is, these general bounds speak to the quality of the experiment or the sensing schema itself, rather than to the subsequent signal processing. If the sensing scheme carries insufficient information about the underlying parameter, then no amount of sophisticated signal processing can extract information that is not there. In other words, if the bound says that the error covariance is larger than specifications require, then the experiment or measurement scheme must be redesigned.

But there is a cautionary note here: the performance bounds we will derive are lower bounds on variance, covariance, and mean-squared error. They are not necessarily tight. So, even if the bound is smaller than specifications require, there is still no guarantee that you will be able to find an estimator that achieves this bound. So the bound may meet variance specifications, but your estimator might not. Peter Schultheiss has called performance bounds *spoiled-sport* results because they establish unwelcome limits on what can be achieved and because they can be used to bring into question performance claims that violate these fundamental bounds.

In this book, measurements are complex-valued and parameters are complex-valued. So the problem is to extract an estimate of a complex parameter from a complex measurement. As we have seen in previous chapters, we need to consider the correlation structure for errors and measurements and their complex conjugates, since this enables us to account for complementary correlation.

In this chapter, we shall consider quadratic performance bounds of the Weiss–Weinstein class, which were introduced by Weiss and Weinstein (1985). The idea behind this class of bounds is to establish a lower bound on the error covariance for a parameter estimator by considering an approximation of the *estimator error* from a *linear* function of a *score function*. This is very sophisticated reasoning: "If I use a linear function of a score function to approximate the estimator error, then maybe I can use the error covariance of this approximation to derive a bound on the error covariance for any estimator." Of course, the bound will depend on the score function that is chosen for the approximation. The idea is then to choose good scores, and among these are the Fisher, Bhattacharyya, Ziv–Zakai, Barankin, and Bobrovsky–Zakai scores. In our extension of this reasoning, we will approximate the error using a *widely linear* function of the score function.

We shall develop a *frequentist* theory of quadratic performance bounding and a *Bayesian* theory. Frequentist bounds are sometimes called *deterministic* bounds and Bayesian bounds are sometimes called *stochastic* bounds. For a *frequentist* the underlying assumption is that an experiment produces a measurement $\mathbf{y}$ that is drawn from a distribution $P_{\boldsymbol{\theta}}(\mathbf{y})$. The parameter $\boldsymbol{\theta}$ of this distribution is unknown, and to be estimated. Repeated trials of the experiment would consist of repeated draws of the measurements $\{\mathbf{y}_i\}$ from the same distribution, with $\boldsymbol{\theta}$ fixed. For a *Bayesian* the underlying assumption is that a random parameter and a random measurement $(\mathbf{y}, \boldsymbol{\theta})$ are drawn from a joint distribution $P(\mathbf{y}, \boldsymbol{\theta})$. Repeated trials of the experiment would consist of repeated draws of the measurement and parameter pairs $(\{\mathbf{y}_i\}, \boldsymbol{\theta})$, with the parameter $\boldsymbol{\theta}$ a single, fixed, random draw. In some cases the repeated draws are $(\{\mathbf{y}_i, \boldsymbol{\theta}_i\})$. For the frequentist and the Bayesian there is a density for the measurement, indexed by the parameter, but for the Bayesian there is also a prior distribution that may concentrate the probability of the parameter. For the frequentist there is none.[1]

The organization of this chapter is as follows. Section 6.1 introduces the frequentist and Bayesian concepts and establishes notation. Section 6.2 discusses quadratic frequentist bounds in general, and Section 6.3 presents the most common frequentist bound, the Cramér–Rao bound. Bayesian bounds are investigated in Section 6.4, and the best-known Bayesian bound, the Fisher–Bayes bound (or "stochastic Cramér–Rao bound"), is derived in Section 6.5. When these bounds are specialized to the multivariate Gaussian model, then the formulae of this chapter reproduce or generalize those of a great number of authors who have computed quadratic bounds for spectrum analysis and beamforming. Finally, Section 6.6 establishes connections and orderings among the bounds.

6.1 Frequentists and Bayesians

Frequentists model the complex measurement as a random vector $\mathbf{y}\colon \Omega \longrightarrow \mathbb{C}^n$ with probability distribution $P_{\boldsymbol{\theta}}(\mathbf{y})$ and probability density function (pdf) $p_{\boldsymbol{\theta}}(\mathbf{y})$. These depend on the parameter $\boldsymbol{\theta}$, which is unknown but known to belong to the set $\boldsymbol{\Theta} \subseteq \mathbb{C}^p$. As explained in Section 2.3, the probability distribution and pdf of a complex random vector are interpreted as the joint distribution and pdf of its real and imaginary parts. Typically, and throughout this chapter, the random vector $\mathbf{y}$ is continuous. If $\mathbf{y}$ contains discrete components, the pdf can be expressed using Dirac δ-functions.

Bayesians regard both the measurement $\mathbf{y}: \Omega \longrightarrow \mathbb{C}^n$ and the parameter $\boldsymbol{\theta}: \Omega \longrightarrow \mathbb{C}^p$ as random vectors defined on the same sample space. They are given a *joint* probability distribution $P(\mathbf{y}, \boldsymbol{\theta})$, and pdf $p(\mathbf{y}, \boldsymbol{\theta})$. The joint pdf may be expressed as

$$p(\mathbf{y}, \boldsymbol{\theta}) = p_{\boldsymbol{\theta}}(\mathbf{y})p(\boldsymbol{\theta}), \tag{6.1}$$

where $p_{\boldsymbol{\theta}}(\mathbf{y}) = p(\mathbf{y}|\boldsymbol{\theta})$ is the conditional measurement pdf and $p(\boldsymbol{\theta})$ is the prior pdf on the random parameter $\boldsymbol{\theta}$.

So, when $\boldsymbol{\theta}$ is deterministic, $p_{\boldsymbol{\theta}}(\mathbf{y})$ denotes a parameterized pdf. When it is random, $p_{\boldsymbol{\theta}}(\mathbf{y})$ denotes a conditional pdf. We do not make any distinction in our notation, since context clarifies which meaning is to be taken.

In our development of performance bounds on estimators of the parameter $\boldsymbol{\theta}$ we will construct *error scores* and *measurement scores*. These are functions of the measurement or of the measurement and parameter. Moreover, later on we shall augment these with their complex conjugates so that we can construct bounds based on widely linear estimates of error scores from measurement scores. We work with the space of square-integrable measurable functions (i.e., their Hermitian second-order moments are finite) of the complex random vector $\mathbf{y}$, or of the complex random vector $(\mathbf{y}, \boldsymbol{\theta})$, and assume all errors and scores to be elements of this space. Of course, a finite Hermitian second-order moment ensures a finite complementary second-order moment.

Definition 6.1. *Throughout this chapter we define*

$$\int_{\mathbb{C}^n} \mathbf{f}(\mathbf{y})d\mathbf{y} \triangleq \int_{\mathbb{R}^{2n}} \mathrm{Re}[\mathbf{f}(\mathbf{y})]d[\mathrm{Re}\,\mathbf{y}]d[\mathrm{Im}\,\mathbf{y}] + \mathrm{j}\int_{\mathbb{R}^{2n}} \mathrm{Im}[\mathbf{f}(\mathbf{y})]d[\mathrm{Re}\,\mathbf{y}]d[\mathrm{Im}\,\mathbf{y}]. \tag{6.2}$$

That is, an integral with respect to a complex variable is the integral with respect to its real and imaginary parts.

The mean value of any function $\mathbf{g}(\mathbf{y})$, computed with respect to the parameterized pdf $p_{\boldsymbol{\theta}}(\mathbf{y})$, is

$$E_{\boldsymbol{\theta}}[\mathbf{g}(\mathbf{y})] = \int_{\mathbb{C}^n} \mathbf{g}(\mathbf{y})p_{\boldsymbol{\theta}}(\mathbf{y})d\mathbf{y}. \tag{6.3}$$

The mean value of any function $\mathbf{g}(\mathbf{y}, \boldsymbol{\theta})$, computed with respect to the joint pdf $p(\mathbf{y}, \boldsymbol{\theta})$, is

$$\begin{aligned} E[\mathbf{g}(\mathbf{y}, \boldsymbol{\theta})] &= \int_{\mathbb{C}^p\times\mathbb{C}^n} \mathbf{g}(\mathbf{y}, \boldsymbol{\theta})p(\mathbf{y}, \boldsymbol{\theta})d\mathbf{y}\, d\boldsymbol{\theta} \\ &= \int_{\mathbb{C}^p} \left[\int_{\mathbb{C}^n} \mathbf{g}(\mathbf{y}, \boldsymbol{\theta})p_{\boldsymbol{\theta}}(\mathbf{y})d\mathbf{y}\right] p(\boldsymbol{\theta})d\boldsymbol{\theta} \\ &= E\{E_{\boldsymbol{\theta}}[\mathbf{g}(\mathbf{y}, \boldsymbol{\theta})]\}. \end{aligned} \tag{6.4}$$

In this notation, $E_{\boldsymbol{\theta}}[\mathbf{g}(\mathbf{y}, \boldsymbol{\theta})]$ is a conditional expectation with respect to the conditional pdf $p_{\boldsymbol{\theta}}(\mathbf{y})$, and the outer E is the expectation over the distribution of $\boldsymbol{\theta}$.

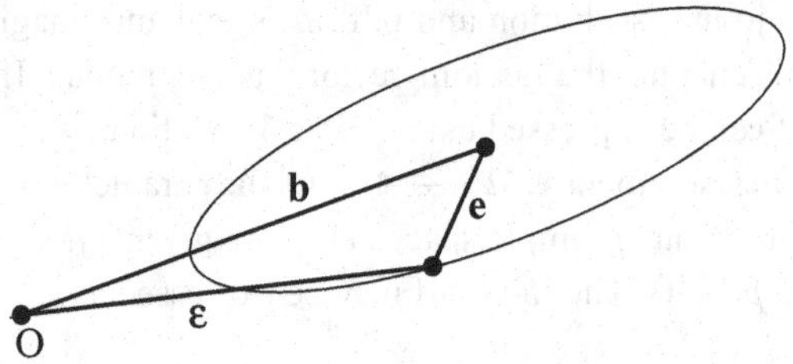

Figure 6.1 Illustration of bias $\mathbf{b}$, error $\boldsymbol{\varepsilon}$, and centered error $\mathbf{e}$. The ellipsoid is the locus of centered errors for which $\mathbf{e}^{\mathrm{H}}\mathbf{Q}^{-1}\mathbf{e}$ is constant.

6.1.1 Bias, error covariance, and mean-squared error

We wish to estimate the parameter $\boldsymbol{\theta}$ with the estimator $\hat{\boldsymbol{\theta}}(\mathbf{y})$. From the **frequentist point of view**, the problem is to determine the pdf $p_{\boldsymbol{\theta}}(\mathbf{y})$, from the set of pdfs $\{p_{\boldsymbol{\theta}}(\mathbf{y}), \boldsymbol{\theta} \in \Theta \subseteq \mathbb{C}^p\}$ that best describes the pdf from which the random measurement $\mathbf{y}$ was drawn. The *error* of the estimator is taken to be

$$\boldsymbol{\varepsilon}_{\boldsymbol{\theta}}(\mathbf{y}) = \hat{\boldsymbol{\theta}}(\mathbf{y}) - \boldsymbol{\theta} \tag{6.5}$$

and the parameter-dependent *frequentist bias* of the estimator error is defined to be

$$\mathbf{b}(\boldsymbol{\theta}) = E_{\boldsymbol{\theta}}[\boldsymbol{\varepsilon}_{\boldsymbol{\theta}}(\mathbf{y})] = E_{\boldsymbol{\theta}}[\hat{\boldsymbol{\theta}}(\mathbf{y})] - \boldsymbol{\theta}. \tag{6.6}$$

The error then has representation $\boldsymbol{\varepsilon}_{\boldsymbol{\theta}}(\mathbf{y}) = \mathbf{e}_{\boldsymbol{\theta}}(\mathbf{y}) + \mathbf{b}(\boldsymbol{\theta})$, where

$$\mathbf{e}_{\boldsymbol{\theta}}(\mathbf{y}) = \boldsymbol{\varepsilon}_{\boldsymbol{\theta}}(\mathbf{y}) - \mathbf{b}(\boldsymbol{\theta}) = \hat{\boldsymbol{\theta}}(\mathbf{y}) - E_{\boldsymbol{\theta}}[\hat{\boldsymbol{\theta}}(\mathbf{y})] \tag{6.7}$$

is called the *centered error score.*

We shall define the function $\boldsymbol{\sigma}_{\boldsymbol{\theta}}(\mathbf{y}) = [\sigma_1(\mathbf{y}), \sigma_2(\mathbf{y}), \ldots, \sigma_m(\mathbf{y})]^{\mathrm{T}}$ to be an m-dimensional vector of complex scores. The mean of the score is $E_{\boldsymbol{\theta}}[\boldsymbol{\sigma}_{\boldsymbol{\theta}}(\mathbf{y})]$ and

$$\mathbf{s}_{\boldsymbol{\theta}}(\mathbf{y}) = \boldsymbol{\sigma}_{\boldsymbol{\theta}}(\mathbf{y}) - E_{\boldsymbol{\theta}}[\boldsymbol{\sigma}_{\boldsymbol{\theta}}(\mathbf{y})] \tag{6.8}$$

is called the *centered measurement score*. Generally we think of the centered measurement score as a judiciously chosen function of the measurement that brings information about the centered error score $\mathbf{e}_{\boldsymbol{\theta}}(\mathbf{y})$.

The frequentist defines the parameter-dependent *mean-squared error matrix* to be

$$\begin{aligned}\mathbf{M}(\boldsymbol{\theta}) &= E_{\boldsymbol{\theta}}[\boldsymbol{\varepsilon}_{\boldsymbol{\theta}}(\mathbf{y})\boldsymbol{\varepsilon}_{\boldsymbol{\theta}}^{\mathrm{H}}(\mathbf{y})] = E_{\boldsymbol{\theta}}[\mathbf{e}_{\boldsymbol{\theta}}(\mathbf{y})\mathbf{e}_{\boldsymbol{\theta}}^{\mathrm{H}}(\mathbf{y})] + \mathbf{b}(\boldsymbol{\theta})\mathbf{b}^{\mathrm{H}}(\boldsymbol{\theta}) \\ &= \mathbf{Q}(\boldsymbol{\theta}) + \mathbf{b}(\boldsymbol{\theta})\mathbf{b}^{\mathrm{H}}(\boldsymbol{\theta}),\end{aligned} \tag{6.9}$$

where $\mathbf{Q}(\boldsymbol{\theta})$ is the frequentist error covariance

$$\mathbf{Q}(\boldsymbol{\theta}) = E_{\boldsymbol{\theta}}[\mathbf{e}_{\boldsymbol{\theta}}(\mathbf{y})\mathbf{e}_{\boldsymbol{\theta}}^{\mathrm{H}}(\mathbf{y})]. \tag{6.10}$$

Here we have exploited the fact that the cross-correlation between $\mathbf{e}_{\boldsymbol{\theta}}(\mathbf{y})$ and $\mathbf{b}(\boldsymbol{\theta})$ is zero. If the frequentist bias of the estimator is zero, then the mean-squared error matrix is the error covariance matrix: $\mathbf{M}(\boldsymbol{\theta}) = \mathbf{Q}(\boldsymbol{\theta})$. The error $\boldsymbol{\varepsilon}_{\boldsymbol{\theta}}(\mathbf{y})$, centered error score $\mathbf{e}_{\boldsymbol{\theta}}(\mathbf{y})$, bias $\mathbf{b}(\boldsymbol{\theta})$, and error covariance $\mathbf{Q}(\boldsymbol{\theta})$ are illustrated in Fig. 6.1.

From the **Bayesian point of view**, we wish to determine the joint pdf $p(\mathbf{y}, \boldsymbol{\theta})$ from the set of pdfs $\{p(\mathbf{y}, \boldsymbol{\theta}), \boldsymbol{\theta} \in \Theta \subseteq \mathbb{C}^p\}$ that best describes the joint pdf from which the random measurement–parameter pair $(\mathbf{y}, \boldsymbol{\theta})$ was drawn. The estimator *error* is

$$\boldsymbol{\varepsilon}(\mathbf{y}, \boldsymbol{\theta}) = \hat{\boldsymbol{\theta}}(\mathbf{y}) - \boldsymbol{\theta} \tag{6.11}$$

and the *Bayes bias* of the estimator error is

$$\mathbf{b} = E[\boldsymbol{\varepsilon}(\mathbf{y}, \boldsymbol{\theta})] = E\{E_{\boldsymbol{\theta}}[\boldsymbol{\varepsilon}_{\boldsymbol{\theta}}(\mathbf{y})]\} = E[\mathbf{b}(\boldsymbol{\theta})] = E[\hat{\boldsymbol{\theta}}(\mathbf{y})] - E[\boldsymbol{\theta}]. \tag{6.12}$$

The Bayes bias is the average of the frequentist bias because the Bayesian and the frequentist define their errors in the same way: $\boldsymbol{\varepsilon}(\mathbf{y}, \boldsymbol{\theta}) = \boldsymbol{\varepsilon}_{\boldsymbol{\theta}}(\mathbf{y})$. The error then has the representation $\boldsymbol{\varepsilon}(\mathbf{y}, \boldsymbol{\theta}) = \mathbf{e}(\mathbf{y}, \boldsymbol{\theta}) + \mathbf{b}$, where the *centered error score* is defined to be

$$\mathbf{e}(\mathbf{y}, \boldsymbol{\theta}) = \boldsymbol{\varepsilon}(\mathbf{y}, \boldsymbol{\theta}) - \mathbf{b} = \hat{\boldsymbol{\theta}}(\mathbf{y}) - E[\hat{\boldsymbol{\theta}}(\mathbf{y})] - (\boldsymbol{\theta} - E[\boldsymbol{\theta}]). \tag{6.13}$$

The measurement score is defined to be the m-dimensional vector of complex scores $\boldsymbol{\sigma}(\mathbf{y}, \boldsymbol{\theta}) = [\sigma_1(\mathbf{y}, \boldsymbol{\theta}), \sigma_2(\mathbf{y}, \boldsymbol{\theta}), \ldots, \sigma_m(\mathbf{y}, \boldsymbol{\theta})]^{\mathrm{T}}$. The mean value of the measurement score is $E[\boldsymbol{\sigma}(\mathbf{y}, \boldsymbol{\theta})] = E\{E_{\boldsymbol{\theta}}[\boldsymbol{\sigma}(\mathbf{y}, \boldsymbol{\theta})]\}$ and the *centered measurement score* is

$$\mathbf{s}(\mathbf{y}, \boldsymbol{\theta}) = \boldsymbol{\sigma}(\mathbf{y}, \boldsymbol{\theta}) - E[\boldsymbol{\sigma}(\mathbf{y}, \boldsymbol{\theta})]. \tag{6.14}$$

The mean $E[\boldsymbol{\sigma}(\mathbf{y}, \boldsymbol{\theta})]$ is generally not $E\{E_{\boldsymbol{\theta}}[\boldsymbol{\sigma}_{\boldsymbol{\theta}}(\mathbf{y})]\}$ because the Bayesian will define a score $\boldsymbol{\sigma}(\mathbf{y}, \boldsymbol{\theta})$ that differs from the frequentist score $\boldsymbol{\sigma}_{\boldsymbol{\theta}}(\mathbf{y})$, in order to account for prior information about the parameter $\boldsymbol{\theta}$. Again we think of scores as judiciously chosen functions of the measurement that bring information about the true parameter.

The Bayesian defines the mean-squared error matrix for the estimator $\hat{\boldsymbol{\theta}}(\mathbf{y})$ to be $\mathbf{M} = E[\boldsymbol{\varepsilon}(\mathbf{y}, \boldsymbol{\theta})\boldsymbol{\varepsilon}^{\mathrm{H}}(\mathbf{y}, \boldsymbol{\theta})] = E\{E_{\boldsymbol{\theta}}[\boldsymbol{\varepsilon}_{\boldsymbol{\theta}}(\mathbf{y})\boldsymbol{\varepsilon}_{\boldsymbol{\theta}}^{\mathrm{H}}(\mathbf{y})]\} = E[\mathbf{M}(\boldsymbol{\theta})]$, which is the average over the prior distribution on $\boldsymbol{\theta}$ of the frequentist mean-squared error matrix $\mathbf{M}(\boldsymbol{\theta})$. This average may be written

$$\mathbf{M} = E[\boldsymbol{\varepsilon}(\mathbf{y}, \boldsymbol{\theta})\boldsymbol{\varepsilon}^{\mathrm{H}}(\mathbf{y}, \boldsymbol{\theta})] = E[\mathbf{e}(\mathbf{y}, \boldsymbol{\theta})\mathbf{e}^{\mathrm{H}}(\mathbf{y}, \boldsymbol{\theta})] + \mathbf{b}\mathbf{b}^{\mathrm{H}} = \mathbf{Q} + \mathbf{b}\mathbf{b}^{\mathrm{H}}, \tag{6.15}$$

where $\mathbf{Q}$ is the Bayes error covariance

$$\mathbf{Q} = E[\mathbf{e}(\mathbf{y}, \boldsymbol{\theta})\mathbf{e}^{\mathrm{H}}(\mathbf{y}, \boldsymbol{\theta})]. \tag{6.16}$$

Note that the Bayes bias and error covariance matrix are independent of the parameter $\boldsymbol{\theta}$. If the Bayes bias of the estimator is zero, then the mean-squared error matrix is the error covariance matrix: $\mathbf{M} = \mathbf{Q}$.

6.1.2 Connection between frequentist and Bayesian approaches

We have established that the Bayes bias $\mathbf{b} = E[\mathbf{b}(\boldsymbol{\theta})]$ is the average of the frequentist bias and the Bayes mean-squared error matrix $\mathbf{M} = E[\mathbf{M}(\boldsymbol{\theta})]$ is the average of the frequentist mean-squared error matrix. But what about the Bayes error covariance $\mathbf{Q}$? Is it the average over the distribution of the parameter $\boldsymbol{\theta}$ of the frequentist error covariance $\mathbf{Q}(\boldsymbol{\theta})$? The answer (surprisingly) is no, as the following argument shows.

The Bayesian's centered error score may be written in terms of the frequentist's centered error score as $\mathbf{e}(\mathbf{y}, \boldsymbol{\theta}) = \mathbf{e}_{\boldsymbol{\theta}}(\mathbf{y}) + (\mathbf{b}(\boldsymbol{\theta}) - \mathbf{b})$. In this representation, $\mathbf{e}_{\boldsymbol{\theta}}(\mathbf{y})$ and

$\mathbf{b}(\boldsymbol{\theta}) - \mathbf{b}$ are conditionally and unconditionally uncorrelated, so as a consequence the Bayesian's error covariance may be written in terms of the frequentist's parameter-dependent error covariance as

$$\mathbf{Q} = E[\mathbf{Q}(\boldsymbol{\theta})] + \mathbf{R}_{bb}, \tag{6.17}$$

where $\mathbf{R}_{bb} = E[(\mathbf{b}(\boldsymbol{\theta}) - \mathbf{b})(\mathbf{b}(\boldsymbol{\theta}) - \mathbf{b})^{\mathrm{H}}]$ is the *covariance of the parameter-dependent bias* $\mathbf{b}(\boldsymbol{\theta})$. If the frequentist bias of the estimator is constant at $\mathbf{b}(\boldsymbol{\theta}) = \mathbf{b}$, then $\mathbf{R}_{bb} = \mathbf{0}$ and the Bayes error covariance is the average over the distribution of the parameter $\boldsymbol{\theta}$ of the frequentist error covariance. In other words, we have the following result.

Result 6.1. *The frequentist error covariance can be averaged for the Bayes error covariance to produce the identity*

$$\mathbf{Q} = E[\mathbf{Q}(\boldsymbol{\theta})] \tag{6.18}$$

if and only if *the frequentist bias of the estimator is constant.*

This generally is a property of linear minimum-variance distortionless response (LMVDR) estimators, also called best linear unbiased estimators (BLUEs) – for which $\mathbf{b}(\boldsymbol{\theta}) = \mathbf{0}$ – but not of MMSE estimators. The following example is illuminating.

Example 6.1. Consider the linear signal-plus-noise model

$$\mathbf{y} = \mathbf{H}\boldsymbol{\theta} + \mathbf{n}, \tag{6.19}$$

where $\mathbf{H} \in \mathbb{C}^{n\times p}$, $\boldsymbol{\theta} \in \mathbb{C}^p$ is zero-mean proper Gaussian with covariance matrix $\mathbf{R}_{\theta\theta}$, $\mathbf{n}$ is zero-mean proper Gaussian with covariance matrix $\mathbf{R}_{nn}$, and $\boldsymbol{\theta}$ and $\mathbf{n}$ are independent. Following Section 5.6, the LMVDR estimator (also called BLUE), which in this case is also the maximum-likelihood (ML) estimator, is

$$\hat{\boldsymbol{\theta}}_{\mathrm{LMVDR}}(\mathbf{y}) = (\mathbf{H}^{\mathrm{H}}\mathbf{R}_{nn}^{-1}\mathbf{H})^{-1}\mathbf{H}^{\mathrm{H}}\mathbf{R}_{nn}^{-1}\mathbf{y} = \boldsymbol{\theta} + (\mathbf{H}^{\mathrm{H}}\mathbf{R}_{nn}^{-1}\mathbf{H})^{-1}\mathbf{H}^{\mathrm{H}}\mathbf{R}_{nn}^{-1}\mathbf{n}, \tag{6.20}$$

which ignores the prior distribution on $\boldsymbol{\theta}$. Consequently the frequentist bias is $\mathbf{b}(\boldsymbol{\theta}) = \mathbf{0}$ and the Bayes bias is $\mathbf{b} = \mathbf{0}$. The centered frequentist error is $\mathbf{e}_{\boldsymbol{\theta}}(\mathbf{y}) = (\mathbf{H}^{\mathrm{H}}\mathbf{R}_{nn}^{-1}\mathbf{H})^{-1}\mathbf{H}^{\mathrm{H}}\mathbf{R}_{nn}^{-1}\mathbf{n}$. The frequentist error covariance is then $\mathbf{Q}(\boldsymbol{\theta}) = (\mathbf{H}^{\mathrm{H}}\mathbf{R}_{nn}^{-1}\mathbf{H})^{-1}$ and this is also the Bayes error covariance $\mathbf{Q}$:

$$\mathbf{Q}_{\mathrm{LMVDR}} = (\mathbf{H}^{\mathrm{H}}\mathbf{R}_{nn}^{-1}\mathbf{H})^{-1} = E[\mathbf{Q}_{\mathrm{LMVDR}}(\boldsymbol{\theta})] = \mathbf{Q}_{\mathrm{LMVDR}}(\boldsymbol{\theta}). \tag{6.21}$$

The MMSE estimator accounts for the distribution of $\boldsymbol{\theta}$:

$$\begin{aligned}\hat{\boldsymbol{\theta}}_{\mathrm{MMSE}}(\mathbf{y}) &= (\mathbf{H}^{\mathrm{H}}\mathbf{R}_{nn}^{-1}\mathbf{H} + \mathbf{R}_{\theta\theta}^{-1})^{-1}\mathbf{H}^{\mathrm{H}}\mathbf{R}_{nn}^{-1}\mathbf{y} \\ &= (\mathbf{H}^{\mathrm{H}}\mathbf{R}_{nn}^{-1}\mathbf{H} + \mathbf{R}_{\theta\theta}^{-1})^{-1}\mathbf{H}^{\mathrm{H}}\mathbf{R}_{nn}^{-1}\mathbf{H}\boldsymbol{\theta} + (\mathbf{H}^{\mathrm{H}}\mathbf{R}_{nn}^{-1}\mathbf{H} + \mathbf{R}_{\theta\theta}^{-1})^{-1}\mathbf{H}^{\mathrm{H}}\mathbf{R}_{nn}^{-1}\mathbf{n}.\end{aligned} \tag{6.22}$$

The frequentist bias of this estimator is $\mathbf{b}(\boldsymbol{\theta}) = (\mathbf{H}^{\mathrm{H}}\mathbf{R}_{nn}^{-1}\mathbf{H} + \mathbf{R}_{\theta\theta}^{-1})^{-1}\mathbf{H}^{\mathrm{H}}\mathbf{R}_{nn}^{-1}\mathbf{H}\boldsymbol{\theta} - \boldsymbol{\theta}$ and the Bayes bias is $\mathbf{b} = \mathbf{0}$. The frequentist error covariance of this estimator is $\mathbf{Q}(\boldsymbol{\theta}) = (\mathbf{H}^{\mathrm{H}}\mathbf{R}_{nn}^{-1}\mathbf{H} + \mathbf{R}_{\theta\theta}^{-1})^{-1}\mathbf{H}^{\mathrm{H}}\mathbf{R}_{nn}\mathbf{H}(\mathbf{H}^{\mathrm{H}}\mathbf{R}_{nn}^{-1}\mathbf{H} + \mathbf{R}_{\theta\theta}^{-1})^{-1}$. The Bayes error covariance is the sum of the expectation of the frequentist error covariance $\mathbf{Q}(\boldsymbol{\theta})$ and the covariance of the parameter-dependent bias $\mathbf{R}_{bb} = E[\mathbf{b}(\boldsymbol{\theta})\mathbf{b}^{\mathrm{H}}(\boldsymbol{\theta})]$. But, since the frequentist error covariance is independent of $\boldsymbol{\theta}$, we may write

$$\begin{aligned}\mathbf{Q}_{\mathrm{MMSE}} &= \mathbf{Q}_{\mathrm{MMSE}}(\boldsymbol{\theta}) + \mathbf{R}_{bb} = (\mathbf{H}^{\mathrm{H}}\mathbf{R}_{nn}^{-1}\mathbf{H} + \mathbf{R}_{\theta\theta}^{-1})^{-1} \\ &\leq (\mathbf{H}^{\mathrm{H}}\mathbf{R}_{nn}^{-1}\mathbf{H})^{-1} = \mathbf{Q}_{\mathrm{LMVDR}}(\boldsymbol{\theta}) = \mathbf{Q}_{\mathrm{LMVDR}}. \end{aligned} \tag{6.23}$$

There are lessons here: the MMSE estimator has smaller Bayes error covariance than the LMVDR estimator; it also has smaller frequentist error covariance. Its only vice is that it is not frequentist unbiased. So, from the point of view of Bayes error covariance, frequentist unbiasedness is no virtue. In fact, it is uncommon for the frequentist bias of an estimator of a random parameter to be zero, but common for the Bayes bias to be zero.

6.1.3 Extension to augmented errors

These results extend to the characterization of augmented error scores $\underline{\mathbf{e}}^{\mathrm{T}} = [\mathbf{e}^{\mathrm{T}}, \mathbf{e}^{\mathrm{H}}]$. The augmented Bayes bias remains the average of the augmented frequentist bias, which is to say that $\underline{\mathbf{b}} = E[\underline{\mathbf{b}}(\boldsymbol{\theta})]$, the augmented Bayes mean-squared error matrix remains the average of the augmented frequentist mean-squared error matrix, $\underline{\mathbf{M}} = E[\underline{\mathbf{M}}(\boldsymbol{\theta})]$, and the Bayes and frequentist augmented error covariances remain related as

$$\underline{\mathbf{Q}} = E[\underline{\mathbf{Q}}(\boldsymbol{\theta})] + \underline{\mathbf{R}}_{bb}, \tag{6.24}$$

where $\underline{\mathbf{Q}}(\boldsymbol{\theta}) = E_{\boldsymbol{\theta}}[\underline{\mathbf{e}}(\mathbf{y})\underline{\mathbf{e}}^{\mathrm{H}}(\mathbf{y})]$ is the augmented frequentist error covariance matrix and $\underline{\mathbf{R}}_{bb} = E[(\underline{\mathbf{b}}(\boldsymbol{\theta}) - \underline{\mathbf{b}})(\underline{\mathbf{b}}(\boldsymbol{\theta}) - \underline{\mathbf{b}})^{\mathrm{H}}]$ is the augmented covariance matrix for the frequentist bias.

6.2 Quadratic frequentist bounds

Let us begin our discussion of quadratic *frequentist* covariance bounds by considering the Hermitian version of the story, where complementary covariances are ignored in the bounding procedure. Then we will extend these bounds by accounting for complementary covariances and find that this tightens the bounds.

Here is the idea: from the measurement $\mathbf{y}$ is computed the estimator $\hat{\boldsymbol{\theta}}(\mathbf{y})$, with frequentist error $\boldsymbol{\varepsilon}_{\boldsymbol{\theta}}(\mathbf{y}) = \hat{\boldsymbol{\theta}}(\mathbf{y}) - \boldsymbol{\theta}$ and centered error score $\mathbf{e}_{\boldsymbol{\theta}}(\mathbf{y}) = \boldsymbol{\varepsilon}_{\boldsymbol{\theta}}(\mathbf{y}) - \mathbf{b}(\boldsymbol{\theta})$. We would like to approximate this centered error score by using a linear function of the centered measurement score $\mathbf{s}_{\boldsymbol{\theta}}(\mathbf{y})$.

6.2.1 The virtual two-channel experiment and the quadratic frequentist bound

Consider the virtual two-channel estimation problem of Fig. 6.2, which is reminiscent of the virtual two-channel problems developed in Section 5.3. In Fig. 6.2, the centered error score $\mathbf{e}_{\boldsymbol{\theta}}(\mathbf{y})$ is to be estimated from the centered measurement score $\mathbf{s}_{\boldsymbol{\theta}}(\mathbf{y})$. The first

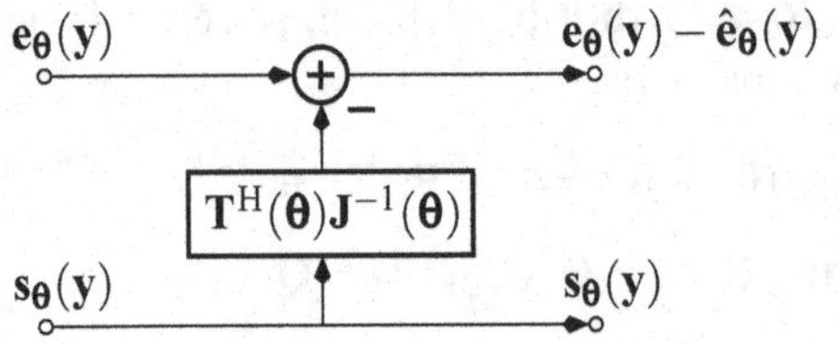

Figure 6.2 Two-channel representation of centered error $\mathbf{e}_{\boldsymbol{\theta}}(\mathbf{y})$ and centered score $\mathbf{s}_{\boldsymbol{\theta}}(\mathbf{y})$.

step in the derivation of quadratic performance bounds is to define the *composite* vector of the error score and the measurement score, $[\mathbf{e}_{\boldsymbol{\theta}}^{\mathrm{T}}(\mathbf{y}), \mathbf{s}_{\boldsymbol{\theta}}^{\mathrm{T}}(\mathbf{y})]^{\mathrm{T}}$. The covariance matrix for this composite vector is then

$$\mathbb{R}_{es}(\boldsymbol{\theta}) = E_{\boldsymbol{\theta}}\left\{\begin{bmatrix}\mathbf{e}_{\boldsymbol{\theta}}(\mathbf{y})\\ \mathbf{s}_{\boldsymbol{\theta}}(\mathbf{y})\end{bmatrix}\begin{bmatrix}\mathbf{e}_{\boldsymbol{\theta}}^{\mathrm{H}}(\mathbf{y}) & \mathbf{s}_{\boldsymbol{\theta}}^{\mathrm{H}}(\mathbf{y})\end{bmatrix}\right\} = \begin{bmatrix}\mathbf{Q}(\boldsymbol{\theta}) & \mathbf{T}^{\mathrm{H}}(\boldsymbol{\theta})\\ \mathbf{T}(\boldsymbol{\theta}) & \mathbf{J}(\boldsymbol{\theta})\end{bmatrix}. \tag{6.25}$$

In this equation the covariance matrices $\mathbf{Q}(\boldsymbol{\theta})$, $\mathbf{T}(\boldsymbol{\theta})$, and $\mathbf{J}(\boldsymbol{\theta})$ are defined as

$$\begin{aligned}\mathbf{Q}(\boldsymbol{\theta}) &= E_{\boldsymbol{\theta}}[\mathbf{e}_{\boldsymbol{\theta}}(\mathbf{y})\mathbf{e}_{\boldsymbol{\theta}}^{\mathrm{H}}(\mathbf{y})],\\ \mathbf{T}(\boldsymbol{\theta}) &= E_{\boldsymbol{\theta}}[\mathbf{s}_{\boldsymbol{\theta}}(\mathbf{y})\mathbf{e}_{\boldsymbol{\theta}}^{\mathrm{H}}(\mathbf{y})],\\ \mathbf{J}(\boldsymbol{\theta}) &= E_{\boldsymbol{\theta}}[\mathbf{s}_{\boldsymbol{\theta}}(\mathbf{y})\mathbf{s}_{\boldsymbol{\theta}}^{\mathrm{H}}(\mathbf{y})].\end{aligned} \tag{6.26}$$

The covariance matrix $\mathbf{Q}(\boldsymbol{\theta})$ is the frequentist covariance for the centered error score, $\mathbf{T}(\boldsymbol{\theta})$ is the cross-correlation between the centered error score and the centered measurement score, which is often called the *sensitivity* or *expansion-coefficient matrix*, and $\mathbf{J}(\boldsymbol{\theta})$ is the covariance matrix for the centered measurement score, which is always called the *information matrix*. We shall assume it to be nonsingular, since there is no good reason to carry around linearly dependent scores. In the case of a Fisher score, to be discussed in Section 6.3, the information is *Fisher information*.

The LMMSE estimator of the centered error score from the centered measurement score is $\hat{\mathbf{e}}_{\boldsymbol{\theta}}(\mathbf{y}) = \mathbf{T}^{\mathrm{H}}(\boldsymbol{\theta})\mathbf{J}^{-1}(\boldsymbol{\theta})\mathbf{s}_{\boldsymbol{\theta}}(\mathbf{y})$. The nonnegative error covariance matrix for this estimator is the Schur complement $\mathbf{Q}(\boldsymbol{\theta}) - \mathbf{T}^{\mathrm{H}}(\boldsymbol{\theta})\mathbf{J}^{-1}(\boldsymbol{\theta})\mathbf{T}(\boldsymbol{\theta})$. From this follows the frequentist bound:

Result 6.2. *The error covariance matrix is bounded by the general quadratic frequentist bound*

$$\mathbf{Q}(\boldsymbol{\theta}) \geq \mathbf{T}^{\mathrm{H}}(\boldsymbol{\theta})\mathbf{J}^{-1}(\boldsymbol{\theta})\mathbf{T}(\boldsymbol{\theta}), \tag{6.27}$$

and the mean-squared error is bounded as

$$\mathbf{M}(\boldsymbol{\theta}) \geq \mathbf{T}^{\mathrm{H}}(\boldsymbol{\theta})\mathbf{J}^{-1}(\boldsymbol{\theta})\mathbf{T}(\boldsymbol{\theta}) + \mathbf{b}(\boldsymbol{\theta})\mathbf{b}^{\mathrm{H}}(\boldsymbol{\theta}). \tag{6.28}$$

This argument could have been simplified by saying that the Hermitian covariance matrix $\mathbb{R}_{es}(\boldsymbol{\theta})$ for the composite error score and measurement score is positive semidefinite, so the Schur complement must be positive semidefinite. But this argument would not have revealed the essence of quadratic performance bounding, which is to *linearly estimate centered error scores from centered measurement scores.*

As noted by Weinstein and Weiss (1988), the Battacharrya, Ziv–Zakai, Barankin, and Bobrovsky–Zakai bounds all have this quadratic structure. Only the choice of score distinguishes one bound from another. Moreover, each quadratic bound is invariant with respect to nonsingular linear transformation of the score function $\mathbf{s}$. That is, the transformed score $\mathbf{Ls}$ transforms the sensitivity matrix $\mathbf{T}$ to $\mathbf{LT}$ and the information matrix $\mathbf{J}$ to $\mathbf{LJL}^{\mathrm{H}}$, leaving the bound invariant.

Definition 6.2. *An estimator is said to be* efficient *with respect to a given score if the bound in* (6.27) *is achieved with equality:* $\mathbf{Q}(\boldsymbol{\theta}) = \mathbf{T}^{\mathrm{H}}(\boldsymbol{\theta})\mathbf{J}^{-1}(\boldsymbol{\theta})\mathbf{T}(\boldsymbol{\theta})$.

That is, the estimated error score equals the actual error score:

$$\mathbf{e}_{\boldsymbol{\theta}}(\mathbf{y}) = \mathbf{T}^{\mathrm{H}}(\boldsymbol{\theta})\mathbf{J}^{-1}(\boldsymbol{\theta})\mathbf{s}_{\boldsymbol{\theta}}(\mathbf{y}). \tag{6.29}$$

This is also invariant with respect to linear transformation $\mathbf{L}$. That is, an estimator is efficient if, element-by-element, its centered error score lies in the subspace spanned by the elements of the centered measurement score. The corresponding representation for the estimator itself is $\hat{\boldsymbol{\theta}}(\mathbf{y}) - \boldsymbol{\theta} = \mathbf{T}^{\mathrm{H}}(\boldsymbol{\theta})\mathbf{J}^{-1}(\boldsymbol{\theta})\mathbf{s}_{\boldsymbol{\theta}}(\mathbf{y})$.

One score is better than another if its quadratic bound is larger than the other. So the search goes on for a better bound, or in fact a better score.

Good and bad scores

We shall not give the details here, but McWhorter and Scharf (1993c) showed that a given measurement score can be improved, in the sense that it improves the quadratic frequentist bound, by "Rao–Blackwellizing" it. To "Rao–Blackwellize" is to compute the expected value of the centered measurement score, conditioned on the frequentist sufficient statistic for the parameter. In other words, to be admissible a score must be a function of the sufficient statistic for the parameter to be estimated. Moreover, additional score functions never hurt, which is to say that the quadratic performance bound is never loosened with the addition of additional scores.

6.2.2 Projection-operator and integral-operator representations of quadratic frequentist bounds

Let us consider the variance bound for one component of the parameter vector $\boldsymbol{\theta}$. Call it θ. The corresponding component of the estimator $\hat{\boldsymbol{\theta}}(\mathbf{y})$ is called $\hat{\theta}(\mathbf{y})$, and the corresponding component of the centered error score is $e_\theta(\mathbf{y}) = \hat{\theta}(\mathbf{y}) - E_{\boldsymbol{\theta}}[\hat{\theta}(\mathbf{y})]$. Without loss of generality, assume that scores have been transformed into a coordinate system where the information matrix is $\mathbf{J}(\theta) = \mathbf{I}$. The composite covariance matrix for $[e_\theta(\mathbf{y}), \mathbf{s}_\theta^{\mathrm{T}}(\mathbf{y})]^{\mathrm{T}}$ is

$$\mathbb{R}_{es}(\theta) = E_{\boldsymbol{\theta}}\left\{\begin{bmatrix} e_\theta(\mathbf{y}) \\ \mathbf{s}_\theta(\mathbf{y}) \end{bmatrix}\begin{bmatrix} e_\theta^*(\mathbf{y}) & \mathbf{s}_\theta^{\mathrm{H}}(\mathbf{y}) \end{bmatrix}\right\} = \begin{bmatrix} Q(\theta) & \mathbf{t}^{\mathrm{H}}(\theta) \\ \mathbf{t}(\theta) & \mathbf{I} \end{bmatrix}, \tag{6.30}$$

where $\mathbf{t}(\theta) = E_{\boldsymbol{\theta}}[\mathbf{s}_\theta(\mathbf{y})e_\theta^*(\mathbf{y})]$ is the m-dimensional column vector of expansion coefficients in this transformed coordinate system. Then the invariant quadratic frequentist

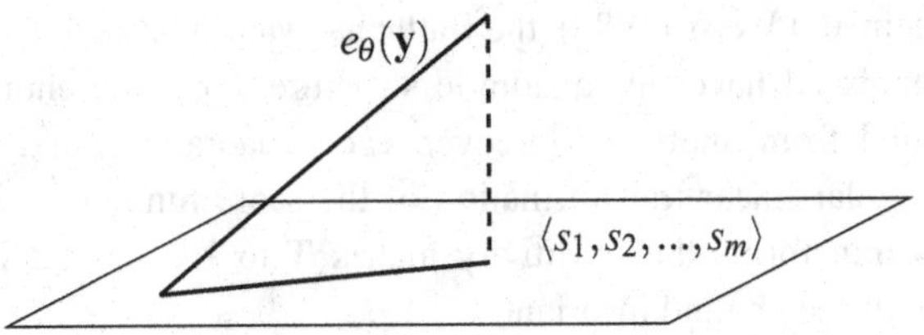

Figure 6.3 Projection of the error $e_\theta(\mathbf{y})$ onto the Hilbert space of centered error scores $s_1, s_2, \ldots, s_m$.

bound may be written

$$Q(\theta) \geq \mathbf{t}^{\mathrm{H}}(\theta)\mathbf{t}(\theta) = \sum_{i=1}^{m} |t_i(\theta)|^2, \tag{6.31}$$

where $t_i(\theta)$ is the ith coordinate of the m-dimensional expansion-coefficient vector $\mathbf{t}(\theta)$. The score random variables are uncorrelated with variance unity, so we may write the previous equation as

$$Q(\theta) \geq E_{\boldsymbol{\theta}}\left[\sum_{i=1}^{m}\sum_{j=1}^{m} t_i s_i^* t_j^* s_j\right] = E_{\boldsymbol{\theta}}|\Pi_s e_\theta(\mathbf{y})|^2, \tag{6.32}$$

where the operator $\Pi_s e_\theta(\mathbf{y})$ acts like this:

$$\Pi_s e_\theta(\mathbf{y}) = \sum_{i=1}^{m} t_i^* s_i = \sum_{i=1}^{m} E_{\boldsymbol{\theta}}[s_i^*(\mathbf{y}) e_\theta(\mathbf{y})] s_i(\mathbf{y}). \tag{6.33}$$

This projection operator projects the centered error score $e_\theta(\mathbf{y})$ onto each element of the centered measurement score (or onto the Hilbert space of centered measurement scores), as illustrated in Fig. 6.3. Thus, for the bound (6.31) to be tight, the actual error score $e_\theta(\mathbf{y})$ must lie close to the subspace spanned by the m centered measurement scores. This interpretation makes it easy to see that the elements of the vector $\mathbf{t}(\theta)$ are just the coefficients of the true error $e_\theta(\mathbf{y})$ in the basis $\mathbf{s}_\theta(\mathbf{y})$. That is, $t_i(\theta) = E_{\boldsymbol{\theta}}[s_i(\mathbf{y}) e_\theta^*(\mathbf{y})]$.

There is also a kernel representation of this bound due to McWhorter and Scharf (1993c). Write the projection as

$$\begin{aligned} \Pi_s e_\theta(\mathbf{y}) &= \sum_{i=1}^{m} t_i^* s_i = \int e_\theta(\mathbf{y}') \sum_{i=1}^{m} s_i(\mathbf{y}) s_i^*(\mathbf{y}') p_\theta(\mathbf{y}') d\mathbf{y}' \\ &= \int e_\theta(\mathbf{y}') K_\theta(\mathbf{y}, \mathbf{y}') p_\theta(\mathbf{y}') d\mathbf{y}'. \end{aligned} \tag{6.34}$$

So, for a tight bound, $\Pi_s e_\theta(\mathbf{y})$ must be nearly the actual error $e_\theta(\mathbf{y})$, which is to say that the actual error $e_\theta(\mathbf{y})$ must nearly be an element of the reproducing kernel Hilbert space defined by the kernel

$$K_\theta(\mathbf{y}, \mathbf{y}') = \sum_{i=1}^{m} s_i(\mathbf{y}) s_i^*(\mathbf{y}'). \tag{6.35}$$

6.2.3 Extension of the quadratic frequentist bound to improper errors and scores

The essential step in the extension of quadratic frequentist bounds to improper error and measurement scores is to define the composite vector of *augmented* errors and scores, $[\underline{\mathbf{e}}^{\mathrm{T}}, \underline{\mathbf{s}}^{\mathrm{T}}]^{\mathrm{T}}$. The covariance matrix for this composite vector is

$$E_{\boldsymbol{\theta}}\left\{\begin{bmatrix}\underline{\mathbf{e}}_{\boldsymbol{\theta}}(\mathbf{y})\\ \underline{\mathbf{s}}_{\boldsymbol{\theta}}(\mathbf{y})\end{bmatrix}\begin{bmatrix}\underline{\mathbf{e}}_{\boldsymbol{\theta}}^{\mathrm{H}}(\mathbf{y}) & \underline{\mathbf{s}}_{\boldsymbol{\theta}}^{\mathrm{H}}(\mathbf{y})\end{bmatrix}\right\} = \begin{bmatrix}\underline{\mathbf{Q}}(\boldsymbol{\theta}) & \underline{\mathbf{T}}^{\mathrm{H}}(\boldsymbol{\theta})\\ \underline{\mathbf{T}}(\boldsymbol{\theta}) & \underline{\mathbf{J}}(\boldsymbol{\theta})\end{bmatrix}. \tag{6.36}$$

In this equation the constituent matrices are defined as

$$\underline{\mathbf{Q}}(\boldsymbol{\theta}) = E_{\boldsymbol{\theta}}[\underline{\mathbf{e}}_{\boldsymbol{\theta}}(\mathbf{y})\underline{\mathbf{e}}_{\boldsymbol{\theta}}^{\mathrm{H}}(\mathbf{y})] = \begin{bmatrix}\mathbf{Q}(\boldsymbol{\theta}) & \widetilde{\mathbf{Q}}(\boldsymbol{\theta})\\ \widetilde{\mathbf{Q}}^{*}(\boldsymbol{\theta}) & \mathbf{Q}^{*}(\boldsymbol{\theta})\end{bmatrix},$$

$$\underline{\mathbf{T}}(\boldsymbol{\theta}) = E_{\boldsymbol{\theta}}[\underline{\mathbf{s}}_{\boldsymbol{\theta}}(\mathbf{y})\underline{\mathbf{e}}_{\boldsymbol{\theta}}^{\mathrm{H}}(\mathbf{y})] = \begin{bmatrix}\mathbf{T}(\boldsymbol{\theta}) & \widetilde{\mathbf{T}}(\boldsymbol{\theta})\\ \widetilde{\mathbf{T}}^{*}(\boldsymbol{\theta}) & \mathbf{T}^{*}(\boldsymbol{\theta})\end{bmatrix}, \tag{6.37}$$

$$\underline{\mathbf{J}}(\boldsymbol{\theta}) = E_{\boldsymbol{\theta}}[\underline{\mathbf{s}}_{\boldsymbol{\theta}}(\mathbf{y})\underline{\mathbf{s}}_{\boldsymbol{\theta}}^{\mathrm{H}}(\mathbf{y})] = \begin{bmatrix}\mathbf{J}(\boldsymbol{\theta}) & \widetilde{\mathbf{J}}(\boldsymbol{\theta})\\ \widetilde{\mathbf{J}}^{*}(\boldsymbol{\theta}) & \mathbf{J}^{*}(\boldsymbol{\theta})\end{bmatrix}.$$

The covariance matrix $\widetilde{\mathbf{T}}(\boldsymbol{\theta}) = E_{\boldsymbol{\theta}}[\mathbf{s}_{\boldsymbol{\theta}}(\mathbf{y})\mathbf{e}_{\boldsymbol{\theta}}^{\mathrm{T}}(\mathbf{y})]$ is the complementary sensitivity or expansion-coefficient matrix, and $\widetilde{\mathbf{J}}(\boldsymbol{\theta}) = E_{\boldsymbol{\theta}}[\mathbf{s}_{\boldsymbol{\theta}}(\mathbf{y})\mathbf{s}_{\boldsymbol{\theta}}^{\mathrm{T}}(\mathbf{y})]$ is the complementary covariance matrix for the centered measurement score, which is called the complementary information matrix. Furthermore, $\underline{\mathbf{Q}}(\boldsymbol{\theta})$ is the augmented error covariance matrix, $\underline{\mathbf{T}}(\boldsymbol{\theta})$ is the augmented sensitivity or expansion-coefficient matrix, and $\underline{\mathbf{J}}(\boldsymbol{\theta})$ is the augmented information matrix. Obviously, we are interested in bounding $\mathbf{Q}(\boldsymbol{\theta})$ rather than $\underline{\mathbf{Q}}(\boldsymbol{\theta})$, but it is convenient to derive the bounds on $\underline{\mathbf{Q}}(\boldsymbol{\theta})$ first, from which one can immediately obtain the bound on $\mathbf{Q}(\boldsymbol{\theta})$.

The covariance matrix for the composite vector of augmented error score and augmented measurement score is positive semidefinite, so, assuming the information matrix to be nonsingular, we obtain the bound $\underline{\mathbf{Q}}(\boldsymbol{\theta}) \geq \underline{\mathbf{T}}^{\mathrm{H}}(\boldsymbol{\theta})\underline{\mathbf{J}}^{-1}(\boldsymbol{\theta})\underline{\mathbf{T}}(\boldsymbol{\theta})$. From this, we can read out the northwest block of $\underline{\mathbf{Q}}(\boldsymbol{\theta})$ to obtain the following result.

Result 6.3. *The frequentist error covariance is bounded as*

$$\mathbf{Q}(\boldsymbol{\theta}) \geq \begin{bmatrix}\mathbf{T}^{\mathrm{H}}(\boldsymbol{\theta}) & \widetilde{\mathbf{T}}^{\mathrm{T}}(\boldsymbol{\theta})\end{bmatrix}\underline{\mathbf{J}}^{-1}(\boldsymbol{\theta})\begin{bmatrix}\mathbf{T}(\boldsymbol{\theta})\\ \widetilde{\mathbf{T}}^{*}(\boldsymbol{\theta})\end{bmatrix}. \tag{6.38}$$

The quadratic form on the right-hand side is the general widely linear-quadratic frequentist bound on the error covariance matrix for improper error and measurement scores. Of course the underlying idea is that $\underline{\mathbf{Q}}(\boldsymbol{\theta}) - \underline{\mathbf{T}}^{\mathrm{H}}(\boldsymbol{\theta})\underline{\mathbf{J}}^{-1}(\boldsymbol{\theta})\underline{\mathbf{T}}(\boldsymbol{\theta})$ is the augmented error covariance matrix of the *widely linear* estimator of the error score from the measurement score:

$$\hat{\mathbf{e}}_{\boldsymbol{\theta}}(\mathbf{y}) = \begin{bmatrix}\mathbf{T}^{\mathrm{H}}(\boldsymbol{\theta}) & \widetilde{\mathbf{T}}^{\mathrm{T}}(\boldsymbol{\theta})\end{bmatrix}\underline{\mathbf{J}}^{-1}(\boldsymbol{\theta})\underline{\mathbf{s}}_{\boldsymbol{\theta}}(\mathbf{y}). \tag{6.39}$$

The bound in Result 6.3 is tighter than the bound in Result 6.2, but for proper measurement scores (i.e., $\widetilde{\mathbf{J}}(\boldsymbol{\theta}) = \mathbf{0}$) and cross-proper measurement and error scores (i.e., $\widetilde{\mathbf{T}}(\boldsymbol{\theta}) = \mathbf{0}$) the bounds are identical.

A widely linear estimator is efficient if equality holds in (6.38). This is the case *if and only if* the estimator of the centered error score equals the actual error, meaning that the estimator has the representation

$$\hat{\boldsymbol{\theta}}(\mathbf{y}) - \boldsymbol{\theta} = \begin{bmatrix} \mathbf{T}^{\mathrm{H}}(\boldsymbol{\theta}) & \widetilde{\mathbf{T}}^{\mathrm{T}}(\boldsymbol{\theta}) \end{bmatrix} \underline{\mathbf{J}}^{-1}(\boldsymbol{\theta})\underline{\mathbf{s}}_{\boldsymbol{\theta}}(\mathbf{y}). \tag{6.40}$$

6.3 Fisher score and the Cramér–Rao bound

Our results for quadratic frequentist bounds so far are general. To make them applicable we need to consider a concrete score for which $\mathbf{T}(\boldsymbol{\theta})$ and $\mathbf{J}(\boldsymbol{\theta})$ can be computed. For this we choose the Fisher score and compute its associated Cramér–Rao bound.

Definition 6.3. *The* Fisher score *is defined as*

$$\boldsymbol{\sigma}_{\boldsymbol{\theta}}(\mathbf{y}) = \left[\frac{\partial}{\partial \boldsymbol{\theta}} \log p_{\boldsymbol{\theta}}(\mathbf{y})\right]^{\mathrm{H}} = \left[\frac{\partial}{\partial \theta_1^*} \log p_{\boldsymbol{\theta}}(\mathbf{y}), \ldots, \frac{\partial}{\partial \theta_p^*} \log p_{\boldsymbol{\theta}}(\mathbf{y})\right]^{\mathrm{T}}, \tag{6.41}$$

where the partial derivatives are Wirtinger derivatives *as discussed in Appendix 2.*

The ith component of the Fisher score is

$$\sigma_i(\mathbf{y}) = \frac{\partial}{\partial(\operatorname{Re}\theta_i)} \log p_{\boldsymbol{\theta}}(\mathbf{y}) + \mathrm{j}\frac{\partial}{\partial(\operatorname{Im}\theta_i)} \log p_{\boldsymbol{\theta}}(\mathbf{y}). \tag{6.42}$$

Thus, the Fisher score is a p-dimensional complex-valued column vector. The notation $(\partial/\partial\boldsymbol{\theta})p_{\boldsymbol{\theta}}(\mathbf{y})$ means $(\partial/\partial\boldsymbol{\phi})p_{\boldsymbol{\phi}}(\mathbf{y})$, evaluated at $\boldsymbol{\phi} = \boldsymbol{\theta}$. We shall demonstrate shortly that the expected value of the Fisher score is $E_{\boldsymbol{\theta}}[\boldsymbol{\sigma}_{\boldsymbol{\theta}}(\mathbf{y})] = \mathbf{0}$, so that Definition 6.3 also defines the centered measurement score $\mathbf{s}_{\boldsymbol{\theta}}(\mathbf{y})$.

The Fisher score has a number of properties that make it a compelling statistic for inferring the value of a parameter and for bounding the error covariance of any estimator for that parameter. We list these properties and annotate them here.

1. We may write the partial derivative as

$$\frac{\partial}{\partial \theta_i^*} \log p_{\boldsymbol{\theta}}(\mathbf{y}) = \frac{1}{p_{\boldsymbol{\theta}}(\mathbf{y})} \frac{\partial}{\partial \theta_i^*} p_{\boldsymbol{\theta}}(\mathbf{y}), \tag{6.43}$$

which is a normalized measure of the sensitivity of the pdf $p_{\boldsymbol{\theta}}(\mathbf{y})$ to variations in the parameter θ_i. Large sensitivity is valued, and this will be measured by the variance of the score.

2. The Fisher score is a zero-mean random variable. This statistical property is consistent with maximum-likelihood procedures that search for maxima of the log-likelihood function $\log p_{\boldsymbol{\theta}}(\mathbf{y})$ by searching for the zeros of $(\partial/\partial\boldsymbol{\theta})\log p_{\boldsymbol{\theta}}(\mathbf{y})$. Functions with steep slopes are valued, so we value a zero-mean score with large variance.

3. The cross-correlation between the centered Fisher score and the centered error score is the expansion-coefficient matrix

$$\mathbf{T}(\boldsymbol{\theta}) = \mathbf{I} + \left[\frac{\partial}{\partial \boldsymbol{\theta}} \mathbf{b}(\boldsymbol{\theta})\right]^{\mathrm{H}}, \tag{6.44}$$

where as before $\mathbf{b}(\boldsymbol{\theta})$ is the frequentist bias of the estimator. The (i, j)th element of the matrix $(\partial/\partial\boldsymbol{\theta})\mathbf{b}(\boldsymbol{\theta})$ is $(\partial/\partial\theta_j)b_i(\boldsymbol{\theta})$. When the estimator is unbiased, then $\mathbf{T} = \mathbf{I}$, which is to say that the measurement score is perfectly aligned with the error score. The effect of bias is to misalign them.

4. The Fisher information matrix is also the expected Hessian of the score function (up to a minus sign):

$$\mathbf{J}_F(\boldsymbol{\theta}) = E_{\boldsymbol{\theta}}\left\{\left[\frac{\partial}{\partial\boldsymbol{\theta}}\log p_{\boldsymbol{\theta}}(\mathbf{y})\right]^H \frac{\partial}{\partial\boldsymbol{\theta}}\log p_{\boldsymbol{\theta}}(\mathbf{y})\right\}$$

$$= -E_{\boldsymbol{\theta}}\left\{\frac{\partial}{\partial\boldsymbol{\theta}}\left[\frac{\partial}{\partial\boldsymbol{\theta}}\log p_{\boldsymbol{\theta}}(\mathbf{y})\right]^H\right\}. \tag{6.45}$$

The first of these four properties is interpretive only. The next three are interpretive and analytical. Let's prove them.

2. The expected value of the Fisher score is

$$E_{\boldsymbol{\theta}}[\boldsymbol{\sigma}_{\boldsymbol{\theta}}(\mathbf{y})] = \int\left[\frac{\partial}{\partial\boldsymbol{\theta}}\log p_{\boldsymbol{\theta}}(\mathbf{y})\right]^H p_{\boldsymbol{\theta}}(\mathbf{y})d\mathbf{y} = \int\left[\frac{\partial}{\partial\boldsymbol{\theta}}p_{\boldsymbol{\theta}}(\mathbf{y})\right]^H d\mathbf{y} = \mathbf{0}. \tag{6.46}$$

3. The frequentist bias $\mathbf{b}(\boldsymbol{\theta})$ of the parameter estimator $\hat{\boldsymbol{\theta}}(\mathbf{y})$ may be differentiated with respect to $\boldsymbol{\theta}$ as follows:

$$\frac{\partial}{\partial\boldsymbol{\theta}}\mathbf{b}(\boldsymbol{\theta}) = \frac{\partial}{\partial\boldsymbol{\theta}}\int[\hat{\boldsymbol{\theta}}(\mathbf{y}) - \boldsymbol{\theta}]p_{\boldsymbol{\theta}}(\mathbf{y})d\mathbf{y}$$

$$= \int[\hat{\boldsymbol{\theta}}(\mathbf{y}) - \boldsymbol{\theta}]\left[\frac{\partial}{\partial\boldsymbol{\theta}}\log p_{\boldsymbol{\theta}}(\mathbf{y})\right]p_{\boldsymbol{\theta}}(\mathbf{y})d\mathbf{y} - \mathbf{I}$$

$$= E_{\boldsymbol{\theta}}[\boldsymbol{\varepsilon}_{\boldsymbol{\theta}}(\mathbf{y})\boldsymbol{\sigma}_{\boldsymbol{\theta}}^H(\mathbf{y})] - \mathbf{I} = \mathbf{T}^H(\boldsymbol{\theta}) - \mathbf{I}. \tag{6.47}$$

In the last line, we have used $E_{\boldsymbol{\theta}}[\boldsymbol{\sigma}_{\boldsymbol{\theta}}(\mathbf{y})] = \mathbf{0}$.

4. Consider the $p \times p$ Hessian matrix

$$\frac{\partial}{\partial\boldsymbol{\theta}}\left[\frac{\partial}{\partial\boldsymbol{\theta}}\log p_{\boldsymbol{\theta}}(\mathbf{y})\right]^H = \frac{\partial}{\partial\boldsymbol{\theta}}\left[\left(\frac{\partial}{\partial\boldsymbol{\theta}}p_{\boldsymbol{\theta}}(\mathbf{y})\right)^H \frac{1}{p_{\boldsymbol{\theta}}(\mathbf{y})}\right]$$

$$= \frac{\partial}{\partial\boldsymbol{\theta}}\left[\frac{\partial}{\partial\boldsymbol{\theta}}p_{\boldsymbol{\theta}}(\mathbf{y})\right]^H \frac{1}{p_{\boldsymbol{\theta}}(\mathbf{y})} - \left[\frac{\partial}{\partial\boldsymbol{\theta}}\log p_{\boldsymbol{\theta}}(\mathbf{y})\right]^H \frac{\partial}{\partial\boldsymbol{\theta}}\log p_{\boldsymbol{\theta}}(\mathbf{y}). \tag{6.48}$$

The expectation of this matrix is

$$E_{\boldsymbol{\theta}}\left\{\frac{\partial}{\partial\boldsymbol{\theta}}\left[\frac{\partial}{\partial\boldsymbol{\theta}}\log p_{\boldsymbol{\theta}}(\mathbf{y})\right]^H\right\} = \mathbf{0} - E_{\boldsymbol{\theta}}[\mathbf{s}_{\boldsymbol{\theta}}(\mathbf{y})\mathbf{s}_{\boldsymbol{\theta}}^H(\mathbf{y})] = -\mathbf{J}_F(\boldsymbol{\theta}). \tag{6.49}$$

Result 6.4. *The quadratic frequentist bound for the Fisher score is the Cramér–Rao bound:*

$$\mathbf{Q}(\boldsymbol{\theta}) \geq \left[\mathbf{I} + \frac{\partial}{\partial\boldsymbol{\theta}}\mathbf{b}(\boldsymbol{\theta})\right]\mathbf{J}_F^{-1}(\boldsymbol{\theta})\left[\mathbf{I} + \frac{\partial}{\partial\boldsymbol{\theta}}\mathbf{b}(\boldsymbol{\theta})\right]^H. \tag{6.50}$$

When the frequentist bias is zero, then the Cramér–Rao bound is

$$\mathbf{Q}(\boldsymbol{\theta}) \geq \mathbf{J}_{\mathrm{F}}^{1}(\boldsymbol{\theta}), \tag{6.51}$$

where $\mathbf{J}_{\mathrm{F}}(\boldsymbol{\theta})$ *is the Fisher information.*

This bound is sometimes also called the *deterministic* Cramér–Rao bound, in order to differentiate it from the corresponding Bayesian (i.e., stochastic) bound in Result 6.7. It is the most celebrated of all quadratic frequentist bounds.

If repeated measurements carry information about one fixed and deterministic parameter $\boldsymbol{\theta}$ through the product pdf $\prod_{i=1}^{M} p_{\boldsymbol{\theta}}(\mathbf{y}_i)$, then the sensitivity matrix remains fixed and the Fisher information matrix scales with M. Consequently the Cramér–Rao bound decreases as M^{-1}.

6.3.1 Nuisance parameters

There are many ways to show the effect of nuisance parameters on error bounds, and some of these are given by Scharf (1991, pp. 231–233). But perhaps the easiest, if not most general, way to establish the effect is this. Begin with the Fisher information matrix $\mathbf{J}(\boldsymbol{\theta})$ and the corresponding Cramér–Rao bound for frequentist unbiased estimators, $\mathbf{Q}(\boldsymbol{\theta}) \geq \mathbf{J}^{-1}(\boldsymbol{\theta})$. (In order to simplify the notation we do not subscript $\mathbf{J}$ as $\mathbf{J}_{\mathrm{F}}$.) The (i, i)th element of $\mathbf{J}(\boldsymbol{\theta})$ is $J_{ii}(\boldsymbol{\theta})$ and the (i, i)th element of $\mathbf{J}^{-1}(\boldsymbol{\theta})$ is denoted by $(\mathbf{J}^{-1})_{ii}(\boldsymbol{\theta})$.

From the definition of a score we see that $J_{ii}(\boldsymbol{\theta})$ is the Fisher information for the ith element of $\boldsymbol{\theta}$, when only the ith parameter in the parameter vector $\boldsymbol{\theta}$ is unknown. The Cauchy–Schwarz inequality says that $(\mathbf{y}^{\mathrm{H}}\mathbf{J}\mathbf{y})(\mathbf{x}^{\mathrm{H}}\mathbf{J}\mathbf{x}) \geq |\mathbf{y}^{\mathrm{H}}\mathbf{J}\mathbf{x}|^2$. Choose $\mathbf{x} = \mathbf{u}_k$ and $\mathbf{y} = \mathbf{J}^{-1}\mathbf{u}_k$, with $\mathbf{u}_k$ the kth Euclidean basis vector. Then $(\mathbf{J}^{-1})_{ii}(\boldsymbol{\theta})J_{ii}(\boldsymbol{\theta}) \geq 1$, or $(\mathbf{J}^{-1})_{ii}(\boldsymbol{\theta}) \geq 1/J_{ii}(\boldsymbol{\theta})$. These results actually generalize to show that any r-by-r-dimensional submatrix of the p-by-p inverse $\mathbf{J}^{-1}(\boldsymbol{\theta})$ is more positive definite than the inverse of the corresponding r-by-r Fisher matrix $\mathbf{J}(\boldsymbol{\theta})$. So nuisance parameters increase the Cramér–Rao bound.

6.3.2 The Cramér–Rao bound in the proper multivariate Gaussian model

In Result 2.5, the pdf for a proper complex Gaussian random variable $\mathbf{y}\colon \Omega \longrightarrow \mathbb{C}^n$ was shown to be

$$p_{\boldsymbol{\theta}}(\mathbf{y}) = \frac{1}{\pi^n \det \mathbf{R}_{yy}} \exp\{-(\mathbf{y} - \boldsymbol{\mu}_y)^{\mathrm{H}}\mathbf{R}_{yy}^{-1}(\mathbf{y} - \boldsymbol{\mu}_y)\}, \tag{6.52}$$

where $\boldsymbol{\mu}_y$ is the mean and $\mathbf{R}_{yy}$ the Hermitian covariance matrix of $\mathbf{y}$. For our purposes we assume that the mean value and the Hermitian covariance matrix are both functions of the unknown parameter vector $\boldsymbol{\theta}$, even though we have not made this dependence explicit in the notation. We shall assume that we have drawn M independent copies of the random vector $\mathbf{y}$ from the pdf $p_{\boldsymbol{\theta}}(\mathbf{y})$, so that the logarithm of the joint pdf of

$\mathbf{Y} = [\mathbf{y}_1, \mathbf{y}_2, \ldots, \mathbf{y}_M]$ is

$$\log\left(\prod_{i=1}^{M} p_{\boldsymbol{\theta}}(\mathbf{y}_i)\right) = -Mn\log\pi - M\log\det\mathbf{R}_{yy} - M\,\mathrm{tr}\,(\mathbf{R}_{yy}^{-1}\mathbf{S}_{yy}), \tag{6.53}$$

where $\mathbf{S}_{yy} = M^{-1}\sum_{i=1}^{M}(\mathbf{y}_i - \boldsymbol{\mu}_y)(\mathbf{y}_i - \boldsymbol{\mu}_y)^{\mathrm{H}}$ is the sample Hermitian covariance matrix.

Using the results of Appendix 2 for Wirtinger derivatives, in particular the results for differentiating logarithms and traces in Section A2.2, we may express the jth element of the centered measurement score $\mathbf{s}_{\boldsymbol{\theta}}(\mathbf{Y})$ as

$$\begin{aligned} s_j(\mathbf{Y}) &= -M\frac{\partial}{\partial\theta_j^*}\log\det\mathbf{R}_{yy} - M\frac{\partial}{\partial\theta_j^*}\mathrm{tr}(\mathbf{R}_{yy}^{-1}\mathbf{S}_{yy}) \\ &= -M\,\mathrm{tr}\left[\mathbf{R}_{yy}^{-1}\frac{\partial}{\partial\theta_j^*}\mathbf{R}_{yy}\right] + M\,\mathrm{tr}\left[\mathbf{R}_{yy}^{-1}\left(\frac{\partial}{\partial\theta_j^*}\mathbf{R}_{yy}\right)\mathbf{R}_{yy}^{-1}\mathbf{S}_{yy}\right] - M\,\mathrm{tr}\left[\mathbf{R}_{yy}^{-1}\frac{\partial}{\partial\theta_j^*}\mathbf{S}_{yy}\right]. \end{aligned} \tag{6.54}$$

It is a simple matter to show that $\partial\mathbf{S}_{yy}/\partial\theta_j^*$ has mean-value zero. So to compute the Hessian term $-E_{\boldsymbol{\theta}}[(\partial/\partial\theta_i)s_j(\mathbf{Y})]$ we can ignore any terms that involve a first partial derivative of $\mathbf{S}_{yy}$. The net result, after a few lines of algebra, is

$$\begin{aligned} J_{\mathrm{F},ij} &= -E_{\boldsymbol{\theta}}\left[\frac{\partial}{\partial\theta_i}s_j(\mathbf{Y})\right] \\ &= M\,\mathrm{tr}\left[\mathbf{R}_{yy}^{-1}\left(\frac{\partial}{\partial\theta_i}\mathbf{R}_{yy}\right)\mathbf{R}_{yy}^{-1}\left(\frac{\partial}{\partial\theta_j^*}\mathbf{R}_{yy}\right)\right] + M\,\mathrm{tr}\left[\mathbf{R}_{yy}^{-1}E_{\boldsymbol{\theta}}\left(\frac{\partial}{\partial\theta_i}\frac{\partial}{\partial\theta_j^*}\mathbf{S}_{yy}\right)\right] \\ &= M\,\mathrm{tr}\left[\mathbf{R}_{yy}^{-1}\left(\frac{\partial}{\partial\theta_i}\mathbf{R}_{yy}\right)\mathbf{R}_{yy}^{-1}\left(\frac{\partial}{\partial\theta_j^*}\mathbf{R}_{yy}\right)\right] + M\left[\frac{\partial}{\partial\theta_i}\boldsymbol{\mu}_y\right]^{\mathrm{H}}\mathbf{R}_{yy}^{-1}\left[\frac{\partial}{\partial\theta_j^*}\boldsymbol{\mu}_y\right] \\ &\quad + M\left[\frac{\partial}{\partial\theta_j^*}\boldsymbol{\mu}_y\right]^{\mathrm{H}}\mathbf{R}_{yy}^{-1}\left[\frac{\partial}{\partial\theta_i}\boldsymbol{\mu}_y\right] = J_{\mathrm{F},ji}^*. \end{aligned} \tag{6.55}$$

This is the general formula for the (i, j)th element of the Fisher information matrix in the proper multivariate Gaussian experiment that brings information about $\boldsymbol{\theta}$ in its mean and covariance. The real version of this result dates at least to Slepian (1954) and the complex version to Bangs (1971). There are a few special cases. If the covariance matrix $\mathbf{R}_{yy}$ is independent of $\boldsymbol{\theta}$ then the first term vanishes. If the mean $\boldsymbol{\mu}_y$ is independent of $\boldsymbol{\theta}^*$ then the second term vanishes, and if it is independent of $\boldsymbol{\theta}$, the third term vanishes.

6.3.3 The separable linear statistical model and the geometry of the Cramér–Rao bound

To gain geometrical insight into the Fisher matrix and the Cramér–Rao bound, we now apply the results of the previous subsection to parameter estimation in the linear model $\mathbf{y} = \mathbf{H}\boldsymbol{\theta} + \mathbf{n}$, where $\mathbf{n}$ is a zero-mean proper Gaussian with covariance $\mathbf{R}_{nn}$ independent

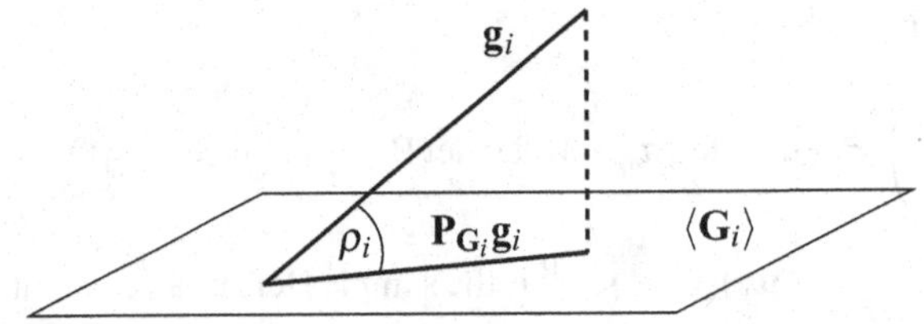

Figure 6.4 Geometry of the Cramér–Rao bound in the separable statistical model with multivariate Gaussian errors; the variance is large when mode $\mathbf{g}_i$ lies near the subspace $\mathbf{G}_i$ of other modes.

of $\boldsymbol{\theta}$. We shall find that it is the sensitivity of noise-free measurements to small variations in parameters that determines the performance of an estimator.

The partial derivatives are $\partial \boldsymbol{\mu}_y / \partial \theta_i = \mathbf{h}_i$, where $\mathbf{h}_i$ is the ith column of $\mathbf{H}$ and the Fisher matrix is $\mathbf{J}_{\mathrm{F}} = M\mathbf{H}^{\mathrm{H}}\mathbf{R}_{nn}^{-1}\mathbf{H}$. The Cramér–Rao bound for frequentist-unbiased estimators is thus

$$\mathbf{Q}(\boldsymbol{\theta}) \geq \frac{1}{M}(\mathbf{H}^{\mathrm{H}}\mathbf{R}_{nn}^{-1}\mathbf{H})^{-1}. \tag{6.56}$$

With the definition $\mathbf{G} = \mathbf{R}_{nn}^{-1/2}\mathbf{H}$, the (i, i)th element may be written as

$$Q_{ii}(\boldsymbol{\theta}) \geq \frac{1}{M}\frac{1}{\mathbf{g}_i^{\mathrm{H}}\mathbf{g}_i}\frac{\mathbf{g}_i^{\mathrm{H}}\mathbf{g}_i}{\mathbf{g}_i^{\mathrm{H}}(\mathbf{I}-\mathbf{P}_{\mathbf{G}_i})\mathbf{g}_i} = \frac{1}{M}\frac{1}{\mathbf{g}_i^{\mathrm{H}}\mathbf{g}_i}\frac{1}{\sin^2\rho_i}, \tag{6.57}$$

where $\mathbf{P}_{\mathbf{G}_i}$ is the projection onto the subspace $\langle \mathbf{G}_i \rangle$ spanned by all but the ith mode in $\mathbf{G}$, ρ_i is the angle that the mode vector $\mathbf{g}_i$ makes with this subspace, and $\mathbf{g}_i^{\mathrm{H}}(\mathbf{I}-\mathbf{P}_{\mathbf{G}_i})\mathbf{g}_i/(\mathbf{g}_i^{\mathrm{H}}\mathbf{g}_i)$ is the sine-squared of this angle. Thus, as illustrated in Fig. 6.4, the lower bound on the variance in estimating θ_i is a large multiple of $(M\mathbf{g}_i^{\mathrm{H}}\mathbf{g}_i)^{-1}$ when the ith mode can be linearly approximated with the other modes in $\mathbf{G}_i$. For closely spaced modes, only a large number of independent samples or a large value of $\mathbf{g}_i^{\mathrm{H}}\mathbf{g}_i$ – producing a large output signal-to-noise ratio $M\mathbf{g}_i^{\mathrm{H}}\mathbf{g}_i$ – can produce a small lower bound. With low output signal-to-noise ratio and closely spaced modes, any estimator of θ_i will be poor, meaning that the resolution in amplitude of the ith mode will be poor. This result generalizes to mean-value vectors more general than $\mathbf{H}\boldsymbol{\theta}$, on replacing $\mathbf{h}_i$ with $\partial \boldsymbol{\mu}/\partial \theta_i$.

Example 6.2. Let the noise covariance matrix in the proper multivariate Gaussian model be $\mathbf{R}_{nn} = \sigma^2\mathbf{I}$ and the matrix $\mathbf{H} = [A\boldsymbol{\psi}_1, A\boldsymbol{\psi}_2]$ with $\boldsymbol{\psi}_k = [1, \mathrm{e}^{\mathrm{j}\phi_k}, \ldots, \mathrm{e}^{\mathrm{j}(n-1)\phi_k}]^{\mathrm{T}}$. We call $\boldsymbol{\psi}_k$ a complex exponential mode with mode angle ϕ_k and A a complex mode amplitude. The Cramér–Rao bound is

$$Q_{ii}(\boldsymbol{\theta}) \geq \frac{1}{M}\frac{1}{n|A|^2/\sigma^2}\frac{1}{1-l_n^2(\phi_1-\phi_2)}, \tag{6.58}$$

where

$$0 \leq l_n^2(\phi) = \frac{1}{n^2}\frac{\sin^2(n\phi/2)}{\sin^2(\phi/2)} \leq 1 \tag{6.59}$$

is the Lanczos kernel. In this bound, $\mathrm{snr} = |A|^2/\sigma^2$ is the per-sample or input signal-to-noise ratio, $\mathrm{SNR} = n\mathrm{snr}$ is the output signal-to-noise ratio and $1 - l_n^2(\phi_1 - \phi_2)$ is the sine-squared of the angle between the subspaces $\langle \boldsymbol{\psi}_1 \rangle$ and $\langle \boldsymbol{\psi}_2 \rangle$. This bound is $(M\mathrm{SNR})^{-1}$ at $\phi_1 - \phi_2 = 2\pi/n$, and this angle difference is called the *Rayleigh limit to resolution*. It governs much of optics, radar, sonar, and geophysics, even though it is quite conservative in the sense that many independent samples or large input SNR can override the aperture effects of the Lanczos kernel.[2]

This example bounds the error covariance matrix for estimating the linear parameters in the separable linear model, not the nonlinear parameters that would determine the matrix $\mathbf{H}(\boldsymbol{\theta})$. Typically these parameters would be frequency, wavenumber, delay, and so on. The Cramér–Rao bound for these parameters then depends on terms like $\partial \mathbf{h}_i/\partial \theta_j^*$.[3]

6.3.4 Extension of Fisher score and the Cramér–Rao bound to improper errors and scores

In order to extend the Cramér–Rao bound to improper errors and scores, we need only compute the complementary expansion-coefficient matrix $\widetilde{\mathbf{T}}(\boldsymbol{\theta}) = E[\mathbf{s}_{\boldsymbol{\theta}}(\mathbf{y})\mathbf{e}_{\boldsymbol{\theta}}^{\mathrm{T}}(\mathbf{y})]$ and complementary Fisher information matrix $\tilde{\mathbf{J}}(\boldsymbol{\theta}) = E[\mathbf{s}_{\boldsymbol{\theta}}(\mathbf{y})\mathbf{s}_{\boldsymbol{\theta}}^{\mathrm{T}}(\mathbf{y})]$. To this end, consider the following conjugate partial derivative of the bias:

$$\begin{aligned}\frac{\partial}{\partial \boldsymbol{\theta}^*}\mathbf{b}(\boldsymbol{\theta}) &= \frac{\partial}{\partial \boldsymbol{\theta}^*}\int [\hat{\boldsymbol{\theta}}(\mathbf{y}) - \boldsymbol{\theta}]p_{\boldsymbol{\theta}}(\mathbf{y})d\mathbf{y} \\ &= \int [\hat{\boldsymbol{\theta}}(\mathbf{y}) - \boldsymbol{\theta}]\left[\frac{\partial}{\partial \boldsymbol{\theta}^*}\log p_{\boldsymbol{\theta}}(\mathbf{y})\right]p_{\boldsymbol{\theta}}(\mathbf{y})d\mathbf{y} - \mathbf{0} = \widetilde{\mathbf{T}}^{\mathrm{T}}(\boldsymbol{\theta}).\end{aligned} \tag{6.60}$$

This produces the identity

$$\widetilde{\mathbf{T}}(\boldsymbol{\theta}) = \left[\frac{\partial}{\partial \boldsymbol{\theta}}\mathbf{b}^*(\boldsymbol{\theta})\right]^{\mathrm{H}} = \left[\frac{\partial}{\partial \boldsymbol{\theta}^*}\mathbf{b}(\boldsymbol{\theta})\right]^{\mathrm{T}}. \tag{6.61}$$

The (i, j)th element of the matrix $(\partial/\partial\boldsymbol{\theta}^*)\mathbf{b}(\boldsymbol{\theta})$ is $(\partial/\partial\theta_j^*)b_i(\boldsymbol{\theta})$. Importantly, for frequentist unbiased estimators, the complementary expansion-coefficient matrix is zero, and the augmented expansion-coefficient matrix is $\underline{\mathbf{T}}(\boldsymbol{\theta}) = \mathbf{I}$.

For the complementary Fisher information, consider the $p \times p$ Hessian

$$\begin{aligned}\frac{\partial}{\partial \boldsymbol{\theta}}\left[\frac{\partial}{\partial \boldsymbol{\theta}}\log p_{\boldsymbol{\theta}}(\mathbf{y})\right]^{\mathrm{T}} &= \frac{\partial}{\partial \boldsymbol{\theta}}\left[\left(\frac{\partial}{\partial \boldsymbol{\theta}}p_{\boldsymbol{\theta}}(\mathbf{y})\right)^{\mathrm{T}}\frac{1}{p_{\boldsymbol{\theta}}(\mathbf{y})}\right] \\ &= \frac{\partial}{\partial \boldsymbol{\theta}}\left[\frac{\partial}{\partial \boldsymbol{\theta}}p_{\boldsymbol{\theta}}(\mathbf{y})\right]^{\mathrm{T}}\frac{1}{p_{\boldsymbol{\theta}}(\mathbf{y})} - \left[\frac{\partial}{\partial \boldsymbol{\theta}}\log p_{\boldsymbol{\theta}}(\mathbf{y})\right]^{\mathrm{T}}\frac{\partial}{\partial \boldsymbol{\theta}}\log p_{\boldsymbol{\theta}}(\mathbf{y}).\end{aligned} \tag{6.62}$$

Taking expectations, we find the identity

$$\tilde{\mathbf{J}}_{\mathrm{F}}(\boldsymbol{\theta}) = E_{\boldsymbol{\theta}}[\mathbf{s}_{\boldsymbol{\theta}}(\mathbf{y})\mathbf{s}_{\boldsymbol{\theta}}^{\mathrm{T}}(\mathbf{y})] = -E_{\boldsymbol{\theta}}\left\{\frac{\partial}{\partial \boldsymbol{\theta}}\left[\frac{\partial}{\partial \boldsymbol{\theta}}\log p_{\boldsymbol{\theta}}(\mathbf{y})\right]^{\mathrm{T}}\right\}, \tag{6.63}$$

which is the complementary dual to (6.45). The augmented Cramér–Rao bound for improper error and measurement scores is the bound of Result 6.3, applied to the Fisher score.

Result 6.5. *For frequentist-unbiased estimators, the widely linear Cramér–Rao bound is*

$$\mathbf{Q}(\boldsymbol{\theta}) \geq [\mathbf{J}_{\mathrm{F}}(\boldsymbol{\theta}) - \tilde{\mathbf{J}}_{\mathrm{F}}(\boldsymbol{\theta})\mathbf{J}_{\mathrm{F}}^{-*}(\boldsymbol{\theta})\tilde{\mathbf{J}}_{\mathrm{F}}^{*}(\boldsymbol{\theta})]^{-1}. \tag{6.64}$$

Since $[\mathbf{J}_{\mathrm{F}}(\boldsymbol{\theta}) - \tilde{\mathbf{J}}_{\mathrm{F}}(\boldsymbol{\theta})\mathbf{J}_{\mathrm{F}}^{-*}(\boldsymbol{\theta})\tilde{\mathbf{J}}_{\mathrm{F}}^{*}(\boldsymbol{\theta})]^{-1} \geq \mathbf{J}_{\mathrm{F}}^{-1}(\boldsymbol{\theta})$, the widely linear bound is *tighter* than the standard Cramér–Rao bound in Result 6.4, which ignores complementary Fisher information. Result 6.5 was essentially given by van den Bos (1994b).

Note that Results 6.3 and 6.5 assume that $\underline{\mathbf{J}}_{\mathrm{F}}(\boldsymbol{\theta})$ is nonsingular. An important case in which $\underline{\mathbf{J}}_{\mathrm{F}}(\boldsymbol{\theta})$ is always singular is when the parameter $\boldsymbol{\theta}$ is *real*. For real $\boldsymbol{\theta}$, we have $\mathbf{J}_{\mathrm{F}}(\boldsymbol{\theta}) = \tilde{\mathbf{J}}_{\mathrm{F}}(\boldsymbol{\theta})$, so there is no need to compute the complementary Fisher matrix. In this case, widely linear estimation of the error from the measurement score is equivalent to linear estimation, and the widely linear Cramér–Rao bound is the standard Cramér–Rao bound: $\mathbf{Q}(\boldsymbol{\theta}) \geq \mathbf{J}_{\mathrm{F}}^{-1}(\boldsymbol{\theta})$. However, *this bound still depends on whether or not* $\mathbf{y}$ *is proper because even the Hermitian Fisher matrix* $\mathbf{J}_{\mathrm{F}}(\boldsymbol{\theta})$ *depends on the complementary correlation of* $\mathbf{y}$, as the following subsection illustrates.

6.3.5 The Cramér–Rao bound in the improper multivariate Gaussian model

It is straightforward to extend the results of Section 6.3.2 to improper Gaussians. Let $\mathbf{y}\colon \Omega \longrightarrow \mathbb{C}^n$ be Gaussian with augmented mean $\underline{\boldsymbol{\mu}}_y$ and augmented covariance matrix $\underline{\mathbf{R}}_{yy}$. Assuming as before that we have drawn M independent copies of $\mathbf{y}$, the (i, j)th element of the Hermitian Fisher matrix is

$$J_{\mathrm{F},ij} = \frac{M}{2}\operatorname{tr}\left[\underline{\mathbf{R}}_{yy}^{-1}\left(\frac{\partial}{\partial\theta_i}\underline{\mathbf{R}}_{yy}\right)\underline{\mathbf{R}}_{yy}^{-1}\left(\frac{\partial}{\partial\theta_j^*}\underline{\mathbf{R}}_{yy}\right)\right] + M\left[\frac{\partial}{\partial\theta_i}\underline{\boldsymbol{\mu}}_y\right]^{\mathrm{H}}\underline{\mathbf{R}}_{yy}^{-1}\left[\frac{\partial}{\partial\theta_j^*}\underline{\boldsymbol{\mu}}_y\right] = J_{\mathrm{F},ji}^*, \tag{6.65}$$

where the second summand is real. The (i, j)th element of the complementary Fisher matrix is

$$\tilde{J}_{\mathrm{F},ij} = \frac{M}{2}\operatorname{tr}\left[\underline{\mathbf{R}}_{yy}^{-1}\left(\frac{\partial}{\partial\theta_i}\underline{\mathbf{R}}_{yy}\right)\underline{\mathbf{R}}_{yy}^{-1}\left(\frac{\partial}{\partial\theta_j}\underline{\mathbf{R}}_{yy}\right)\right] + M\left[\frac{\partial}{\partial\theta_i}\underline{\boldsymbol{\mu}}_y\right]^{\mathrm{H}}\underline{\mathbf{R}}_{yy}^{-1}\left[\frac{\partial}{\partial\theta_j}\underline{\boldsymbol{\mu}}_y\right] = \tilde{J}_{\mathrm{F},ji}, \tag{6.66}$$

and again the second summand is real. The case of real $\boldsymbol{\theta}$, $\mathbf{J}_{\mathrm{F}}(\boldsymbol{\theta}) = \tilde{\mathbf{J}}_{\mathrm{F}}(\boldsymbol{\theta})$, has been treated by Delmas and Abeida (2004). They have shown that, when $\boldsymbol{\theta}$ is real, the Cramér–Rao bound $\mathbf{Q}(\boldsymbol{\theta}) \geq \mathbf{J}_{\mathrm{F}}^{-1}(\boldsymbol{\theta})$ is actually *lowered* if $\mathbf{y}$ is improper rather than proper. That is, for a given Hermitian covariance matrix $\mathbf{R}_{yy}$, the Hermitian Fisher matrix $\mathbf{J}_{\mathrm{F}}(\boldsymbol{\theta})$ is more positive definite if $\widetilde{\mathbf{R}}_{yy} \neq \mathbf{0}$ than if $\widetilde{\mathbf{R}}_{yy} = \mathbf{0}$. At first, this may seem like a contradiction of the statement made after Result 6.5 that improper errors and scores *tighten* (i.e., increase) the Cramér–Rao bound. So what is going on?

There are two different effects at work. On the one hand, the actual mean-squared error in widely linearly estimating $\boldsymbol{\theta}$ (real or complex) from $\mathbf{y}$ is *reduced* when the degree of impropriety of $\mathbf{y}$ is increased while the Hermitian covariance matrix $\mathbf{R}_{yy}$ is kept constant. This is reflected by the *lowered* Cramér–Rao bound. On the other hand, if $\boldsymbol{\theta}$ is complex, a widely linear estimate of the error score from the measurement score is a more accurate predictor of the actual mean-squared error for estimating $\boldsymbol{\theta}$ from $\mathbf{y}$ than is a linear estimate of the error score from the measurement score. This is reflected by a *tightening* Cramér–Rao bound. It depends on the problem which one of these two effects is stronger.

Finally, we note that (6.65) and (6.66) do not cover the maximally improper (also called *strictly noncircular* or *rectilinear*) case in which $\underline{\mathbf{R}}_{yy}$ is singular. This was addressed by Römer and Haardt (2007).

6.3.6 Fisher score and Cramér–Rao bounds for functions of parameters

Sometimes it is a function of the parameter $\boldsymbol{\theta}$ that is to be estimated rather than the parameter itself. What can we say about the Fisher matrix in this case? Let $\mathbf{w} = \mathbf{g}(\boldsymbol{\theta})$ be a continuous bijection from $\mathbb{C}^p$ to $\mathbb{C}^p$, with inverse $\boldsymbol{\theta} = \mathbf{g}^{-1}(\mathbf{w})$. In the linear case, $\mathbf{w} = \mathbf{H}\boldsymbol{\theta}$ and $\boldsymbol{\theta} = \mathbf{H}^{-1}\mathbf{w}$, and in the widely linear case, $\underline{\mathbf{w}} = \underline{\mathbf{H}}\,\underline{\boldsymbol{\theta}}$ and $\underline{\boldsymbol{\theta}} = \underline{\mathbf{H}}^{-1}\underline{\mathbf{w}}$. Define the Jacobian matrix

$$\underline{\mathbf{F}}(\mathbf{w}) = \frac{\partial \underline{\boldsymbol{\theta}}}{\partial \underline{\mathbf{w}}} = \begin{bmatrix} \mathbf{F}(\mathbf{w}) & \widetilde{\mathbf{F}}(\mathbf{w}) \\ \widetilde{\mathbf{F}}^*(\mathbf{w}) & \mathbf{F}^*(\mathbf{w}) \end{bmatrix} = \begin{bmatrix} \dfrac{\partial \boldsymbol{\theta}}{\partial \mathbf{w}} & \dfrac{\partial \boldsymbol{\theta}}{\partial \mathbf{w}^*} \\ \left(\dfrac{\partial \boldsymbol{\theta}}{\partial \mathbf{w}^*}\right)^* & \left(\dfrac{\partial \boldsymbol{\theta}}{\partial \mathbf{w}}\right)^* \end{bmatrix}. \tag{6.67}$$

In the linear case, $\mathbf{F}(\mathbf{w}) = \mathbf{H}^{-1}$ and $\widetilde{\mathbf{F}} = \mathbf{0}$. In the widely linear case, $\underline{\mathbf{F}}(\mathbf{w}) = \underline{\mathbf{H}}^{-1}$. The pdf parameterized by $\mathbf{w}$ is taken to be $p_{\boldsymbol{\theta}}(\mathbf{y})$ at $\boldsymbol{\theta} = \mathbf{g}^{-1}(\mathbf{w})$. Using the chain rule for non-holomorphic functions (A2.23), the Fisher score for $\mathbf{w}$ is then

$$\begin{aligned} \mathbf{s}_{\mathbf{w}}(\mathbf{y}) = \left[\frac{\partial}{\partial \mathbf{w}} \log p_{\mathbf{w}}(\mathbf{y})\right]^{\mathrm{H}} &= \left[\frac{\partial}{\partial \boldsymbol{\theta}} \log p_{\boldsymbol{\theta}}(\mathbf{y})\mathbf{F}(\mathbf{w}) + \frac{\partial}{\partial \boldsymbol{\theta}^*} \log p_{\boldsymbol{\theta}}(\mathbf{y})\widetilde{\mathbf{F}}^*(\mathbf{w})\right]^{\mathrm{H}} \\ &= \mathbf{F}^{\mathrm{H}}(\mathbf{w})\mathbf{s}_{\boldsymbol{\theta}}(\mathbf{y}) + \widetilde{\mathbf{F}}^{\mathrm{T}}(\mathbf{w})\mathbf{s}^*_{\boldsymbol{\theta}}(\mathbf{y}) \text{ at } \boldsymbol{\theta} = \mathbf{g}(\mathbf{w}), \end{aligned}$$

or, equivalently,

$$\underline{\mathbf{s}}_{\mathbf{w}}(\mathbf{y}) = \underline{\mathbf{F}}^{\mathrm{H}}(\mathbf{w})\underline{\mathbf{s}}_{\boldsymbol{\theta}}(\mathbf{y}) \text{ at } \boldsymbol{\theta} = \mathbf{g}(\mathbf{w}), \tag{6.68}$$

where $\underline{\mathbf{s}}_{\boldsymbol{\theta}}(\mathbf{y})$ is the augmented Fisher score for $\boldsymbol{\theta}$. If $\mathbf{g}$ is widely linear, then $\underline{\mathbf{s}}_{\mathbf{w}}(\mathbf{y}) = \underline{\mathbf{H}}^{-\mathrm{H}}\underline{\mathbf{s}}_{\boldsymbol{\theta}}(\mathbf{y})$.

The centered error in estimating $\mathbf{w}$ is $\mathbf{e}_{\mathbf{w}}(\mathbf{y}) = \mathbf{g}^{-1}(\mathbf{e}_{\boldsymbol{\theta}}(\mathbf{y}))$. In the widely linear case, we have $\underline{\mathbf{e}}_{\mathbf{w}}(\mathbf{y}) = \underline{\mathbf{H}}\,\underline{\mathbf{e}}_{\boldsymbol{\theta}}(\mathbf{y})$. The expansion-coefficient matrix cannot be modeled generally, but for widely linear $\mathbf{g}$ the augmented version is

$$\underline{\mathbf{T}}(\mathbf{w}) = E[\underline{\mathbf{s}}_{\mathbf{w}}(\mathbf{y})\underline{\mathbf{e}}^{\mathrm{H}}_{\mathbf{w}}(\mathbf{y})] = \underline{\mathbf{H}}^{-\mathrm{H}}\underline{\mathbf{T}}(\boldsymbol{\theta})\underline{\mathbf{H}}^{\mathrm{H}} \text{ at } \underline{\boldsymbol{\theta}} = \underline{\mathbf{H}}^{-1}\underline{\mathbf{w}}. \tag{6.69}$$

The augmented Fisher matrix for $\mathbf{w}$ is

$$\underline{\mathbf{J}}_F(\mathbf{w}) = \underline{\mathbf{F}}^H(\mathbf{w})\underline{\mathbf{J}}_F(\boldsymbol{\theta})\underline{\mathbf{F}}(\mathbf{w}) \text{ at } \boldsymbol{\theta} = \mathbf{g}^{-1}(\mathbf{w}). \tag{6.70}$$

If $\mathbf{g}$ is widely linear, this is $\underline{\mathbf{J}}_F(\mathbf{w}) = \underline{\mathbf{H}}^{-H}\underline{\mathbf{J}}_F(\boldsymbol{\theta})\underline{\mathbf{H}}^{-1}$ at $\underline{\boldsymbol{\theta}} = \underline{\mathbf{H}}^{-1}\underline{\mathbf{w}}$. So, in the general case we can characterize the Fisher matrix and in the widely linear case we can characterize the Cramér–Rao bound:

$$\underline{\mathbf{Q}}(\mathbf{w}) \geq \underline{\mathbf{T}}^H(\mathbf{w})\underline{\mathbf{J}}_F^{-1}(\mathbf{w})\underline{\mathbf{T}}(\mathbf{w}) = \underline{\mathbf{H}}\underline{\mathbf{T}}^H(\boldsymbol{\theta})\underline{\mathbf{J}}_F^{-1}(\boldsymbol{\theta})\underline{\mathbf{T}}(\boldsymbol{\theta})\underline{\mathbf{H}}^H \text{ at } \underline{\boldsymbol{\theta}} = \underline{\mathbf{H}}^{-1}\underline{\mathbf{w}}. \tag{6.71}$$

If the original estimator was frequentist unbiased, then $\underline{\mathbf{T}}(\boldsymbol{\theta}) = \mathbf{I}$ and the bound is

$$\underline{\mathbf{Q}}(\mathbf{w}) \geq \underline{\mathbf{H}}\,\underline{\mathbf{J}}_F^{-1}(\boldsymbol{\theta})\underline{\mathbf{H}}^H \text{ at } \underline{\boldsymbol{\theta}} = \underline{\mathbf{H}}^{-1}\underline{\mathbf{w}}. \tag{6.72}$$

To obtain a bound on $\mathbf{Q}(\mathbf{w})$, we read out the northwest block of $\underline{\mathbf{Q}}(\mathbf{w})$. The simplifications in the strictly linear case should be obvious.

6.4 Quadratic Bayesian bounds

As in the development of quadratic frequentist bounds, the idea of quadratic *Bayesian* bounds is this: from the measurement $\mathbf{y}$ is computed the estimator $\hat{\boldsymbol{\theta}}(\mathbf{y})$, with error $\boldsymbol{\varepsilon}(\mathbf{y}, \boldsymbol{\theta}) = \hat{\boldsymbol{\theta}}(\mathbf{y}) - \boldsymbol{\theta}$ and centered error score $\mathbf{e}(\mathbf{y}, \boldsymbol{\theta}) = \boldsymbol{\varepsilon}(\mathbf{y}, \boldsymbol{\theta}) - \mathbf{b}$. We would like to approximate this centered error score by using a linear or widely linear function of the centered measurement score $\mathbf{s}(\mathbf{y}, \boldsymbol{\theta}) = \boldsymbol{\sigma}(\mathbf{y}, \boldsymbol{\theta}) - E[\boldsymbol{\sigma}(\mathbf{y}, \boldsymbol{\theta})]$. Thus, the virtual two-channel estimation problem of Fig. 6.2 still holds. The centered error score $\mathbf{e}(\mathbf{y}, \boldsymbol{\theta})$ is to be estimated from the centered measurement score $\mathbf{s}(\mathbf{y}, \boldsymbol{\theta})$.

We will from the outset discuss the *widely linear* story, and the simplifications in the linear case should be obvious. The *composite* vector of *augmented* centered errors and scores $[\underline{\mathbf{e}}^T, \underline{\mathbf{s}}^T]^T$ has covariance matrix

$$E\left\{\begin{bmatrix}\underline{\mathbf{e}}(\mathbf{y}, \boldsymbol{\theta}) \\ \underline{\mathbf{s}}(\mathbf{y}, \boldsymbol{\theta})\end{bmatrix}\begin{bmatrix}\underline{\mathbf{e}}^H(\mathbf{y}, \boldsymbol{\theta}) & \underline{\mathbf{s}}^H(\mathbf{y}, \boldsymbol{\theta})\end{bmatrix}\right\} = \begin{bmatrix}\underline{\mathbf{Q}} & \underline{\mathbf{T}}^H \\ \underline{\mathbf{T}} & \underline{\mathbf{J}}\end{bmatrix}. \tag{6.73}$$

In this equation the covariance matrices $\underline{\mathbf{Q}}$, $\underline{\mathbf{T}}$, and $\underline{\mathbf{J}}$ are defined as

$$\begin{aligned}\underline{\mathbf{Q}} &= E[\underline{\mathbf{e}}(\mathbf{y}, \boldsymbol{\theta})\underline{\mathbf{e}}^H(\mathbf{y}, \boldsymbol{\theta})], \\ \underline{\mathbf{T}} &= E[\underline{\mathbf{s}}(\mathbf{y}, \boldsymbol{\theta})\underline{\mathbf{e}}^H(\mathbf{y}, \boldsymbol{\theta})], \\ \underline{\mathbf{J}} &= E[\underline{\mathbf{s}}(\mathbf{y}, \boldsymbol{\theta})\underline{\mathbf{s}}^H(\mathbf{y}, \boldsymbol{\theta})],\end{aligned} \tag{6.74}$$

where $\underline{\mathbf{Q}}$ is the augmented error covariance for the estimator $\hat{\boldsymbol{\theta}}(\mathbf{y})$, $\underline{\mathbf{T}}$ is the augmented sensitivity or expansion-coefficient matrix, and $\underline{\mathbf{J}}$ is the augmented information matrix associated with the centered measurement score $\mathbf{s}(\mathbf{y}, \boldsymbol{\theta})$. In the case of the Fisher–Bayes score, to be discussed in Section 6.5, the information is *Fisher–Bayes information*.

The covariance matrix for the composite augmented error and score is positive semidefinite, so, assuming the augmented information matrix to be nonsingular, we obtain the following bound.

Result 6.6. *The augmented error covariance is bounded by the general augmented quadratic Bayesian bound*

$$\underline{\mathbf{Q}} \geq \underline{\mathbf{T}}^{\mathrm{H}}\underline{\mathbf{J}}^{-1}\underline{\mathbf{T}}. \tag{6.75}$$

The augmented mean-squared error matrix is bounded by

$$\underline{\mathbf{M}} \geq \underline{\mathbf{T}}^{\mathrm{H}}\underline{\mathbf{J}}^{-1}\underline{\mathbf{T}} + \underline{\mathbf{b}}\underline{\mathbf{b}}^{\mathrm{H}}. \tag{6.76}$$

The bounds on $\mathbf{Q}$ and $\mathbf{M}$ are obtained by reading out the northwest block of $\underline{\mathbf{Q}}$ and $\underline{\mathbf{M}}$. Only the choice of score distinguishes one bound from another. Moreover, all quadratic Bayes bounds are invariant with respect to nonsingular widely linear transformation of the score function $\mathbf{s}$. That is, the transformed score $\underline{\mathbf{L}}\,\underline{\mathbf{s}}$ transforms the augmented sensitivity matrix $\underline{\mathbf{T}}$ to $\underline{\mathbf{L}}\,\underline{\mathbf{T}}$ and the augmented information matrix $\underline{\mathbf{J}}$ to $\underline{\mathbf{L}}\,\underline{\mathbf{J}}\,\underline{\mathbf{L}}^{\mathrm{H}}$, leaving the bound invariant.

Definition 6.4. *A widely linear estimator will be said to be* Bayes-efficient *with respect to a given score if the bound in (6.75) is achieved with equality.*

A necessary and sufficient condition for efficiency is

$$\underline{\hat{\boldsymbol{\theta}}}(\mathbf{y}) - \underline{\boldsymbol{\theta}} - (E[\underline{\hat{\boldsymbol{\theta}}}] - E[\underline{\boldsymbol{\theta}}]) = \underline{\mathbf{T}}^{\mathrm{H}}\underline{\mathbf{J}}^{-1}\underline{\mathbf{s}}(\mathbf{y}, \boldsymbol{\theta}) \tag{6.77}$$

which is also invariant with respect to widely linear transformation $\underline{\mathbf{L}}$. That is, an estimator is efficient if, element-by-element, the centered error scores lie in the widely linear subspace spanned by the elements of the centered measurement score.

The minimum achievable error covariance will be the covariance of the conditional mean estimator, which is unbiased. Thus, the Fisher–Bayes bound will be the greatest achievable quadratic lower bound if and only if the conditional mean estimator can be written as $E[\underline{\boldsymbol{\theta}}|\mathbf{y}] = \underline{\mathbf{T}}^{\mathrm{H}}\underline{\mathbf{J}}^{-1}\underline{\mathbf{s}}(\mathbf{y}, \boldsymbol{\theta}) + \underline{\boldsymbol{\theta}}$. This result follows from the fact that the MMSE estimator has zero mean, which is to say that $E[\hat{\boldsymbol{\theta}}(\mathbf{y})] = E[\boldsymbol{\theta}]$.

The comments in Section 6.2.1, regarding good and bad scores, and Section 6.2.2 apply almost directly to Bayesian bounds, with one cautionary note: the Bayes expansion coefficient $t_i = E[s_i(\mathbf{y}, \theta)e^*(\mathbf{y}, \theta)]$ is *not* generally $E[t_i(\theta)] = E\{E_\theta[t_i(\theta)]\}$, since the frequentist's score $\mathbf{s}_\theta(\mathbf{y})$ is not generally the Bayesian's score $\mathbf{s}(\mathbf{y}, \theta)$, which will include prior information about the parameter θ.

6.5 Fisher–Bayes score and Fisher–Bayes bound

Our results for quadratic Bayes bounds so far are general. To make them applicable, we need to consider a concrete score for which the augmented covariances $\underline{\mathbf{T}}$ and $\underline{\mathbf{J}}$ can be computed. Following the development of the Fisher score, we choose the *Fisher–Bayes score* and compute its associated *Fisher–Bayes bound.*

6.5.1 Fisher–Bayes score and information

Definition 6.5. *The* Fisher–Bayes measurement score *is defined as*

$$\boldsymbol{\sigma}(\mathbf{y},\boldsymbol{\theta}) = \left[\frac{\partial}{\partial\boldsymbol{\theta}}\log p(\mathbf{y},\boldsymbol{\theta})\right]^{\mathrm{H}} = \left[\frac{\partial}{\partial\theta_1^*}\log p(\mathbf{y},\boldsymbol{\theta}),\ldots,\frac{\partial}{\partial\theta_p^*}\log p(\mathbf{y},\boldsymbol{\theta})\right]^{\mathrm{T}}. \tag{6.78}$$

As throughout this chapter, all partial derivatives are Wirtinger derivatives as defined in Appendix 2. Thus, the Fisher–Bayes score is a p-dimensional complex column vector. The Fisher–Bayes score is the sum of the Fisher score and the prior score,

$$\begin{aligned}\boldsymbol{\sigma}(\mathbf{y},\boldsymbol{\theta}) &= \left[\frac{\partial}{\partial\boldsymbol{\theta}}\log p(\mathbf{y},\boldsymbol{\theta})\right]^{\mathrm{H}} \\ &= \left[\frac{\partial}{\partial\boldsymbol{\theta}}\log p_{\boldsymbol{\theta}}(\mathbf{y})\right]^{\mathrm{H}} + \left[\frac{\partial}{\partial\boldsymbol{\theta}}\log p(\boldsymbol{\theta})\right]^{\mathrm{H}} = \boldsymbol{\sigma}_{\boldsymbol{\theta}}(\mathbf{y}) + \boldsymbol{\sigma}(\boldsymbol{\theta}),\end{aligned} \tag{6.79}$$

where $\boldsymbol{\sigma}_{\boldsymbol{\theta}}(\mathbf{y})$ is the frequentist score and $\boldsymbol{\sigma}(\boldsymbol{\theta})$ is a new score that scores the prior density. The pdf $p(\mathbf{y},\boldsymbol{\theta})$ is the joint pdf for the measurement $\mathbf{y}$ and parameter $\boldsymbol{\theta}$. Since the expected value of Fisher–Bayes score is $\mathbf{0}$, Definition 6.5 also defines the centered measurement score $\mathbf{s}(\mathbf{y},\boldsymbol{\theta})$.

There are several properties of the Fisher–Bayes score that make it a compelling statistic for inferring the value of a parameter and for bounding the mean-squared error of any estimator for that parameter. In the following list, properties 1, 2, and 4 are analogous to the properties of the Fisher score listed in Section 6.3, but property 3 is different.

1. We may write the partial derivative as

$$\frac{\partial}{\partial\theta_i^*}\log p(\mathbf{y},\boldsymbol{\theta}) = \frac{1}{p(\mathbf{y},\boldsymbol{\theta})}\frac{\partial}{\partial\theta_i^*}p(\mathbf{y},\boldsymbol{\theta}), \tag{6.80}$$

which is a normalized measure of the sensitivity of the pdf $p(\mathbf{y},\boldsymbol{\theta})$ to variations in the parameter θ_i. Large sensitivity is valued, and this will be measured by the variance of the score.

2. The Fisher–Bayes score is a zero-mean random variable. This statistical property is consistent with maximum-likelihood procedures that search for maxima of the log-likelihood function $\log p(\mathbf{y},\boldsymbol{\theta})$ by searching for zeros of $(\partial/\partial\boldsymbol{\theta})\log p(\mathbf{y},\boldsymbol{\theta})$. Functions with steep slopes are valued, so we value a zero-mean score with large variance.
3. The expansion-coefficient matrix is $\mathbf{T}=\mathbf{I}$, and the complementary expansion-coefficient matrix is $\widetilde{\mathbf{T}}=\mathbf{0}$, thus $\underline{\mathbf{T}}=\mathbf{I}$. So, for the Fisher–Bayes measurement score, the centered measurement score is *perfectly aligned* with the centered error score.
4. The Fisher–Bayes information matrix is also the expected Hessian of the score function (up to a minus sign):

$$\begin{aligned}\mathbf{J}_{\mathrm{FB}} = E\left[\mathbf{s}(\mathbf{y},\boldsymbol{\theta})\mathbf{s}^{\mathrm{H}}(\mathbf{y},\boldsymbol{\theta})\right] &= E\left\{\left[\frac{\partial}{\partial\boldsymbol{\theta}}\log p(\mathbf{y},\boldsymbol{\theta})\right]^{\mathrm{H}}\frac{\partial}{\partial\boldsymbol{\theta}}\log p(\mathbf{y},\boldsymbol{\theta})\right\} \\ &= -E\left\{\frac{\partial}{\partial\boldsymbol{\theta}}\left[\frac{\partial}{\partial\boldsymbol{\theta}}\log p(\mathbf{y},\boldsymbol{\theta})\right]^{\mathrm{H}}\right\}.\end{aligned} \tag{6.81}$$

Analogously, the complementary Fisher–Bayes information matrix is

$$\tilde{\mathbf{J}}_{\mathrm{FB}} = E[\mathbf{s}(\mathbf{y},\boldsymbol{\theta})\mathbf{s}^{\mathrm{T}}(\mathbf{y},\boldsymbol{\theta})] = -E\left\{\frac{\partial}{\partial\boldsymbol{\theta}}\left[\frac{\partial}{\partial\boldsymbol{\theta}}\log p(\mathbf{y},\boldsymbol{\theta})\right]^{\mathrm{T}}\right\}. \tag{6.82}$$

The proofs of properties 2 and 4 for the Fisher–Bayes score are not substantially different than the corresponding proofs for the Fisher score, so we omit them. Property 3 is proved as follows. The bias $\mathbf{b}$ of the parameter estimator $\hat{\boldsymbol{\theta}}$ is differentiated with respect to $\boldsymbol{\theta}$ as

$$\begin{aligned}\mathbf{0} &= \left[\frac{\partial}{\partial\boldsymbol{\theta}}\mathbf{b}\right]^{\mathrm{H}} = \left[\frac{\partial}{\partial\boldsymbol{\theta}}\iint(\hat{\boldsymbol{\theta}}(\mathbf{y})-\boldsymbol{\theta})p(\mathbf{y},\boldsymbol{\theta})d\mathbf{y}\,d\boldsymbol{\theta}\right]^{\mathrm{H}}\\ &= \iint\left[\frac{\partial}{\partial\boldsymbol{\theta}}\log p(\mathbf{y},\boldsymbol{\theta})\right]^{\mathrm{H}}[\hat{\boldsymbol{\theta}}(\mathbf{y})-\boldsymbol{\theta}]^{\mathrm{H}}p(\mathbf{y},\boldsymbol{\theta})d\mathbf{y}\,d\boldsymbol{\theta} - \mathbf{I} = \mathbf{T}-\mathbf{I},\end{aligned} \tag{6.83}$$

which produces the identity $\mathbf{T} = \mathbf{I}$. On the other hand, the complementary expansion-coefficient matrix is derived analogously by considering

$$\begin{aligned}\mathbf{0} &= \left[\frac{\partial}{\partial\boldsymbol{\theta}}\mathbf{b}^*\right]^{\mathrm{H}} = \left[\frac{\partial}{\partial\boldsymbol{\theta}}\iint(\hat{\boldsymbol{\theta}}(\mathbf{y})-\boldsymbol{\theta})^*p(\mathbf{y},\boldsymbol{\theta})d\mathbf{y}\,d\boldsymbol{\theta}\right]^{\mathrm{H}}\\ &= \iint\left[\frac{\partial}{\partial\boldsymbol{\theta}}\log p_{\boldsymbol{\theta}}(\mathbf{y})\right]^{\mathrm{H}}[\hat{\boldsymbol{\theta}}(\mathbf{y})-\boldsymbol{\theta}]^{\mathrm{T}}p(\mathbf{y},\boldsymbol{\theta})d\mathbf{y}\,d\boldsymbol{\theta} - \mathbf{0} = \tilde{\mathbf{T}}.\end{aligned} \tag{6.84}$$

This shows that $\tilde{\mathbf{T}} = \mathbf{0}$, which means that the error and Fisher–Bayes score are always cross-proper. This is a consequence of the way the Fisher–Bayes score is defined. In summary, the augmented expansion-coefficient matrix is $\underline{\mathbf{T}} = \mathbf{I}$.

The Fisher–Bayes matrix $\underline{\mathbf{J}}_{\mathrm{FB}}$ is the sum of the expected Fisher matrix and the expected prior information matrix, since the frequentist score and prior score are uncorrelated:

$$\underline{\mathbf{J}}_{\mathrm{FB}} = E[\underline{\mathbf{J}}_{\mathrm{F}}(\boldsymbol{\theta})] + E[\underline{\mathbf{J}}_{\mathrm{P}}(\boldsymbol{\theta})]. \tag{6.85}$$

Thus, the Fisher–Bayes information is not simply the average of the Fisher information, although, for diffuse priors or a plenitude of measurements, it is approximately so.

6.5.2 Fisher–Bayes bound

The *Fisher–Bayes bound* on the augmented error covariance matrix is Result 6.6 with Fisher–Bayes score: $\underline{\mathbf{Q}} \geq \underline{\mathbf{J}}_{\mathrm{FB}}^{-1}$. From it, we obtain the bound on the error covariance matrix.

Result 6.7. *The widely linear* Fisher–Bayes bound *on the error covariance matrix is*

$$\mathbf{Q} \geq (\mathbf{J}_{\mathrm{FB}} - \tilde{\mathbf{J}}_{\mathrm{FB}}\mathbf{J}_{\mathrm{FB}}^{-*}\tilde{\mathbf{J}}_{\mathrm{FB}}^{*})^{-1}. \tag{6.86}$$

The Fisher–Bayes bound is sometimes also called the *stochastic Cramér–Rao bound.* In a number of aspects it behaves essentially like the "deterministic" Cramér–Rao bound in Result 6.4.

- The widely linear Fisher–Bayes bound is tighter than the linear Fisher–Bayes bound ignoring complementary Fisher–Bayes information because $(\mathbf{J}_{\mathrm{FB}} - \tilde{\mathbf{J}}_{\mathrm{FB}}\mathbf{J}_{\mathrm{FB}}^{-*}\tilde{\mathbf{J}}_{\mathrm{FB}}^{*})^{-1} \geq \mathbf{J}_{\mathrm{FB}}^{-1}$. However, $\mathbf{J}_{\mathrm{FB}}$ itself depends on whether or not $\mathbf{y}$ is proper.
- If M repeated measurements carry information about one fixed parameter $\boldsymbol{\theta}$ through the product pdf $\prod_{i=1}^{M} p(\mathbf{y}_i, \boldsymbol{\theta})$, the Fisher–Bayes bound decreases as M^{-1}.
- The influence of nuisance parameters on the Fisher–Bayes bound remains unchanged from Section 6.3.1: nuisance parameters increase the Fisher–Bayes bound.
- A tractable formula for Fisher–Bayes bounds for functions of parameters is obtained for the widely linear case, in which the development of Section 6.3.6 proceeds essentially unchanged.

Suppose that measurement and parameter have a jointly Gaussian distribution $p(\mathbf{y}, \boldsymbol{\theta}) = p_{\boldsymbol{\theta}}(\mathbf{y})p(\boldsymbol{\theta})$, and $\boldsymbol{\theta}$ has augmented covariance matrix $\underline{\mathbf{R}}_{\theta\theta}$. The new term added to the Fisher information matrix to produce the Fisher–Bayes information matrix is the prior information $\underline{\mathbf{J}}_{\mathrm{P}} = \underline{\mathbf{R}}_{\theta\theta}^{-1}$, which is independent of the mean of $\boldsymbol{\theta}$. Consequently the augmented Fisher–Bayes bound is

$$\underline{\mathbf{Q}} \geq (\underline{\mathbf{J}}_{\mathrm{F}} + \underline{\mathbf{R}}_{\theta\theta}^{-1})^{-1}, \tag{6.87}$$

where $\underline{\mathbf{J}}_{\mathrm{F}} = E[\underline{\mathbf{J}}_{\mathrm{F}}(\boldsymbol{\theta})]$ and the elements of the augmented Fisher matrix $\underline{\mathbf{J}}_{\mathrm{F}}(\boldsymbol{\theta})$ have been derived in Section 6.3.5.

6.6 Connections and orderings among bounds

Generally, we would like to connect Bayesian bounds to frequentist bounds, and particularly the Fisher–Bayes bound to the Cramér–Rao bound. In order to do so, we appeal to the fundamental identity

$$\mathbf{Q} = E[\mathbf{Q}(\boldsymbol{\theta})] + \mathbf{R}_{bb}. \tag{6.88}$$

Thus, there is the Bayesian bound

$$\mathbf{Q} \geq E[\mathbf{T}^{\mathrm{H}}(\boldsymbol{\theta})\mathbf{J}^{-1}(\boldsymbol{\theta})\mathbf{T}(\boldsymbol{\theta})] + E[(\mathbf{b}(\boldsymbol{\theta}) - \mathbf{b})(\mathbf{b}(\boldsymbol{\theta}) - \mathbf{b})^{\mathrm{H}}]. \tag{6.89}$$

This is a covariance bound on any estimator whose frequentist bias is $\mathbf{b}(\boldsymbol{\theta})$ and Bayes bias is $\mathbf{b}$. If the frequentist bias is constant at $\mathbf{b}$ – as for LMVDR (BLUE) estimators, for which $\mathbf{b}(\boldsymbol{\theta}) = \mathbf{0}$ – then $\mathbf{R}_{bb} = \mathbf{0}$, and the bound is $\mathbf{Q} \geq E[\mathbf{T}^{\mathrm{H}}(\boldsymbol{\theta})\mathbf{J}^{-1}(\boldsymbol{\theta})\mathbf{T}(\boldsymbol{\theta})]$. If the score is the Fisher score, and the bias is constant, then $\mathbf{T}(\boldsymbol{\theta}) = \mathbf{I}$ and the bound is

$$\mathbf{Q} \geq E[\mathbf{J}_{\mathrm{F}}^{-1}(\boldsymbol{\theta})]. \tag{6.90}$$

This is a bound on any estimator of a random parameter with the property that its frequentist and Bayes bias are constant, as in LMVDR estimators. We know this bound is larger than the Fisher–Bayes bound, because $\mathbf{J}_{\mathrm{FB}} = E[\mathbf{J}_{\mathrm{F}}(\boldsymbol{\theta})] + E[\mathbf{J}_{\mathrm{P}}(\boldsymbol{\theta})]$, so we have the ordering

$$E[\mathbf{J}_{\mathrm{F}}^{-1}(\boldsymbol{\theta})] \geq \mathbf{J}_{\mathrm{FB}}^{-1}. \tag{6.91}$$

This ordering makes sense because the right-hand side is a bound on estimators for which there is no constraint on the frequentist bias, as in MMSE estimation, and the left-hand side is a bound on estimators for which there is the constraint that the frequentist bias be constant. The constraint increases error covariance, as we have seen when comparing the variance of LMVDR and MMSE estimators.

When there is a plenitude of data or when the prior is so diffuse that $E[\mathbf{J}_{\mathrm{P}}(\boldsymbol{\theta})] \ll E[\mathbf{J}_{\mathrm{F}}(\boldsymbol{\theta})]$, then $E[\mathbf{J}_{\mathrm{F}}(\boldsymbol{\theta})]$ approximates $\mathbf{J}_{\mathrm{FB}}$ from below and consequently $(E[\mathbf{J}_{\mathrm{F}}(\boldsymbol{\theta})])^{-1} \geq \mathbf{J}_{\mathrm{FB}}^{-1}$. Then, by Jensen's inequality, $E[\mathbf{J}_{\mathrm{F}}^{-1}(\boldsymbol{\theta})] \geq (E[\mathbf{J}_{\mathrm{F}}(\boldsymbol{\theta})])^{-1}$, so we may squeeze $(E[\mathbf{J}_{\mathrm{F}}(\boldsymbol{\theta})])^{-1}$ between the two bounds to obtain the ordering (cf. Van Trees and Bell (2007))

$$E[\mathbf{J}_{\mathrm{F}}^{-1}(\boldsymbol{\theta})] \geq (E[\mathbf{J}_{\mathrm{F}}(\boldsymbol{\theta})])^{-1} \geq \mathbf{J}_{\mathrm{FB}}^{-1}. \tag{6.92}$$

To reiterate, the rightmost inequality follows from the fact that the Fisher–Bayes information matrix is more positive semidefinite than the expected value of the Fisher information matrix. The leftmost follows from Jensen's inequality.

Why *should* this ordering hold? The leftmost bound is a bound for estimators that are frequentist *and* Bayes unbiased, and frequentist unbiasedness is not generally a desirable property for an estimator of a random parameter. So this extra property tends to increase mean-squared error. The middle term ignores prior information about the random parameter, so it is only a bound in the limit as the Fisher–Bayes information matrix converges to the expected Fisher information matrix. The rightmost bound enforces no constraints on the frequentist or Bayes bias. The leftmost bound we tend to associate with LMVDR and ML estimators, the middle term we associate with the limiting performance of MMSE estimators when the prior is diffuse or there is a plenitude of measurements, and the rightmost bound we tend to associate with MMSE estimators. In fact, in our examples we have seen that the LMVDR and ML estimators achieve the leftmost bound and the MMSE estimator achieves the rightmost. Of course, it is not possible or meaningful to fit the frequentist Cramér–Rao bound into this ordering, except when $\mathbf{J}_{\mathrm{F}}(\boldsymbol{\theta})$ is independent of $\boldsymbol{\theta}$, in which case $\mathbf{J}_{\mathrm{F}}^{-1} = E[\mathbf{J}_{\mathrm{F}}^{-1}] = (E[\mathbf{J}_{\mathrm{F}}])^{-1}$. Then $\mathbf{J}_{\mathrm{F}}^{-1} \geq \mathbf{J}_{\mathrm{FB}}^{-1}$.

Notes

1 There is a vast literature on frequentist and Bayesian bounds. This literature is comprehensively represented, with original results by the editors, in Van Trees and Bell (2007). We make no attempt in this chapter to cite a representative subset of theoretical and applied papers in this edited volume, or in the published literature at large, since to do so would risk offense to the authors of the papers excluded.

This chapter addresses performance bounds for the entire class of quadratic frequentist and Bayesian bounds, for proper and improper complex parameters and measurements. Our method has been to cast the problem of quadratic performance bounding in the framework of error score estimation from measurement scores in a virtual two-channel experiment. This point of view is consistent with the point of view used in Chapter 5 and it allows us to extend the reasoning of Weinstein and Weiss (1988) to widely linear estimators.

We have not covered the derivation of performance bounds under parameter constraints. Representative published works are those by Gorman and Hero (1990), Marzetta (1993), and Stoica and Ng (1998).

2 The examples we give in this chapter are not representative of a great number of problems in signal processing where mode parameters like frequency, wavenumber, direction of arrival, and so on are to be estimated. In these problems the Cramér–Rao and Fisher–Bayes bounds are tight at high SNR and very loose at low SNR. Consequently they do not predict performance breakdown at a threshold SNR. In an effort to analyze the onset of breakdown, many alternative scores have been proposed. Even the Cramér–Rao and Fisher–Bayes bounds can be adapted to this problem by using a method of intervals, wherein a mixture of the Cramér–Rao or Fisher–Bayes bound and a constant bound representative of total breakdown is averaged. The averaging probability is the probability of error lying outside the interval wherein the likelihood is quadratic, and it may be computed for many problems. This method, which has been used to effect by Richmond (2006), is remarkably accurate. So we might say that quadratic performance bounding, augmented with a method of intervals, is a powerful way to model error covariances from high SNR, through the threshold region of low SNR where the quadratic bound by itself is not tight, to the region of total breakdown.

3 The Cramér–Rao bounds for these nonlinear parameters have been treated extensively in the literature, and are well represented by Van Trees and Bell (2007). McWhorter and Scharf (1993b) discuss concentration ellipsoids for error covariances when estimating the sum and difference frequencies of closely spaced modes within the matrix $\mathbf{H}(\boldsymbol{\theta})$. See also the discussion of the geometry of the Cramér–Rao bound by McWhorter and Scharf (1993a).

7 Detection

Detection is the electrical engineer's term for the statistician's *hypothesis testing*. The problem is to determine which of two or more competing models best describes experimental measurements. If the competition is between two models, then the detection problem is a binary detection problem. Such problems apply widely to communication, radar, and sonar. But even a binary problem can be composite, which is to say that one or both of the hypotheses may consist of a set of models. We shall denote by H_0 the hypothesis that the underlying model, or set of models, is M_0 and by H_1 the hypothesis that it is M_1.

There are two main lines of development for detection theory: Neyman–Pearson and Bayes. The Neyman–Pearson theory is a frequentist theory that assigns no prior probability of occurrence to the competing models. Bayesian theory does. Moreover, the measure of optimality is different. To a frequentist the game is to maximize the detection probability under the constraint that the false-alarm probability is not greater than a prespecified value. To a Bayesian the game is to assign costs to incorrect decisions, and then to minimize the average (or Bayes) cost. The solution in any case is to evaluate the *likelihood* of the measurement under each hypothesis, and to choose the model whose likelihood is higher. Well – not quite. It is the *likelihood ratio* that is evaluated, and when this ratio exceeds a threshold, determined either by the false-alarm rate or by the Bayes cost, one or other of the hypotheses is accepted.

It is commonplace for each of the competing models to contain unknown *nuisance parameters* and unknown *decision* parameters. Only the decision parameters are tested, but the nuisance parameters complicate the computation of likelihoods and defeat attempts at optimality. There are generally two strategies: enforce a condition of *invariance*, which limits competition to detectors that share these invariances, or invoke a principle of likelihood that replaces nuisance parameters by their maximum-likelihood estimates. These lead to *generalized likelihood ratios*. In many cases the two approaches produce identical tests. So, in summary, the likelihood-ratio statistic is the gold standard. It plays the same role for binary hypothesis testing as the conditional mean estimator plays in MMSE estimation.

The theory of Neyman–Pearson and Bayes detection we shall review is quite general.[1] It reveals functions of the measurement that may include arbitrary functions of the complex measurement $\mathbf{y}$, including its complex conjugate $\mathbf{y}^*$. However, in order to demonstrate the role of widely linear and widely quadratic forms in the theory of hypothesis testing, we shall concentrate on hypothesis testing within the proper and improper Gaussian measurement models.

Throughout this chapter we assume that measurements have pdfs, which are allowed to contain Dirac δ-functions. Thus, all of our results hold for mixed random variables with discrete and continuous, and even singular, components.

7.1 Binary hypothesis testing

A hypothesis is a statement about the distribution of a measurement $\mathbf{y} \in \mathcal{Y}$. Typically the hypothesis is coded by a parameter $\boldsymbol{\theta} \in \boldsymbol{\Theta}$ that indexes the pdf $p_{\boldsymbol{\theta}}(\mathbf{y})$ for the measurement random vector $\mathbf{y}: \Omega \longrightarrow \mathbb{C}^n$. We might say the parameter *modulates* the measurement in the sense that the value of $\boldsymbol{\theta}$ modulates the distribution of the measurement. Often the parameter modulates the mean vector, as in communications, but sometimes it modulates the covariance matrix, as in oceanography and optics, where a frequency–wavenumber spectrum determines a covariance matrix.

We shall be interested in two competing hypotheses, H_0 and H_1, and therefore we address ourselves to *binary hypothesis testing*, as opposed to *multiple hypothesis testing*. The hypothesis H_0 is sometimes called the *hypothesis* and the hypothesis H_1 is sometimes called the *alternative*, so we are testing a hypothesis against its alternative. Each of the hypotheses may be *simple*, in which case the pdf of the measurement is $p_{\boldsymbol{\theta}_i}(\mathbf{y})$ under hypothesis H_i. Or each of them may be composite, in which case the distribution of the measurement is $p_{\boldsymbol{\theta}}(\mathbf{y})$, $\boldsymbol{\theta} \in \boldsymbol{\Theta}_i$, under hypothesis H_i. The parameter sets $\boldsymbol{\Theta}_i$ and the measurement space $\mathbf{y} \in \mathcal{Y}$ may be quite general.

Definition 7.1. *A binary test of H_1 versus H_0 is a* test statistic, *or* threshold detector $\phi: \mathcal{Y} \rightarrow \{0, 1\}$ *of the form*

$$\phi(\mathbf{y}) = \begin{cases} 1, & \mathbf{y} \in \mathcal{Y}_1, \\ 0, & \mathbf{y} \in \mathcal{Y}_0. \end{cases} \tag{7.1}$$

We might say that ϕ is a one-bit quantizer of the random vector $\mathbf{y}$. It sorts measurements into two sets: the detection region $\mathcal{Y}_1$ and its complement region $\mathcal{Y}_0$. Such a test will make two types of errors: missed detections and false alarms. That is, it will sometimes classify a measurement drawn from the pdf $p_{\boldsymbol{\theta}}(\mathbf{y})$, $\boldsymbol{\theta} \in \boldsymbol{\Theta}_1$, as a measurement drawn from the pdf $p_{\boldsymbol{\theta}}(\mathbf{y})$, $\boldsymbol{\theta} \in \boldsymbol{\Theta}_0$. This is a missed detection. Or it might classify a measurement drawn from the pdf $p_{\boldsymbol{\theta}}(\mathbf{y})$, $\boldsymbol{\theta} \in \boldsymbol{\Theta}_0$, as a measurement drawn from the pdf $p_{\boldsymbol{\theta}}(\mathbf{y})$, $\boldsymbol{\theta} \in \boldsymbol{\Theta}_1$. This is a false alarm.

For a simple hypothesis and simple alternative the *detection probability* P_{D} and the *false-alarm probability* P_{F} of a test are defined as follows:

$$P_{\mathrm{D}} = P_{\boldsymbol{\theta}_1}(\phi = 1) = E_{\boldsymbol{\theta}_1}[\phi], \tag{7.2}$$

$$P_{\mathrm{F}} = P_{\boldsymbol{\theta}_0}(\phi = 1) = E_{\boldsymbol{\theta}_0}[\phi]. \tag{7.3}$$

The subscripts on the probability and expectation operators denote probability and expectation computed under the pdf of the measurement $p_{\theta_i}(\mathbf{y})$. We shall generalize these definitions for testing composite hypotheses in due course.

Detection probability and false-alarm probability are engineering terms for the statistician's *power* and *size*. Both terms are evocative, with the statistician advocating for a test that is small but powerful, like a compact drill motor, and the engineer advocating for a test that has high detection probability and low false-alarm probability, like a good radar detector on one of the authors' Mazda RX-7.

7.1.1 The Neyman–Pearson lemma

In binary hypothesis testing of a simple alternative versus a simple hypothesis, the test statistic ϕ is a Bernoulli random variable $B[p]$, with $p = P_D$ under H_1 and $p = P_F$ under H_0. Ideally, it would have the property $P_D = 1$ and $P_F = 0$, but these are unattainable for non-degenerate tests. So we aim to maximize the detection probability under the constraint that the false-alarm probability be no greater than the design value of $P_F = \alpha$. In other words, it should have maximum power for its small size. We first state the *Neyman–Pearson (NP) lemma* without proof.[2]

Result 7.1. *(Neyman–Pearson) Among all competing binary tests ϕ' of a simple alternative versus a simple hypothesis, with probability of false alarm less than or equal to α, none has larger detection probability than the* likelihood-ratio test

$$\phi(\mathbf{y}) = \begin{cases} 1, & l(\mathbf{y}) > \eta, \\ \gamma, & l(\mathbf{y}) = \eta, \\ 0, & l(\mathbf{y}) < \eta. \end{cases} \tag{7.4}$$

In this equation, $0 \le l(\mathbf{y}) < \infty$ is the real-valued likelihood ratio

$$l(\mathbf{y}) = \frac{p_{\theta_1}(\mathbf{y})}{p_{\theta_0}(\mathbf{y})} \tag{7.5}$$

and the threshold parameters (η, γ) are chosen to satisfy the following constraint on the false-alarm probability:

$$P_{\theta_0}[l(\mathbf{y}) > \eta] + \gamma P_{\theta_0}[l(\mathbf{y}) = \eta] = \alpha. \tag{7.6}$$

When $\phi(\mathbf{y}) = \gamma$, we select H_1 with probability γ, and H_0 with probability $1 - \gamma$.

The likelihood ratio l is a highly compressed function of the measurement, returning a nonnegative number, and the test function ϕ returns what might be called a one-bit quantized version of the likelihood ratio. Any monotonic function of the likelihood ratio l will do, so in multivariate Gaussian theory it is the logarithm of l, or the *log-likelihood ratio* L, that is used in place of the likelihood-ratio.

There is a cautionary note about the reading of these equations. The test statistic ϕ and the likelihood ratio l are functions of the measurement random variable, making ϕ

and l *random variables*. When evaluated at a particular realization or measurement, they are test values and likelihood-ratio values.

7.1.2 Bayes detectors

The only thing that distinguishes a Bayes detector from a Neyman–Pearson (NP) detector is the choice of threshold η. It is chosen to minimize the Bayes risk, which is defined to be $\pi_1 C_{10}(1 - P_D) + (1 - \pi_1)C_{01} P_F$, where π_1 is the prior probability that hypothesis H_1 will be in force, C_{10} is the cost of choosing H_0 when H_1 is in force, and $1 - P_D$ is the probability of doing so. Similarly, C_{01} is the cost of choosing H_1 when H_0 is in force, and P_F is the probability of doing so.

Assume that the Bayes test sorts measurements into the set $S \subset \mathbb{C}^n$, where H_1 is accepted, and the complement set, where H_0 is accepted. Then the Bayes risk may be written as

$$\pi_1 C_{10}(1 - P_D) + (1 - \pi_1)C_{01} P_F = \pi_1 C_{10} + \int_S [(1 - \pi_1)C_{01} p_{\boldsymbol{\theta}_0}(\mathbf{y}) - \pi_1 C_{10} p_{\boldsymbol{\theta}_1}(\mathbf{y})]d\mathbf{y}. \tag{7.7}$$

As in Chapter 6, this integral with respect to a complex variable is interpreted as the integral with respect to its real and imaginary parts:

$$\int f(\mathbf{y})d\mathbf{y} \triangleq \int f(\mathbf{y})d[\operatorname{Re}\mathbf{y}]d[\operatorname{Im}\mathbf{y}]. \tag{7.8}$$

The Bayes risk is minimized by choosing S in (7.7) to be the set

$$S = \{\mathbf{y}\colon \pi_1 C_{10} p_{\boldsymbol{\theta}_1}(\mathbf{y}) > (1 - \pi_1)C_{01} p_{\boldsymbol{\theta}_0}(\mathbf{y})\} = \left\{\mathbf{y}\colon l(\mathbf{y}) > \frac{(1 - \pi_1)C_{01}}{\pi_1 C_{10}} = \eta\right\}. \tag{7.9}$$

This is just the likelihood-ratio test with a particular threshold η. The only difference between a Bayes test and an NP test is that their thresholds are different.

7.1.3 Adaptive Neyman–Pearson and empirical Bayes detectors

It is often the case that only a subset of the parameters is to be tested, leaving the others as nuisance parameters. In these cases the principles of NP or Bayes detection may be expanded to say that unknown nuisance parameters are replaced by their maximum-likelihood estimators or their Bayes estimators, under each of the two hypotheses. The resulting detectors are called *adaptive*, *empirical*, or *generalized likelihood-ratio* detectors. The literature on them is comprehensive, and we make no attempt to review it here. Some examples are discussed in the context of testing for correlation structure in Section 4.5. Later, in Section 7.6, we outline a different story, which is based on invariance with respect to nuisance parameters. This story also leads to detectors that are adaptive.

7.2 Sufficiency and invariance

The test function ϕ is a Bernoulli random variable. Its probability distribution depends on the parameter $\boldsymbol{\theta}$ according to $p = P_D$ under H_1 and $p = P_F$ under H_0. However, the value of ϕ, given its likelihood ratio l, is 0 if $l < \eta$, γ if $l = \eta$, and 1 if $l > \eta$, independently of $\boldsymbol{\theta}$. Thus, according to the Fisher–Neyman factorization theorem, the likelihood-ratio statistic is a *sufficient statistic* for testing H_1 versus H_0.

Definition 7.2. *We call a statistic* $\mathbf{t}(\mathbf{y})$ *sufficient if the pdf* $p_{\boldsymbol{\theta}}(\mathbf{y})$ *may be factored as* $p_{\boldsymbol{\theta}}(\mathbf{y}) = a(\mathbf{y})q_{\boldsymbol{\theta}}(\mathbf{t})$*, where* $q_{\boldsymbol{\theta}}(\mathbf{t})$ *is the pdf for the sufficient statistic.*

In this case the likelihood ratio equals the likelihood ratio for the sufficient statistic $\mathbf{t}$:

$$l(\mathbf{y}) = \frac{p_{\boldsymbol{\theta}_1}(\mathbf{y})}{p_{\boldsymbol{\theta}_0}(\mathbf{y})} = \frac{q_{\boldsymbol{\theta}_1}(\mathbf{t})}{q_{\boldsymbol{\theta}_0}(\mathbf{t})} = l'(\mathbf{t}). \tag{7.10}$$

We may thus rewrite the NP lemma in terms of the sufficient statistic:

$$\phi(\mathbf{t}) = \begin{cases} 1, & l'(\mathbf{t}) > \eta, \\ \gamma, & l'(\mathbf{t}) = \eta, \\ 0, & l'(\mathbf{t}) < \eta. \end{cases} \tag{7.11}$$

In this equation, (η, γ) are chosen to satisfy the false-alarm probability constraint:

$$P_{\boldsymbol{\theta}_0}[l'(\mathbf{t}) > \eta] + \gamma P_{\boldsymbol{\theta}_0}[l'(\mathbf{t}) = \eta] = \alpha. \tag{7.12}$$

The statistic $\mathbf{t}$ is often a scalar function of the original measurement, and its likelihood can often be inverted with a monotonic function for the statistic itself. Then the test statistic is a *threshold test* that simply compares the sufficient statistic against a threshold.

Definition 7.3. *A sufficient statistic is* minimal, *meaning that it is most memory efficient, if it is a function of every other sufficient statistic. A sufficient statistic is an* invariant statistic *with respect to the transformation group G if* $\phi(g(\mathbf{t})) = \phi(\mathbf{t})$ *for all* $g \in G$*. The composition rule for any two transformations within G is* $g_1 \circ g_2(\mathbf{t}) = g_1(g_2(\mathbf{t}))$.

7.3 Receiver operating characteristic

It is obvious that no NP detector will have a worse $(P_{\mathrm{D}}, P_{\mathrm{F}})$ pair than (α, α), since the test $\phi' = \alpha$ achieves this performance and, according to the NP lemma, cannot be better than the NP likelihood-ratio test. Moreover, from the definitions of $(P_{\mathrm{D}}, P_{\mathrm{F}})$, an increase in the threshold η will decrease the false-alarm probability and decrease the detection probability. So a plot of achievable $(P_{\mathrm{D}}, P_{\mathrm{F}})$ pairs, called an ROC curve for *receiver operation characteristic*, appears as the convex curve in Fig. 7.1.

Why is it convex? Consider two points on the ROC curve. Draw a straight line between them, to find an operating point for some suboptimum detector. At the P_{F} for this detector there is a point on the ROC curve for the NP detector that cannot lie below it. Thus the ROC curve for the NP detector is convex.

Every detector will have an operating point somewhere within the convex set bounded by the ROC curve for the NP detector and the (α, α) line. One such operating point is

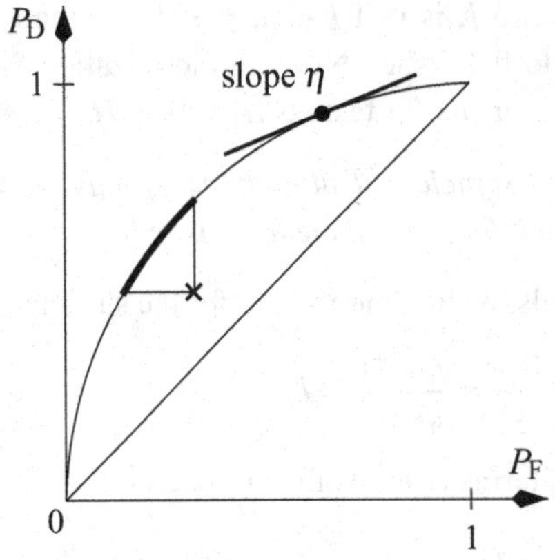

Figure 7.1 Geometry of the ROC for a Neyman–Pearson detector.

illustrated with a cross in Fig. 7.1. Any suboptimum test with this operating point may be improved with an NP test whose operating point lies on the arc subtended by the wedge that lies to the northwest of the cross. To improve a test is to decrease its P_F for a fixed P_D along the horizontal line of the wedge, increase its P_D for a fixed P_F along the vertical line of the wedge, or decrease P_F and increase P_D by heading northwest to the ROC line.

Recall the identity (7.10) for a sufficient statistic **t**. If the sufficient statistic is the likelihood itself, that is $\mathbf{t} = l(\mathbf{y})$, (7.10) shows that the pdf for likelihood under the two hypotheses is connected as follows:

$$q_{\boldsymbol{\theta}_1}(l) = l q_{\boldsymbol{\theta}_0}(l). \tag{7.13}$$

The probability of detection and probability of false alarm may be written

$$1 - P_D = \int_0^\eta q_{\boldsymbol{\theta}_1}(l)\mathrm{d}l = \int_0^\eta l q_{\boldsymbol{\theta}_0}(l)\mathrm{d}l, \tag{7.14}$$

$$1 - P_F = \int_0^\eta q_{\boldsymbol{\theta}_0}(l)\mathrm{d}l. \tag{7.15}$$

Differentiate each of these formulae with respect to η to find

$$\frac{\partial}{\partial\eta}P_D = -\eta q_{\boldsymbol{\theta}_0}(\eta) \tag{7.16}$$

$$\frac{\partial}{\partial\eta}P_F = -q_{\boldsymbol{\theta}_0}(\eta) \tag{7.17}$$

and therefore

$$\frac{\mathrm{d}P_D}{\mathrm{d}P_F} = \eta. \tag{7.18}$$

Thus, at any operating point on the ROC curve, the derivative of the curve is the threshold η required to operate there.[3]

7.4 Simple hypothesis testing in the improper Gaussian model

In the context of this book, where an augmented second-order theory that includes Hermitian and complementary covariances forms the basis for inference, the most important example is the problem of testing in the multivariate Gaussian model. We may test for mean-value and/or covariance structure.

7.4.1 Uncommon means and common covariance

The first problem is to test the alternative H_1 that an improper Gaussian measurement $\mathbf{y}$ has mean value $\boldsymbol{\mu}_1$ versus the hypothesis H_0 that it has mean value $\boldsymbol{\mu}_0$. Under both hypotheses the measurement $\mathbf{y}$ has Hermitian correlation $\mathbf{R} = E_{\boldsymbol{\theta}_0}[(\mathbf{y} - \boldsymbol{\mu}_0)(\mathbf{y} - \boldsymbol{\mu}_0)^{\mathrm{H}}] = E_{\boldsymbol{\theta}_1}[(\mathbf{y} - \boldsymbol{\mu}_1)(\mathbf{y} - \boldsymbol{\mu}_1)^{\mathrm{H}}]$ and symmetric complementary covariance $\widetilde{\mathbf{R}} = E_{\boldsymbol{\theta}_0}[(\mathbf{y} - \boldsymbol{\mu}_0)(\mathbf{y} - \boldsymbol{\mu}_0)^{\mathrm{T}}] = E_{\boldsymbol{\theta}_1}[(\mathbf{y} - \boldsymbol{\mu}_1)(\mathbf{y} - \boldsymbol{\mu}_1)^{\mathrm{T}}]$. These covariances are coded into the augmented covariance matrix

$$\underline{\mathbf{R}} = \begin{bmatrix} \mathbf{R} & \widetilde{\mathbf{R}} \\ \widetilde{\mathbf{R}}^* & \mathbf{R}^* \end{bmatrix}. \tag{7.19}$$

For convenience, we drop the subscripts on covariance matrices ($\mathbf{R}_{yy}$ etc.) in this chapter. Following Result 2.4, we may write down the pdf of the improper Gaussian vector $\mathbf{y}$ under hypothesis H_i:

$$p_{\boldsymbol{\theta}_i}(\mathbf{y}) = \frac{1}{\pi^n \det^{1/2} \underline{\mathbf{R}}} \exp\left\{-\tfrac{1}{2}(\underline{\mathbf{y}} - \underline{\boldsymbol{\mu}}_i)^{\mathrm{H}} \underline{\mathbf{R}}^{-1} (\underline{\mathbf{y}} - \underline{\boldsymbol{\mu}}_i)\right\}. \tag{7.20}$$

Of course, this pdf also covers the proper case where $\underline{\mathbf{R}}$ is block-diagonal. After some algebra, the real-valued log-likelihood ratio may be expressed as the following widely linear function of $\mathbf{y}$:

$$\begin{aligned} L = \log\left(\frac{p_{\boldsymbol{\theta}_1}(\mathbf{y})}{p_{\boldsymbol{\theta}_0}(\mathbf{y})}\right) &= -\tfrac{1}{2}(\underline{\mathbf{y}} - \underline{\boldsymbol{\mu}}_1)^{\mathrm{H}} \underline{\mathbf{R}}^{-1} (\underline{\mathbf{y}} - \underline{\boldsymbol{\mu}}_1) + \tfrac{1}{2}(\underline{\mathbf{y}} - \underline{\boldsymbol{\mu}}_0)^{\mathrm{H}} \underline{\mathbf{R}}^{-1} (\underline{\mathbf{y}} - \underline{\boldsymbol{\mu}}_0) \\ &= (\underline{\boldsymbol{\mu}}_1 - \underline{\boldsymbol{\mu}}_0)^{\mathrm{H}} \underline{\mathbf{R}}^{-1} (\underline{\mathbf{y}} - \underline{\mathbf{y}}_0) \end{aligned} \tag{7.21}$$

with

$$\underline{\mathbf{y}}_0 = \tfrac{1}{2}\left(\underline{\boldsymbol{\mu}}_1 + \underline{\boldsymbol{\mu}}_0\right). \tag{7.22}$$

The log-likelihood ratio may be written as the inner product

$$L = \left[\underline{\mathbf{R}}^{-1}(\underline{\boldsymbol{\mu}}_1 - \underline{\boldsymbol{\mu}}_0)\right]^{\mathrm{H}} (\underline{\mathbf{y}} - \underline{\mathbf{y}}_0) = 2\,\mathrm{Re}\,\{\mathbf{w}^{\mathrm{H}}(\mathbf{y} - \mathbf{y}_0)\}, \tag{7.23}$$

with

$$\underline{\mathbf{w}} = \begin{bmatrix} \mathbf{w} \\ \mathbf{w}^* \end{bmatrix} = \underline{\mathbf{R}}^{-1}(\underline{\boldsymbol{\mu}}_1 - \underline{\boldsymbol{\mu}}_0). \tag{7.24}$$

Taking the inner product in (7.23) can be interpreted as *coherent matched filtering*. In the proper case, the matching vector $\mathbf{w}$ simplifies to $\mathbf{w} = \mathbf{R}^{-1}(\boldsymbol{\mu}_1 - \boldsymbol{\mu}_0)$.

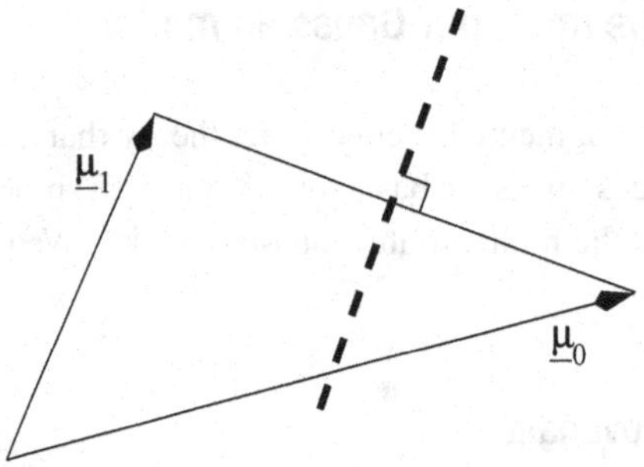

Figure 7.2 Geometrical interpretation of a coherent matched filter. The thick dashed line is the locus of measurements such that $[\underline{\mathbf{R}}^{-1}(\underline{\boldsymbol{\mu}}_1 - \underline{\boldsymbol{\mu}}_0)]^{\mathrm{H}}(\underline{\mathbf{y}} - \underline{\mathbf{y}}_0) = \mathbf{0}$.

The centered complex measurement $\underline{\mathbf{y}} - \underline{\mathbf{y}}_0$ is resolved onto the line $\langle \underline{\mathbf{R}}^{-1}(\underline{\boldsymbol{\mu}}_1 - \underline{\boldsymbol{\mu}}_0)\rangle$ to produce the *real* log-likelihood ratio L, which is compared against a threshold. The geometry of the log-likelihood ratio is illustrated in Fig. 7.2: we choose lines parallel to the thick dashed line in order to get the desired probability of false alarm P_{F}.

But how do we determine the threshold to get the desired P_{F}? We will see that the deflection, or output signal-to-noise ratio, plays a key role here.

Definition 7.4. *The* deflection *for a likelihood-ratio test with log-likelihood ratio L is defined as*

$$d = \frac{[E_{\boldsymbol{\theta}_1}(L) - E_{\boldsymbol{\theta}_0}(L)]^2}{\operatorname{var}_{\boldsymbol{\theta}_0}(L)}, \tag{7.25}$$

where $\operatorname{var}_{\boldsymbol{\theta}_0}(L)$ *denotes the variance of L under H_0.*

We will first show that L given by (7.23) results in the deflection

$$d = (\underline{\boldsymbol{\mu}}_1 - \underline{\boldsymbol{\mu}}_0)^{\mathrm{H}} \underline{\mathbf{R}}^{-1} (\underline{\boldsymbol{\mu}}_1 - \underline{\boldsymbol{\mu}}_0), \tag{7.26}$$

which is simply the *output signal-to-noise ratio*. The mean of L is $d/2$ under H_1 and $-d/2$ under H_0, and its variance is d under both hypotheses. Thus, the log-likelihood ratio statistic L is distributed as a real Gaussian random variable with mean value $\pm d/2$ and variance d. Thus, L has deflection, or output signal-to-noise ratio, d. In the general non-Gaussian case, the deflection is much easier to compute than a complete ROC curve.

Now the false-alarm probability for the detector ϕ that compares L against a threshold η is

$$P_{\mathrm{F}} = 1 - \int_{-\infty}^{\eta} \frac{1}{\sqrt{2\pi d}} \exp\left\{-\frac{1}{2d}(L + d/2)^2\right\} \mathrm{d}L = 1 - \Phi\left(\frac{\eta + d/2}{\sqrt{d}}\right), \tag{7.27}$$

where Φ is the probability distribution function of a zero-mean, variance-one, real Gaussian random variable. A similar calculation shows the probability of detection to be

$$P_{\mathrm{D}} = 1 - \int_{-\infty}^{\eta} \frac{1}{\sqrt{2\pi d}} \exp\left\{-\frac{1}{2d}(L - d/2)^2\right\} \mathrm{d}L = 1 - \Phi\left(\frac{\eta - d/2}{\sqrt{d}}\right). \tag{7.28}$$

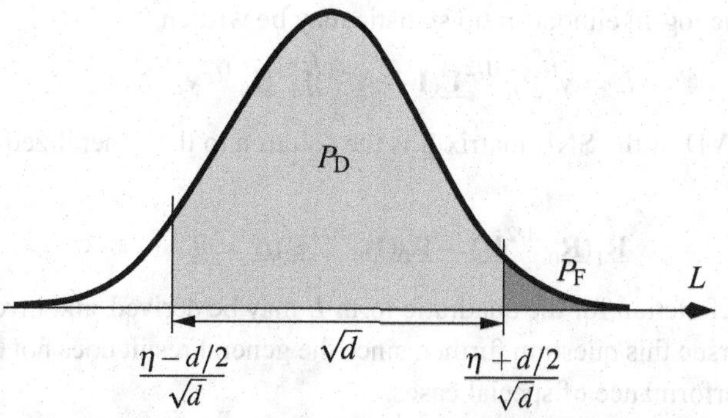

Figure 7.3 The false-alarm probability P_F (dark gray) and detection probability P_D (light and dark gray) are tail probabilities of a normalized Gaussian. They are determined by the threshold η and the deflection d.

So, by choosing the threshold η, any design value of false-alarm probability may be achieved, and this choice of threshold determines the detection probability as well. In Fig. 7.3 the argument is made that the normalized zero-mean, variance-one, Gaussian density determines performance. The designer is given a string of length $\sqrt{d}$, whose head end is placed at location $(\eta + d/2)/\sqrt{d}$, to achieve the false-alarm probability P_F, and whose tail end falls at location $(\eta - d/2)/\sqrt{d}$, to determine the detection probability P_D. The length of the string is $\sqrt{d}$, which is the output *voltage SNR*, so obviously the larger the SNR the better.

7.4.2 Common mean and uncommon covariances

When the mean of an improper complex random vector is common to both the hypothesis and the alternative, but the covariances are uncommon, we might as well assume that the mean is zero, since the measurement can always be centered. Then the pdf for the measurement $\mathbf{y}: \Omega \longrightarrow \mathbb{C}^n$ under hypothesis H_i is

$$p_{\boldsymbol{\theta}_i}(\mathbf{y}) = \frac{1}{\pi^n \det^{1/2} \underline{\mathbf{R}}_i} \exp\left\{-\tfrac{1}{2}\underline{\mathbf{y}}^{\mathrm{H}} \underline{\mathbf{R}}_i^{-1} \underline{\mathbf{y}}\right\} \tag{7.29}$$

where $\underline{\mathbf{R}}_i$ is the augmented covariance matrix under hypothesis H_i. For a measured $\mathbf{y}$ this is also the likelihood that the covariance model $\underline{\mathbf{R}}_i$ would have produced it. The log-likelihood ratio for comparing the likelihood of alternative H_1 with the likelihood of hypothesis H_0 is then the real widely quadratic form

$$L = \underline{\mathbf{y}}^{\mathrm{H}}(\underline{\mathbf{R}}_0^{-1} - \underline{\mathbf{R}}_1^{-1})\underline{\mathbf{y}} = \underline{\mathbf{y}}^{\mathrm{H}} \underline{\mathbf{R}}_0^{-\mathrm{H}/2}(\mathbf{I} - \underline{\mathbf{S}}^{-1})\underline{\mathbf{R}}_0^{-1/2}\underline{\mathbf{y}}, \tag{7.30}$$

where $\underline{\mathbf{S}} = \underline{\mathbf{R}}_0^{-1/2} \underline{\mathbf{R}}_1 \underline{\mathbf{R}}_0^{-\mathrm{H}/2}$ is the augmented *signal-to-noise ratio matrix*. The transformed measurement $\underline{\mathbf{R}}_0^{-1/2}\underline{\mathbf{y}}$ has augmented covariance matrix $\mathbf{I}$ under hypothesis H_0 and augmented covariance matrix $\underline{\mathbf{S}}$ under hypothesis H_1. Thus L is the log-likelihood ratio for testing that the widely linearly transformed measurement, extracted as the top half of $\underline{\mathbf{R}}_0^{-1/2}\underline{\mathbf{y}}$, is *proper and white* versus the alternative that it is improper with augmented covariance $\underline{\mathbf{S}}$. When the SNR matrix is given its augmented EVD $\underline{\mathbf{S}} = \underline{\mathbf{U}}\,\underline{\boldsymbol{\Lambda}}\,\underline{\mathbf{U}}^{\mathrm{H}}$

(cf. Result 3.1), the log-likelihood-ratio statistic may be written

$$L = \underline{\mathbf{y}}^{\mathrm{H}} \underline{\mathbf{R}}_0^{-\mathrm{H}/2} \underline{\mathbf{U}}(\mathbf{I} - \underline{\mathbf{\Lambda}}^{-1}) \underline{\mathbf{U}}^{\mathrm{H}} \underline{\mathbf{R}}_0^{-1/2} \underline{\mathbf{y}}. \tag{7.31}$$

The augmented EVD of the SNR matrix $\underline{\mathbf{S}}$ is the solution to the generalized eigenvalue problem

$$\underline{\mathbf{R}}_1 (\underline{\mathbf{R}}_0^{-\mathrm{H}/2} \underline{\mathbf{U}}) - \underline{\mathbf{R}}_0 (\underline{\mathbf{R}}_0^{-\mathrm{H}/2} \underline{\mathbf{U}}) \underline{\mathbf{\Lambda}} = \mathbf{0}. \tag{7.32}$$

The characteristic function for the quadratic form L may be derived, and inverted for its pdf. We do not pursue this question further, since the general result does not bring much insight into the performance of special cases.

7.4.3 Comparison between linear and widely linear detection

We would now like to compare the performance of linear and widely linear detection. For this, we consider a special case of testing for a common mean and uncommon covariances. We wish to detect whether the Gaussian signal $\mathbf{x}\colon \Omega \longrightarrow \mathbb{C}^n$ is present in additive white Gaussian noise $\mathbf{n}\colon \Omega \longrightarrow \mathbb{C}^n$. Under H_0, the measurement is $\mathbf{y} = \mathbf{n}$, and under H_1, it is $\mathbf{y} = \mathbf{x} + \mathbf{n}$. The signal $\mathbf{x}$ has augmented covariance matrix $\underline{\mathbf{R}}_{xx}$ and the noise is white and proper with augmented covariance matrix $\underline{\mathbf{R}}_{nn} = N_0\mathbf{I}$. The signal and noise both have zero mean, and they are uncorrelated. Thus, the covariance matrix under H_0 is $\underline{\mathbf{R}}_0 = N_0\mathbf{I}$, and under H_1 it is $\underline{\mathbf{R}}_1 = \underline{\mathbf{R}}_{xx} + N_0\mathbf{I}$.

Let $\underline{\mathbf{R}}_{xx} = \underline{\mathbf{U}}\,\underline{\mathbf{\Lambda}}\,\underline{\mathbf{U}}^{\mathrm{H}}$ denote the augmented EVD of $\underline{\mathbf{R}}_{xx}$. Note that $\underline{\mathbf{\Lambda}}$ is now the augmented eigenvalue matrix of $\underline{\mathbf{R}}_{xx}$, whereas before in Section 7.4.2, $\underline{\mathbf{\Lambda}}$ denoted the augmented eigenvalue matrix of $\underline{\mathbf{S}} = \underline{\mathbf{R}}_0^{-1/2} \underline{\mathbf{R}}_1 \underline{\mathbf{R}}_0^{-\mathrm{H}/2} = \underline{\mathbf{R}}_1/N_0 = \underline{\mathbf{R}}_{xx}/N_0 + \mathbf{I}$. Therefore, the augmented eigenvalue matrix of $\underline{\mathbf{S}}$ is now $\underline{\mathbf{\Lambda}}/N_0 + \mathbf{I}$, and the log-likelihood ratio (7.31) is

$$\begin{aligned} L &= \frac{1}{N_0} \underline{\mathbf{y}}^{\mathrm{H}} \underline{\mathbf{U}} [\mathbf{I} - (\underline{\mathbf{\Lambda}}/N_0 + \mathbf{I})^{-1}] \underline{\mathbf{U}}^{\mathrm{H}} \underline{\mathbf{y}} \\ &= \frac{1}{N_0} \underline{\mathbf{y}}^{\mathrm{H}} \underline{\mathbf{U}}\,\underline{\mathbf{\Lambda}} (\underline{\mathbf{\Lambda}} + N_0\mathbf{I})^{-1} \underline{\mathbf{U}}^{\mathrm{H}} \underline{\mathbf{y}}. \end{aligned} \tag{7.33}$$

It is worthwhile to point out that L can also be written as

$$L = \underline{\mathbf{y}}^{\mathrm{H}} (\underline{\mathbf{R}}_{xy} \underline{\mathbf{R}}_{yy}^{-1} \underline{\mathbf{y}}) = \underline{\mathbf{y}}^{\mathrm{H}} \underline{\hat{\mathbf{x}}} = 2\,\mathrm{Re}\,\{\mathbf{y}^{\mathrm{H}} \hat{\mathbf{x}}\}, \tag{7.34}$$

where $\underline{\mathbf{R}}_{xy} = \underline{\mathbf{R}}_{xx}$ and $\underline{\mathbf{R}}_{yy} = \underline{\mathbf{R}}_1 = \underline{\mathbf{R}}_{xx} + N_0\mathbf{I}$, so that $\hat{\mathbf{x}}$ is the *widely linear* MMSE estimate of $\mathbf{x}$ from $\mathbf{y}$. Hence, the log-likelihood detector for the signal-plus-white-noise setup is really a cascade of a WLMMSE filter and a matched filter.

In the proper case, the augmented covariance matrix $\underline{\mathbf{R}}_{xx}$ is block-diagonal, $\underline{\mathbf{R}}_{xx} = \mathbf{Diag}(\mathbf{R}_{xx}, \mathbf{R}_{xx}^*)$, and the eigenvalues of $\underline{\mathbf{R}}_{xx}$ occur in pairs. Let $\mathbf{M} = \mathbf{Diag}(\mu_1, \ldots, \mu_n)$ be the eigenvalue matrix of the Hermitian covariance matrix $\mathbf{R}_{xx}$. The log-likelihood ratio is now

$$L = \frac{2}{N_0}\,\mathrm{Re}\,\{\mathbf{y}^{\mathrm{H}} \mathbf{U} \mathbf{M} (\mathbf{M} + N_0\mathbf{I})^{-1} \mathbf{U}^{\mathrm{H}} \mathbf{y}\} = 2\,\mathrm{Re}\,\{\mathbf{y}^{\mathrm{H}} \hat{\mathbf{x}}\}, \tag{7.35}$$

where $\hat{\mathbf{x}}$ is now the *linear* MMSE estimate of $\mathbf{x}$ from $\mathbf{y}$.

So what performance advantage does widely linear processing offer over linear processing? Answering this question proceeds along the lines of Section 5.4.2. The performance criterion we choose here is the deflection d. Schreier *et al.* (2005) computed the deflection for the improper signal-plus-white-noise detection scenario:

$$d = \frac{\left(\sum_{i=1}^{2n} \frac{\lambda_i^2}{\lambda_i + N_0}\right)^2}{2N_0^2 \sum_{i=1}^{2n} \left(\frac{\lambda_i}{\lambda_i + N_0}\right)^2}. \tag{7.36}$$

Here, $\{\lambda_i\}_{i=1}^{2n}$ are the eigenvalues of the augmented covariance matrix $\underline{\mathbf{R}}_{xx}$. In order to evaluate the maximum performance advantage of widely linear over linear processing, we need to maximize the deflection for fixed $\mathbf{R}_{xx}$ and varying $\widetilde{\mathbf{R}}_{xx}$. Using Result A3.3, it can be shown that the deflection is a Schur-convex function of the eigenvalues $\{\lambda_i\}$. Therefore, maximizing the deflection requires maximum spread of the $\{\lambda_i\}$ in the sense of majorization. According to Result 3.7, this is achieved for

$$\lambda_i = 2\mu_i, \quad i = 1, \ldots, n, \quad \text{and} \quad \lambda_i = 0, \quad i = n+1, \ldots, 2n, \tag{7.37}$$

where $\{\mu_i\}_{i=1}^{n}$ are the eigenvalues of the Hermitian covariance matrix $\mathbf{R}_{xx}$. By plugging this into (7.36), we obtain the maximum deflection achieved by *widely linear* processing,

$$\max_{\widetilde{\mathbf{R}}_{xx}} d = \frac{2\left(\sum_{i=1}^{n} \frac{\mu_i^2}{2\mu_i + N_0}\right)^2}{N_0^2 \sum_{i=1}^{n} \left(\frac{\mu_i}{2\mu_i + N_0}\right)^2}. \tag{7.38}$$

On the other hand, *linear* processing implicitly assumes that $\widetilde{\mathbf{R}}_{xx} = \mathbf{0}$, in which case the deflection is

$$d_{\widetilde{\mathbf{R}}_{xx}=\mathbf{0}} = \frac{\left(\sum_{i=1}^{n} \frac{\mu_i^2}{\mu_i + N_0}\right)^2}{N_0^2 \sum_{i=1}^{n} \left(\frac{\mu_i}{\mu_i + N_0}\right)^2}. \tag{7.39}$$

Thus, the maximum performance advantage, as measured by deflection, is

$$\max_{N_0} \frac{\max_{\widetilde{\mathbf{R}}_{xx}} d}{d_{\widetilde{\mathbf{R}}_{xx}=\mathbf{0}}} = 2, \tag{7.40}$$

which is attained for $N_0 \longrightarrow 0$ or $N_0 \longrightarrow \infty$. This bound was derived by Schreier *et al.* (2005). We note that, if (7.37) is not satisfied, the maximum performance advantage, which is then less than a factor of 2, occurs for some noise level $N_0 > 0$.

The factor of 2 is a very conservative performance advantage bound because it assumes the worst-case scenario of *white* additive noise. If the noise is colored, the performance difference between widely linear and linear processing can be much larger. One such example is the detection of a real message $\mathbf{x}$ in purely imaginary noise $\mathbf{n}$. This is a trivial detection problem, but only if the widely linear operation $\mathrm{Re}\,\mathbf{y}$ is admitted.

7.5 Composite hypothesis testing and the Karlin–Rubin theorem

So far we have tested a simple alternative H_1 versus a simple hypothesis H_0 and found the NP detector to be most powerful among all competitors whose false-alarm rates do not exceed the false-alarm rate of the NP detector. Now we would like to generalize the notion of most powerful to the notion of *uniformly most powerful* (UMP). That is, we would like to consider composite alternative and composite hypothesis and argue that an NP test is uniformly most powerful, which is to say most powerful against all competitors for every possible pair of parameters $(\boldsymbol{\theta}_0 \in \boldsymbol{\Theta}_0, \boldsymbol{\theta}_1 \in \boldsymbol{\Theta}_1)$. This is too much to ask, so we restrict ourselves to a *scalar real-valued parameter* θ, and ask for a test that would be most powerful for every pair $(\theta_0 \in \Theta_0, \theta_1 \in \Theta_1)$. Moreover, we shall restrict the sets Θ_0 and Θ_1 to the form $\Theta_0 = \{\theta\colon \theta \leq 0\}$ and $\Theta_1 = \{\theta\colon \theta > 0\}$, or $\Theta_0 = \{0\}$ and $\Theta_1 = \{\theta\colon \theta > 0\}$.

Definition 7.5. *A detector ϕ is* uniformly most powerful (UMP) *for testing $H_1\colon \theta \in \Theta_1$ versus $H_0\colon \theta \in \Theta_0$ if its detection probability is greater at every value of $\theta \in \Theta_1$ than that of every competitor ϕ' whose false-alarm probability is no greater than that of ϕ. That is*

$$\sup_{\theta\in\Theta_0} E_\theta[\phi'] \leq \sup_{\theta\in\Theta_0} E_\theta[\phi] = \alpha, \tag{7.41}$$

$$E_\theta[\phi'] \leq E_\theta[\phi], \quad \forall \theta \in \Theta_1. \tag{7.42}$$

A scalar sufficient statistic, such as the likelihood ratio itself, has been identified. Assume that the likelihood ratio for this statistic, namely $q_{\theta_1}(t)/q_{\theta_0}(t)$ is monotonically increasing in t for all $\theta_1 > \theta_0$. Then the test function ϕ may be replaced by the *threshold test*

$$\phi(t) = \begin{cases} 1, & t > t_0, \\ \gamma, & t = t_0, \\ 0, & t < t_0 \end{cases} \tag{7.43}$$

with (t_0, γ) chosen to satisfy the false-alarm probability. That is

$$P_{\theta_0}[t > t_0] + \gamma P_{\theta_0}[t = t_0] = \alpha. \tag{7.44}$$

The Karlin–Rubin theorem, which we shall not prove, says the threshold test ϕ is uniformly most powerful for testing $\{\theta \leq 0\}$ versus $\{\theta > 0\}$ or for testing $\{\theta = 0\}$ versus $\{\theta > 0\}$. Moreover, the false-alarm probability is computed at $\theta = 0$.

Example 7.1. Let's consider a variation on Section 7.4.1. The measurement is $\mathbf{y} = \boldsymbol{\mu}\theta + \mathbf{n}$, where the signal $\boldsymbol{\mu}$ is deterministic complex, θ is real, and the Gaussian noise $\mathbf{n}$ has zero mean and augmented covariance matrix $\underline{\mathbf{R}}_{nn}$. As usual we work with the augmented measurement $\underline{\mathbf{y}} = \underline{\boldsymbol{\mu}}\theta + \underline{\mathbf{n}}$.

First, we whiten the augmented measurement to obtain $\underline{\mathbf{v}} = \underline{\boldsymbol{\kappa}}\theta + \underline{\mathbf{w}}$, where $\underline{\mathbf{v}} = \underline{\mathbf{R}}_{nn}^{-1/2}\underline{\mathbf{y}}$, $\underline{\boldsymbol{\kappa}} = \underline{\mathbf{R}}_{nn}^{-1/2}\underline{\boldsymbol{\mu}}$, and $\underline{\mathbf{w}} = \underline{\mathbf{R}}_{nn}^{-1/2}\underline{\mathbf{n}}$. Then $\mathbf{v}$ is proper and white, but the pdf may still be written in augmented form:

$$p_\theta(\mathbf{v}) = \frac{1}{\pi^n}\exp\left\{-\tfrac{1}{2}(\underline{\mathbf{v}} - \underline{\boldsymbol{\kappa}}\theta)^{\mathrm{H}}(\underline{\mathbf{v}} - \underline{\boldsymbol{\kappa}}\theta)\right\}. \tag{7.45}$$

Thus, a sufficient statistic is $t = \underline{\boldsymbol{\kappa}}^{\mathrm{H}}\underline{\mathbf{v}} = 2\,\mathrm{Re}\,\{\boldsymbol{\kappa}^{\mathrm{H}}\mathbf{v}\}$, which is a real Gaussian with mean value $\theta\underline{\boldsymbol{\kappa}}^{\mathrm{H}}\underline{\boldsymbol{\kappa}} = 2\theta\,\mathrm{Re}\,\{\boldsymbol{\kappa}^{\mathrm{H}}\boldsymbol{\kappa}\}$ and variance $\underline{\boldsymbol{\kappa}}^{\mathrm{H}}\underline{\boldsymbol{\kappa}} = 2\mathrm{Re}\,\{\boldsymbol{\kappa}^{\mathrm{H}}\boldsymbol{\kappa}\}$. It is easy to check that the likelihood ratio $l(t) = q_{\theta_1}(t)/q_{\theta_0}(t)$ increases monotonically in t for all $\theta_1 > \theta_0$. Thus the Karlin–Rubin theorem applies and the test

$$\phi(t) = \begin{cases} 1, & t > t_0, \\ \gamma, & t = t_0, \\ 0, & t < t_0 \end{cases} \tag{7.46}$$

is uniformly most powerful for testing $\{\theta \le 0\}$ versus $\{\theta > 0\}$ or for testing $\{\theta = 0\}$ versus $\{\theta > 0\}$. The threshold may be set to achieve the desired false alarm probability at $\theta = 0$. The probability of detection then depends on the scale of θ:

$$P_{\mathrm{F}} = 1 - \Phi\left(\frac{t_0}{\sqrt{d}}\right) \quad \text{and} \quad P_{\mathrm{D}} = 1 - \Phi\left(\frac{t_0 - \theta d}{\sqrt{d}}\right). \tag{7.47}$$

As before, Φ is the distribution function of a zero-mean, variance-one, real Gaussian random variable. For $\theta = 0$ under H_0, the deflection is $d = \theta^2\underline{\boldsymbol{\kappa}}^{\mathrm{H}}\underline{\boldsymbol{\kappa}} = \theta^2\underline{\boldsymbol{\mu}}^{\mathrm{H}}\underline{\mathbf{R}}_{nn}^{-1}\underline{\boldsymbol{\mu}}$. Figure 7.3 continues to apply.

In the proper case, the sufficient statistic is $t = \underline{\boldsymbol{\kappa}}^{\mathrm{H}}\underline{\mathbf{v}} = 2\,\mathrm{Re}\,\{\boldsymbol{\mu}^{\mathrm{H}}\mathbf{R}_{nn}^{-1/2}\mathbf{y}\}$. The term inside the real part operation is proper complex with real mean $\theta\boldsymbol{\mu}^{\mathrm{H}}\mathbf{R}_{nn}^{-1}\boldsymbol{\mu}$ and variance $\boldsymbol{\mu}^{\mathrm{H}}\mathbf{R}_{nn}^{-1}\boldsymbol{\mu}$. So twice the real part has mean $2\theta\boldsymbol{\mu}^{\mathrm{H}}\mathbf{R}_{nn}^{-1}\boldsymbol{\mu}$ and variance $2\boldsymbol{\mu}^{\mathrm{H}}\mathbf{R}_{nn}^{-1}\boldsymbol{\mu}$. The resulting deflection is $d = 2\theta^2\boldsymbol{\mu}^{\mathrm{H}}\mathbf{R}_{nn}^{-1}\boldsymbol{\mu}$.

7.6 Invariance in hypothesis testing

There are many problems in detection theory where a channel introduces impairments to a measurement. Examples would be the addition of bias or the introduction of gain. Typically, even the most exquisite testing and calibration cannot ensure that these are, respectively, zero and one. Thus there are unknown bias and gain terms that cannot be cancelled out or equalized. Moreover, the existence of these terms in the measurement defeats our attempts to find NP tests that are most powerful or uniformly most powerful. The idea behind invariance in the theory of hypothesis testing is then to make an end-run by requiring that any detector we design be *invariant* with respect to these impairments. This is sensible. However, in order to develop a theory that captures the essence of this reasoning, we must be clear about what would constitute an *invariant hypothesis-testing problem* and its corresponding *invariant test*.[4]

The complex random measurement $\mathbf{y}$ is drawn from the pdf $p_{\boldsymbol{\theta}}(\mathbf{y})$. We are to test the alternative H_1: $\boldsymbol{\theta} \in \boldsymbol{\Theta}_1$ versus the hypothesis H_0: $\boldsymbol{\theta} \in \boldsymbol{\Theta}_0$. If g is a transformation in the group G, where the group operation is composition, $g_1 \circ g_2(\mathbf{y}) = g_1(g_2(\mathbf{y}))$, then the transformed measurement $\mathbf{t} = g(\mathbf{y})$ will have pdf $q_{\boldsymbol{\theta}}(\mathbf{t})$. If this pdf is $q_{\boldsymbol{\theta}}(\mathbf{t}) = p_{\boldsymbol{\theta}'}(\mathbf{t})$, then we say the transformation group G leaves the measurement distribution invariant, since only the parameterization of the probability law has changed, not the law itself. If the induced transformation on the parameter $\boldsymbol{\theta}$ leaves the sets $\boldsymbol{\Theta}_1$ and $\boldsymbol{\Theta}_0$ invariant, which is to say that $g'(\boldsymbol{\Theta}_i) = \boldsymbol{\Theta}_i$, then the hypothesis-testing problem is said to be invariant-G.

Let us first recall the definition of a maximal invariant. A statistic $\mathbf{t}(\mathbf{y})$ is said to be a *maximal invariant statistic* if $\mathbf{t}(\mathbf{y}_2) = \mathbf{t}(\mathbf{y}_1)$ *if and only if* $\mathbf{y}_2 = g(\mathbf{y}_1)$ for some $g \in G$. In this way, $\mathbf{t}$ sorts measurements into orbits or equivalence classes where $\mathbf{t}$ is constant and on this orbit all measurements are within a transformation g of each other. On this orbit any other invariant statistic will also be constant, making it a function of the maximal invariant statistic. It is useful to note that an invariant test is uniformly most powerful (UMP) if it is more powerful than every other test that is invariant-G.

We illustrate the application of invariance in detection by solving a problem that has attracted a great deal of attention in the radar, sonar, and communications literature, namely matched subspace detection. This problem has previously been addressed for real-valued and proper complex-valued measurements. The solution we give extends the result to improper complex-valued measurements.[5]

7.6.1 Matched subspace detector

Let $\mathbf{y}$: $\Omega \longrightarrow \mathbb{C}^n$ denote a complex measurement of the form

$$\mathbf{y} = \mathbf{H}\boldsymbol{\theta} + \mathbf{n} \tag{7.48}$$

with $\mathbf{H} \in \mathbb{C}^{n\times p}$ and $\boldsymbol{\theta} \in \mathbb{C}^p$, $p < n$. We say that the signal component of the measurement is an element of the p-dimensional subspace $\langle\mathbf{H}\rangle$. The noise is taken, for the time being, to be *proper* with Hermitian covariance matrix $\mathbf{R}$. We wish to test the alternative H_1: $\boldsymbol{\theta} \neq \mathbf{0}$ versus the hypothesis H_0: $\boldsymbol{\theta} = \mathbf{0}$. But we note that $\boldsymbol{\theta} = \mathbf{0}$ *if and only if* $\boldsymbol{\theta}^{\mathrm{H}}\mathbf{M}\boldsymbol{\theta} = 0$ for any positive definite Hermitian matrix $\mathbf{M} = \mathbf{F}^{\mathrm{H}}\mathbf{F}$. So we might as well be testing H_1: $\boldsymbol{\theta}^{\mathrm{H}}\mathbf{F}^{\mathrm{H}}\mathbf{F}\boldsymbol{\theta} > 0$ versus H_0: $\boldsymbol{\theta}^{\mathrm{H}}\mathbf{F}^{\mathrm{H}}\mathbf{F}\boldsymbol{\theta} = 0$, for a matrix $\mathbf{F}$ of our choice.

We may whiten the measurement with the nonsingular transformation $\mathbf{v} = \mathbf{R}^{-1/2}\mathbf{y}$ to return the measurement model

$$\mathbf{v} = \mathbf{G}\boldsymbol{\theta} + \mathbf{w}, \tag{7.49}$$

where $\mathbf{G} = \mathbf{R}^{-1/2}\mathbf{H}$, and $\mathbf{w} = \mathbf{R}^{-1/2}\mathbf{n}$ is proper and white noise. Without loss of generality, we may even replace this model with the model

$$\mathbf{v} = \mathbf{U}\boldsymbol{\phi} + \mathbf{w}, \tag{7.50}$$

where the matrix $\mathbf{U} \in \mathbb{C}^{n\times p}$ is a unitary basis for the subspace $\langle\mathbf{G}\rangle$, and the parameter $\boldsymbol{\phi} = \mathbf{F}\boldsymbol{\theta}$ is a reparameterization of $\boldsymbol{\theta}$. So the hypothesis test can be written equivalently as the alternative H_1: $\boldsymbol{\phi}^{\mathrm{H}}\boldsymbol{\phi} > 0$ versus the hypothesis H_0: $\boldsymbol{\phi}^{\mathrm{H}}\boldsymbol{\phi} = 0$.

At this point the pdf of the whitened measurement $\mathbf{v}$ may be expressed as

$$p_{\boldsymbol{\phi}}(\mathbf{v}) = \frac{1}{\pi^n} \exp\{(\mathbf{v} - \mathbf{U}\boldsymbol{\phi})^{\mathrm{H}}(\mathbf{v} - \mathbf{U}\boldsymbol{\phi})\}. \tag{7.51}$$

The statistic $\mathbf{t} = \mathbf{U}^{\mathrm{H}}\mathbf{v}$ is sufficient for $\boldsymbol{\phi}$ and the statistic $\mathbf{t}^{\mathrm{H}}\mathbf{t}$ may be written as $\mathbf{t}^{\mathrm{H}}\mathbf{t} = \mathbf{v}^{\mathrm{H}}\mathbf{U}\mathbf{U}^{\mathrm{H}}\mathbf{v} = \mathbf{v}^{\mathrm{H}}\mathbf{P}_{\mathrm{U}}\mathbf{v}$. The projection onto the subspace $\langle\mathbf{U}\rangle$ is denoted $\mathbf{P}_{\mathrm{U}} = \mathbf{U}\mathbf{U}^{\mathrm{H}}$, and the projection onto the subspace orthogonal to $\langle\mathbf{U}\rangle$ is $\mathbf{P}_{\mathrm{U}}^{\perp} = \mathbf{I} - \mathbf{U}\mathbf{U}^{\mathrm{H}}$. The statistic $\mathbf{t}^{\mathrm{H}}\mathbf{t}$ is invariant with respect to rotations $g(\mathbf{v}) = (\mathbf{U}\mathbf{Q}\mathbf{U}^{\mathrm{H}} + \mathbf{P}_{\mathrm{U}}^{\perp})\mathbf{v}$ of the whitened measurement $\mathbf{v}$ in the subspace $\langle\mathbf{U}\rangle$, where $\mathbf{Q} \in \mathbb{C}^{p\times p}$ is a rotation matrix. This rotation leaves the noise white and proper and rotates the signal $\mathbf{U}\boldsymbol{\phi}$ to the signal $\mathbf{U}\mathbf{Q}\boldsymbol{\phi}$, thus rotating the parameter $\boldsymbol{\phi}$ to the parameter $\boldsymbol{\phi}' = \mathbf{Q}\boldsymbol{\phi}$. So the pdf for $\mathbf{v}$ is invariant-$\mathbf{Q}$, with transformed parameter $\mathbf{Q}\boldsymbol{\phi}$. Moreover, $\boldsymbol{\phi}^{\mathrm{H}}\mathbf{Q}^{\mathrm{H}}\mathbf{Q}\boldsymbol{\phi} = \boldsymbol{\phi}^{\mathrm{H}}\boldsymbol{\phi}$, leaving the hypothesis-testing problem invariant.

In summary, the hypothesis-testing problem is invariant with respect to rotations of the measurement in the subspace $\langle\mathbf{U}\rangle$, and the *matched subspace detector* statistic

$$L = \mathbf{t}^{\mathrm{H}}\mathbf{t} = \mathbf{v}^{\mathrm{H}}\mathbf{P}_{\mathrm{U}}\mathbf{v} \tag{7.52}$$

is invariant-$\mathbf{Q}$. If the complex measurements $\mathbf{y}$ are Gaussian, this detector statistic is the sum of squares of p independent complex Gaussian random variables, each of variance 1, making it distributed as a $\chi^2_{2p}(\boldsymbol{\phi}^{\mathrm{H}}\boldsymbol{\phi})$ random variable, which is to say a chi-squared random variable with $2p$ degrees of freedom and non-centrality parameter $\boldsymbol{\phi}^{\mathrm{H}}\boldsymbol{\phi}$. This pdf has a monotonic likelihood ratio, so a threshold test of L is a uniformly most powerful invariant test of H_1 versus H_0. The test statistic L matches to the subspace $\langle\mathbf{U}\rangle$ and then computes its energy in this subspace, earning the name *noncoherent matched subspace detector*. This energy may be compared against a threshold because the scale of the additive noise is known a priori. Therefore large values of energy in the subspace $\langle\mathbf{U}\rangle$ are unlikely to be realized with noise only in the measurement.

Actually, rotations are not the largest set of transformations under which the statistic L is invariant. It is invariant also with respect to translations of $\mathbf{v}$ in the subspace perpendicular to the subspace $\langle\mathbf{U}\rangle$. The sufficient statistic remains sufficient, the hypothesis problem remains invariant, and the quadratic form in the projection operator remains invariant. So the set of invariances for the noncoherent matched subspace detector is the cylindrical set illustrated in Fig. 7.4.

Example 7.2. For $p = 1$, the matrix $\mathbf{U}$ is a unit vector $\boldsymbol{\psi}$, and the measurement model is

$$\mathbf{v} = \boldsymbol{\psi}\phi + \mathbf{w}. \tag{7.53}$$

The noncoherent matched subspace detector is

$$L = |\boldsymbol{\psi}^{\mathrm{H}}\mathbf{v}|^2, \tag{7.54}$$

which is invariant with respect to rotations of the whitened measurement in the subspace $\langle\boldsymbol{\psi}\rangle$ and to bias in the subspace orthogonal to $\boldsymbol{\psi}$. A rotation of the measurement amounts

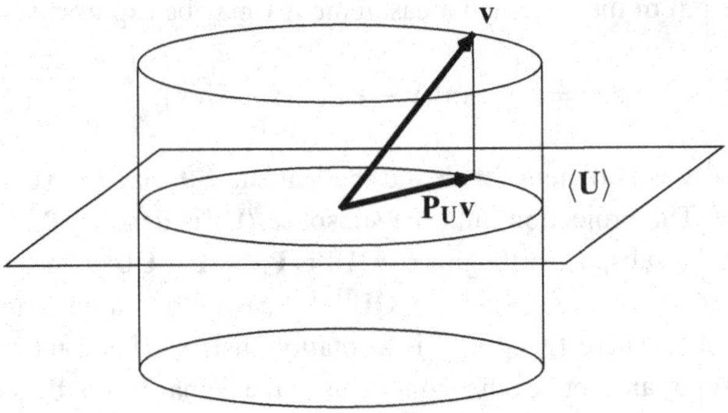

Figure 7.4 The cylinder is the set of transformations of the whitened measurement that leave the matched subspace detector invariant, and the energy in the subspace $\langle \mathbf{U} \rangle$ constant.

to the transformation

$$g(\mathbf{v}) = (\boldsymbol{\psi} e^{j\beta} \boldsymbol{\psi}^H + \mathbf{P}_{\boldsymbol{\psi}}^{\perp})\mathbf{v} = \boldsymbol{\psi} e^{j\beta} \phi + \mathbf{w}', \tag{7.55}$$

where the rotated noise $\mathbf{w}'$ remains white and proper and the rotated signal is just a true phase rotation of the original signal. This is the natural transformation to be invariant to when nothing is known a priori about the amplitude and phase of the parameter ϕ.

Let us now extend our study of this problem to the case in which the measurement noise $\mathbf{n}$ is improper with Hermitian covariance $\mathbf{R}$ and complementary covariance $\widetilde{\mathbf{R}}$. To this end, we construct the augmented measurement $\underline{\mathbf{y}}$ and whiten it with the square root of the augmented covariance matrix $\underline{\mathbf{R}}^{-1/2}$. The resulting augmented measurement $\underline{\mathbf{v}} = \underline{\mathbf{R}}^{-1/2}\underline{\mathbf{y}}$ has identity augmented covariance, so $\mathbf{v}$ is white and proper, and a widely linear function of the measurement $\mathbf{y}$. However, by whitening the measurement we have turned the linear model $\mathbf{y} = \mathbf{H}\boldsymbol{\theta} + \mathbf{n}$ into the *widely linear* model

$$\underline{\mathbf{v}} = \underline{\mathbf{G}}\,\underline{\boldsymbol{\theta}} + \underline{\mathbf{w}}, \tag{7.56}$$

where

$$\underline{\mathbf{G}} = \underline{\mathbf{R}}^{-1/2} \begin{bmatrix} \mathbf{H} & \mathbf{0} \\ \mathbf{0} & \mathbf{H}^* \end{bmatrix} \tag{7.57}$$

and $\mathbf{w}$ is proper and white noise. This model is then replaced with

$$\underline{\mathbf{v}} = \underline{\mathbf{U}}\,\underline{\boldsymbol{\phi}} + \underline{\mathbf{w}}, \tag{7.58}$$

where the augmented matrix $\underline{\mathbf{U}}$ is a widely unitary basis for the widely linear (i.e., real) subspace $\langle \underline{\mathbf{G}} \rangle$, which satisfies $\underline{\mathbf{U}}^H\underline{\mathbf{U}} = \mathbf{I}$, and $\underline{\boldsymbol{\phi}}$ is a reparameterization of $\underline{\boldsymbol{\theta}}$. It is a straightforward adaptation of prior arguments to argue that the real test statistic

$$L = \tfrac{1}{2}\underline{\mathbf{v}}^H\mathbf{P}_{\underline{\mathbf{U}}}\,\underline{\mathbf{v}} = \tfrac{1}{2}\underline{\mathbf{v}}^H\underline{\mathbf{U}}\,\underline{\mathbf{U}}^H\underline{\mathbf{v}} \tag{7.59}$$

is invariant with respect to rotations in the widely linear (i.e., real) subspace $\langle \underline{\mathbf{U}} \rangle$ and to bias in the widely linear (i.e., real) subspace orthogonal to $\langle \underline{\mathbf{U}} \rangle$. The statistic L may also

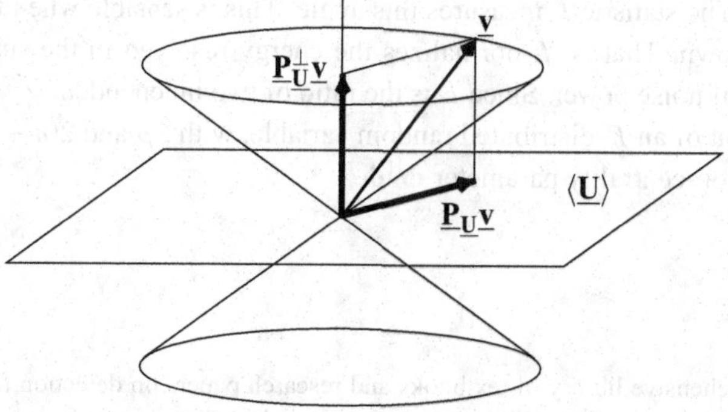

Figure 7.5 The double cone is the set of transformations of the whitened measurement that leave the CFAR matched subspace detector invariant, and leave the angle that the measurement makes with the subspace $\langle\underline{\mathbf{U}}\rangle$ constant.

be written in terms of **v** and the components of

$$\underline{\mathbf{U}} = \begin{bmatrix} \mathbf{U}_1 & \mathbf{U}_2 \\ \mathbf{U}_2^* & \mathbf{U}_1^* \end{bmatrix} \tag{7.60}$$

as

$$L = \mathbf{v}^{\mathrm{H}}(\mathbf{U}_1\mathbf{U}_1^{\mathrm{H}} + \mathbf{U}_2\mathbf{U}_2^{\mathrm{H}})\mathbf{v} + \operatorname{Re}\{\mathbf{v}^{\mathrm{T}}(\mathbf{U}_1^*\mathbf{U}_2^{\mathrm{H}} + \mathbf{U}_2^*\mathbf{U}_1^{\mathrm{H}})\mathbf{v}\}. \tag{7.61}$$

In the proper case, $\mathbf{U}_2 = \mathbf{0}$ and $\mathbf{U}_1 = \mathbf{U}$, so L simplifies to (7.52). If the complex measurement **y** is Gaussian, the statistic L is distributed as a $\chi^2_{2p}(\boldsymbol{\phi}^{\mathrm{H}}\boldsymbol{\phi})$ random variable. This is the same distribution as in the proper case – the fact that L is coded as a real-valued *widely quadratic* form does not change its distribution.

7.6.2 CFAR matched subspace detector

The arguments for a constant-false-alarm-rate (CFAR) matched subspace detector follow along lines very similar to the lines of argument for the matched subspace detector. The essential difference is this: the *scale* of the improper noise is unknown, so the Hermitian covariance and the complementary covariance are known only up to a nonnegative scale. That is, $\sigma^2\mathbf{R}$ and $\sigma^2\widetilde{\mathbf{R}}$ are the Hermitian and complementary covariances, with $\mathbf{R}$ and $\widetilde{\mathbf{R}}$ known, but $\sigma^2 > 0$ unknown. We ask for invariance with respect to one more transformation, namely scaling of the measurement by a nonzero constant, which would have the effect of scaling the variance of the noise and scaling the value of the subspace parameter $\boldsymbol{\phi}$, while leaving the hypothesis-testing problem invariant. For proper complex noise, the statistic

$$L = \frac{\underline{\mathbf{v}}^{\mathrm{H}}\mathbf{P}_{\underline{\mathbf{U}}}\,\underline{\mathbf{v}}}{\underline{\mathbf{v}}^{\mathrm{H}}\mathbf{P}_{\underline{\mathbf{U}}}^{\perp}\,\underline{\mathbf{v}}} \tag{7.62}$$

is invariant with respect to rotations in the widely linear (i.e., real) subspaces $\langle\underline{\mathbf{U}}\rangle$ and $\langle\underline{\mathbf{U}}\rangle^{\perp}$, and to scaling by a nonzero constant. The set of measurements $g(\mathbf{v})$ with respect to which the statistic L is invariant is illustrated in Fig. 7.5. This invariant set demonstrates that it is the *angle* that a measurement makes with a subspace that matters, not its energy in that subspace. The statistic L measures this angle. This is sensible when the scale of the noise is unknown. That is, L normalizes the energy resolved in the subspace $\langle \underline{\mathbf{U}} \rangle$ with an estimate of noise power. Since L is the ratio of two independent χ^2 statistics, it is within a constant of an F-distributed random variable, with $2p$ and $2(n-p)$ degrees of freedom, and non-centrality parameter $\boldsymbol{\phi}^{\mathrm{H}}\boldsymbol{\phi}$.

Notes

1 There is a comprehensive library of textbooks and research papers on detection theory, which we make no attempt to catalog. What we have reviewed in this chapter is just a sample of what seems relevant to detection in the improper Gaussian model, a topic that has not received much attention in the literature. Even regarding that, we have said nothing about extensions of detection theory to a theory of adaptive detection that aims to resolve hypotheses when the subspace is a priori unknown or the covariance matrix of the noise is unknown. The study of these problems, in the case of proper noise, began with the seminal work of Reed *et al.* (1974), Kelly (1986), and Robey *et al.* (1992), and has seen its most definitive conclusions in the work of Kraut and Scharf and their collaborators (see Kraut *et al.* (2001)) and Conte and his collaborators (see Conte *et al.* (2001)).

2 The Neyman–Pearson lemma, the Fisher–Neyman factorization theorem, and the Karlin–Rubin theorem are proved in most books on hypothesis testing and decision theory. Proofs are contained, for instance, in Ferguson (1967). These fundamental theorems are used in Chapter 4 of Scharf (1991) to derive a number of uniformly most powerful and uniformly most powerful invariant detectors for *real-valued* signals.

3 This result is originally due to Van Trees (see Van Trees (2001), where it is proved using a moment-matching argument).

4 The theory of invariance in hypothesis testing is covered comprehensively by Ferguson (1967) and applied to problems of bias and scale by Scharf (1991).

5 The results in this chapter for matched subspace detectors generalize the results of Scharf (1991) and Scharf and Friedlander (1994) from real-valued signals to complex-valued signals that may be proper or improper. Moreover, the adaptive versions of these matched subspace detectors, termed adaptive subspace detectors, may be derived, along the lines of papers by Kraut *et al.* (2001) and Conte *et al.* (2001). Adaptive subspace detectors use sample covariance matrices. These ideas can be generalized further to the matched direction detectors introduced by Besson *et al.* (2005).

There is also a body of results addressed to the problem of testing for covariance structure. In this case the hypothesis is composite, consisting of covariance matrices of specified structure, and the alternative is composite, consisting of all Hermitian covariance matrices. A selection of these problems has been addressed in Section 4.5.

Part III

Complex random processes

8 Wide-sense stationary processes

The remaining chapters of the book deal with complex-valued random processes. In this chapter, we discuss wide-sense stationary (WSS) signals. In Chapter 9, we look at nonstationary signals, and in Chapter 10, we treat cyclostationary signals, which are an important subclass of nonstationary signals.

Our discussion of WSS signals continues the preliminary exposition given in Section 2.6. WSS processes have shift-invariant second-order statistics, which leads to the definition of a time-invariant power spectral density (PSD) – an intuitively pleasing idea. For improper signals, the PSD needs to be complemented by the complementary power spectral density (C-PSD), which is generally complex-valued. In Section 8.1, we will see that WSS processes allow an easy characterization of all possible PSD/C-PSD pairs and also a spectral representation of the process itself. Section 8.2 discusses widely linear shift-invariant filtering, with an application to analytic and complex baseband signals. We also introduce the noncausal widely linear minimum mean-squared error, or Wiener, filter for estimating a message signal from a noisy measurement. In order to find the causal approximation of the Wiener filter, we need to adapt existing spectral factorization algorithms to the improper case. This is done in Section 8.3, where we build causal synthesis, analysis, and Wiener filters for improper WSS vector-valued time series.

Section 8.4 introduces rotary-component and polarization analysis, which are widely used in a number of research areas, ranging from optics, geophysics, meteorology, and oceanography to radar. These techniques are usually applied to deterministic signals, but we present them in a more general stochastic framework. The idea is to represent a two-dimensional signal in the complex plane as a superposition of ellipses, which can be analyzed in terms of their shape and orientation. Each ellipse is the sum of a counterclockwise and a clockwise rotating phasor, called the rotary components. If there is complete coherence between the rotary components (i.e., they are linearly dependent), then the signal is completely polarized.

The chapter is concluded with a brief exposition of higher-order statistics of Nth-order stationary signals in Section 8.5, where we focus on higher-order moment spectra and the principal domains of analytic signals.

8.1 Spectral representation and power spectral density

Consider a zero-mean wide-sense stationary (WSS) continuous-time complex-valued random process $x(t) = u(t) + jv(t)$, which is composed from the two real random processes $u(t)$ and $v(t)$ defined for all $t \in \mathbb{R}$. Let us first recall a few key definitions and results from Section 2.6. The process $x(t)$ is WSS if and only if both the covariance function $r_{xx}(\tau) = E[x(t+\tau)x^*(t)]$ and the complementary covariance function $\tilde{r}_{xx}(\tau) = E[x(t+\tau)x(t)]$ are independent of t. Equivalently, the auto- and cross-covariance functions of real and imaginary parts, $r_{uu}(\tau) = E[u(t+\tau)u(t)]$, $r_{vv}(\tau) = E[v(t+\tau)v(t)]$, and $r_{uv}(\tau) = E[u(t+\tau)v(t)]$, must all be independent of t. The process $x(t)$ is proper when $\tilde{r}_{xx}(\tau) = 0$ for all τ, which means that $r_{uu}(\tau) \equiv r_{vv}(\tau)$ and $r_{uv}(\tau) \equiv -r_{uv}(-\tau)$, and thus $r_{uv}(0) = 0$.

The Fourier transform of $r_{xx}(\tau)$ is the power spectral density (PSD) $P_{xx}(f)$, and the Fourier transform of $\tilde{r}_{xx}(\tau)$ is the complementary power spectral density (C-PSD) $\widetilde{P}_{xx}(f)$. Result 2.14 characterizes all possible pairs $(P_{xx}(f), \widetilde{P}_{xx}(f))$. This result is easily obtained by considering the augmented signal $\underline{\mathbf{x}}(t) = [x(t), x^*(t)]^{\mathrm{T}}$ whose lag-τ covariance matrix

$$\underline{\mathbf{R}}_{xx}(\tau) = E[\underline{\mathbf{x}}(t+\tau)\underline{\mathbf{x}}^{\mathrm{H}}(t)] = \begin{bmatrix} r_{xx}(\tau) & \tilde{r}_{xx}(\tau) \\ \tilde{r}^*_{xx}(\tau) & r^*_{xx}(\tau) \end{bmatrix} \tag{8.1}$$

is the augmented covariance function of $x(t)$. The Fourier transform of $\underline{\mathbf{R}}_{xx}(\tau)$ is the augmented PSD matrix

$$\underline{\mathbb{P}}_{xx}(f) = \begin{bmatrix} P_{xx}(f) & \widetilde{P}_{xx}(f) \\ \widetilde{P}^*_{xx}(f) & P_{xx}(-f) \end{bmatrix}. \tag{8.2}$$

This matrix is a valid augmented PSD matrix if and only if it is positive semidefinite, which is equivalent to, for all f,

$$P_{xx}(f) \geq 0, \tag{8.3}$$

$$\widetilde{P}_{xx}(f) = \widetilde{P}_{xx}(-f), \tag{8.4}$$

$$|\widetilde{P}_{xx}(f)|^2 \leq P_{xx}(f)P_{xx}(-f). \tag{8.5}$$

Thus, for any pair of L^1-functions $(P_{xx}(f), \widetilde{P}_{xx}(f))$ that satisfy these three conditions, there exists a WSS random process $x(t)$ with PSD $P_{xx}(f)$ and C-PSD $\widetilde{P}_{xx}(f)$. Conversely, the PSD $P_{xx}(f)$ and C-PSD $\widetilde{P}_{xx}(f)$ of any given WSS random process $x(t)$ satisfy the conditions (8.3)–(8.5). The same conditions hold for a discrete-time process $x[k] = u[k] + \mathrm{j}v[k]$.

The PSD and C-PSD of $x(t)$ are connected to the PSDs of real and imaginary parts, $P_{uu}(f)$ and $P_{vv}(f)$, and the cross-PSD between real and imaginary part, $P_{uv}(f)$, through the familiar real-to-complex transformation

$$\mathbf{T} = \begin{bmatrix} 1 & \mathrm{j} \\ 1 & -\mathrm{j} \end{bmatrix}, \qquad \mathbf{T}\mathbf{T}^{\mathrm{H}} = \mathbf{T}^{\mathrm{H}}\mathbf{T} = 2\mathbf{I}, \tag{8.6}$$

as

$$\underline{\mathbb{P}}_{xx}(f) = \mathbf{T} \begin{bmatrix} P_{uu}(f) & P_{uv}(f) \\ P^*_{uv}(f) & P_{vv}(f) \end{bmatrix} \mathbf{T}^{\mathrm{H}}, \tag{8.7}$$

from which we determine

$$P_{xx}(f) = P_{uu}(f) + P_{vv}(f) + 2\,\mathrm{Im}\,P_{uv}(f), \tag{8.8}$$

$$P_{xx}(-f) = P_{uu}(f) + P_{vv}(f) - 2\,\mathrm{Im}\,P_{uv}(f), \tag{8.9}$$

$$\widetilde{P}_{xx}(f) = P_{uu}(f) - P_{vv}(f) + 2\mathrm{j}\,\mathrm{Re}\,P_{uv}(f). \tag{8.10}$$

The PSD and C-PSD provide a statistical description of the spectral properties of $x(t)$. But is there a *spectral representation* of the process $x(t)$ itself? The following result, whose proper version is due to Cramér, is affirmative.

Result 8.1. *Provided that $r_{xx}(\tau)$ is continuous, a complex second-order random process $x(t)$ can be written as*

$$x(t) = \int_{-\infty}^{\infty} \mathrm{e}^{\mathrm{j}2\pi ft}\,\mathrm{d}\xi(f), \tag{8.11}$$

where $\xi(f)$ is a spectral process with orthogonal increments $\mathrm{d}\xi(f)$ *whose second-order moments are $E|\mathrm{d}\xi(f)|^2 = P_{xx}(f)\mathrm{d}f$ and $E[\mathrm{d}\xi(f)\mathrm{d}\xi(-f)] = \widetilde{P}_{xx}(f)\mathrm{d}f$.*

This result bears comment. The expression (8.11) almost looks like an inverse Fourier transform. In fact, if the derivative process $X(f) = \mathrm{d}\xi(f)/\mathrm{d}f$ existed, then $X(f)$ would be the Fourier transform of $x(t)$. We would then be able to replace $\mathrm{d}\xi(f)$ in (8.11) with $X(f)\mathrm{d}f$, turning (8.11) into an actual inverse Fourier transform. However, $X(f)$ can never be a second-order random process with finite second-order moment if $x(t)$ is WSS. This is why we have to resort to the orthogonal increments $\mathrm{d}\xi(f)$. This issue is further clarified in Section 9.2.

The process $\xi(f)$ has orthogonal increments, which means that $E[\mathrm{d}\xi(f_1)\mathrm{d}\xi^*(f_2)] = 0$ and $E[\mathrm{d}\xi(f_1)\mathrm{d}\xi(-f_2)] = 0$ for $f_1 \neq f_2$. For $f_1 = f_2 = f$, $E|\mathrm{d}\xi(f)|^2 = P_{xx}(f)\mathrm{d}f$ and $E[\mathrm{d}\xi(f)\mathrm{d}\xi(-f)] = \widetilde{P}_{xx}(f)\mathrm{d}f$. The minus sign in the expression for $\widetilde{P}_{xx}(f)$ is owed to the fact that the spectral process corresponding to $x^*(t)$ is $\xi^*(-f)$:

$$x^*(t) = \left(\int_{-\infty}^{\infty} \mathrm{e}^{\mathrm{j}2\pi ft}\,\mathrm{d}\xi(f)\right)^* = \int_{-\infty}^{\infty} \mathrm{e}^{\mathrm{j}2\pi ft}\,\mathrm{d}\xi^*(-f). \tag{8.12}$$

The process $x(t)$ can be decomposed into two orthogonal and WSS parts as $x(t) = x_\mathrm{c}(t) + x_\mathrm{d}(t)$, where $x_\mathrm{c}(t)$ belongs to a continuous spectral process $\xi_\mathrm{c}(f)$ and $x_\mathrm{d}(t)$ to a purely discontinuous spectral process $\xi_\mathrm{d}(f)$. The spectral representation of $x_\mathrm{d}(t)$ reduces to a countable sum

$$x_\mathrm{d}(t) = \int_{-\infty}^{\infty} \mathrm{e}^{\mathrm{j}2\pi ft}\,\mathrm{d}\xi_d(f) = \sum_n C_{f_n}\mathrm{e}^{\mathrm{j}2\pi f_n t}, \tag{8.13}$$

where $\{f_n\}$ is the set of discontinuities of $\xi(f)$. The random variables

$$C_{f_n} = \lim_{\varepsilon\to 0+} [\xi_\mathrm{d}(f_n + \varepsilon) - \xi_\mathrm{d}(f_n - \varepsilon)] \tag{8.14}$$

are orthogonal in the sense that $E[C_{f_n}C^*_{f_m}] = 0$ and $E[C_{f_n}C_{-f_m}] = 0$ for $n \neq m$. At frequency f_n, the process $x_\mathrm{d}(t)$ has power $E|C_{f_n}|^2$ and complementary power $E[C_{f_n}C_{-f_n}]$, and thus unbounded power spectral *density*. The discontinuities $\{f_n\}$ correspond to periodic components of the correlation function $r_{xx}(\tau)$. The process $x_c(t)$ with continuous spectral process $\xi_c(f)$ has zero power at all frequencies (but nonzero power spectral *density*, unless $x_c(t) \equiv 0$).

Example 8.1. Consider the ellipse

$$x(t) = C_+ e^{j2\pi f_0 t} + C_- e^{-j2\pi f_0 t}, \tag{8.15}$$

where C_+ and C_- are two complex random variables. The spectral process $\xi(f)$ is purely discontinuous:

$$\xi(f) = \begin{cases} 0, & f < -f_0, \\ C_-, & -f_0 \le f < f_0, \\ C_- + C_+, & f_0 \le f. \end{cases} \tag{8.16}$$

In Example 1.5 we found that $x(t)$ is WSS if and only if $E[C_+ C_-^*] = 0$, $E\, C_+^2 = 0$, and $E\, C_-^2 = 0$, in which case the covariance function is

$$r_{xx}(\tau) = E|C_+|^2 e^{j2\pi f_0 \tau} + E|C_-|^2 e^{-j2\pi f_0 \tau} \tag{8.17}$$

and the complementary covariance function is

$$\tilde{r}_{xx}(\tau) = 2E[C_+ C_-]\cos(2\pi f_0 \tau). \tag{8.18}$$

The covariance and complementary covariance are periodic functions of τ. The power of $x(t)$ at frequency f_0 is $E|C_+|^2$, and at $-f_0$ it is $E|C_-|^2$. The complementary power at $\pm f_0$ is $E[C_+ C_-]$. The PSD and C-PSD are thus unbounded line spectra:

$$P_{xx}(f) = E\,|C_+|^2 \delta(f - f_0) + E|C_-|^2 \delta(f + f_0), \tag{8.19}$$

$$\tilde{P}_{xx}(f) = E[C_+ C_-][\delta(f - f_0) + \delta(f + f_0)]. \tag{8.20}$$

Aliasing in the spectral representation of a discrete-time process $x[k]$ will be discussed in the context of nonstationary processes in Section 9.2.

8.2 Filtering

The output of a widely linear shift-invariant (WLSI) filter with impulse response $(h_1(t), h_2(t))$ and input $x(t)$ is

$$y(t) = (h_1 * x)(t) + (h_2 * x^*)(t) = \int_{-\infty}^{\infty} [h_1(t-\tau)x(\tau) + h_2(t-\tau)x^*(\tau)]d\tau. \tag{8.21}$$

If $x(t)$ and $y(t)$ have spectral representations with orthogonal increments $d\xi(f)$ and $d\upsilon(f)$, respectively, they are connected as

$$d\upsilon(f) = H_1(f)d\xi(f) + H_2(f)d\xi^*(-f). \tag{8.22}$$

This may be written in augmented notation as

$$\begin{bmatrix} \mathrm{d}\upsilon(f) \\ \mathrm{d}\upsilon^*(-f) \end{bmatrix} = \begin{bmatrix} H_1(f) & H_2(f) \\ H_2^*(-f) & H_1^*(-f) \end{bmatrix} \begin{bmatrix} \mathrm{d}\xi(f) \\ \mathrm{d}\xi^*(-f) \end{bmatrix}. \tag{8.23}$$

We call

$$\underline{\mathbb{H}}(f) = \begin{bmatrix} H_1(f) & H_2(f) \\ H_2^*(-f) & H_1^*(-f) \end{bmatrix} \tag{8.24}$$

the *augmented frequency-response matrix* of the WLSI filter. The relationship between the augmented PSD matrices of $x(t)$ and $y(t)$ is

$$\underline{\mathbb{P}}_{yy}(f) = \underline{\mathbb{H}}(f)\underline{\mathbb{P}}_{xx}(f)\underline{\mathbb{H}}^{\mathrm{H}}(f). \tag{8.25}$$

In the following, we will take a look at three important (W)LSI filters: the filter constructing the analytic signal using the Hilbert transform, the noncausal Wiener filter, and the causal Wiener filter.

8.2.1 Analytic and complex baseband signals

In Section 1.4, we discussed the role of analytic and equivalent complex baseband signals in communications. In this section, we will investigate properties of *random* analytic and complex baseband signals. Let $x(t) = u(t) + \mathrm{j}\hat{u}(t)$ denote the analytic signal constructed from the real signal $u(t)$ and its Hilbert transform $\hat{u}(t)$. As discussed in Section 1.4.2, a Hilbert transformer is a linear shift-invariant (LSI) filter with impulse response $h(t) = 1/(\pi t)$ and frequency response $H(f) = -\mathrm{j}\,\mathrm{sgn}(f)$. Producing the analytic signal $x(t)$ from the real signal $u(t)$ is therefore an LSI operation with augmented frequency-response matrix

$$\underline{\mathbb{H}}(f) = \begin{bmatrix} 1 + \mathrm{j}H(f) & 0 \\ 0 & 1 - \mathrm{j}H^*(-f) \end{bmatrix} = \begin{bmatrix} 1 + \mathrm{j}H(f) & 0 \\ 0 & 1 - \mathrm{j}H(f) \end{bmatrix}, \tag{8.26}$$

where the second equality is due to $H(f) = H^*(-f)$. The augmented PSD matrix of the *real* input signal $u(t)$ is

$$\underline{\mathbb{P}}_{uu}(f) = \begin{bmatrix} P_{uu}(f) & P_{uu}(f) \\ P_{uu}(f) & P_{uu}(f) \end{bmatrix}, \tag{8.27}$$

which is rank-deficient for all frequencies f. The augmented PSD matrix of the analytic signal $x(t)$ is

$$\underline{\mathbb{P}}_{xx}(f) = \underline{\mathbb{H}}(f)\underline{\mathbb{P}}_{uu}(f)\underline{\mathbb{H}}^{\mathrm{H}}(f). \tag{8.28}$$

With the mild assumption that $u(t)$ has no DC component, $P_{uu}(0) = 0$, we obtain the PSD of $x(t)$

$$\begin{aligned} P_{xx}(f) &= P_{uu}(f) + H(f)P_{uu}(f)H^*(f) + 2\,\mathrm{Im}\,\{P_{uu}(f)H^*(f)\} \\ &= 2P_{uu}(f)(1 + \mathrm{sgn}(f)) = 4\Gamma(f)P_{uu}(f), \end{aligned} \tag{8.29}$$

where $\Gamma(f)$ is the unit-step function. The C-PSD is

$$\widetilde{P}_{xx}(f) = P_{uu}(f) - H(f)P_{uu}(f)H^*(f) + 2\text{j}\,\text{Re}\,\{P_{uu}(f)H^*(f)\} = 0. \tag{8.30}$$

So the analytic signal constructed from a WSS real signal has zero PSD for negative frequencies and four times the PSD of the real signal for positive frequencies, and it is *proper*. Propriety also follows from the bound (8.5) since

$$|\widetilde{P}_{xx}(f)|^2 \leq P_{xx}(f)P_{xx}(-f) = 0. \tag{8.31}$$

There is also a close interplay between wide-sense stationarity and propriety for equivalent complex baseband signals. As in Section 1.4, let $p(t) = \text{Re}\,\{x(t)\text{e}^{\text{j}2\pi f_0 t}\}$ denote a real passband signal obtained by complex modulation of the complex baseband signal $x(t)$ with bandwidth $\Omega < f_0$. It is easy to see that a complex-modulated signal $x(t)\text{e}^{\text{j}2\pi f_0 t}$ is proper if and only if $x(t)$ is proper. If $x(t)\text{e}^{\text{j}2\pi f_0 t}$ is the analytic signal obtained from $p(t)$, we find the following.

Result 8.2. *A real passband signal $p(t)$ is WSS if and only if the equivalent complex baseband signal $x(t)$ is WSS and proper.*

An alternative way to prove this is to express the covariance function of $p(t)$ as

$$r_{pp}(t,\tau) = \text{Re}\big\{r_{xx}(t,\tau)\text{e}^{\text{j}2\pi f_0 \tau}\big\} + \text{Re}\big\{\tilde{r}_{xx}(t,\tau)\text{e}^{\text{j}2\pi f_0(2t+\tau)}\big\}. \tag{8.32}$$

If $r_{pp}(t,\tau)$ is to be independent of t, we need the covariance function $r_{xx}(t,\tau)$ to be independent of t and the complementary covariance function $\tilde{r}_{xx}(t,\tau) \equiv 0$.

We note a subtle difference between analytic signals and complex baseband signals: while there are no improper WSS analytic signals, improper WSS complex baseband signals do exist. However, the real passband signal produced from an improper WSS complex baseband signal is *cyclostationary* rather than WSS. Result 8.2 is important for communications because passband thermal noise is modeled as WSS. Hence, its complex baseband representation is WSS and proper.

8.2.2 Noncausal Wiener filter

The celebrated Wiener filter produces a linear estimate $\hat{x}(t)$ of a *message signal* $x(t)$ from an *observation* (or measurement) $y(t)$ based on the PSD of $y(t)$, denoted $P_{yy}(f)$, and the cross-PSD between $x(t)$ and $y(t)$, denoted $P_{xy}(f)$. This assumes that $x(t)$ and $y(t)$ are jointly WSS. The estimate $\hat{x}(t)$ is optimal in the sense that it minimizes the mean-squared error $E\,|\hat{x}(t) - x(t)|^2$, which is independent of t because $x(t)$ and $y(t)$ are jointly WSS. The frequency response of the noncausal Wiener filter is

$$H(f) = \frac{P_{xy}(f)}{P_{yy}(f)}. \tag{8.33}$$

If $y(t) = x(t) + n(t)$, with uncorrelated noise $n(t)$ of PSD $P_{nn}(f)$, then the frequency response is

$$H(f) = \frac{P_{xx}(f)}{P_{xx}(f) + P_{nn}(f)}. \tag{8.34}$$

This has an intuitively appealing interpretation: the filter attenuates frequency components where the noise is strong compared with the signal.

The extension of the linear Wiener filter to the widely linear Wiener filter for improper complex signals presents no particular difficulties. The augmented frequency-response matrix of the noncausal widely linear Wiener filter is

$$\underline{\mathbb{H}}(f) = \underline{\mathbb{P}}_{xy}(f)\underline{\mathbb{P}}_{yy}^{-1}(f). \tag{8.35}$$

Using the matrix-inversion lemma (A1.42) to invert the 2×2 matrix $\underline{\mathbb{P}}_{yy}(f)$, we can derive explicit formulae for $H_1(f)$ and $H_2(f)$, similarly to the vector case in Result 5.3.

Example 8.2. If $y(t) = x(t) + n(t)$, where $n(t)$ is proper, white noise, uncorrelated with $x(t)$, with PSD $P_{nn}(f) = N_0$, we obtain

$$H_1(f) = \frac{P_{xx}(f)(P_{xx}(f) + N_0) - |\widetilde{P}_{xx}(f)|^2}{(P_{xx}(f) + N_0)^2 - |\widetilde{P}_{xx}(f)|^2},$$

$$H_2(f) = \frac{\widetilde{P}_{xx}(f)N_0}{(P_{xx}(f) + N_0)^2 - |\widetilde{P}_{xx}(f)|^2}.$$

It is easy to see that, for $\widetilde{P}_{xx}(f) = 0$, the frequency response simplifies to

$$H_1(f) = \frac{P_{xx}(f)}{P_{xx}(f) + N_0} \quad \text{and} \quad H_2(f) = 0.$$

8.3 Causal Wiener filter

The Wiener filter derived above is not suitable for real-time applications because it is noncausal. In this section, we find the causal Wiener filter for an improper WSS vector-valued time series. This requires the spectral factorization of an augmented PSD matrix. We will see that it is straightforward to apply existing spectral factorization algorithms to the improper case, following Spurbeck and Schreier (2007).

8.3.1 Spectral factorization

So far, we have dealt with scalar continuous-time processes. Consider now the zero-mean WSS *vector-valued discrete-time* process $\mathbf{x}[k] = \mathbf{u}[k] + \mathrm{j}\mathbf{v}[k]$ with *matrix-valued* covariance function

$$\mathbf{R}_{xx}[\kappa] = E\mathbf{x}[k+\kappa]\mathbf{x}^{\mathrm{H}}[k] \tag{8.36}$$

and *matrix-valued* complementary covariance function

$$\widetilde{\mathbf{R}}_{xx}[\kappa] = E\mathbf{x}[k+\kappa]\mathbf{x}^{\mathrm{T}}[k]. \tag{8.37}$$

The augmented matrix covariance function is

$$\underline{\mathbf{R}}_{xx}[\kappa] = E\underline{\mathbf{x}}[k+\kappa]\underline{\mathbf{x}}^{\mathrm{H}}[k] \tag{8.38}$$

for $\underline{\mathbf{x}}[k] = [\mathbf{x}^{\mathrm{T}}[k], \mathbf{x}^{\mathrm{H}}[k]]^{\mathrm{T}}$. It is positive semidefinite for $\kappa = 0$ and $\underline{\mathbf{R}}_{xx}[\kappa] = \underline{\mathbf{R}}_{xx}^{\mathrm{H}}[-\kappa]$. The augmented PSD matrix $\underline{\mathbb{P}}_{xx}(\theta)$ is the discrete-time Fourier transform (DTFT) of $\underline{\mathbf{R}}_{xx}[\kappa]$. However, in order to spectrally factor $\underline{\mathbb{P}}_{xx}(\theta)$, we need to work with the z-transform

$$\underline{\mathbb{P}}_{xx}(z) = \begin{bmatrix} \mathbb{P}_{xx}(z) & \widetilde{\mathbb{P}}_{xx}(z) \\ \widetilde{\mathbb{P}}_{xx}^{*}(z^{*}) & \mathbb{P}_{xx}^{*}(z^{*}) \end{bmatrix} \tag{8.39}$$

instead. It satisfies the symmetry property $\underline{\mathbb{P}}_{xx}(z) = \underline{\mathbb{P}}_{xx}^{\mathrm{H}}(z^{-*})$, and thus $\mathbb{P}_{xx}(z) = \mathbb{P}_{xx}^{\mathrm{H}}(z^{-*})$ and $\widetilde{\mathbb{P}}_{xx}(z) = \widetilde{\mathbb{P}}_{xx}^{\mathrm{T}}(z^{-1})$. We indulge in a minor abuse of notation by writing $\underline{\mathbb{P}}_{xx}(\theta) = \underline{\mathbb{P}}_{xx}(z)|_{z=e^{j\theta}}$. The augmented PSD matrix $\underline{\mathbb{P}}_{xx}(\theta)$ is positive semidefinite and Hermitian: $\underline{\mathbb{P}}_{xx}(\theta) = \underline{\mathbb{P}}_{xx}^{\mathrm{H}}(\theta)$.

We wish to factor $\underline{\mathbb{P}}_{xx}(z)$ as

$$\underline{\mathbb{P}}_{xx}(z) = \underline{\mathbb{A}}_{xx}(z)\underline{\mathbb{A}}_{xx}^{\mathrm{H}}(z^{-*}), \tag{8.40}$$

where $\underline{\mathbb{A}}_{xx}(z)$ is minimum phase, meaning that both $\underline{\mathbb{A}}_{xx}(z)$ and $\underline{\mathbb{A}}_{xx}^{-1}(z)$ are causal and stable.[1] This means that both the coloring filter $\underline{\mathbb{A}}_{xx}(z)$ and the whitening filter $\underline{\mathbb{A}}_{xx}^{-1}(z)$ have all their zeros and poles inside the unit circle. Moreover, since $\underline{\mathbb{A}}_{xx}(z)$ must represent a WLSI filter, it needs to satisfy the pattern

$$\underline{\mathbb{A}}_{xx}(z) = \begin{bmatrix} \mathbb{A}_{xx}(z) & \widetilde{\mathbb{A}}_{xx}(z) \\ \widetilde{\mathbb{A}}_{xx}^{*}(z^{*}) & \mathbb{A}_{xx}^{*}(z^{*}) \end{bmatrix}. \tag{8.41}$$

We first determine $\mathbb{P}_{zz}(z)$ of the real vector time series $\mathbf{z}[k] = [\mathbf{u}^{\mathrm{T}}[k], \mathbf{v}^{\mathrm{T}}[k]]^{\mathrm{T}}$ as $\mathbb{P}_{zz}(z) = \frac{1}{4}\mathbf{T}^{\mathrm{H}}\underline{\mathbb{P}}_{xx}(z)\mathbf{T}$. We will assume that $\underline{\mathbb{P}}_{xx}(z)$ and thus also $\mathbb{P}_{zz}(z)$ are rational, which means that every element can be written as

$$[\mathbb{P}_{zz}]_{ij}(z) = \frac{b_{ij}(z)}{a_{ij}(z)}, \tag{8.42}$$

where the polynomials $b_{ij}(z)$ and $a_{ij}(z)$ are coprime with real coefficients. We further require that no $a_{ij}(z)$ have roots on the unit circle. Now we factor $\mathbb{P}_{zz}(z)$ into a matrix of polynomials $\boldsymbol{\beta}(z)$ divided by a single denominator polynomial $\alpha(z)$:

$$\mathbb{P}_{zz}(z) = \frac{1}{\alpha(z)}\boldsymbol{\beta}(z). \tag{8.43}$$

This factorization is achieved by finding the polynomial $\alpha(z)$ that is the least common multiple of all the denominator polynomials $\{a_{ij}(z)\}$. The polynomial $\alpha(z)$ may be iteratively computed using Euclid's algorithm for finding the greatest common divisor of two polynomials (see, e.g., Blahut (1985)). The elements of $\boldsymbol{\beta}(z)$ are then

$$\beta_{ij}(z) = b_{ij}(z)\frac{\alpha(z)}{a_{ij}(z)}. \tag{8.44}$$

Now we factor the polynomial $\alpha(z)$ to obtain a minimum-phase polynomial $\alpha_0(z)$ such that $\alpha(z) = \alpha_0(z)\alpha_0(z^{-1})$. Next, we can utilize an efficient algorithm such as the one provided by Jezek and Kucera (1985) or Spurbeck and Mullis (1998) to factor the polynomial matrix $\boldsymbol{\beta}(z) = \boldsymbol{\beta}_0(z)\boldsymbol{\beta}_0^{\mathrm{T}}(z^{-1})$ with $\boldsymbol{\beta}_0(z)$ minimum phase. Note that $\alpha_0(z^{-1}) = \alpha_0^*(z^{-*})$ and $\boldsymbol{\beta}_0^{\mathrm{T}}(z^{-1}) = \boldsymbol{\beta}_0^{\mathrm{H}}(z^{-*})$ because the polynomials in $\alpha(z)$ and $\boldsymbol{\beta}(z)$ have real coefficients. Then, $\boldsymbol{\beta}_0(z)/\alpha_0(z)$ is a rational minimum-phase coloring filter for the real time series $\mathbf{z}[k]$. Thus,

$$\underline{\mathbb{A}}_{xx}(z) = \frac{1}{\sqrt{2}\alpha_0(z)}\mathbf{T}\boldsymbol{\beta}_0(z)\mathbf{T}^{\mathrm{H}} \tag{8.45}$$

is a rational minimum-phase coloring filter for the complex time series $\mathbf{x}[k]$. It is easy to verify that (8.40) holds for $\underline{\mathbb{A}}_{xx}(z)$ given by (8.45).

8.3.2 Causal synthesis, analysis, and Wiener filters

We can now causally synthesize a WSS vector-valued time series $\mathbf{x}[k]$ with desired PSD matrix $\mathbb{P}_{xx}(\theta)$ and C-PSD matrix $\widetilde{\mathbb{P}}_{xx}(\theta)$ from a complex white vector sequence $\mathbf{w}[k]$. To do this, we factor the augmented PSD matrix as $\underline{\mathbb{P}}_{xx}(z) = \underline{\mathbb{A}}_{xx}(z)\underline{\mathbb{A}}_{xx}^{\mathrm{H}}(z^{-*})$ as described above. The z-transform of the synthesized vector time series $\mathbf{x}[k] \longleftrightarrow \mathbf{X}(z)$ is then

$$\mathbf{X}(z) = \mathbb{A}_{xx}(z)\mathbf{W}(z) + \widetilde{\mathbb{A}}_{xx}(z)\mathbf{W}^*(z^*), \tag{8.46}$$

where the input $\mathbf{w}[k] \longleftrightarrow \mathbf{W}(z)$ is a complex white and proper vector time series with $\mathbb{P}_{ww}(\theta) = \mathbf{I}$ and $\widetilde{\mathbb{P}}_{ww}(\theta) = \mathbf{0}$.

We can also causally whiten (analyze) a WSS vector time series $\mathbf{x}[k]$. In augmented notation, the z-transform of the filter output is

$$\underline{\mathbf{W}}(z) = \underline{\mathbb{A}}_{xx}^{-1}(z)\underline{\mathbf{X}}(z). \tag{8.47}$$

The augmented transfer function of the noncausal Wiener filter that provides the widely linear MMSE estimate $\hat{\mathbf{x}}[k]$ of $\mathbf{x}[k]$ from measurements $\mathbf{y}[k]$ is

$$\underline{\mathbb{H}}_{\mathrm{nc}}(z) = \underline{\mathbb{P}}_{xy}(z)\underline{\mathbb{P}}_{yy}^{-1}(z). \tag{8.48}$$

With the factorization $\underline{\mathbb{P}}_{yy}(z) = \underline{\mathbb{A}}_{yy}(z)\underline{\mathbb{A}}_{yy}^{\mathrm{H}}(z^{-*})$, the augmented transfer function of the causal Wiener filter is then

$$\underline{\mathbb{H}}_{\mathrm{c}}(z) = \left[\underline{\mathbb{P}}_{xy}(z)\underline{\mathbb{A}}_{yy}^{-\mathrm{H}}(z^{-*})\right]_+ \underline{\mathbb{A}}_{yy}^{-1}(z), \tag{8.49}$$

where $[\cdot]_+$ denotes the causal part. The z-transform of the estimate is given in augmented notation by $\widehat{\underline{\mathbf{X}}}(z) = \underline{\mathbb{H}}_{\mathrm{c}}(z)\underline{\mathbf{Y}}(z)$.

8.4 Rotary-component and polarization analysis

Rotary-component and polarization analysis are widely used in many research areas, including optics, geophysics, meteorology, oceanography, and radar.[2] The idea is to represent a two-dimensional signal in the complex plane as an integral of ellipses, and

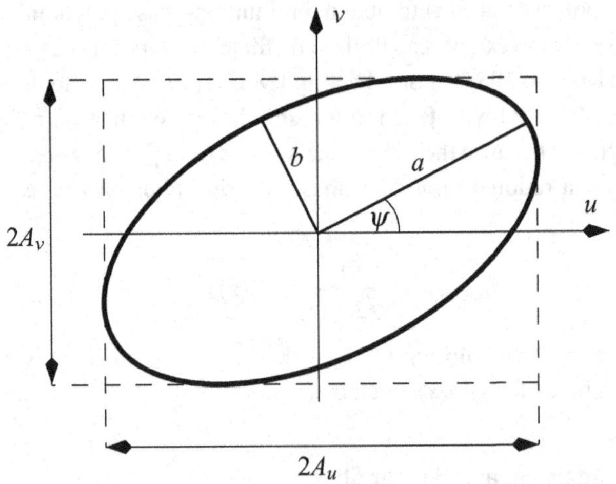

Figure 8.1 The ellipse traced out by the monochromatic and deterministic signal $x(t)$.

each ellipse as the sum of a counterclockwise and a clockwise rotating phasor, which are called the *rotary components*. From a mathematical point of view, each ellipse is the widely linear minimum mean-squared error approximation of the signal from its rotary components at a given frequency f. From a practical point of view, the analysis of ellipse properties (such as its shape and orientation) and properties of the rotary components (such as their coherence) carries information about the underlying physics.

8.4.1 Rotary components

In order to illustrate the fundamental ideas of rotary-component analysis, we begin with a *monochromatic* and *deterministic* bivariate signal.[3] The most general such signal is described as

$$\begin{aligned} u(t) &= A_u \cos(2\pi f_0 t + \theta_u), \\ v(t) &= A_v \cos(2\pi f_0 t + \theta_v), \end{aligned} \tag{8.50}$$

where A_u and A_v are two given nonnegative amplitudes and θ_u and θ_v are two given phase offsets. As shown in Fig. 8.1, this signal moves periodically around an ellipse, which is inscribed in a rectangle whose sides are parallel to the u- and v-axes and have lengths $2A_u$ and $2A_v$. The description (8.50) is said to be a decomposition of this ellipse into its *linearly polarized components* $u(t)$ and $v(t)$. A complex representation equivalent to (8.50) is

$$x(t) = u(t) + \mathrm{j}v(t) = \underbrace{A_+ \mathrm{e}^{\mathrm{j}\theta_+} \mathrm{e}^{\mathrm{j}2\pi f_0 t}}_{x_+(t)} + \underbrace{A_- \mathrm{e}^{-\mathrm{j}\theta_-} \mathrm{e}^{-\mathrm{j}2\pi f_0 t}}_{x_-(t)}. \tag{8.51}$$

This is the sum of the counterclockwise (CCW) turning phasor $x_+(t)$ and the clockwise (CW) turning phasor $x_-(t)$, which are called the *rotary components* or *circularly polarized components*. The ellipse itself is called the *polarization ellipse*.

The real description (8.50) and complex description (8.51) can be related through their PSD matrices. Restricted to nonnegative frequencies f, the PSD matrix of the vector $\mathbf{z}(t) = [u(t), v(t)]^{\mathrm{T}}$ is

$$\mathbb{P}_{zz}(f) = \tfrac{1}{4}\begin{bmatrix} A_u^2 & A_u A_v \mathrm{e}^{\mathrm{j}\theta} \\ A_u A_v \mathrm{e}^{-\mathrm{j}\theta} & A_v^2 \end{bmatrix} \delta(f - f_0), \tag{8.52}$$

with $\theta = \theta_u - \theta_v$. Since $\mathbb{P}_{zz}(f) = \mathbb{P}_{zz}^*(-f)$, the PSDs of positive and negative frequencies are trivially related and the restriction to nonnegative frequencies f presents no loss of information. The augmented PSD matrix for the complex description is

$$\underline{\mathbb{P}}_{xx}(f) = \begin{bmatrix} A_+^2 & A_+ A_- \mathrm{e}^{\mathrm{j}2\psi} \\ A_+ A_- \mathrm{e}^{-\mathrm{j}2\psi} & A_-^2 \end{bmatrix} \delta(f - f_0), \tag{8.53}$$

with $2\psi = \theta_+ - \theta_-$. The PSD matrices $\mathbb{P}_{zz}(f)$ and $\underline{\mathbb{P}}_{xx}(f)$ are connected through the real-to-complex transformation $\mathbf{T}$ as

$$\underline{\mathbb{P}}_{xx}(f) = \mathbf{T}\mathbb{P}_{zz}(f)\mathbf{T}^{\mathrm{H}}, \tag{8.54}$$

from which we obtain the following expressions:

$$A_+^2 + A_-^2 = \frac{A_u^2 + A_v^2}{2}, \tag{8.55}$$

$$A_+^2 - A_-^2 = A_u A_v \sin\theta, \tag{8.56}$$

$$\tan(2\psi) = \frac{2A_u A_v \cos\theta}{A_u^2 - A_v^2} = \frac{\operatorname{Im}\{A_+ A_- \mathrm{e}^{\mathrm{j}2\psi}\}}{\operatorname{Re}\{A_+ A_- \mathrm{e}^{\mathrm{j}2\psi}\}}. \tag{8.57}$$

Thus, the ellipse is parameterized by either (A_u, A_v, θ) or (A_+, A_-, ψ). If the reader suspects that the latter will turn out to be more useful, then he or she is right.

We have already determined in Section 1.3 that $\psi = (\theta_+ - \theta_-)/2$ is the angle between the major axis and the u-axis, simply referred to as the *orientation* of the ellipse. Moreover, we found that $2a = 2(A_+ + A_-)$ is the length of the major axis and $2b = 2|A_+ - A_-|$ the length of the minor axis. From this, we obtain the area of the ellipse

$$ab\pi = (A_+ + A_-)|A_+ - A_-|\pi = |A_+^2 - A_-^2|\pi. \tag{8.58}$$

The numerical eccentricity

$$\frac{\sqrt{a^2 - b^2}}{a} = \frac{2\sqrt{A_+ A_-}}{A_+ + A_-} \tag{8.59}$$

is seen to be the ratio of geometric mean to arithmetic mean of A_+ and A_-. In optics, it is common to introduce another angle χ through

$$\sin(2\chi) = \pm\frac{2ab}{a^2 + b^2} = \frac{A_+^2 - A_-^2}{A_+^2 + A_-^2}, \quad -\frac{\pi}{4} \le \chi \le \frac{\pi}{4}, \tag{8.60}$$

where the expression in the middle is the ratio of geometric mean to arithmetic mean of a^2 and b^2, and its sign indicates the rotation direction of the ellipse: CCW for "+" and CW for "−." Because

$$\tan\chi = \pm\frac{b}{a} = \frac{A_+ - A_-}{A_+ + A_-} \tag{8.61}$$

we may say that χ characterizes the scale-invariant shape of the ellipse.

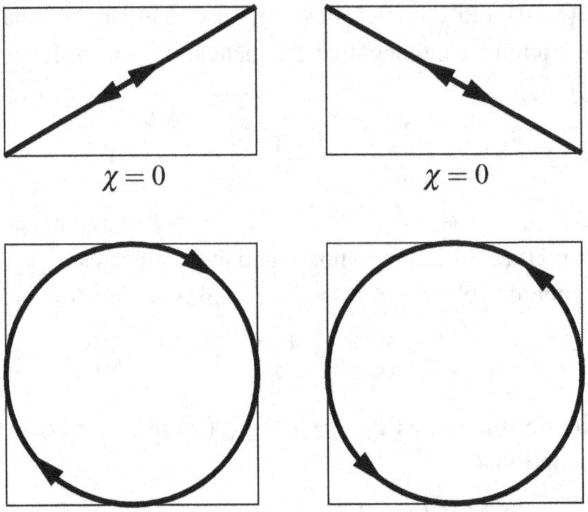

Figure 8.2 Linear and circular shape-polarization.

- For $-\pi/4 \leq \chi < 0$, $x(t)$ traces out the ellipse in the *clockwise* direction, and $x(t)$ is called *CW polarized* or *right(-handed) polarized.* If $\chi = -\pi/4$, then the ellipse degenerates into a circle because $A_+ = 0$, and $x(t)$ is called *CW circularly polarized* or *right-circularly polarized.*
- For $0 < \chi \leq \pi/4$, $x(t)$ traces out the ellipse in the *counterclockwise* direction, and $x(t)$ is called *CCW polarized* or *left(-handed) polarized.* If $\chi = \pi/4$, the ellipse becomes a circle because $A_- = 0$, and $x(t)$ is *CCW circularly polarized* or *left-circularly polarized.*
- For $\chi = 0$, the ellipse degenerates into a line because $A_+ = A_-$, and $x(t)$ is called *linearly polarized.* The angle of the line is the ellipse orientation ψ.

Linear and circular polarization is illustrated in Fig. 8.2. Unfortunately, the term polarization has two quite different meanings in the literature. The type of polarization discussed here is "shape polarization," which characterizes the shape (elliptical, circular, or linear) and rotation direction of the ellipse. There is another type of polarization, to be discussed in Section 8.4.3, which is not related to shape polarization.

8.4.2 Rotary components of random signals

Now we will generalize the discussion of the previous subsection to a WSS random signal $x(t)$. Using the spectral representation from Result 8.1, we write $x(t)$ as

$$x(t) = \int_0^\infty d\xi(f)e^{j2\pi ft} + d\xi(-f)e^{-j2\pi ft}. \tag{8.62}$$

This creates the minor technical issue that $d\xi(0)$ is counted twice in the integral, which can be remedied by scaling $d\xi(0)$ by a factor of $1/2$. Throughout this section, we will let f denote a nonnegative frequency. The representation (8.62) is the superposition of ellipses, and one ellipse

$$\varepsilon_f(t) = \underbrace{\mathrm{d}\xi(f)\mathrm{e}^{\mathrm{j}2\pi ft}}_{\varepsilon_{f_+}(t)} + \underbrace{\mathrm{d}\xi(-f)\mathrm{e}^{-\mathrm{j}2\pi ft}}_{\varepsilon_{f_-}(t)} \tag{8.63}$$

can be constructed for a given frequency $f \geq 0$. We emphasize that the ellipse $\varepsilon_f(t)$ and the *rotary components* $\varepsilon_{f_+}(t)$ and $\varepsilon_{f_-}(t)$ are WSS random processes. The rotary components are each individually proper, but the ellipse $\varepsilon_f(t)$ is generally improper.

Interpretation of the random ellipse

There is a powerful interpretation of the ellipse $\varepsilon_f(t)$. Assume that we wanted to linearly estimate the signal $x(t)$ from the rotary components at a given frequency f, i.e., from $\varepsilon_{f_+}(t) = \mathrm{d}\xi(f)\mathrm{e}^{\mathrm{j}2\pi ft}$ and $\varepsilon_{f_-}(t) = \mathrm{d}\xi(-f)\mathrm{e}^{-\mathrm{j}2\pi ft}$. Because $\mathrm{d}\xi(f)$ and $\mathrm{d}\xi^*(-f)$ are generally correlated, meaning that $\widetilde{P}_{xx}(f)\mathrm{d}f = E[\mathrm{d}\xi(f)\mathrm{d}\xi(-f)]$ may be nonzero, we have the foresight to use a *widely linear* estimator of the form

$$\begin{aligned}\hat{x}_f(t) = {} & [H_1(f)\mathrm{d}\xi(f) + H_2(f)\mathrm{d}\xi^*(-f)]\mathrm{e}^{\mathrm{j}2\pi ft} \\ & + H_1(-f)\mathrm{d}\xi(-f) + H_2(-f)\mathrm{d}\xi^*(f)]\mathrm{e}^{-\mathrm{j}2\pi ft}.\end{aligned} \tag{8.64}$$

So, for a fixed frequency f, we need to determine $H_1(f)$, $H_1(-f)$, $H_2(f)$, and $H_2(-f)$ such that $E|\hat{x}_f(t) - x(t)|^2$ (which is independent of t since $x(t)$ is WSS) is minimized.

To solve this problem, we first note that the vector $[\mathrm{d}\xi(f), \mathrm{d}\xi^*(-f)]$ is uncorrelated with the vector $[\mathrm{d}\xi(-f), \mathrm{d}\xi^*(f)]$ because $x(t)$ is WSS. This allows us to determine the pair $(H_1(f), H_2(f))$ independently from the pair $(H_1(-f), H_2(-f))$. In other words, the WLMMSE estimate $\hat{x}_f(t)$ is the sum of

- the LMMSE estimate of $x(t)$ from $\varepsilon_{f_+}(t) = \mathrm{d}\xi(f)\mathrm{e}^{\mathrm{j}2\pi ft}$ and $\varepsilon^*_{f_-}(t) = \mathrm{d}\xi^*(-f)\mathrm{e}^{\mathrm{j}2\pi ft}$, and
- the LMMSE estimate of $x(t)$ from $\varepsilon^*_{f_+}(t) = \mathrm{d}\xi^*(f)\mathrm{e}^{-\mathrm{j}2\pi ft}$ and $\varepsilon_{f_-}(t) = \mathrm{d}\xi(-f)\mathrm{e}^{-\mathrm{j}2\pi ft}$.

The first is

$$\begin{aligned}& \left[E(x(t)\mathrm{d}\xi^*(f)\mathrm{e}^{-\mathrm{j}2\pi ft}) \quad E(x(t)\mathrm{d}\xi(-f)\mathrm{e}^{-\mathrm{j}2\pi ft})\right] \\ & \quad \times \begin{bmatrix} E|\mathrm{d}\xi(f)|^2 & E(\mathrm{d}\xi(f)\mathrm{d}\xi(-f)) \\ E(\mathrm{d}\xi^*(f)\mathrm{d}\xi^*(-f)) & E|\mathrm{d}\xi(-f)|^2 \end{bmatrix}^{-1} \begin{bmatrix} \varepsilon_{f_+}(t) \\ \varepsilon^*_{f_-}(t) \end{bmatrix} \\ & = \left[P_{xx}(f)\mathrm{d}f \quad \widetilde{P}_{xx}(f)\mathrm{d}f\right] \begin{bmatrix} P_{xx}(f)\mathrm{d}f & \widetilde{P}_{xx}(f)\mathrm{d}f \\ \widetilde{P}^*_{xx}(f)\mathrm{d}f & P_{xx}(-f)\mathrm{d}f \end{bmatrix}^{-1} \begin{bmatrix} \varepsilon_{f_+}(t) \\ \varepsilon^*_{f_-}(t) \end{bmatrix}\end{aligned} \tag{8.65}$$

$$= \frac{1}{P_{xx}(f)P_{xx}(-f) - |\widetilde{P}_{xx}(f)|^2} \begin{bmatrix} P_{xx}(f)P_{xx}(-f) - |\widetilde{P}_{xx}(f)|^2 \\ -P_{xx}(f)\widetilde{P}_{xx}(f) + \widetilde{P}_{xx}(f)P_{xx}(f) \end{bmatrix}^{\mathrm{T}} \begin{bmatrix} \varepsilon_{f_+}(t) \\ \varepsilon^*_{f_-}(t) \end{bmatrix} \tag{8.66}$$

$$= \varepsilon_{f_+}(t). \tag{8.67}$$

In (8.65) we have used

$$E(x(t)\mathrm{d}\xi^*(f)\mathrm{e}^{-\mathrm{j}2\pi ft}) = E\left[\left(\int_{-\infty}^{\infty} \mathrm{d}\xi(\nu)\mathrm{e}^{\mathrm{j}2\pi\nu t}\right)\mathrm{d}\xi^*(f)\mathrm{e}^{-\mathrm{j}2\pi ft}\right]$$
$$= E|\mathrm{d}\xi(f)|^2 = P_{xx}(f)\mathrm{d}f, \tag{8.68}$$
$$E(x(t)\mathrm{d}\xi(-f)\mathrm{e}^{-\mathrm{j}2\pi ft}) = E\left[\left(\int_{-\infty}^{\infty} \mathrm{d}\xi(\nu)\mathrm{e}^{\mathrm{j}2\pi\nu t}\right)\mathrm{d}\xi(-f)\mathrm{e}^{-\mathrm{j}2\pi ft}\right]$$
$$= E(\mathrm{d}\xi(f)\mathrm{d}\xi(-f)) = \widetilde{P}_{xx}(f)\mathrm{d}f, \tag{8.69}$$

which hold because $\xi(f)$ has orthogonal increments, i.e., $E[\mathrm{d}\xi(\nu)\mathrm{d}\xi^*(f)] = 0$ and $E[\mathrm{d}\xi(\nu)\mathrm{d}\xi(-f)] = 0$ for $\nu \neq f$.

A completely analogous computation to (8.66)–(8.67) shows that the LMMSE estimate of $x(t)$ from $\varepsilon_{f_+}^*(t)$ and $\varepsilon_{f_-}(t)$ is simply $\varepsilon_{f_-}(t)$. All of this taken together means that

$$\hat{x}_f(t) = \varepsilon_f(t) = \varepsilon_{f_+}(t) + \varepsilon_{f_-}(t), \tag{8.70}$$

i.e., $H_1(f) = H_1(-f) = 1$ and $H_2(f) = H_2(-f) = 0$. This says that the ellipse $\varepsilon_f(t)$ in (8.63) is actually the best widely linear estimate of $x(t)$ that may be constructed from the rotary components at a given frequency f. Moreover, the widely linear estimate turns out to be strictly linear. That is, even if the rotary components $\varepsilon_{f_+}(t)$ and $\varepsilon_{f_-}(t)$ have nonzero complementary correlation, it cannot be exploited to build a widely linear estimator with smaller mean-squared error than a strictly linear estimator.

Statistical properties of the random ellipse

The second-order statistical properties of the WSS random ellipse $\varepsilon_f(t)$ can be *approximately* determined from the augmented PSD matrix $\underline{\mathbf{P}}_{xx}(f)$. From (8.58), we approximate the expected area of the random ellipse as

$$\pi|P_{xx}(f) - P_{xx}(-f)|\mathrm{d}f. \tag{8.71}$$

This is a crude approximation because it involves exchanging the order of the expectation operator and the absolute value. The expected rotation direction of the ellipse is given by the sign of $P_{xx}(f) - P_{xx}(-f)$, where "+" indicates CCW and "−" CW direction. The angle $\psi(f)$, obtained as half the phase of the C-PSD through

$$\tan(2\psi(f)) = \frac{\mathrm{Im}\,\widetilde{P}_{xx}(f)}{\mathrm{Re}\,\widetilde{P}_{xx}(f)}, \tag{8.72}$$

gives an approximation of the expected ellipse orientation. Similarly, $\chi(f)$ obtained through

$$\sin(2\chi(f)) = \frac{P_{xx}(f) - P_{xx}(-f)}{P_{xx}(f) + P_{xx}(-f)} \tag{8.73}$$

approximates the expected ellipse shape. The approximations (8.72) and (8.73) are derived from (8.57) and (8.60), respectively, by applying the expectation operator to the numerator and denominator separately, thus ignoring the fact that these are generally correlated, and exchanging the order of the expectation operator and the sin or tan function. However, it was shown by Rubin-Delanchy and Walden (2008) that, in the *Gaussian* case, $\psi(f)$ is the mean ellipse orientation – not just an approximation.

The properties of the ellipse all depend on the PSD and C-PSD, which we have defined as ensemble averages. In most practical applications, only one realization of a random process is available. Assuming that the random process is ergodic (which is often a reasonable assumption), all ensemble averages can be replaced by time averages.

8.4.3 Polarization and coherence

Definition 8.1. *Let $x(t)$ be a WSS complex random process $x(t)$ with augmented PSD matrix $\underline{\mathbf{P}}_{xx}(f)$, and let $\Lambda_1(f)$ and $\Lambda_2(f)$ denote the two real eigenvalues of $\underline{\mathbf{P}}_{xx}(f)$, assuming $\Lambda_1(f) \geq \Lambda_2(f)$. The* degree of polarization *of $x(t)$ at frequency f is defined as*

$$\Phi(f) = \frac{\Lambda_1(f) - \Lambda_2(f)}{\Lambda_1(f) + \Lambda_2(f)}, \tag{8.74}$$

which satisfies $0 \leq \Phi(f) \leq 1$. If $\Phi(f) = 0$, then $x(t)$ is called unpolarized *at frequency f. If $\Phi(f) = 1$, then $x(t)$ is called* completely polarized *at frequency f.*

We need to emphasize again that there are two meanings of the term polarization: one is this definition and the other is "shape polarization" as discussed in Section 8.4.1. A signal that is polarized in the sense of Definition 8.1 does not have to be polarized in the sense that the shape of the ellipse degenerates into a line or circle, which is called linear or circular polarization. Similarly, a linearly or circularly polarized signal does not have to be polarized in the sense of Definition 8.1 either.

The degree of polarization has an intricate relationship with and is in fact sometimes confused with the *magnitude-squared coherence* between the CCW and CW rotary components. "Coherence" is a synonymous term for correlation coefficient, but, in the frequency domain, "coherence" is much more commonly used than "correlation coefficient." In the terminology of Chapter 4, the coherence defined in the following would be called a *reflectional correlation coefficient.*

Definition 8.2. *The* complex coherence *between CCW and CW rotary components is*

$$\widetilde{\rho}_{xx}(f) = \frac{E[d\xi(f)d\xi(-f)]}{\sqrt{E|d\xi(f)|^2 E\,|d\xi(-f)|^2}} = \frac{\widetilde{P}_{xx}(f)}{\sqrt{P_{xx}(f)P_{xx}(-f)}}. \tag{8.75}$$

If either $P_{xx}(f) = 0$ or $P_{xx}(-f) = 0$, then we define $\widetilde{\rho}_{xx}(f) = 1$.

We usually consider only the magnitude-squared coherence $|\widetilde{\rho}_{xx}(f)|^2$, which satisfies $|\widetilde{\rho}_{xx}(f)|^2 \leq 1$. If and only if $x(t)$ is completely polarized at f, then $\Lambda_2(f) = 0$, $|\widetilde{P}_{xx}(f)|^2 = P_{xx}(f)P_{xx}(-f)$, and thus det $\underline{\mathbf{P}}_{xx}(f) = 0$. We thus find the following connection between the degree of polarization and the magnitude-squared coherence.

Result 8.3. *A WSS complex random process $x(t)$ is completely polarized at frequency f, i.e., $\Phi(f) = 1$, if and only if $\widetilde{\rho}_{xx}(f) = 1$. The corresponding ellipse $\varepsilon_f(t)$ has complete* *coherence between its CCW and CW rotary components. Provided that $P_{xx}(-f) \neq 0$, we have $d\xi(f) = c_f\, d\xi^*(-f)$ with probability 1, where the complex constant c_f is*

$$c_f = \frac{\widetilde{P}_{xx}(f)}{P_{xx}(-f)} = \frac{P_{xx}(f)}{\widetilde{P}^*_{xx}(f)}. \tag{8.76}$$

If $x(t)$ is completely polarized at f, *all* sample functions of the random ellipse $\varepsilon_f(t)$ turn in the same direction, which is either clockwise – if $-\pi/4 \le \chi(f) < 0$, $|c_f| < 1$ with c_f given by (8.76) – or counterclockwise – if $0 < \chi(f) \le \pi/4$, $|c_f| > 1$. Monochromatic deterministic signals are always completely polarized, since their augmented PSD matrix is always rank-deficient. Moreover, all analytic signals are completely polarized for all f, as is evident from $\det \underline{\mathbf{P}}_{xx}(f) \equiv 0$ because $P_{xx}(f)P_{xx}(-f) \equiv 0$.

However, $\widetilde{\rho}_{xx}(f) = 0$, i.e., $\widetilde{P}_{xx}(f) = 0$, is only a necessary but not sufficient condition for a signal to be unpolarized at f, which requires $\Lambda_1(f) = \Lambda_2(f)$. In fact, for $\widetilde{\rho}_{xx}(f) = 0$ the degree of polarization takes on a particularly simple form:

$$\begin{aligned} \Phi(f) &= \left| \frac{P_{xx}(f) - P_{xx}(-f)}{P_{xx}(f) + P_{xx}(-f)} \right| \\ &= |\sin(2\chi(f))|\,. \end{aligned} \tag{8.77}$$

The additional condition of equal power in the rotary components, $P_{xx}(f) = P_{xx}(-f)$, for a signal to be unpolarized at f can thus be visualized as a requirement that the random ellipse $\varepsilon_f(t)$ have no preferred rotation direction.

We can find an expression for the degree of polarization $\Phi(f)$ by using the formula (A1.13) for the two eigenvalues of the 2×2 augmented PSD matrix:

$$\Lambda_{1,2}(f) = \tfrac{1}{2}\, \mathrm{tr}\, \underline{\mathbf{P}}_{xx}(f) \pm \tfrac{1}{2}\sqrt{\mathrm{tr}^2 \underline{\mathbf{P}}_{xx}(f) - 4 \det \underline{\mathbf{P}}_{xx}(f)}. \tag{8.78}$$

By inserting this expression into (8.74), we then obtain

$$\Phi(f) = \sqrt{1 - \frac{4 \det \underline{\mathbf{P}}_{xx}(f)}{\mathrm{tr}^2 \underline{\mathbf{P}}_{xx}(f)}}, \tag{8.79}$$

without the need for an explicit eigenvalue decomposition of $\underline{\mathbf{P}}_{xx}(f)$.

For a given frequency f, it is possible to decompose any WSS signal $x(t)$ into a WSS completely polarized signal $p(t)$ and a WSS unpolarized signal $n(t)$ (but this decomposition is generally different for different frequencies).[4] This is achieved by expanding the eigenvalue matrix as

$$\mathbf{\Lambda}_{xx}(f) = \mathbf{\Lambda}_{pp}(f) + \mathbf{\Lambda}_{nn}(f) = \begin{bmatrix} \Lambda_1(f) - \Lambda_2(f) & 0 \\ 0 & 0 \end{bmatrix} + \begin{bmatrix} \Lambda_2(f) & 0 \\ 0 & \Lambda_2(f) \end{bmatrix}. \tag{8.80}$$

The degree of polarization $\Phi(f)$ at frequency f is thus the ratio of the polarized power to the total power:

$$\Phi(f) = \frac{P_{\text{pol}}(f)}{P_{\text{tot}}(f)}. \tag{8.81}$$

The degree of polarization and the magnitude-squared coherence have one more important property.

Result 8.4. *The degree of polarization $\Phi(f)$ and the magnitude-squared coherence $|\tilde{\rho}_{xx}(f)|^2$ are both invariant under coordinate rotation, i.e., $x(t)$ and $x(t)\mathrm{e}^{\mathrm{j}\phi}$ have the same $\Phi(f)$ and $|\tilde{\rho}_{xx}(f)|^2$ for a fixed real angle ϕ.*

This is easy to see. If $x(t)$ is replaced with $x(t)\mathrm{e}^{\mathrm{j}\phi}$, the PSD and C-PSD transform as

$$P_{xx}(f) \longrightarrow P_{xx}(f),$$
$$\tilde{P}_{xx}(f) \longrightarrow \tilde{P}_{xx}(f)\mathrm{e}^{\mathrm{j}2\phi}.$$

That is, $P_{xx}(f)$ and $|\tilde{P}_{xx}(f)|$ are invariant under coordinate rotation. It follows that the ellipse shape $\chi(f)$, the degree of polarization $\Phi(f)$, and the magnitude-squared coherence $|\tilde{\rho}_{xx}(f)|^2$ are all invariant under coordinate rotation. On the other hand, the ellipse orientation $\psi(f)$ is *covariant* under coordinate rotation, i.e., it transforms as $\psi(f) \longrightarrow \mathrm{e}^{\mathrm{j}\phi}\psi(f)$.

The invariance property of $|\tilde{\rho}_{xx}(f)|^2$ is another key advantage of the rotary component versus the Cartesian description. It is obviously also possible to define the Cartesian coherence

$$\rho_{uv}(f) = \frac{P_{uv}(f)}{\sqrt{P_{uu}(f)P_{vv}(f)}} \tag{8.82}$$

as a normalized cross-spectrum between the u- and v-components. However, the Cartesian magnitude-squared coherence $|\rho_{uv}(f)|^2$ does depend on the orientation of the u- and v-axes, thus limiting its usefulness.

8.4.4 Stokes and Jones vectors

George G. Stokes introduced a set of four parameters to characterize the state of polarization of partially polarized light. These parameters are closely related to the PSD and C-PSD.

Definition 8.3. *The* Stokes vector $\boldsymbol{\Sigma}(f) = [\Sigma_0(f), \Sigma_1(f), \Sigma_2(f), \Sigma_3(f)]^{\mathrm{T}}$ *is defined as*

$$\Sigma_0(f) = \mathrm{Ev}\, P_{xx}(f) = \tfrac{1}{2}[P_{xx}(f) + P_{xx}(-f)],$$
$$\Sigma_1(f) = \mathrm{Re}\, \tilde{P}_{xx}(f),$$
$$\Sigma_2(f) = \mathrm{Im}\, \tilde{P}_{xx}(f),$$
$$\Sigma_3(f) = \mathrm{Od}\, P_{xx}(f) = \tfrac{1}{2}[P_{xx}(f) - P_{xx}(-f)],$$

where Ev $P_{xx}(f)$ *and* Od $P_{xx}(f)$ *denote the even and odd parts of* $P_{xx}(f)$.

The four real-valued parameters $(\Sigma_0(f), \Sigma_1(f), \Sigma_2(f), \Sigma_3(f))$ are an equivalent parameterization of $(P_{xx}(f), P_{xx}(-f), \widetilde{P}_{xx}(f))$, where $P_{xx}(\pm f)$ is real and $\widetilde{P}_{xx}(f)$ complex. Hence, the polarization-ellipse properties may be determined from $\boldsymbol{\Sigma}(f)$. The polarized power is

$$P_{\text{pol}}(f) = \sqrt{\Sigma_1^2(f) + \Sigma_2^2(f) + \Sigma_3^2(f)} = \sqrt{|\widetilde{P}_{xx}(f)|^2 + \text{Od}^2 P_{xx}(f)}, \tag{8.83}$$

and the total power is

$$P_{\text{tot}}(f) = \Sigma_0(f) = \text{Ev}\, P_{xx}(f). \tag{8.84}$$

The degree of polarization (8.79) can thus be expressed as

$$\Phi(f) = \frac{\sqrt{|\widetilde{P}_{xx}(f)|^2 + \text{Od}^2 P_{xx}(f)}}{\text{Ev}\, P_{xx}(f)}. \tag{8.85}$$

An unpolarized signal has $\Sigma_1(f) = \Sigma_2(f) = \Sigma_3(f) = 0$, and a completely polarized signal has $\Sigma_1^2(f) + \Sigma_2^2(f) + \Sigma_3^2(f) = \Sigma_0^2(f)$. Thus, three real-valued parameters suffice to characterize a completely polarized signal. Which three parameters to choose is, of course, somewhat arbitrary. The most common way uses $(P_{uu}(f), P_{vv}(f), \arg P_{uv}(f))$, which are combined in the *Jones vector*

$$\mathbf{j}(f) = \begin{bmatrix} j_1(f) \\ j_2(f) \end{bmatrix} = \begin{bmatrix} P_{uu}(f) \\ P_{vv}(f) e^{\text{j} \arg P_{uv}(f)} \end{bmatrix}. \tag{8.86}$$

We note that only the angle of the cross-PSD $P_{uv}(f)$ between the real part $u(t)$ and the imaginary part $v(t)$ is specified. The magnitude $|P_{uv}(f)|$ is determined by the assumption that the signal is completely polarized, which means that the determinant of the PSD matrix $\mathbf{P}_{zz}(f)$ is zero:

$$\det \mathbf{P}_{zz}(f) = \det \begin{bmatrix} P_{uu}(f) & P_{uv}(f) \\ P_{uv}^*(f) & P_{vv}(f) \end{bmatrix} = 0 \Longleftrightarrow |P_{uv}(f)|^2 = P_{uu}(f) P_{vv}(f). \tag{8.87}$$

The reason why the Jones vector uses the Cartesian coordinate system rather than the rotary components is simply historical. It would be just as easy (and maybe more insightful) to instead use the three parameters $(P_{xx}(f), P_{xx}(-f), \arg \widetilde{P}_{xx}(f))$, and fix the magnitude of the C-PSD as $|\widetilde{P}_{xx}(f)|^2 = P_{xx}(f) P_{xx}(-f)$.

On the other hand, we may also express the Stokes vector in the Cartesian coordinate system as $\Sigma_0(f) = P_{uu}(f) + P_{vv}(f)$, $\Sigma_1(f) = P_{uu}(f) - P_{vv}(f)$, $\Sigma_2(f) = 2\,\text{Re}\, P_{uv}(f)$, and $\Sigma_3(f) = 2\,\text{Im}\, P_{uv}(f)$. This allows us to determine equivalent Jones and Stokes vectors for fully polarized signals. Remember that the Jones vectors cannot describe partially polarized or unpolarized signals.

The effect of optical elements on Stokes vectors can be described by multiplication from the left with a real-valued 4×4 matrix. This is called the *Mueller calculus*. For completely polarized light, there exists an analog calculus for manipulation of Jones vectors, which uses complex-valued 2×2 matrices. This is called the *Jones calculus*.[5]

Example 8.3. The following table lists some states of polarization and corresponding Jones and Stokes vectors:

Polarization	Jones vector	Stokes vector
Horizontal linear	$[1, 0]$	$[1, 1, 0, 0]$
Vertical linear	$[0, 1]$	$[1, -1, 0, 0]$
Linear at 45°	$[\frac{1}{2}, \frac{1}{2}]$	$[1, 0, 1, 0]$
Right-hand (CW) circular	$[\frac{1}{2}, -\mathrm{j}/2]$	$[1, 0, 0, -1]$
Left-hand (CCW) circular	$[\frac{1}{2}, \mathrm{j}/2]$	$[1, 0, 0, 1]$

The effects of two exemplary optical elements on Jones and Stokes vectors are described by pre-multiplying the vectors with the following matrices:

Optical element	Jones matrix	Mueller matrix
Horizontal linear polarizer	$\begin{bmatrix} 1 & 0 \\ 0 & 0 \end{bmatrix}$	$\frac{1}{2}\begin{bmatrix} 1 & 1 & 0 & 0 \\ 1 & 1 & 0 & 0 \\ 0 & 0 & 0 & 0 \\ 0 & 0 & 0 & 0 \end{bmatrix}$
Left-circular (CCW) polarizer	$\frac{1}{2}\begin{bmatrix} 1 & -\mathrm{j} \\ \mathrm{j} & 1 \end{bmatrix}$	$\frac{1}{2}\begin{bmatrix} 1 & 0 & 0 & 1 \\ 0 & 0 & 0 & 0 \\ 0 & 0 & 0 & 0 \\ 1 & 0 & 0 & 1 \end{bmatrix}$

In summary, we now have the following alternative characterizations of the statistical properties of the ellipse $\varepsilon_f(t)$:

- the PSD and C-PSD: $(P_{xx}(f), P_{xx}(-f), \widetilde{P}_{xx}(f))$
- the PSDs and cross-PSD of the Cartesian coordinates: $(P_{uu}(f), P_{vv}(f), P_{uv}(f))$
- the PSD, the ellipse orientation (i.e., half the phase of the C-PSD), and the magnitude-squared coherence: $(P_{xx}(f), P_{xx}(-f), \psi(f), |\widetilde{\rho}_{xx}(f)|^2)$
- the Stokes vector $\boldsymbol{\Sigma}(f)$
- the Jones vector $\mathbf{j}(f)$, which assumes a completely polarized signal: $|\widetilde{\rho}_{xx}(f)|^2 = 1$

All of these descriptions are equivalent because each of them can be derived from any other description.

8.4.5 Joint analysis of two signals

Now consider the joint analysis of two complex jointly WSS signals $x(t)$ and $y(t)$. Individually, they are described by the augmented PSD matrices $\underline{\mathbf{P}}_{xx}(f)$ and $\underline{\mathbf{P}}_{yy}(f)$, or any of the equivalent characterizations discussed at the end of the previous subsection.

Their joint properties are described by the augmented cross-PSD matrix

$$\underline{\mathbf{P}}_{xy}(f) = \begin{bmatrix} P_{xy}(f) & \widetilde{P}_{xy}(f) \\ \widetilde{P}^*_{xy}(-f) & P^*_{xy}(-f) \end{bmatrix}, \tag{8.88}$$

which is the Fourier transform of

$$\underline{\mathbf{R}}_{xy}(\tau) = E\left[\underline{\mathbf{x}}(t+\tau)\underline{\mathbf{y}}^{\mathrm{H}}(t)\right] = \begin{bmatrix} r_{xy}(\tau) & \tilde{r}_{xy}(\tau) \\ \tilde{r}^*_{xy}(\tau) & r^*_{xy}(\tau) \end{bmatrix}. \tag{8.89}$$

Denoting the spectral process corresponding to $y(t)$ by $\upsilon(f)$, we can express the cross-PSD and cross-C-PSD as

$$P_{xy}(f)\mathrm{d}f = E[\mathrm{d}\xi(f)\mathrm{d}\upsilon^*(f)], \tag{8.90}$$

$$\widetilde{P}_{xy}(f)\mathrm{d}f = E[\mathrm{d}\xi(f)\mathrm{d}\upsilon(-f)]. \tag{8.91}$$

Instead of using the four complex-valued cross-(C-)PSDs $P_{xy}(f)$, $P_{xy}(-f)$, $\widetilde{P}_{xy}(f)$, and $\widetilde{P}_{xy}(-f)$ directly, we can also define the magnitude-squared coherences

$$|\rho_{xy}(\pm f)|^2 = \frac{|P_{xy}(\pm f)|^2}{P_{xx}(\pm f)P_{yy}(\pm f)}, \tag{8.92}$$

$$|\tilde{\rho}_{xy}(\pm f)|^2 = \frac{|\widetilde{P}_{xy}(\pm f)|^2}{P_{xx}(\pm f)P_{yy}(\mp f)}, \tag{8.93}$$

which are *rotational* and *reflectional* correlation coefficients, respectively. Let $\psi_{xy}(\pm f)$ denote the phase of $P_{xy}(\pm f)$, and $\tilde{\psi}_{xy}(\pm f)$ the phase of $\widetilde{P}_{xy}(\pm f)$. Then the joint analysis of $x(t)$ and $y(t)$ is performed using the quantities

- $(P_{xx}(\pm f), |\rho_{xx}(f)|^2, \psi_{xx}(f))$ to describe the individual properties of $x(t)$,
- $(P_{yy}(\pm f), |\rho_{yy}(f)|^2, \psi_{yy}(f))$ to describe the individual properties of $y(t)$, and
- $(|\rho_{xy}(\pm f)|^2, |\tilde{\rho}_{xy}(\pm f)|^2, \psi_{xy}(\pm f), \tilde{\psi}_{xy}(\pm f))$ to describe the joint properties of $x(t)$ and $y(t)$.

It is also possible to define a *total* coherence between $x(t)$ and $y(t)$ as a correlation coefficient between the vectors $[\mathrm{d}\xi(f), \mathrm{d}\xi^*(-f)]^{\mathrm{T}}$ and $[d\upsilon(f), d\upsilon^*(-f)]^{\mathrm{T}}$. This would proceed along the lines discussed in Chapter 4.

8.5 Higher-order spectra

So far in this chapter, we have considered only second-order properties. We now look at higher-order moments of a complex Nth-order stationary signal $x(t)$ that has moments defined and bounded up to Nth order. For $n \le N$, it is possible to define 2^n different nth-order moment functions, depending on where complex conjugate operators are placed:

$$m_{x,\diamond}(\boldsymbol{\tau}) = E\left[x^{\diamond_n}(t)\prod_{i=1}^{n-1} x^{\diamond_i}(t+\tau_i)\right]. \tag{8.94}$$

In this equation, $\boldsymbol{\tau} = [\tau_1, \ldots, \tau_{n-1}]^{\mathrm{T}}$ is a vector of time lags, and $\diamond = [\diamond_n, \diamond_1, \diamond_2, \ldots, \diamond_{n-1}]$ has elements $\diamond_i$ that are either 1 or the conjugating star $*$. For example, if $\diamond = [1, 1, *]$, then $m_{x,\diamond}(\boldsymbol{\tau}) = m_{xxx^*}(\tau_1, \tau_2) = E[x(t)x(t+\tau_1)x^*(t+\tau_2)]$. The number of stars in $\diamond$ will be denoted by q and the number of 1s by $n-q$. If $x(t)$ is Nth-order stationary, then all moments up to order N are functions of $\boldsymbol{\tau}$ only. It is possible that some moments do not depend on t, whereas others do. Such a process is not Nth-order stationary.

For $n = 2$, $m_{x,\diamond}(\boldsymbol{\tau})$ is a correlation function (Hermitian or complementary), for $n = 3$ it is a *bicorrelation* function, and for $n = 4$ it is a *tricorrelation* function. Many of the 2^n moments obtained for the 2^n possible conjugation patterns are redundant. First of all, moments with the same number of stars are equivalent because they are related to each other through a simple coordinate transformation. For example,

$$m_{xxx^*}(\tau_1, \tau_2) = m_{x^*xx}(\tau_1 - \tau_2, -\tau_2). \tag{8.95}$$

Secondly, moments with q stars are related to moments with $n-q$ stars through complex conjugation and possibly an additional coordinate transformation. For instance,

$$m_{xxx^*}(\tau_1, \tau_2) = m^*_{x^*xx^*}(\tau_2, \tau_1). \tag{8.96}$$

We call two moments *equivalent* if they can be expressed in terms of each other, and *distinct* if they cannot be expressed in terms of each other. There are $\lfloor n/2 \rfloor + 1$ distinct nth-order moment functions, where $\lfloor n/2 \rfloor$ denotes the greatest integer less than or equal to $n/2$. In general, all distinct nth-order moment functions are required for an nth-order description of $x(t)$.

Higher-order statistical analysis often uses cumulants rather than moments. For simplicity, we consider only moments here. However, the extension to cumulants is straightforward, using the connection between cumulants and moments given in Section 2.5.

8.5.1 Moment spectra and principal domains

We assume that the spectral process $\xi(f)$ corresponding to $x(t)$ is an Nth-order random function with moments defined and bounded up to Nth order. In the following, let $\diamond_i f_i$ denote $-f_i$ if $\diamond_i$ is the conjugating star $*$. The $(n-1)$-dimensional Fourier transform of $m_{x,\diamond}(\boldsymbol{\tau})$ is the nth-order moment spectrum $M_{x,\diamond}(\mathbf{f})$, with $\mathbf{f} = [f_1, f_2, \ldots, f_{n-1}]^{\mathrm{T}}$:

$$M_{x,\diamond}(\mathbf{f}) = \int_{\mathbb{R}^{n-1}} m_{x,\diamond}(\boldsymbol{\tau}) \mathrm{e}^{-\mathrm{j}2\pi \mathbf{f}^{\mathrm{T}}\boldsymbol{\tau}} \, \mathrm{d}^{n-1}\boldsymbol{\tau}. \tag{8.97}$$

This can be expressed in terms of the increments of the spectral process $\xi(f)$ as

$$M_{x,\diamond}(\mathbf{f})\mathrm{d}^{n-1}\mathbf{f} = E\left[\mathrm{d}\xi^{\diamond_n}(-\diamond_n \mathbf{f}^{\mathrm{T}}\mathbf{1}) \prod_{i=1}^{n-1} \mathrm{d}\xi^{\diamond_i}(\diamond_i f_i)\right], \tag{8.98}$$

with $\mathbf{1} = [1, 1, \ldots, 1]^{\mathrm{T}}$ and $\mathbf{f}^{\mathrm{T}}\mathbf{1} = f_1 + f_2 + \cdots + f_{n-1}$. The relationship (8.98) will be further illuminated in Section 9.5. For $n = 2$, $M_{x,\diamond}(\mathbf{f})$ is a power spectrum (the PSD or C-PSD), for $n = 3$ it is a *bispectrum*, and for $n = 4$ it is a *trispectrum*. Which ones of equivalent spectra to use is mainly a matter of taste, but each spectrum comes with its own symmetry properties. It is sufficient to compute a moment spectrum over a smaller region, called the *principal domain*, because the complete spectrum can be reconstructed from its symmetry properties. For complex signals, each moment spectrum with a different conjugation pattern has its own symmetries and therefore its own principal domain.

Example 8.4. We consider the bispectrum of a lowpass signal $x(t)$ that is limited to the frequency band $[-f_{\max}, f_{\max}]$. The two distinct third-order correlations we choose are $m_{xxx}(\tau_1, \tau_2)$ and $m_{x^*xx}(\tau_1, \tau_2)$. The corresponding frequency-domain expressions are

$$M_{xxx}(f_1, f_2)\mathrm{d}f_1\,\mathrm{d}f_2 = E[\mathrm{d}\xi(f_1)\mathrm{d}\xi(f_2)\mathrm{d}\xi(-f_1 - f_2)],$$

$$M_{x^*xx}(f_1, f_2)\mathrm{d}f_1\,\mathrm{d}f_2 = E[\mathrm{d}\xi(f_1)\mathrm{d}\xi(f_2)\mathrm{d}\xi^*(f_1 + f_2)].$$

We first look at $M_{xxx}(f_1, f_2)$. The support of $M_{xxx}(f_1, f_2)$ is the overlap of the support of $\mathrm{d}\xi(f_1)\mathrm{d}\xi(f_2)$, the square in Fig. 8.3(a), with the support of $\mathrm{d}\xi(-f_1 - f_2)$, the northwest to southeast corridor between the two solid 45° lines. The lines of symmetry are $f_2 = f_1$, $f_2 = -2f_1$, and $f_2 = -\frac{1}{2}f_1$. Thus, the principal domain of $M_{xxx}(f_1, f_2)$ is the gray region in Fig. 8.3(a). Now consider $M_{x^*xx}(f_1, f_2)$. This bispectrum has only one line of symmetry, which is $f_2 = f_1$. Thus, its principal domain is the gray polygon in Fig. 8.3(b).

If $x(t)$ is real, the increments of its spectral process satisfy the Hermitian symmetry $\mathrm{d}\xi(f) = \mathrm{d}\xi^*(-f)$. This means that the bispectrum of a real signal has many more symmetry relations, and thus a smaller principal domain. For comparison, the principal domain of the bispectrum for a real signal is shown in Fig. 8.3(c).[6]

8.5.2 Analytic signals

Now consider the WSS analytic signal $x(t) = u(t) + \mathrm{j}\hat{u}(t)$, where $\hat{u}(t)$ is the Hilbert transform of the real signal $u(t)$. The increments of the spectral processes corresponding to $x(t)$ and $u(t)$ are related by $\mathrm{d}\xi(f) = 2\Gamma(f)\mathrm{d}\eta(f)$, where $\Gamma(f)$ is the Heaviside unit-step function. We thus find that the nth-order moment spectra of complex $x(t)$ are connected to the nth-order moment spectrum of its corresponding real signal $u(t)$ as

$$M_{x,\diamond}(\mathbf{f}) = 2^n\Gamma(-\diamond_n\,\mathbf{f}^{\mathrm{T}}\mathbf{1})\prod_{i=1}^{n-1}\Gamma(\diamond_i f_i)M_u(\mathbf{f}). \tag{8.99}$$

This shows that $M_{x,\diamond}(\mathbf{f})$ cuts out regions of $M_u(\mathbf{f})$. Therefore, on its nonzero domain, $M_{x,\diamond}(\mathbf{f})$ also inherits the symmetries of $M_u(\mathbf{f})$. The conjugation pattern determines the selected region and thus the symmetry properties. For example, the pattern $\diamond = [*, 1, 1, \ldots, 1]$ selects the orthant $f_1 \geq 0, f_2 \geq 0, \ldots, f_{n-1} \geq 0$.

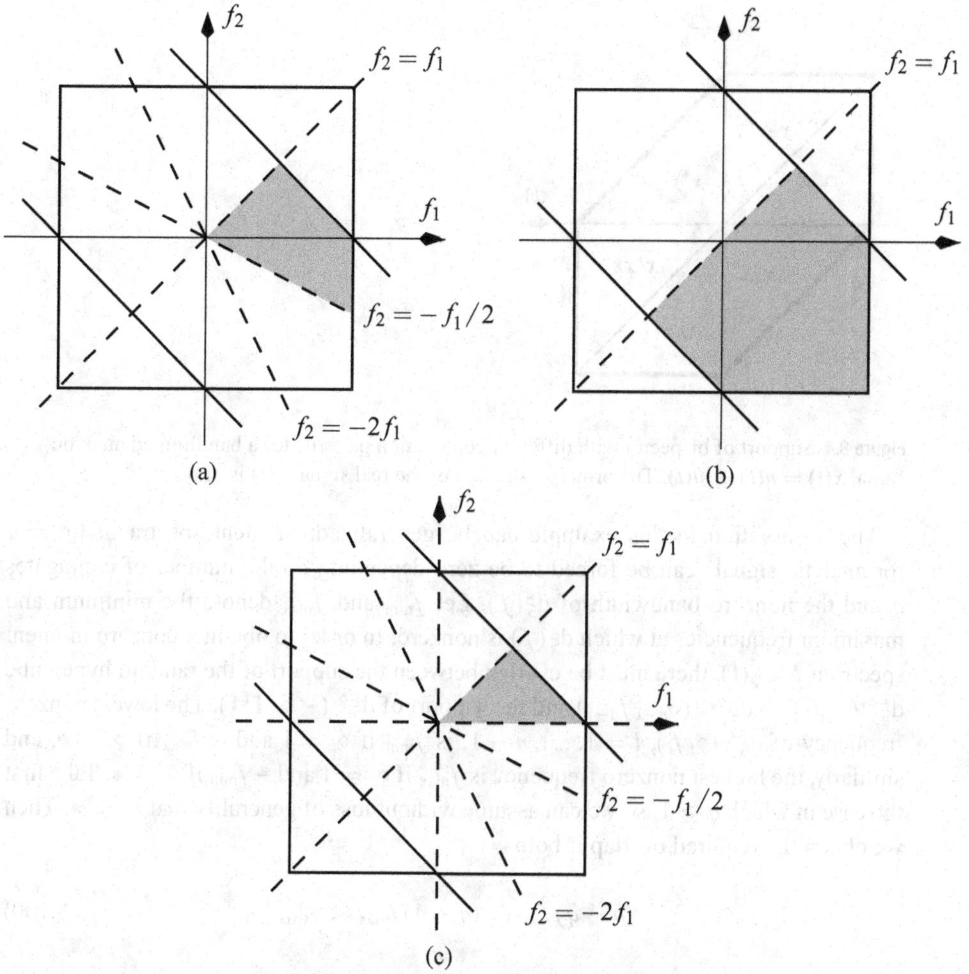

Figure 8.3 Principal domains (gray) of different bispectra: (a) $M_{xxx}(f_1, f_2)$, (b) $M_{x^*xx}(f_1, f_2)$, and (c) for a real signal. Dashed lines are lines of symmetries.

Example 8.5. Figure 8.4 shows the support of bispectra of a bandlimited analytic signal $x(t)$ with different conjugation patterns. All the conjugation patterns shown are equivalent, meaning that every bispectrum with one or two stars ($q = 1, 2$) contains the same information. Yet since $M_{x^*xx}(f_1, f_2)$ covers the principal domain of $M_{xxx}(f_1, f_2)$ in its most common definition, gray in the figure, one could regard x^*xx as the canonical starring pattern.

Because $\Gamma(f_1)\Gamma(f_2)\Gamma(-f_1 - f_2) \equiv 0$ and $\Gamma(-f_1)\Gamma(-f_2)\Gamma(f_1 + f_2) \equiv 0$, the support for $M_{xxx}(f_1, f_2)$ and $M_{x^*x^*x^*}(f_1, f_2)$ is empty. In other words, if $x(t)$ is a WSS analytic signal, bispectra with $q = 0$ or $q = 3$ are identically zero.

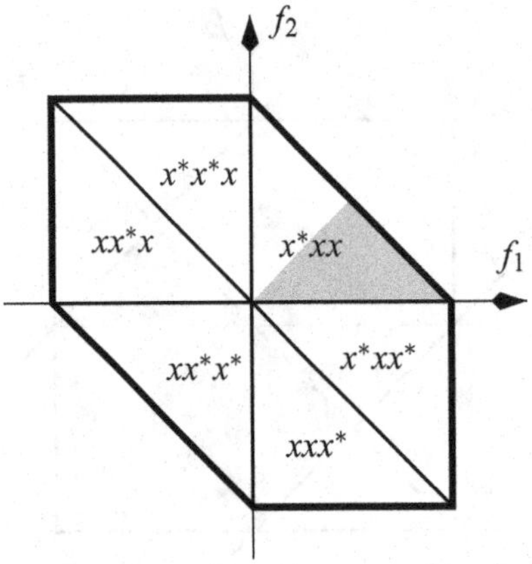

Figure 8.4 Support of bispectra with different conjugation patterns for a bandlimited analytic signal $x(t) = u(t) + j\hat{u}(t)$. The principal domain of the real signal $u(t)$ is gray.

The observation in this example can be generalized: moment spectra of order n for analytic signals can be forced to be zero depending on the number of conjugates q and the nonzero bandwidth of $d\xi(f)$. Let $f_{\min}$ and $f_{\max}$ denote the minimum and maximum frequencies at which $d\xi(f)$ is nonzero. In order to obtain a nonzero moment spectrum $M_{x,\diamond}(\mathbf{f})$, there must be overlap between the support of the random hypercube $d\xi^{\diamond_1}(\diamond_1 f_1)\cdots d\xi^{\diamond_{n-1}}(\diamond_{n-1} f_{n-1})$ and the support of $d\xi^{\diamond_n}(-\diamond_n \mathbf{f}^T\mathbf{1})$. The lowest nonzero frequency of $d\xi^{\diamond_i}(\diamond_i f_i)$, $i = 1, \ldots, n-1$, is $f_{\min}$ if $\diamond_i = 1$ and $-f_{\max}$ if $\diamond_i = *$, and similarly, the highest nonzero frequency is $f_{\max}$ if $\diamond_i = 1$ and $-f_{\min}$ if $\diamond_i = *$. Take first the case in which $q \geq 1$, so we can assume without loss of generality that $\diamond_n = *$. Then we obtain the required overlap if both

$$(n-q)f_{\min} - (q-1)f_{\max} < f_{\max} \tag{8.100}$$

and

$$(n-q)f_{\max} - (q-1)f_{\min} > f_{\min}. \tag{8.101}$$

Now, if $q = 0$, then $\diamond_n = 1$ and we require

$$(n-1)f_{\min} < -f_{\min} \tag{8.102}$$

and

$$(n-1)f_{\max} > -f_{\max}, \tag{8.103}$$

which shows that (8.100) and (8.101) also hold for $q = 0$. Since one of the inequalities (8.100) and (8.101) will always be trivially satisfied, a simple *necessary* (not sufficient) condition for a nonzero $M_{x,\diamond}(\mathbf{f})$ is

$$\begin{aligned} (n-q)f_{\min} &< qf_{\max}, \quad \text{if } 2q \leq n, \\ (n-q)f_{\max} &> qf_{\min}, \quad \text{if } 2q > n. \end{aligned} \tag{8.104}$$

If either $q = 0$ or $q = n$, this condition requires $f_{\min} < 0$, which is impossible for an analytic signal. Thus, moment spectra of analytic signals with $q = 0$ or $q = n$ must be identically zero.[7]

Example 8.6. For a WSS analytic signal $x(t)$, evaluating (8.100) and (8.101) for $n = 4$ confirms that trispectra with $q = 0$ or $q = 4$ are zero. A necessary condition for trispectra with $q = 1$ or $q = 3$ to be nonzero is $3f_{\min} < f_{\max}$, and trispectra with $q = 2$ cannot be identically zero for a nonzero signal.

For a real WSS signal $u(t)$, we may thus calculate the fourth-order moment $Eu^4(t) = m_{xxxx}(0, 0, 0)$ from the analytic signal $x(t) = u(t) + j\hat{u}(t)$ as

$$Eu^4(t) = \tfrac{1}{16}E[x(t) + x^*(t)]^4 = \tfrac{1}{2}\,\mathrm{Re}\,\{E[x^3(t)x^*(t)]\} + \tfrac{3}{8}E|x(t)|^4,$$

taking into account that $Ex^4(t) = 0$ and $Ex^{*4}(t) = 0$. If $3f_{\min} > f_{\max}$, tricorrelations with $q = 1$ or $q = 3$ must also be identically zero, in which case

$$Eu^4(t) = \tfrac{3}{8}E|x(t)|^4.$$

The term $E[x^3(t)x^*(t)]$ is equal to the integral of a trispectrum for one particular conjugation pattern with $q = 1$ or $q = 3$, and $E\,|x(t)|^4$ is the integral of a trispectrum for one particular conjugation pattern with $q = 2$.

Reversing the inequalities in (8.104) leads to a sufficient condition for a zero moment spectrum. Let us assume that $2q \le n$, which constitutes no loss of generality since moment spectra with q and $n - q$ conjugates are equivalent. If (8.104) holds for $q - 1$ and some fixed order n, it also holds for q. Thus, if trispectra with q conjugates are zero, so are trispectra with $q - 1$ conjugates. We may therefore ask up to which order N a signal with given $f_{\min}$ and $f_{\max}$ is circular. Remember that a signal is Nth-order circular (Nth-order proper) if its only nonzero moment functions up to order N have an equal number of conjugated and nonconjugated terms, and hence all odd-order moments (up to order N) are zero. Moreover, if a signal is circular up to order $N = 2m - 1$, then it is also circular for order $N = 2m$. Therefore, N can be assumed even without loss of generality. If $x(t)$ is Nth-order circular, all moment spectra with $q \le N/2 - 1$ are zero but (8.104) does not hold for $q = N/2 - 1$. This immediately leads to

$$N \le 2\left\lfloor \frac{f_{\max} + f_{\min}}{f_{\max} - f_{\min}} \right\rfloor, \tag{8.105}$$

where $\lfloor\cdot\rfloor$ again denotes the floor function.[8]

Notes

1 The multivariate spectral factorization problem was first addressed by Wiener and Masani (1957, 1958), who provided the proof that a causal factorization exists for continuous-time spectral density matrices. Further developments were reported by Youla (1961) and Rozanov

(1963), who provide general algorithms that perform spectral factorization on rational spectral-density matrices for continuous-time WSS processes. Several spectral-factorization algorithms for discrete-time spectral-density matrices have been proposed by Davis (1963), Wilson (1972), and Jezek and Kucera (1985).

2 The seminal paper that first investigated the coherence properties of partially polarized light is Wolf (1959), and an extended discussion is provided by Born and Wolf (1999). The rotary component method was introduced in oceanography by Gonella (1972) and Mooers (1973). Other key papers that deal with polarization and coherence are Jones (1979) and Samson (1980).

3 The presentation in Section 8.4.1 follows Schreier (2008b). This paper is ©IEEE, and portions are used with permission. A similar discussion is also provided by Lilly and Gascard (2006).

4 There are other interesting signal decompositions for a given frequency f besides the one into a completely polarized and an unpolarized signal. Rubin-Delanchy and Walden (2008) present a decomposition into two ellipses whose orientations are orthogonal, but with the same aspect ratio. The magnitude-squared coherence between CCW and CW rotary components controls the relative influences of the two ellipses. If $|\widetilde{\rho}_{xx}(f)|^2 = 1$, then the second ellipse vanishes. If $|\widetilde{\rho}_{xx}(f)|^2 = 0$, then the two ellipses have equal influence.

5 The Stokes parameters were introduced by Stokes (1852), and the Jones parameters and calculus by Jones (1941). Mueller calculus was invented by Hans Müller (whose last name is commonly transcribed as "Mueller") at MIT in the early 1940s. Interestingly, the first time Müller published work using his calculus was only several years later, in Mueller (1948).

6 The symmetry relations of the bispectrum for real signals are detailed by Pflug *et al.* (1992) and Molle and Hinich (1995). The bispectrum of complex signals has been discussed by Jouny and Moses (1992), but they did not consider the bispectrum with no conjugated terms. Higher-order spectra with all distinct conjugation patterns were discussed by Schreier and Scharf (2006b).

7 The principal domains of the bispectrum and trispectrum of real signals were derived by Pflug *et al.* (1992). Necessary conditions for nonzero moment spectra of analytic signals were given by Picinbono (1994) and Amblard *et al.* (1996b). This discussion was extended by Izzo and Napolitano (1997) to necessary conditions for nonzero spectra of equivalent complex baseband signals.

8 A thorough discussion of circular random signals, including the interplay between circularity and stationarity, is provided by Picinbono (1994). The result that connects Nth-order circularity to the bandwidth of analytic signals is also due to Picinbono (1994), and was derived by Amblard *et al.* (1996b) as well.

9 Nonstationary processes

Wide-sense stationary (WSS) processes admit a spectral representation (see Result 8.1) in terms of the Fourier basis, which allows a frequency interpretation. The transform-domain description of a WSS signal $x(t)$ is a spectral process $\xi(f)$ with *orthogonal* increments $d\xi(f)$. For nonstationary signals, we have to sacrifice either the Fourier basis, and thus its frequency interpretation, or the orthogonality of the transform-domain representation. We will discuss both possibilities.

The Karhunen–Loève (KL) expansion uses an orthonormal basis other than the Fourier basis but retains the orthogonality of the transform-domain description. The KL expansion is applied to a continuous-time signal of finite duration, which means that its transform-domain description is a *countably* infinite number of orthogonal random coefficients. This is analogous to the Fourier series, which produces a countably infinite number of Fourier coefficients, as opposed to the Fourier transform, which is applied to an infinite-duration continuous-time signal. The KL expansion presented in Section 9.1 takes into account the complementary covariance of an improper signal. It can be considered the continuous-time equivalent of the eigenvalue decomposition of improper random vectors discussed in Section 3.1.

An alternative approach is the Cramér–Loève (CL) spectral representation, which retains the Fourier basis and its frequency interpretation but sacrifices the orthogonality of the increments $d\xi(f)$. As discussed in Section 9.2, the increments $d\xi(f)$ of the spectral process of an improper signal can have nonzero Hermitian correlation and complementary correlation between different frequencies. Starting from the CL representation, we introduce energy and power spectral densities for nonstationary signals. We then discuss the CL representation for analytic signals and discrete-time signals.

Yet another description, which allows deep insights into the time-varying nature of nonstationary signals, is possible in the joint time–frequency domain. In Section 9.3, we focus our attention on the Rihaczek distribution, which is a member of Cohen's class of bilinear time–frequency distributions. The Rihaczek distribution is not as widely used as, for instance, the Wigner–Ville distribution, but it possesses a compelling property: it is an inner product between the spectral increments and the time-domain process at a given point in the time–frequency plane. This property leads to an evocative geometrical interpretation. It is also the basis for extending the rotary-component and polarization analysis presented in the previous chapter to nonstationary signals. This is discussed in Section 9.4. Finally, Section 9.5 presents a short exposition of higher-order statistics for nonstationary signals.

9.1 Karhunen–Loève expansion

We first consider a representation for finite-length continuous-time signals. Like the Fourier series, this representation produces a countably infinite number of uncorrelated random coefficients. However, because it uses an orthonormal basis other than the Fourier basis it generally does not afford these random coefficients a frequency-domain interpretation. The improper KL expansion is the continuous-time equivalent of the finite-dimensional improper eigenvalue decomposition, which is given in Result 3.1.[1]

Result 9.1. *Suppose that* $\{x(t),\ 0 \le t \le T\}$ *is a zero-mean second-order complex random process with augmented covariance function* $\underline{\mathbf{R}}_{xx}(t_1, t_2)$*, where both the covariance function* $r_{xx}(t_1, t_2) = E[x(t_1)x^*(t_2)]$ *and the complementary covariance function* $\tilde{r}_{xx}(t_1, t_2) = E[x(t_1)x(t_2)]$ *are continuous on* $[0, T] \times [0, T]$*. Then* $\underline{\mathbf{R}}_{xx}(t_1, t_2)$ *can be expanded in the* Mercer series

$$\underline{\mathbf{R}}_{xx}(t_1, t_2) = \sum_{n=1}^{\infty} \underline{\mathbf{\Phi}}_n(t_1)\underline{\mathbf{\Lambda}}_n\underline{\mathbf{\Phi}}_n^{\mathrm{H}}(t_2), \tag{9.1}$$

which converges uniformly in t_1 *and* t_2*. The augmented eigenvalue matrix* $\underline{\mathbf{\Lambda}}_n$ *is real and it contains two nonnegative eigenvalues* λ_{2n-1} *and* λ_{2n}*:*

$$\underline{\mathbf{\Lambda}}_n = \frac{1}{2}\begin{bmatrix} \lambda_{2n-1} + \lambda_{2n} & \lambda_{2n-1} - \lambda_{2n} \\ \lambda_{2n-1} - \lambda_{2n} & \lambda_{2n-1} + \lambda_{2n} \end{bmatrix}. \tag{9.2}$$

The augmented eigenfunction matrix

$$\underline{\mathbf{\Phi}}_n(t) = \begin{bmatrix} \phi_n(t) & \widetilde{\phi}_n(t) \\ \widetilde{\phi}_n^*(t) & \phi_n^*(t) \end{bmatrix} \tag{9.3}$$

satisfies the orthogonality condition

$$\int_0^T \underline{\mathbf{\Phi}}_n^{\mathrm{H}}(t)\underline{\mathbf{\Phi}}_m(t)\mathrm{d}t = \mathbf{I}\,\delta_{nm}. \tag{9.4}$$

The matrices $\underline{\mathbf{\Lambda}}_n$ *and* $\underline{\mathbf{\Phi}}_n(t)$ *are found as the solutions to the equation*

$$\underline{\mathbf{\Phi}}_n(t_1)\underline{\mathbf{\Lambda}}_n = \int_0^T \underline{\mathbf{R}}_{xx}(t_1, t_2)\underline{\mathbf{\Phi}}_n(t_2)\mathrm{d}t_2. \tag{9.5}$$

Then $x(t)$ *can be represented by the* Karhunen–Loève (KL) expansion

$$\underline{\mathbf{x}}(t) = \sum_{n=1}^{\infty} \underline{\mathbf{\Phi}}_n(t)\underline{\mathbf{x}}_n \Leftrightarrow x(t) = \sum_{n=1}^{\infty} \phi_n(t)x_n + \widetilde{\phi}_n(t)x_n^*, \tag{9.6}$$

where equality holds in the mean-square sense, and convergence is uniform in t*. The complex KL coefficients* x_n *are given by*

$$\underline{\mathbf{x}}_n = \int_0^T \underline{\mathbf{\Phi}}_n^{\mathrm{H}}(t)\underline{\mathbf{x}}(t)\mathrm{d}t \Leftrightarrow x_n = \int_0^T \left[\phi_n^*(t)x(t) + \widetilde{\phi}_n(t)x^*(t)\right]\mathrm{d}t. \tag{9.7}$$

The KL coefficients are improper, with covariance and complementary covariance

$$E(x_n x_m^*) = \tfrac{1}{2}(\lambda_{2n-1} + \lambda_{2n})\delta_{nm}, \tag{9.8}$$

$$E(x_n x_m) = \tfrac{1}{2}(\lambda_{2n-1} - \lambda_{2n})\delta_{nm}. \tag{9.9}$$

The proof of this result provides some interesting insights. Let $\mathbb{C}_*^2$ be the image of $\mathbb{R}^2$ under the map

$$\mathbf{T} = \begin{bmatrix} 1 & \mathrm{j} \\ 1 & -\mathrm{j} \end{bmatrix}. \tag{9.10}$$

The space of augmented square-integrable functions defined on $[0, T]$ is denoted by $L^2([0, T], \mathbb{C}_*^2)$. This space is $\mathbb{R}$-linear (i.e., $\mathbb{C}$-widely linear) but not $\mathbb{C}$-linear.

Using the results of Kelly and Root (1960) for vector-valued random processes, we can write down the KL expansion for an augmented signal. Let the assumptions be as in the statement of Result 9.1. Then the augmented covariance matrix $\underline{\mathbf{R}}_{xx}(t_1, t_2)$ can be expanded in the uniformly convergent series (called *Mercer's expansion*)

$$\underline{\mathbf{R}}_{xx}(t_1, t_2) = \sum_{n=1}^{\infty} \lambda_n \underline{\mathbf{f}}_n(t_1)\underline{\mathbf{f}}_n^{\mathrm{H}}(t_2), \tag{9.11}$$

where $\{\lambda_n\}_{n=1}^{\infty}$ are the nonnegative scalar eigenvalues and $\{\underline{\mathbf{f}}_n(t) = [f_n(t), f_n^*(t)]^{\mathrm{T}}\}$ are the corresponding orthonormal augmented eigenfunctions. Each $\underline{\mathbf{f}}_n(t)$ is $L^2([0, T], \mathbb{C}_*^2)$. Eigenvalues and eigenfunctions are obtained as solutions to the integral equation

$$\lambda_n \underline{\mathbf{f}}_n(t_1) = \int_0^T \underline{\mathbf{R}}_{xx}(t_1, t_2)\underline{\mathbf{f}}_n(t_2)\mathrm{d}t_2, \quad 0 \le t_1 \le T, \tag{9.12}$$

where the augmented eigenfunctions $\underline{\mathbf{f}}_n(t)$ are orthonormal in $L^2([0, T], \mathbb{C}_*^2)$:

$$\langle \underline{\mathbf{f}}_n(t), \underline{\mathbf{f}}_m(t)\rangle = \int_0^T \underline{\mathbf{f}}_n^{\mathrm{H}}(t)\underline{\mathbf{f}}_m(t)\mathrm{d}t = 2\,\mathrm{Re}\int_0^T f_n^*(t) f_m(t)\mathrm{d}t = \delta_{nm}. \tag{9.13}$$

Then $\underline{\mathbf{x}}(t) \Leftrightarrow x(t)$ can be represented by the series

$$\underline{\mathbf{x}}(t) = \sum_{n=1}^{\infty} u_n \underline{\mathbf{f}}_n(t) \Leftrightarrow x(t) = \sum_{n=1}^{\infty} u_n f_n(t), \tag{9.14}$$

where equality holds in the mean-square sense and convergence is uniform in t. The KL coefficients are

$$u_n = \langle \underline{\mathbf{f}}_n(t), \underline{\mathbf{x}}(t)\rangle = \int_0^T \underline{\mathbf{f}}_n^{\mathrm{H}}(t)\underline{\mathbf{x}}(t)\mathrm{d}t = 2\,\mathrm{Re}\int_0^T f_n^*(t)x(t)\mathrm{d}t. \tag{9.15}$$

The surprising result here is that these coefficients are *real* scalars with covariance

$$E(u_n u_m) = \lambda_n \delta_{nm}. \tag{9.16}$$

The reason why the coefficients are real is found in (9.13). It shows that the functions $f_n(t)$ do not have to be orthogonal in $L^2([0, T], \mathbb{C})$ to ensure that the augmented functions $\underline{\mathbf{f}}_n(t)$ be orthogonal in $L^2([0, T], \mathbb{C}_*^2)$. In fact, there are *twice* as many orthogonal augmented functions $\underline{\mathbf{f}}_n(t)$ in $L^2([0, T], \mathbb{C}_*^2)$ as there are orthogonal functions $f_n(t)$ in $L^2([0, T], \mathbb{C})$.

That's why we have been able to reduce the dimension of the internal description by a factor of 2 (real rather than complex KL coefficients).

From (9.11) it is not clear how the improper version of Mercer's expansion specializes to its proper version. To make this connection apparent, we rewrite (9.11) by combining terms with $2n-1$ and $2n$ as

$$\begin{aligned}\underline{\mathbf{R}}_{xx}(t_1,t_2) &= \sum_{n=1}^{\infty}\left\{\left([\underline{\mathbf{f}}_{2n-1}(t_1),\underline{\mathbf{f}}_{2n}(t_1)]\frac{\mathbf{T}^{\mathrm{H}}}{\sqrt{2}}\right)\left(\frac{\mathbf{T}}{\sqrt{2}}\begin{bmatrix}\lambda_{2n-1} & 0\\ 0 & \lambda_{2n}\end{bmatrix}\frac{\mathbf{T}^{\mathrm{H}}}{\sqrt{2}}\right)\right.\\ &\quad\left.\times\left(\frac{\mathbf{T}}{\sqrt{2}}\begin{bmatrix}\underline{\mathbf{f}}^{\mathrm{H}}_{2n-1}(t_2)\\ \underline{\mathbf{f}}^{\mathrm{H}}_{2n}(t_2)\end{bmatrix}\right)\right\}\\ &= \sum_{n=1}^{\infty}\begin{bmatrix}\phi_n(t_1) & \widetilde{\phi}_n(t_1)\\ \widetilde{\phi}^*_n(t_1) & \phi^*_n(t_1)\end{bmatrix}\begin{bmatrix}\frac{1}{2}(\lambda_{2n-1}+\lambda_{2n}) & \frac{1}{2}(\lambda_{2n-1}-\lambda_{2n})\\ \frac{1}{2}(\lambda_{2n-1}-\lambda_{2n}) & \frac{1}{2}(\lambda_{2n-1}+\lambda_{2n})\end{bmatrix}\begin{bmatrix}\phi^*_n(t_2) & \widetilde{\phi}_n(t_2)\\ \widetilde{\phi}^*_n(t_2) & \phi_n(t_2)\end{bmatrix}\\ &= \sum_{n=1}^{\infty}\underline{\mathbf{\Phi}}_n(t_1)\underline{\mathbf{\Lambda}}_n\underline{\mathbf{\Phi}}^{\mathrm{H}}_n(t_2),\end{aligned} \tag{9.17}$$

where

$$\phi_n(t) = \frac{1}{\sqrt{2}}[f_{2n-1}(t) - \mathrm{j}f_{2n}(t)], \tag{9.18}$$

$$\widetilde{\phi}_n(t) = \frac{1}{\sqrt{2}}[f_{2n-1}(t) + \mathrm{j}f_{2n}(t)]. \tag{9.19}$$

Thus, the latent representation is now given by *complex* KL coefficients

$$x_n = \frac{1}{\sqrt{2}}(u_{2n-1} + \mathrm{j}u_{2n}), \tag{9.20}$$

$$\underline{\mathbf{x}}_n = \begin{bmatrix}x_n\\ x^*_n\end{bmatrix} = \frac{\mathbf{T}}{\sqrt{2}}\begin{bmatrix}u_{2n-1}\\ u_{2n}\end{bmatrix} = \int_0^T \underline{\mathbf{\Phi}}^{\mathrm{H}}_n(t)\underline{\mathbf{x}}(t)\mathrm{d}t. \tag{9.21}$$

For these coefficients we find, because of (9.16),

$$\begin{aligned}E(x_nx_m) &= \tfrac{1}{2}[E(u_{2n-1}u_{2m-1}) - E(u_{2n}u_{2m}) + \mathrm{j}E(u_{2n-1}u_{2m}) + \mathrm{j}E(u_{2n}u_{2m-1})]\\ &= \tfrac{1}{2}(\lambda_{2n-1}-\lambda_{2n})\delta_{nm}\end{aligned} \tag{9.22}$$

and $E(x_nx^*_m) = \frac{1}{2}(\lambda_{2n-1}+\lambda_{2n})\delta_{nm}$. This completes the proof.

If $x(t)$ is *proper*, $\tilde{r}_{xx}(t_1,t_2) \equiv 0$, the KL expansion simplifies because $\lambda_{2n-1} = \lambda_{2n}$ and $\widetilde{\phi}_n(t) \equiv 0$ for all n, and the KL coefficients x_n are proper. In the proper case, the Mercer series expands the *Hermitian* covariance function as

$$r_{xx}(t_1,t_2) = \sum_{n=1}^{\infty}\mu_n\phi_n(t_1)\phi^*_n(t_2). \tag{9.23}$$

Here $\{\mu_n\}$ are the eigenvalues of the kernel $r_{xx}(t_1,t_2)$, which are related to the eigenvalues of the diagonal kernel $\underline{\mathbf{R}}_{xx}(t_1,t_2)$ as $\mu_n = \lambda_{2n-1} = \lambda_{2n}$.

In the proper or improper case, finding the eigenvalues and eigenfunctions of a linear integral equation with arbitrary kernel in (9.12) ranges from difficult to impossible. An alternative is the Rayleigh–Ritz technique, which numerically solves operator equations. The Rayleigh–Ritz technique is discussed by Chen *et al.* (1997); Navarro-Moreno *et al.* (2006) have applied this technique to obtain approximate series expansions of stochastic processes.

Example 9.1. To gain more insight into the improper KL expansion, consider the following communications example. Suppose we want to detect a *real* waveform $u(t)$ that is transmitted over a channel that rotates it by some random phase ϕ and adds complex white Gaussian noise $n(t)$. The observations are then given by

$$y(t) = u(t)e^{j\phi} + n(t), \tag{9.24}$$

where we assume pairwise mutual independence of $u(t)$, $n(t)$, and ϕ. Furthermore, we denote the rotated signal by $x(t) = u(t)e^{j\phi}$. Its covariance is given by $r_{xx}(t_1, t_2) = E[u(t_1)u(t_2)]$ and its complementary covariance is $\tilde{r}_{xx}(t_1, t_2) = E[u(t_1)u(t_2)] \cdot Ee^{j2\phi}$.

There are two important special cases. If the phase ϕ is uniformly distributed, then $\tilde{r}_{xx}(t_1, t_2) \equiv 0$ and detection is *noncoherent*. The eigenvalues of the augmented covariance of $x(t)$ satisfy $\lambda_{2n-1} = \lambda_{2n} = \mu_n$. On the other hand, if ϕ is known, then $\tilde{r}_{xx}(t_1, t_2) \equiv e^{j2\phi} r_{xx}(t_1, t_2)$ and detection is *coherent*. If we order the eigenvalues appropriately, we have $\lambda_{2n-1} = 2\mu_n$ and $\lambda_{2n} = 0$. Therefore, the coherent case is the most improper case under the power constraint $\lambda_{2n-1} + \lambda_{2n} = 2\mu_n$. These comments are clarified by noting that

$$\tfrac{1}{2}\lambda_{2n-1} = E\{x(t)\mathrm{Re}\, x_n\}, \qquad \tfrac{j}{2}\lambda_{2n} = E\{x(t)\mathrm{Im}\, x_n\}. \tag{9.25}$$

Thus, λ_{2n-1} measures the covariance between the real part of the observable coordinate x_n and the continuous-time signal, and λ_{2n} does so for the imaginary part.

In the noncoherent version of (9.24), these two covariances are equal, suggesting that the information is carried equally in the real and imaginary parts of x_n. In the coherent version, $\lambda_{2n-1} = 2\mu_n$, $\lambda_{2n} = 0$ shows that the information is carried exclusively in the real part of x_n, making $\mathrm{Re}\, x_n$ a *sufficient statistic* for the decision on $x(t)$. Therefore, in the coherent problem, WL processing amounts to considering only the real part of the internal description. The more interesting applications of WL filtering, however, lie between the coherent and the noncoherent case, being characterized by a nonuniform phase distribution, or in adaptive realizations of coherent algorithms.

9.1.1 Estimation

The KL expansion can be used to solve the following problem: widely linearly estimate a nonstationary improper complex zero-mean random signal $x(t)$ with augmented covariance $\underline{\mathbf{R}}_{xx}(t_1, t_2)$ in complex white (i.e., proper) noise $n(t)$ with power-spectral density N_0. The observations are

$$y(t) = x(t) + n(t), \quad 0 \le t \le T, \tag{9.26}$$

and the noise will be assumed to be uncorrelated with the signal. We are looking for a widely linear (WL) estimator $\hat{\underline{\mathbf{x}}}(t) \Leftrightarrow \hat{x}(t)$ of the form

$$\hat{\underline{\mathbf{x}}}(t) = \int_0^T \underline{\mathbf{H}}(t, v)\underline{\mathbf{y}}(v)\mathrm{d}v, \tag{9.27}$$

with augmented filter impulse response

$$\underline{\mathbf{H}}(t, v) = \begin{bmatrix} h_1(t, v) & h_2(t, v) \\ h_2^*(t, v) & h_1^*(t, v) \end{bmatrix}. \tag{9.28}$$

Thus,

$$\hat{x}(t) = \int_0^T h_1(t, v)y(v)\mathrm{d}v + \int_0^T h_2(t, v)y^*(v)\mathrm{d}v. \tag{9.29}$$

To make this estimator a widely linear minimum mean-squared error (WLMMSE) estimator, we require that it satisfy the orthogonality condition

$$[\underline{\mathbf{x}}(t) - \hat{\underline{\mathbf{x}}}(t)] \perp \underline{\mathbf{y}}(u), \quad \forall (t, u) \in [0, T] \times [0, T], \tag{9.30}$$

which translates to

$$\underline{\mathbf{R}}_{xx}(t, u) = \int_0^T \underline{\mathbf{H}}(t, v)\underline{\mathbf{R}}_{yy}(v, u)\mathrm{d}v. \tag{9.31}$$

Using the KL expansion of Result 9.1 in this equation, we obtain

$$\sum_{n=1}^{\infty} \underline{\mathbf{\Phi}}_n(t)\underline{\mathbf{\Lambda}}_n\underline{\mathbf{\Phi}}_n^{\mathrm{H}}(u) = \int_0^T \underline{\mathbf{H}}(t, v)\left[\sum_{n=1}^{\infty} \underline{\mathbf{\Phi}}_n(v)(\underline{\mathbf{\Lambda}}_n + N_0\mathbf{I})\underline{\mathbf{\Phi}}_n^{\mathrm{H}}(u)\right]\mathrm{d}v. \tag{9.32}$$

We now attempt a solution of the form

$$\underline{\mathbf{H}}(t, v) = \sum_{n=1}^{\infty} \underline{\mathbf{\Phi}}_n(t)\underline{\mathbf{H}}_n\underline{\mathbf{\Phi}}_n^{\mathrm{H}}(v), \tag{9.33}$$

$$\underline{\mathbf{H}}_n = \begin{bmatrix} h_{n,1} & h_{n,2} \\ h_{n,2}^* & h_{n,1}^* \end{bmatrix}. \tag{9.34}$$

Inserting (9.33) into (9.32) we get

$$\sum_{n=1}^{\infty} \underline{\mathbf{\Phi}}_n(t)\underline{\mathbf{\Lambda}}_n\underline{\mathbf{\Phi}}_n^{\mathrm{H}}(u) = \sum_{n=1}^{\infty} \underline{\mathbf{\Phi}}_n(t)\underline{\mathbf{H}}_n(\underline{\mathbf{\Lambda}}_n + N_0\mathbf{I})\underline{\mathbf{\Phi}}_n^{\mathrm{H}}(u), \tag{9.35}$$

which means

$$\underline{\mathbf{H}}_n = \underline{\mathbf{\Lambda}}_n(\underline{\mathbf{\Lambda}}_n + N_0\mathbf{I})^{-1}. \tag{9.36}$$

Thus, the terms of $\underline{\mathbf{H}}_n$ are

$$h_{n,1} = \frac{\lambda_{2n-1}\lambda_{2n} + (N_0/2)(\lambda_{2n-1} + \lambda_{2n})}{\lambda_{2n-1}\lambda_{2n} + N_0(\lambda_{2n-1} + \lambda_{2n}) + N_0^2}, \tag{9.37}$$

$$h_{n,2} = \frac{(N_0/2)(\lambda_{2n-1} - \lambda_{2n})}{\lambda_{2n-1}\lambda_{2n} + N_0(\lambda_{2n-1} + \lambda_{2n}) + N_0^2}. \tag{9.38}$$

If we denote the resolution of the observed signal onto the KL basis functions by

$$\underline{\mathbf{y}}_n = \int_0^T \underline{\mathbf{\Phi}}_n^{\mathrm{H}}(t)\underline{\mathbf{y}}(t)\mathrm{d}t \Leftrightarrow y_n = \int_0^T \left[\phi_n^*(t)y(t) + \widetilde{\phi}_n(t)y^*(t)\right] \mathrm{d}t, \tag{9.39}$$

the WL estimator in (9.27) is

$$\hat{\underline{\mathbf{x}}}(t) = \sum_{n=1}^{\infty} \underline{\mathbf{\Phi}}_n(t)\underline{\mathbf{H}}_n\underline{\mathbf{y}}_n \Leftrightarrow \hat{x}(t) = \sum_{n=1}^{\infty} \hat{x}_n\phi_n(t) + \hat{x}_n^*\widetilde{\phi}_n(t) \tag{9.40}$$

with

$$\hat{x}_n = h_{n,1}y_n + h_{n,2}y_n^*. \tag{9.41}$$

The WLMMSE in the interval $[0, T]$ is

$$\begin{aligned}\text{WLMMSE} &= \tfrac{1}{2}\int_0^T \operatorname{tr} E\{\underline{\mathbf{x}}(t)[\underline{\mathbf{x}}(t) - \hat{\underline{\mathbf{x}}}(t)]^{\mathrm{H}}\}\mathrm{d}t \\ &= \tfrac{1}{2}\int_0^T \operatorname{tr}\left[\underline{\mathbf{R}}_{xx}(t,t) - \int_0^T \underline{\mathbf{H}}(t,v)\underline{\mathbf{R}}_{xx}(t,v)\mathrm{d}v\right]\mathrm{d}t \\ &= \frac{N_0}{2}\sum_{n=1}^{\infty}\frac{\lambda_n}{\lambda_n + N_0}.\end{aligned} \tag{9.42}$$

In the proper case, $\lambda_{2n-1} = \lambda_{2n} = \mu_n$, $\widetilde{\phi}_n(t) \equiv 0$ for all n, and we have

$$h_{n,1} = \frac{\mu_n}{\mu_n + N_0}, \qquad h_{n,2} = 0. \tag{9.43}$$

Thus, the solution simplifies to the LMMSE estimator

$$\hat{x}(t) = \sum_{n=1}^{\infty}\frac{\mu_n}{\mu_n + N_0}y_n\phi_n(t) \tag{9.44}$$

with LMMSE in the interval $[0, T]$ of

$$\text{LMMSE} = N_0\sum_{n=1}^{\infty}\frac{\mu_n}{\mu_n + N_0}. \tag{9.45}$$

Equations (9.42) and (9.45) hide the fact that there are two λ_ns for every μ_n, and thus the sum in (9.42) contains two terms for every one in the sum of (9.45).

The discussion in Section 5.4.2 for the estimation problem $\mathbf{y} = \mathbf{x} + \mathbf{n}$ also applies to the continuous-time version $y(t) = x(t) + n(t)$ discussed here. In particular, Result 5.6 still holds: the maximum performance advantage of WLMMSE over LMMSE processing, in additive white and proper noise, is a factor of 2. This is achieved in the maximally improper case $\lambda_{2n-1} = 2\mu_n$, $\lambda_{2n} = 0$. If the noise is colored or improper, the performance advantage can be arbitrarily large.

9.1.2 Detection

The KL expansion can also be used to solve the following detection problem. We observe a complex signal $y(t)$ over the time interval $0 \le t \le T$. We would like to test the hypotheses

$$\begin{aligned} H_0 &: y(t) = n(t), \\ H_1 &: y(t) = x(t) + n(t), \end{aligned} \tag{9.46}$$

where the noise $n(t)$ is zero-mean complex white (i.e., proper) Gaussian with power spectral density N_0, and the zero-mean complex Gaussian signal $x(t)$ has augmented covariance $\underline{\mathbf{R}}_{xx}(t_1, t_2)$.

Let y_n be the resolution of $y(t)$ onto the KL basis functions as given by (9.39). Now collect the first N of these coefficients in the vector $\mathbf{y} = [y_1, \ldots, y_N]^{\mathrm{T}}$. This leads to a finite-dimensional detection problem exactly like the one discussed in Section 7.4.3. Owing to Grenander's theorem[2] the finite-dimensional log-likelihood ratio converges to the log-likelihood ratio of the infinite-dimensional detection problem as $N \to \infty$. This means that the solution and discussion provided in Section 7.4.3 also apply to the detection problem (9.46). In particular, in additive white and proper Gaussian noise the maximum performance advantage of widely linear processing over linear processing, as measured by deflection, is a factor of 2.

9.2 Cramér–Loève spectral representation

The Cramér–Loève (CL) spectral representation uses the (orthonormal) Fourier basis and thus preserves the frequency interpretation of the transform-domain representation. This, however, comes at the price of *correlated* spectral increments. Signals that have a CL representation are called *harmonizable*. The proper version of the following result is due to Loève (1978).[3]

Result 9.2. *A nonstationary continuous-time signal $x(t)$ can be represented as*

$$x(t) = \int_{-\infty}^{\infty} \mathrm{d}\xi(f) \mathrm{e}^{\mathrm{j}2\pi f t}, \tag{9.47}$$

where $\xi(f)$ is a spectral process with correlated increments

$$E[\mathrm{d}\xi(f_1)\mathrm{d}\xi^*(f_2)] = S_{xx}(f_1, f_2)\mathrm{d}f_1\, \mathrm{d}f_2, \tag{9.48}$$

$$E[\mathrm{d}\xi(f_1)\mathrm{d}\xi(-f_2)] = \widetilde{S}_{xx}(f_1, f_2)\mathrm{d}f_1\, \mathrm{d}f_2, \tag{9.49}$$

if and only if

$$S_{xx}(f_1, f_2) = \int_{\mathbb{R}^2} r_{xx}(t_1, t_2)\mathrm{e}^{-\mathrm{j}2\pi(f_1 t_1 - f_2 t_2)}\, \mathrm{d}t_1\, \mathrm{d}t_2, \tag{9.50}$$

$$\widetilde{S}_{xx}(f_1, f_2) = \int_{\mathbb{R}^2} \widetilde{r}_{xx}(t_1, t_2)\mathrm{e}^{-\mathrm{j}2\pi(f_1 t_1 - f_2 t_2)}\, \mathrm{d}t_1\, \mathrm{d}t_2. \tag{9.51}$$

The representation (9.47) looks like the spectral representation (8.11) in Result 8.1 in the WSS case except that the spectral increments $d\xi(f)$ are now correlated between different frequencies. This correlation is measured by the (Hermitian) *spectral correlation* $S_{xx}(f_1, f_2)$ and *complementary spectral correlation* $\tilde{S}_{xx}(f_1, f_2)$, which are also called dual-frequency spectra or Loève spectra. The spectral correlation and complementary spectral correlation are the two-dimensional Fourier transforms of the time-correlation function $r_{xx}(t_1, t_2) = E[x(t_1)x^*(t_2)]$ and complementary correlation function $\tilde{r}_{xx}(t_1, t_2) = E[x(t_1)x(t_2)]$, respectively. We note, however, that the two-dimensional Fourier transforms in (9.50) and (9.51) are defined with different signs on f_1 and f_2. This is done so that the CL representation simplifies to the spectral representation in Result 8.1 for WSS signals. If necessary, the Fourier integrals (9.50) and (9.51) should be interpreted as generalized Fourier transforms that lead to δ-functions in $S_{xx}(f_1, f_2)$ and $\tilde{S}_{xx}(f_1, f_2)$.

Result 9.3. *The spectral correlation and complementary correlation satisfy the symmetries and bounds*

$$S_{xx}(f_1, f_2) = S^*_{xx}(f_2, f_1), \tag{9.52}$$

$$\tilde{S}_{xx}(f_1, f_2) = \tilde{S}_{xx}(-f_2, -f_1), \tag{9.53}$$

$$|S_{xx}(f_1, f_2)|^2 \leq S_{xx}(f_1, f_1)S_{xx}(f_2, f_2), \tag{9.54}$$

$$|\tilde{S}_{xx}(f_1, f_2)|^2 \leq S_{xx}(f_1, f_1)S_{xx}(-f_2, -f_2). \tag{9.55}$$

These bounds are due to the Cauchy–Schwarz inequality. Result 9.3 is the nonstationary extension of the WSS Result 2.14. However, in contrast to the WSS case, Result 9.3 lists only *necessary conditions*. That is, a pair of functions $(S_{xx}(f_1, f_2), \tilde{S}_{xx}(f_1, f_2))$ that satisfy (9.52)–(9.55) are not necessarily valid spectral correlation and complementary spectral correlation functions. A necessary *and* sufficient condition is that

$$\underline{\mathbb{S}}_{xx}(f_1, f_2) = \begin{bmatrix} S_{xx}(f_1, f_2) & \tilde{S}_{xx}(f_1, f_2) \\ \tilde{S}^*_{xx}(-f_1, -f_2) & S^*_{xx}(-f_1, -f_2) \end{bmatrix}$$

be positive semidefinite in the sense that

$$\int_{\mathbb{R}^2} \mathbf{g}^{\mathrm{H}}(f_1)\underline{\mathbb{S}}_{xx}(f_1, f_2)\mathbf{g}(f_2)\,df_1\,df_2 \geq 0 \tag{9.56}$$

for all continuous functions $\mathbf{g}(f)\colon \mathbb{R} \longrightarrow \mathbb{C}^2$. Unfortunately, it is not easy to interpret this condition as a relation between $S_{xx}(f_1, f_2)$ and $\tilde{S}_{xx}(f_1, f_2)$.

9.2.1 Four-corners diagram

We can gain more insight into the time-varying nature of $x(t)$ by expressing the correlation functions in terms of a *global* (absolute) time variable $t = t_2$ and a *local* (relative) time lag $\tau = t_1 - t_2$. For convenience – but in an abuse of notation – we will reuse the same symbols r and $\tilde{r}$ for the correlation and complementary correlation functions, now

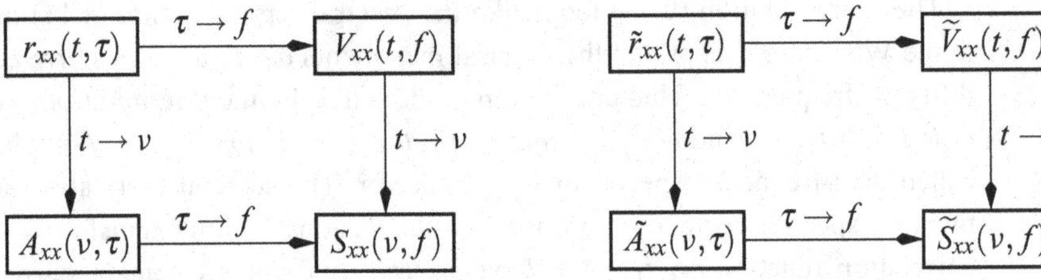

Figure 9.1 Four-corners diagrams showing equivalent second-order characterizations.

expressed in terms of t and τ:

$$r_{xx}(t,\tau) = E[x(t+\tau)x^*(t)], \tag{9.57}$$

$$\tilde{r}_{xx}(t,\tau) = E[x(t+\tau)x(t)]. \tag{9.58}$$

Equivalent representations can be found by applying Fourier transformations to t or τ, as shown in the "four-corners diagrams" in Fig. 9.1. There is a diagram for the Hermitian correlation, and another for the complementary correlation. Each arrow in Fig. 9.1 stands for a Fourier transform in one variable. In contrast to (9.50) and (9.51), these Fourier transforms now use equal signs for both dimensions. The variables t and f are global (absolute) variables, and τ and ν are local (relative) variables. A Fourier transform applied to a local time lag yields a global frequency variable, and a Fourier transform applied to a global time variable yields a local frequency offset.

We have already encountered the spectral correlations in the southeast corners of Fig. 9.1. Again we are reusing the same symbols S and $\widetilde{S}$ for the spectral correlation and complementary correlation, now expressed in terms of $\nu = f_1 - f_2$ and $f = f_1$. The magnitude of the Jacobian determinant of this transformation is 1, i.e., the area of the infinitesimal element $\mathrm{d}\nu\, \mathrm{d}f$ equals the area of $\mathrm{d}f_1\, \mathrm{d}f_2$. Thus,

$$S_{xx}(\nu,f)\mathrm{d}\nu\, \mathrm{d}f = E[\mathrm{d}\xi(f)\mathrm{d}\xi^*(f-\nu)], \tag{9.59}$$

$$\widetilde{S}_{xx}(\nu,f)\mathrm{d}\nu\, \mathrm{d}f = E[\mathrm{d}\xi(f)\mathrm{d}\xi(\nu-f)]. \tag{9.60}$$

The northeast corners of the diagrams contain the *Rihaczek time–frequency representations* $V_{xx}(t,f)$ and $\widetilde{V}_{xx}(t,f)$ in terms of a global time t and global frequency f. We immediately find the expressions

$$V_{xx}(t,f)\mathrm{d}f = E[x^*(t)\mathrm{d}\xi(f)\mathrm{e}^{\mathrm{j}2\pi ft}], \tag{9.61}$$

$$\widetilde{V}_{xx}(t,f)\mathrm{d}f = E[x(t)\mathrm{d}\xi(f)\mathrm{e}^{\mathrm{j}2\pi ft}]. \tag{9.62}$$

Finally, the southwest corners are the *ambiguity functions* $A_{xx}(\nu,\tau)$ and $\tilde{A}_{xx}(\nu,\tau)$, which play a prominent role in radar. The ambiguity functions are representations in terms of a local time τ and local frequency ν.

We remark that there are other ways of defining global/local time and frequency variables. Another definition splits the time lag τ symmetrically as $t_1 = t + \tau/2$ and $t_2 = t - \tau/2$. This alternative definition leads to the frequency variables $f_1 = f + \nu/2$ and $f_2 = f - \nu/2$. Section 9.3 will motivate our choice of definition.[4]

9.2.2 Energy and power spectral densities

Definition 9.1. *If the derivative process* $X(f) = \mathrm{d}\xi(f)/\mathrm{d}f$ *exists, then*

$$E|X(f)|^2 = S_{xx}(0, f) \tag{9.63}$$

is the energy spectral density (ESD) *and*

$$E[X(f)X(-f)] = \widetilde{S}_{xx}(0, f) \tag{9.64}$$

is the complementary energy spectral density (C-ESD).

The ESD shows how the energy of a random process is distributed over frequency. The total energy of $x(t)$ is obtained by integrating the ESD:

$$E_x = \int_{-\infty}^{\infty} S_{xx}(0, f)\mathrm{d}f. \tag{9.65}$$

Wide-sense stationary signals

We will now show that, for a given frequency f, the ESD of a wide-sense stationary (WSS) process is either 0 or ∞. If $x(t)$ is WSS, both $r_{xx}(t, \tau) = r_{xx}(\tau)$ and $\tilde{r}_{xx}(t, \tau) = \tilde{r}_{xx}(\tau)$ are independent of t. For the spectral correlation we find

$$S_{xx}(\nu, f) = \int_{\mathbb{R}^2} r_{xx}(\tau)\mathrm{e}^{-\mathrm{j}2\pi(\nu t + f\tau)}\mathrm{d}t\, \mathrm{d}\tau = P_{xx}(f)\delta(\nu), \tag{9.66}$$

where $P_{xx}(f)$ is the power spectral density (PSD) of $x(t)$. This produces the Wiener–Khinchin relation

$$P_{xx}(f) = \int_{-\infty}^{\infty} r_{xx}(\tau)\mathrm{e}^{-\mathrm{j}2\pi f\tau}\, \mathrm{d}\tau. \tag{9.67}$$

Similarly, the complementary spectral correlation is $\widetilde{S}_{xx}(\nu, f) = \widetilde{P}_{xx}(f)\delta(\nu)$, where $\widetilde{P}_{xx}(f)$ is the complementary power spectral density (C-PSD) of $x(t)$, which is the Fourier transform of $\tilde{r}_{xx}(\tau)$. We see that

$$E[\mathrm{d}\xi(f)\mathrm{d}\xi^*(f - \nu)] = P_{xx}(f)\delta(\nu)\mathrm{d}\nu\, \mathrm{d}f, \tag{9.68}$$

$$E[\mathrm{d}\xi(f)\mathrm{d}\xi(\nu - f)] = \widetilde{P}_{xx}(f)\delta(\nu)\mathrm{d}\nu\, \mathrm{d}f \tag{9.69}$$

and the CL representation of WSS signals simplifies to the spectral representation given in Result 8.1. The line $\nu = 0$ is called the *stationary manifold*. A few remarks on WSS processes are now in order.

- The ESD at any given frequency is either 0 or ∞. Hence, a WSS process $x(t)$ cannot have a second-order Fourier transform $X(f)$.
- The spectral correlation and complementary spectral correlations are zero outside the stationary manifold. Figuratively speaking, "the (C-)ESD is the (C-)PSD sitting on top of a δ-ridge."
- The (C-)PSD itself may contain δ-functions if the signal contains periodic components (see Section 8.1).

What do the remaining two corners of the four-corners diagram look like for WSS signals? The time–frequency distribution of a WSS signal is simply the PSD,

$$V_{xx}(t, f) = \int_{-\infty}^{\infty} r_{xx}(\tau)e^{-j2\pi f\tau}\, d\tau = P_{xx}(f), \tag{9.70}$$

and also $\widetilde{V}_{xx}(t, f) = \widetilde{P}_{xx}(f)$. Finally, the ambiguity function of a WSS signal is

$$A_{xx}(\nu, \tau) = \int_{-\infty}^{\infty} r_{xx}(\tau)e^{-j2\pi\nu t}\, dt = r_{xx}(\tau)\delta(\nu), \tag{9.71}$$

and also $\tilde{A}_{xx}(\nu, \tau) = \tilde{r}_{xx}(\tau)\delta(\nu)$.

Nonstationary signals

There are two types of nonstationary signals:

- *energy signals*, which have nonzero bounded ESD and thus zero PSD; and
- *power signals*, which have nonzero, but not necessarily bounded, PSD and thus unbounded ESD.

WSS signals are always power signals and never energy signals. For energy signals, the ESD and C-ESD are the Fourier transform of *time-integrated* $r_{xx}(t, \tau)$ and $\tilde{r}_{xx}(t, \tau)$, respectively:

$$S_{xx}(0, f) = \int_{-\infty}^{\infty} \left[\int_{-\infty}^{\infty} r_{xx}(t, \tau)dt\right] e^{-j2\pi f\tau}\, d\tau, \tag{9.72}$$

$$\widetilde{S}_{xx}(0, f) = \int_{-\infty}^{\infty} \left[\int_{-\infty}^{\infty} \tilde{r}_{xx}(t, \tau)dt\right] e^{-j2\pi f\tau}\, d\tau. \tag{9.73}$$

Power signals have a WSS component, i.e., $S_{xx}(\nu, f)$ has a δ-ridge on the stationary manifold $\nu = 0$. The definitions of the PSD and C-PSD for WSS signals in Result 8.1 may also be applied to nonstationary signals:

$$E|d\xi(f)|^2 = P_{xx}(f)df \qquad \text{and} \qquad E[d\xi(f)d\xi(-f)] = \widetilde{P}_{xx}(f)df. \tag{9.74}$$

The PSD and C-PSD are the Fourier transforms of *time-averaged* $r_{xx}(t, \tau)$ and $\tilde{r}_{xx}(t, \tau)$:

$$P_{xx}(f) = \int_{-\infty}^{\infty} \left[\lim_{T\to\infty} \frac{1}{T} \int_{-T/2}^{T/2} r_{xx}(t, \tau)dt\right] e^{-j2\pi f\tau} d\tau, \tag{9.75}$$

$$\widetilde{P}_{xx}(f) = \int_{-\infty}^{\infty} \left[\lim_{T\to\infty} \frac{1}{T} \int_{-T/2}^{T/2} \tilde{r}_{xx}(t, \tau)dt\right] e^{-j2\pi f\tau} d\tau. \tag{9.76}$$

The time-averaged correlations, and thus the PSD and C-PSD, vanish for energy signals. It is important to keep in mind that the PSD and C-PSD (or ESD and C-ESD) are only an incomplete second-order characterization of nonstationary signals, since they read out the spectral and complementary spectral correlations on the stationary manifold, and thus ignore everything outside the manifold. For instance, the spectral correlation of cyclostationary signals, discussed in Chapter 10, has δ-ridges parallel to the stationary manifold.

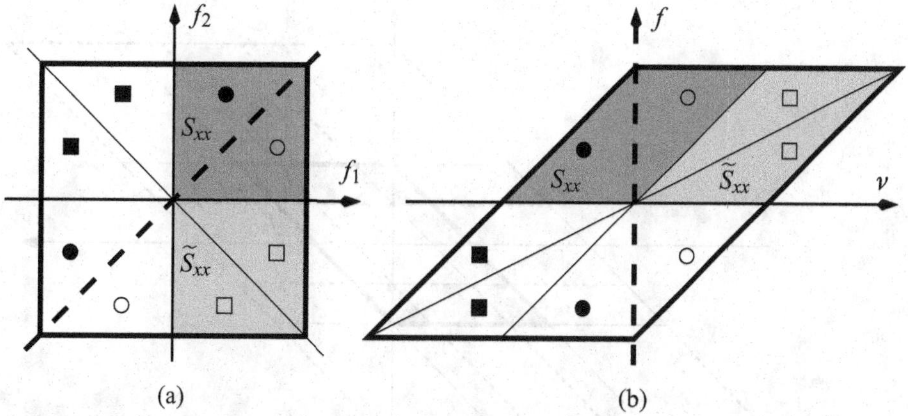

Figure 9.2 Support of the spectral correlation (dark gray) and complementary spectral correlation (light gray) of a bandlimited analytic signal $x(t)$ compared with the support of the spectral correlation of the corresponding real signal $u(t)$ (thick-lined square/parallelogram): (a) uses the global–global frequencies (f_1, f_2) and (b) the local–global frequencies (ν, f). This figure is adapted from Schreier and Scharf (2003b) ©IEEE, and used here with permission.

9.2.3 Analytic signals

Now consider the important special case of the complex analytic signal $x(t) = u(t) + j\hat{u}(t)$, where $\hat{u}(t)$ is the Hilbert transform of $u(t)$. Thus, $d\xi(f) = 2\Gamma(f)d\eta(f)$, where $\Gamma(f)$ is the unit-step function and $\eta(f)$ the spectral process of $u(t)$, whose increments satisfy the Hermitian symmetry $d\eta(f) = d\eta^*(-f)$. In the global–global coordinate system (f_1, f_2), we find the connection

$$S_{xx}(f_1, f_2) = \Gamma(f_1)\Gamma(f_2)S_{uu}(f_1, f_2), \tag{9.77}$$

$$\widetilde{S}_{xx}(f_1, f_2) = \Gamma(f_1)\Gamma(-f_2)S_{uu}(f_1, f_2) \tag{9.78}$$

between the spectral and complementary spectral correlations of $x(t)$ and the spectral correlation of $u(t)$. In the local–global coordinate system (ν, f), this connection becomes

$$S_{xx}(\nu, f) = \Gamma(f)\Gamma(f-\nu)S_{uu}(\nu, f), \tag{9.79}$$

$$\widetilde{S}_{xx}(\nu, f) = \Gamma(f)\Gamma(\nu - f)S_{uu}(\nu, f). \tag{9.80}$$

Thus, the spectral and complementary spectral correlations of $x(t)$ cut out regions from the spectral correlation of $u(t)$. This is illustrated in Fig. 9.2, which shows the support of the spectral correlation (dark gray) and complementary spectral correlation (light gray) of a bandlimited analytic signal $x(t)$ in the two different coordinate systems: global–global in (a) and local–global in (b).

The Hermitian symmetry $d\eta(f) = d\eta^*(-f)$ leads to many symmetry properties of S_{uu}. To illustrate these in Fig. 9.2, the circle and the square represent two exemplary values of the spectral correlation. A filled square or circle is the conjugate of an empty square or circle. On their regions of support, S_{xx} and $\widetilde{S}_{xx}$ inherit these symmetries from

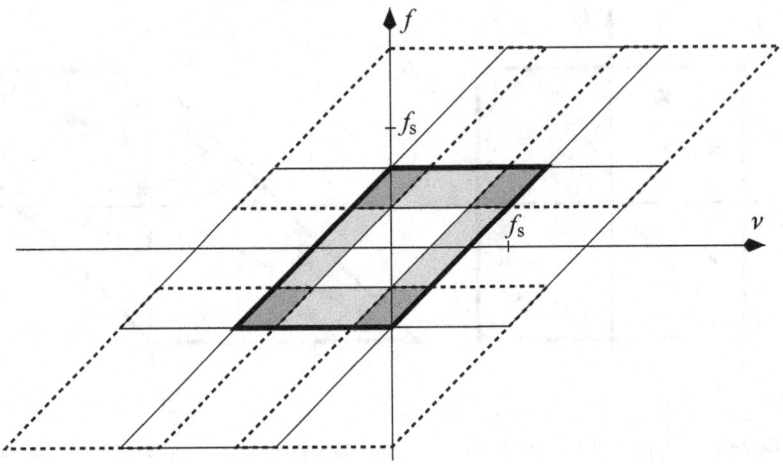

Figure 9.3 Aliasing in the spectral correlation of $x[k]$.

S_{uu}. Figure 9.2 also shows how $\widetilde{S}_{xx}$ complements the information in S_{xx}. It is clearly not possible to reconstruct S_{uu} from S_{xx} alone.

The stationary manifold is the dashed line in Fig. 9.2. This line does not pass through the region of support of $\widetilde{S}_{xx}$. This proves once again that WSS analytic signals – whose support is limited to the stationary manifold – must be proper.

9.2.4 Discrete-time signals

We now investigate what happens if we sample the random process $x(t)$ with sampling frequency $f_s = T_s^{-1}$ to obtain $x[k] = x(kT_s)$. The spectral representation of the sampled process $x[k]$ is

$$\begin{aligned} x[k] &= \int_{-\infty}^{\infty} e^{j2\pi f k T_s}\, d\xi(f) \\ &= \int_{-f_s/2}^{f_s/2} e^{j\frac{2\pi}{f_s} f k} \sum_{m=-\infty}^{\infty} d\xi(f + m f_s). \end{aligned} \tag{9.81}$$

The spectral representation of $x[k]$ is thus subject to aliasing unless $d\xi(f) = 0$ for $|f| > f_s/2$. The spectral correlation of $x[k]$ is

$$S_{xx}^{d}(\nu, f) = \sum_{m=-\infty}^{\infty} \sum_{n=-\infty}^{\infty} S_{xx}(\nu + m f_s, f + n f_s), \tag{9.82}$$

where the letter d stands for discrete-time. The same formula is obtained for the complementary spectral correlation. The term with $m = 0$ and $n = 0$ in (9.82) is called the principal replica of $S_{xx}(\nu, f)$. Figure 9.3 illustrates aliasing for the spectral correlation of a bandlimited, undersampled, signal. The support of the principal replica is the thick-lined parallelogram centered at the origin. Figure 9.3 also shows the replicas that overlap with the principal replica. Areas of overlap are gray. The following points are noteworthy.

- There are regions of the principal replica that overlap with one other replica (light gray) and regions that overlap with three other replicas (dark gray). (This assumes, as in Fig. 9.3, that $d\xi(f) = 0$ for $|f| > f_s$; otherwise the principal replica could overlap with further replicas.)
- In general, replicas with $m \neq 0$ and/or $n \neq 0$ contribute to aliasing in the ESD or PSD along the stationary manifold $\nu = 0$.
- However, the spectral correlation of a WSS signal is zero off the stationary manifold. Therefore, only the replicas with $m = 0$ and $n \neq 0$ can contribute to aliasing in the PSD of a WSS signal.

9.3 Rihaczek time–frequency representation

Let's explore some of the properties of the Rihaczek time–frequency representation,[5] which is comprised of the Hermitian Rihaczek time–frequency distribution (HR-TFD)

$$V_{xx}(t, f) = \int_{-\infty}^{\infty} r_{xx}(t, \tau) e^{-j2\pi f\tau} \, d\tau \tag{9.83}$$

and the complementary Rihaczek time–frequency distribution (CR-TFD)

$$\widetilde{V}_{xx}(t, f) = \int_{-\infty}^{\infty} \tilde{r}_{xx}(t, \tau) e^{-j2\pi f\tau} \, d\tau. \tag{9.84}$$

It is obvious that the Rihaczek time–frequency representation is shift-covariant in both time and frequency:

$$x(t - t_0) \longleftrightarrow V_{xx}(t - t_0, f), \ \widetilde{V}_{xx}(t - t_0, f), \tag{9.85}$$

$$x(t) e^{j2\pi f_0 t} \longleftrightarrow V_{xx}(t, f - f_0), \ \widetilde{V}_{xx}(t, f - f_0). \tag{9.86}$$

Both the HR-TFD and CR-TFD are generally complex-valued. However, the time-marginal of the HR-TFD

$$\int_{-\infty}^{\infty} V_{xx}(t, f) df = r_{xx}(t, 0) = E|x(t)|^2 \geq 0, \tag{9.87}$$

which is the instantaneous power at t, and the frequency-marginal of the HR-TFD

$$\int_{-\infty}^{\infty} V_{xx}(t, f) dt = S_{xx}(0, f) \geq 0, \tag{9.88}$$

which is the energy spectral density at f, are nonnegative.

It is common to interpret time–frequency distributions as distributions of energy or power over time and frequency. However, such an interpretation is fraught with problems because of the following. One would think that the minimum requirements for a distribution to have an interpretation as a distribution of energy or power are that

- it is a bilinear function of the signal (so that it has the right physical units),
- it is covariant with respect to shifts in time and frequency,
- it has the correct time- and frequency-marginals (instantaneous power and energy spectral density, respectively), and
- it is nonnegative.

Yet Wigner's theorem[6] says that such a distribution does not exist. The Rihaczek distribution satisfies the first three properties but it is complex-valued. The Wigner–Ville distribution,[7] which is more popular than the Rihaczek distribution, also satisfies the first three properties but it can take on negative values. Thus, we will not attempt an energy/power-distribution interpretation. Instead we argue for the Rihaczek distribution as a distribution of correlation and then present an evocative geometrical interpretation. This geometrical interpretation is the main reason why we prefer the Rihaczek distribution over the Wigner–Ville distribution.

9.3.1 Interpretation

A key insight is that the HR/CR-TFDs are inner products in the Hilbert space of second-order random variables, between the time-domain signal *at a fixed time instant t* and the frequency-domain representation *at a fixed frequency f*:

$$V_{xx}(t,f)\mathrm{d}f = E[x^*(t)\mathrm{d}\xi(f)\mathrm{e}^{\mathrm{j}2\pi ft}] = \langle x(t), \mathrm{d}\xi(f)\mathrm{e}^{\mathrm{j}2\pi ft}\rangle, \tag{9.89}$$

$$\widetilde{V}_{xx}(t,f)\mathrm{d}f = E[x(t)\mathrm{d}\xi(f)\mathrm{e}^{\mathrm{j}2\pi ft}] = \langle x^*(t), \mathrm{d}\xi(f)\mathrm{e}^{\mathrm{j}2\pi ft}\rangle. \tag{9.90}$$

How should these inner products be interpreted? Let's construct a linear minimum mean-squared error (LMMSE) estimator of the random variable $x(t)$, at fixed t, from the random variable $\mathrm{d}\xi(f)\mathrm{e}^{\mathrm{j}2\pi ft}$, at fixed f:[8]

$$\begin{aligned}\hat{x}_f(t) &= \frac{E\left[x(t)\mathrm{d}\xi^*(f)\mathrm{e}^{-\mathrm{j}2\pi ft}\right]}{E\left[\mathrm{d}\xi(f)\mathrm{e}^{\mathrm{j}2\pi ft}\,\mathrm{d}\xi^*(f)\mathrm{e}^{-\mathrm{j}2\pi ft}\right]}\,\mathrm{d}\xi(f)\mathrm{e}^{\mathrm{j}2\pi ft} \\ &= \frac{V^*_{xx}(t,f)}{S_{xx}(0,f)}\,\frac{\mathrm{d}\xi(f)}{\mathrm{d}f}\,\mathrm{e}^{\mathrm{j}2\pi ft}.\end{aligned} \tag{9.91}$$

The MMSE is thus

$$E|\hat{x}_f(t) - x(t)|^2 = r_{xx}(t,0)(1 - |\rho_{xx}(t,f)|^2), \tag{9.92}$$

where

$$|\rho_{xx}(t,f)|^2 = \frac{|V_{xx}(t,f)|^2}{r_{xx}(t,0)S_{xx}(0,f)} \tag{9.93}$$

is the magnitude-squared *rotational* time–frequency coherence[9] between the time- and frequency-domain descriptions. Similarly, we can linearly estimate $x(t)$ from

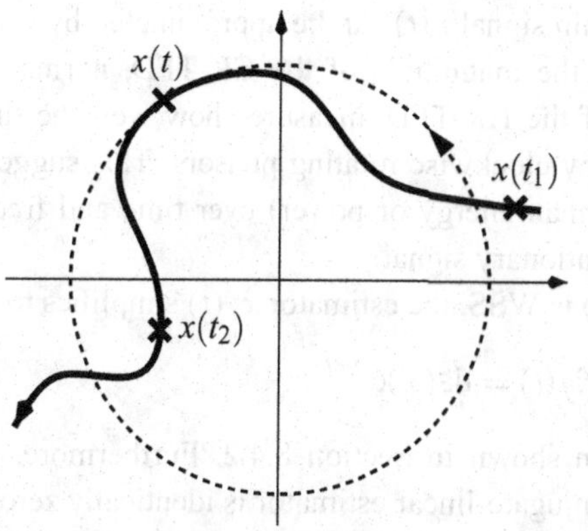

Figure 9.4 The dark solid line shows the time-domain signal $x(t)$ as it moves through the complex plane. At $t' = t$, the counterclockwise turning circle is a perfect approximation.

$d\xi^*(f)e^{-j2\pi ft}$ as

$$\tilde{\hat{x}}_f(t) = \frac{E\left[x(t)d\xi(f)e^{j2\pi ft}\right]}{E\left[d\xi^*(f)e^{-j2\pi ft}d\xi(f)e^{j2\pi ft}\right]} d\xi^*(f)e^{-j2\pi ft}$$
$$= \frac{\widetilde{V}_{xx}(t, f)}{S_{xx}(0, f)} \frac{d\xi^*(f)}{df} e^{-j2\pi ft}, \tag{9.94}$$

and the corresponding magnitude-squared *reflectional* time–frequency coherence is

$$|\widetilde{\rho}_{xx}(t, f)|^2 = \frac{|\widetilde{V}_{xx}(t, f)|^2}{r_{xx}(t, 0)S_{xx}(0, f)}. \tag{9.95}$$

The magnitude-squared time–frequency coherences satisfy $0 \le |\rho_{xx}(t, f)|^2 \le 1$ and $0 \le |\widetilde{\rho}_{xx}(t, f)|^2 \le 1$. It is particularly illuminating to discuss them for the case of *analytic* signals $x(t) = u(t) + j\hat{u}(t)$. Bilinear time–frequency analysis favors the use of analytic signals because there is no interference between positive and negative frequencies due to $d\xi(f) = 0$ for $f < 0$. For a fixed frequency $f > 0$ but varying time t', $d\xi(f)e^{j2\pi ft'}$ is a counterclockwise rotating phasor, and $d\xi^*(f)e^{-j2\pi ft'}$ a clockwise rotating phasor. For fixed (t, f) and varying t', the counterclockwise rotating phasor

$$\hat{x}_f(t') = \frac{V^*_{xx}(t, f)}{S_{xx}(0, f)} \frac{d\xi(f)}{df} e^{j2\pi ft'} \tag{9.96}$$

approximates $x(t)$ at $t' = t$. If $|\rho_{xx}(t, f)|^2 = 1$, this is a perfect approximation. This is illustrated in Fig. 9.4. Similarly, the clockwise rotating phasor

$$\tilde{\hat{x}}_f(t') = \frac{\widetilde{V}_{xx}(t, f)}{S_{xx}(0, f)} \frac{d\xi^*(f)}{df} e^{-j2\pi ft'} \tag{9.97}$$

approximates $x(t)$ at $t' = t$, where $|\widetilde{\rho}_{xx}(t, f)|^2 = 1$ indicates a perfect approximation. Thus, the magnitude of the HR-TFD, normalized by its time- and frequency-marginals, measures how well the time-domain signal $x(t)$ can be approximated by counterclockwise rotating phasors. Similarly, the magnitude of the CR-TFD, normalized by the time- and frequency-marginals of the HR-TFD, measures how well the time-domain signal $x(t)$ can be approximated by clockwise rotating phasors. This suggests that the distribution of coherence (rather than energy or power) over time and frequency is a fundamental descriptor of a nonstationary signal.

Finally we point out that, if $x(t)$ is WSS, the estimator $\hat{x}_f(t)$ simplifies to

$$\hat{x}_f(t) = \mathrm{d}\xi(f)\mathrm{e}^{\mathrm{j}2\pi ft} \tag{9.98}$$

for all t, which has already been shown in Section 8.4.2. Furthermore, since WSS analytic signals are proper, the conjugate-linear estimator is identically zero: $\widetilde{\hat{x}}_f(t) = 0$ for all t. Hence, WSS analytic signals can be estimated only from counterclockwise rotating phasors, whereas nonstationary improper analytic signals may also be estimated from clockwise rotating phasors.

9.3.2 Kernel estimators

The Rihaczek distribution is a mathematical expectation. As a practical matter, it must be estimated. Because such an estimator is likely to be implemented digitally, we present it for a time series $x[k]$. We require that our estimator be a bilinear function of $x[k]$ that is covariant with respect to shifts in time and frequency. Estimators that satisfy these properties constitute Cohen's class.[10] (Since we want to perform time- and frequency-smoothing, we do not expect our estimator to have the correct time- and frequency-marginals.) The discrete-time version of Cohen's class is

$$\widehat{V}_{xx}[k,\theta] = \sum_m \sum_\mu x[k+m+\mu]\phi[m,\mu]x^*[k+m]\mathrm{e}^{-\mathrm{j}\mu\theta}, \tag{9.99}$$

where m is a global and μ a local time variable. The choice of the dual-time kernel $\phi[m,\mu]$ determines the properties of the HR-TFD estimator $\widehat{V}_{xx}[k,\theta)$. Omitting the conjugation in (9.99) gives an estimator of the CR-TFD. It is our objective to design a suitable kernel $\phi[m,\mu]$. A factored kernel that preserves the spirit of the Rihaczek distribution is

$$\phi[m,\mu] = w_1[m+\mu]w_2[\mu]w_3^*[m]. \tag{9.100}$$

In practice, the three windows w_1, w_2, and w_3 might be chosen real and even. By inserting (9.100) into (9.99), we obtain

$$\widehat{V}_{xx}[k,\theta) = \sum_m w_3^*[m]x^*[k+m]\int_{-\pi}^{\pi} W_2(\omega)F_1[k,\theta-\omega)\mathrm{e}^{\mathrm{j}m(\theta-\omega)}\frac{\mathrm{d}\omega}{2\pi}. \tag{9.101}$$

Here $W_2(\omega)$ is the discrete-time Fourier transform (DTFT) of $w_2[\mu]$ and F_1 is the short-time Fourier transform (STFT) of $x[k]$ using window w_1, defined as

$$F_1[k,\theta) = \sum_n w_1[n]x[n+k]\mathrm{e}^{-\mathrm{j}n\theta}. \tag{9.102}$$

The corresponding expression for the CR-TFD estimator is obtained by omitting the two conjugations in (9.101).

In the estimator (9.101), the STFT $F_1[k, \theta - \omega)e^{jm(\theta-\omega)}$ plays the role of $d\xi(\theta)e^{jk\theta}$ in the CL spectral representation. The STFT is averaged over time with window $w_3[k]$ and over frequency with window $W_2(\omega)$. Thus, the three windows play different roles. Window w_1 should be a smooth tapering window to stabilize the STFT F_1, whereas the windows w_3 and W_2 should be localized to concentrate the estimator in time and frequency, respectively. A computationally efficient implementation of this estimator was presented by Hindberg *et al.* (2006). Choosing $\phi[m, \mu] = \delta[m]$, i.e., $w_1[n] = 1$, $w_2[\mu] = 1$, and $w_3[m] = \delta[m]$, produces an instantaneous estimator of the Rihaczek distribution without any time- or frequency-smoothing.

Statistical properties

From (9.99) we find that

$$E\widehat{V}_{xx}[k, \theta) = \sum_m \sum_\mu \phi[m, \mu] r_{xx}[k + m, \mu] e^{-j\mu\theta}$$
$$= \sum_m \int_{-\pi}^{\pi} V_{xx}[m, \omega)\Phi[m - k, \theta - \omega)\frac{d\omega}{2\pi}, \tag{9.103}$$

where $r_{xx}[k, \kappa] = E(x[k + \kappa]x^*[k])$ and $\Phi[m, \omega)$ is the DTFT of $\phi[m, \mu]$ in μ. For the factored kernel (9.100) we have

$$\Phi[m, \omega) = w_3^*[m] \int_{-\pi}^{\pi} W_2(\omega - \nu) W_1(\nu) e^{jm\nu} \frac{d\nu}{2\pi}. \tag{9.104}$$

The mean (9.103) of the HR-TFD estimator is thus a time–frequency-smoothed version of the Rihaczek distribution. The same result holds for the CR-TFD estimator.

In order to derive the second-order properties of the estimator, we assume Gaussian signals. This avoids the appearance of fourth-order terms (cf. Section 2.5.3). An exact expression for the covariance of the HR-TFD estimator is[11]

$$\operatorname{cov}\{\widehat{V}_{xx}[k_1, \theta_1), \widehat{V}_{xx}[k_2, \theta_2)\}$$
$$= \sum_{m_1} \sum_{m_2} \sum_{\mu_1} \sum_{\mu_2} \phi[m_1, \mu_1] \phi^*[m_2, \mu_2] e^{-j(\mu_1\theta_1 - \mu_2\theta_2)}$$
$$\times \{r_{xx}[k_1 + m_1 + \mu_1, k_2 - k_1 + m_2 - m_1 + \mu_2 - \mu_1]$$
$$\times r_{xx}^*[k_1 + m_1, k_2 - k_1 + m_2 - m_1] + \tilde{r}_{xx}[k_1 + m_1, k_2 - k_1 + m_2 - m_1 + \mu_2]$$
$$\times \tilde{r}_{xx}^*[k_1 + m_1 + \mu_1, k_2 - k_1 + m_2 - m_1 - \mu_1]\}.$$

We observe that the covariance of the HR-TFD estimator in general depends on both the Hermitian and the complementary correlation. The covariance of the CR-TFD estimator, on the other hand, depends on the Hermitian correlation only:

$$\begin{aligned}
&\operatorname{cov}\{\widehat{\widetilde{V}}_{xx}[k_1,\theta_1), \widehat{\widetilde{V}}_{xx}[k_2,\theta_2)\} \\
&= \sum_{m_1}\sum_{m_2}\sum_{\mu_1}\sum_{\mu_2} \phi[m_1,\mu_1]\phi^*[m_2,\mu_2]\mathrm{e}^{-\mathrm{j}(\mu_1\theta_1-\mu_2\theta_2)} \\
&\quad \times \{r_{xx}[k_1+m_1+\mu_1, k_2-k_1+m_2-m_1+\mu_2-\mu_1] \\
&\qquad \times r_{xx}[k_1+m_1, k_2-k_1+m_2-m_1] + r_{xx}[k_1+m_1, k_2-k_1+m_2-m_1+\mu_2] \\
&\qquad \times r_{xx}[k_1+m_1+\mu_1, k_2-k_1+m_2-m_1-\mu_1]\}.
\end{aligned}$$

We can derive the following approximate expression for the variance of the HR-TFD estimator, assuming *analytic* and *quasi-stationary* signals whose time duration of stationarity is much greater than the duration of correlation:

$$\begin{aligned}
\operatorname{var}\{\widehat{V}_{xx}[k,\theta)\} \approx \sum_m \int_0^\pi & |\Phi[m-k,\theta-\omega)|^2\,|V_{xx}[k,\omega)|^2 \\
&+ \Phi[m-k,\theta-\omega)\Phi^*[m-k,\theta+\omega)\,|\widetilde{V}_{xx}[k,\omega)|^2\,\frac{\mathrm{d}\omega}{2\pi}.
\end{aligned} \tag{9.105}$$

However, the same simplifying approximation of large duration of stationarity leads to a vanishing CR-TFD estimator because stationary analytic signals have zero complementary correlation. This result should not be taken as an indication that the CR-TFD is not important for analytic signals. Rather it shows that the assumption of quasi-stationarity with large duration of stationarity is a rather crude approximation to the general class of nonstationary signals.

9.4 Rotary-component and polarization analysis

In Section 9.3.1, we determined the LMMSE estimator of the time-domain signal at a fixed time instant t from the frequency-domain representation at a fixed frequency f. This showed us how well the signal can be approximated from clockwise or counterclockwise turning circles (phasors). It thus seems natural to ask how well the signal can be approximated from *ellipses*, which combine the contributions from clockwise and counterclockwise circles. This framework will allow us to extend rotary-component and polarization analysis from the stationary case, described in Section 8.4, to a nonstationary setting.[12] We thus construct a WLMMSE estimator of $x(t)$, for fixed t, from the frequency-domain representation at $+f$ and $-f$, where throughout this entire section f shall denote a fixed nonnegative frequency:

$$\begin{aligned}
\hat{x}_f(t) = {}& W_1(t,f)\mathrm{d}\xi(f)\mathrm{e}^{\mathrm{j}2\pi ft} + W_2(t,-f)\mathrm{d}\xi^*(f)\mathrm{e}^{-\mathrm{j}2\pi ft} \\
&+ W_1(t,-f)\mathrm{d}\xi(-f)\mathrm{e}^{-\mathrm{j}2\pi ft} + W_2(t,f)\mathrm{d}\xi^*(-f)\mathrm{e}^{\mathrm{j}2\pi ft}.
\end{aligned} \tag{9.106}$$

Since (t, f) are fixed, $W_1(t, f)$, $W_1(t, -f)$, $W_2(t, f)$, and $W_2(t, -f)$ are four complex coefficients that are determined such that $E|\hat{x}_f(t) - x(t)|^2$ is minimized. By combining terms with $e^{j2\pi ft}$ and terms with $e^{-j2\pi ft}$, we obtain

$$\hat{x}_f(t) = \underbrace{[W_1(t, f)d\xi(f) + W_2(t, f)d\xi^*(-f)]}_{d\zeta(t, f)} e^{j2\pi ft} + \underbrace{[W_1(t, -f)d\xi(-f) + W_2(t, -f)d\xi^*(f)]}_{d\zeta(t, -f)} e^{-j2\pi ft}. \tag{9.107}$$

For fixed (t, f) but varying t',

$$\varepsilon_{t,f}(t') = \underbrace{d\zeta(t, f)e^{j2\pi ft'}}_{\varepsilon_{t,f_+}(t')} + \underbrace{d\zeta(t, -f)e^{-j2\pi ft'}}_{\varepsilon_{t,f_-}(t')} \tag{9.108}$$

describes an ellipse in the complex time-domain plane. We may say that the varying t' traces out a *local ellipse* $\varepsilon_{t,f}(t')$ whose parameters are "frozen" for a given (t, f). Since $\varepsilon_{t,f}(t) = \hat{x}_f(t)$, the local ellipse $\varepsilon_{t,f}(t')$ provides the best approximation of $x(t)$ at $t' = t$. Following the terminology introduced in Section 8.4.1, the local ellipse $\varepsilon_{t,f}(t')$ is also called the *polarization ellipse* at (t, f), and $\varepsilon_{t,f_+}(t')$ and $\varepsilon_{t,f_-}(t')$ are the *rotary components*.

One ellipse can be constructed for every time–frequency point (t, f). This raises the following obvious question: which ellipses in the time–frequency plane should the analysis focus on? We propose to consider those points (t, f) where the local ellipses provide a good approximation of the nonstationary signal $x(t)$. The quality of the approximation can be measured in terms of a time–frequency coherence, which is introduced in the next subsection.

In the WSS case, we found that the WSS ellipse

$$\varepsilon_f(t) = d\xi(f)e^{j2\pi ft} + d\xi(-f)e^{-j2\pi ft} \tag{9.109}$$

is the WLMMSE estimate of $x(t)$, for *all* t, from the rotary components at frequency f. On comparing the WSS ellipse with the nonstationary solution (9.108), we see that the random variables $d\zeta(t, f)$ and $d\zeta(t, -f)$ now play the roles of the spectral increments $d\xi(f)$ and $d\xi(-f)$ in the WSS ellipse.

In order to determine the optimum coefficients $W_1(t, \pm f)$ and $W_2(t, \pm f)$ in (9.107), we introduce the short-hand notation

$$d\Xi(t, f) = \begin{bmatrix} d\xi(f)e^{j2\pi ft} \\ d\xi^*(-f)e^{j2\pi ft} \\ d\xi(-f)e^{-j2\pi ft} \\ d\xi^*(f)e^{-j2\pi ft} \end{bmatrix}. \tag{9.110}$$

The local ellipse is found as the output of a WLMMSE filter

$$\varepsilon_{t,f}(t') = \mathbf{V}_{x\Xi}(t, f)\mathbf{K}^{\dagger}_{\Xi\Xi}(t, f)\frac{d\Xi(t', f)}{df}, \tag{9.111}$$

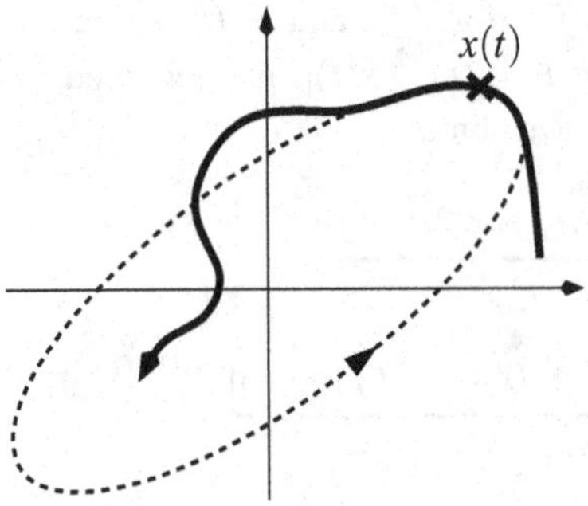

Figure 9.5 The dark solid line shows the time-domain signal $x(t)$ as it moves through the complex plane. At $t' = t$, the ellipse is a perfect approximation.

with

$$\mathbf{V}_{x\Xi}(t, f)\mathrm{d}f = E\left[x(t)\mathrm{d}\boldsymbol{\Xi}^{\mathrm{H}}(t, f)\right]$$
$$= \left[V_{xx}^*(t, f) \quad \widetilde{V}_{xx}(t, -f) \quad V_{xx}^*(t, -f) \quad \widetilde{V}_{xx}(t, f)\right]\mathrm{d}f \tag{9.112}$$

and

$$\mathbf{K}_{\Xi\Xi}(t, f)(\mathrm{d}f)^2 = E\left[\mathrm{d}\boldsymbol{\Xi}(t, f)\mathrm{d}\boldsymbol{\Xi}^{\mathrm{H}}(t, f)\right]. \tag{9.113}$$

In (9.111), $(\cdot)^{\dagger}$ denotes the pseudo-inverse, which is necessary because $\mathbf{K}_{\Xi\Xi}(t, f)$ can be singular. Some background on the pseudo-inverse is given in Section A1.3.2 of Appendix 1. Finding a closed-form solution for the ellipse $\varepsilon_{t,f}(t')$ is tedious because it involves the pseudo-inverse of the 4×4 matrix $\mathbf{K}_{\Xi\Xi}(t, f)$. However, in special cases (i.e., proper, WSS, or analytic signals), $\mathbf{K}_{\Xi\Xi}(t, f)$ has many zero entries and the computations simplify accordingly. In particular, in the WSS case, $\varepsilon_{t,f}(t')$ simplifies to the stationary ellipse $\varepsilon_f(t)$ in (9.109). The analytic case is discussed in some detail below.

9.4.1 Ellipse properties

In order to measure how well the local ellipse $\varepsilon_{t,f}(t')$ approximates $x(t)$ at $t' = t$, we may consult the magnitude-squared *time–frequency coherence*[13]

$$|\bar{\rho}_{xx}(t, f)|^2 = \frac{\mathbf{V}_{x\Xi}(t, f)\mathbf{K}_{\Xi\Xi}^{\dagger}(t, f)\mathbf{V}_{x\Xi}^{\mathrm{H}}(t, f)}{r_{xx}(t, 0)}, \tag{9.114}$$

which is closely related to the approximation error at $t' = t$:

$$E|\varepsilon_{t,f}(t) - x(t)|^2 = r_{xx}(t, 0)(1 - |\bar{\rho}_{xx}(t, f)|^2). \tag{9.115}$$

The magnitude-squared time–frequency coherence satisfies $0 \le |\bar{\rho}_{xx}(t, f)|^2 \le 1$. If $|\bar{\rho}_{xx}(t, f)|^2 = 1$, the ellipse $\varepsilon_{t,f}(t')$ is a perfect approximation of $x(t)$ at $t' = t$. This is illustrated in Fig. 9.5. If $|\bar{\rho}_{xx}(t, f)|^2 = 0$, the best-fit ellipse $\varepsilon_{t,f}(t')$ has zero amplitude. The time–frequency coherence tells us which regions of the time–frequency plane our analysis should focus on: these are the points (t, f) that have magnitude-squared coherence close to 1.

As we have seen above, the random variables $d\zeta(t, f)$ and $d\zeta(t, -f)$ in the nonstationary case play the roles of the spectral increments $d\xi(f)$ and $d\xi(-f)$ in the WSS case. The properties of the local ellipse may therefore be analyzed using the augmented ESD matrix $\underline{\mathbb{J}}_{\varepsilon\varepsilon}(t, f)$ defined by

$$\underline{\mathbb{J}}_{\varepsilon\varepsilon}(t, f)(df)^2 = \begin{bmatrix} J_{\varepsilon\varepsilon}(t, f) & \tilde{J}_{\varepsilon\varepsilon}(t, f) \\ \tilde{J}^*_{\varepsilon\varepsilon}(t, f) & J_{\varepsilon\varepsilon}(t, -f) \end{bmatrix} (df)^2 \tag{9.116}$$

$$= E \begin{bmatrix} |d\zeta(t, f)|^2 & d\zeta(t, f) d\zeta(t, -f) \\ d\zeta^*(t, f) d\zeta^*(t, -f) & |d\zeta(t, -f)|^2 \end{bmatrix}, \tag{9.117}$$

instead of the augmented PSD matrix $\underline{\mathbb{P}}_{xx}(f)$ in the WSS case.[14] Just as in the WSS case discussed in Section 8.4.2, the expected orientation of the ellipse is then approximated by

$$\tan(2\psi(t, f)) = \frac{\operatorname{Im} \tilde{J}_{\varepsilon\varepsilon}(t, f)}{\operatorname{Re} \tilde{J}_{\varepsilon\varepsilon}(t, f)} \tag{9.118}$$

and the area by

$$\pi |J_{\varepsilon\varepsilon}(t, f) - J_{\varepsilon\varepsilon}(t, -f)| (df)^2, \tag{9.119}$$

and its shape is approximately characterized by

$$\sin(2\chi(t, f)) = \frac{J_{\varepsilon\varepsilon}(t, f) - J_{\varepsilon\varepsilon}(t, -f)}{J_{\varepsilon\varepsilon}(t, f) + J_{\varepsilon\varepsilon}(t, -f)}. \tag{9.120}$$

The analysis of polarization and coherence proceeds completely analogously to Section 8.4.3. For instance, as in (8.79), the degree of polarization is

$$\Phi(t, f) = \sqrt{1 - \frac{4 \det \underline{\mathbb{J}}_{\varepsilon\varepsilon}(t, f)}{\operatorname{tr}^2 \underline{\mathbb{J}}_{\varepsilon\varepsilon}(t, f)}}. \tag{9.121}$$

9.4.2 Analytic signals

For analytic signals, these expressions simplify. If $x(t)$ is analytic, $d\xi(-f) = 0$ and the local ellipse becomes

$$\varepsilon_{t,f}(t') = \underbrace{W_1(t, f) d\xi(f)}_{d\zeta(t, f)} e^{j2\pi f t'} + \underbrace{W_2(t, -f) d\xi^*(f)}_{d\zeta(t, -f)} e^{-j2\pi f t'}. \tag{9.122}$$

Thus,

$$\underline{\mathbb{J}}_{\varepsilon\varepsilon}(t, f)(df)^2 = E \begin{bmatrix} W_1(t, f) d\xi(f) \\ W_2^*(t, -f) d\xi(f) \end{bmatrix} \begin{bmatrix} W_1^*(t, f) d\xi^*(f) & W_2(t, -f) d\xi^*(f) \end{bmatrix} \tag{9.123}$$

and then

$$\underline{\mathbb{J}}_{\varepsilon\varepsilon}(t,f) = S_{xx}(0,f)\begin{bmatrix} |W_1(t,f)|^2 & W_1(t,f)W_2(t,-f) \\ W_1^*(t,f)W_2^*(t,-f) & |W_2(t,-f)|^2 \end{bmatrix}. \tag{9.124}$$

From (9.111), we determine the filter coefficients

$$W_1(t,f)\mathrm{d}f = \frac{V_{xx}^*(t,f)S_{xx}(0,f) - \widetilde{V}_{xx}(t,f)\widetilde{S}_{xx}^*(2f,f)\mathrm{e}^{-\mathrm{j}4\pi ft}}{S_{xx}^2(0,f) - |\widetilde{S}_{xx}(2f,f)|^2}, \tag{9.125}$$

$$W_2(t,-f)\mathrm{d}f = \frac{\widetilde{V}_{xx}(t,f)S_{xx}(0,f) - V_{xx}^*(t,f)\widetilde{S}_{xx}(2f,f)\mathrm{e}^{\mathrm{j}4\pi ft}}{S_{xx}^2(0,f) - |\widetilde{S}_{xx}(2f,f)|^2}, \tag{9.126}$$

provided that $S_{xx}^2(0,f) \neq |\widetilde{S}_{xx}(2f,f)|^2$. If $S_{xx}^2(0,f) = |\widetilde{S}_{xx}(2f,f)|^2$, then $\mathrm{d}\xi^*(f) = \mathrm{e}^{\mathrm{j}\alpha}\,\mathrm{d}\xi(f)$ for some real α and the WLMMSE estimator becomes strictly linear. We find for this special case

$$W_1(t,f)\mathrm{d}f = \frac{V_{xx}^*(t,f)}{S_{xx}(0,f)}, \tag{9.127}$$

$$W_2(t,-f)\mathrm{d}f = 0, \tag{9.128}$$

which makes the approximating ellipse a circle.

It is interesting to examine the ellipse shape and polarization, which, for $S_{xx}^2(0,f) \neq |\widetilde{S}_{xx}(2f,f)|^2$, can be done through the angle $\chi(t,f)$ defined by

$$\sin(2\chi(t,f)) = \frac{(|V_{xx}(t,f)|^2 - |\widetilde{V}_{xx}(t,f)|^2)(S_{xx}^2(0,f) - |\widetilde{S}_{xx}(2f,f)|^2)}{D}, \tag{9.129}$$

where

$$\begin{aligned} D = {} & (|V_{xx}(t,f)|^2 + |\widetilde{V}_{xx}(t,f)|^2)(S_{xx}^2(0,f) + |\widetilde{S}_{xx}(2f,f)|^2) \\ & - 4\,\mathrm{Re}\left\{V_{xx}^*(t,f)\widetilde{V}_{xx}^*(t,f)S_{xx}(0,f)\widetilde{S}_{xx}(2f,f)\mathrm{e}^{\mathrm{j}4\pi ft}\right\}. \end{aligned}$$

If $x(t)$ is proper at time t and frequency f, $\widetilde{V}_{xx}(t,f) = 0$ and $\widetilde{S}_{xx}(2f,f) = 0$, then $\chi(t,f) = \pi/4$. This says that a proper analytic signal is *counterclockwise circularly polarized*. On the other hand, if $S_{xx}^2(0,f) = |\widetilde{S}_{xx}(2f,f)|^2$, then $\mathrm{d}\xi^*(f) = \mathrm{e}^{\mathrm{j}\alpha}\,\mathrm{d}\xi(f)$ and thus $|V_{xx}(t,f)|^2 = |\widetilde{V}_{xx}(t,f)|^2$. In this case, the signal $x(t)$ can be regarded as maximally improper at frequency f. A maximally improper analytic signal has $\chi(t,f) = \pi/4$ and is therefore also *counterclockwise circularly polarized.*

Since $\det \underline{\mathbb{J}}_{\varepsilon\varepsilon}(t,f) = 0$ for all (t,f), all analytic signals are *completely polarized.* Yet, while $|\widetilde{S}_{xx}(2f,f)|^2 \leq S_{xx}^2(0,f)$, the magnitude of the HR-TFD does not provide an upper bound on the magnitude of the CR-TFD, i.e., $|\widetilde{V}_{xx}(t,f)|^2 \nleq |V_{xx}(t,f)|^2$. Moreover, $|V_{xx}(t,f)|^2 = |\widetilde{V}_{xx}(t,f)|^2$ does not imply $S_{xx}^2(0,f) = |\widetilde{S}_{xx}(2f,f)|^2$. Therefore it is possible that $x(t)$ is clockwise polarized at (t,f), i.e., $\chi(t,f) < 0$, provided that the signal is "sufficiently improper" at (t,f). This result may seem surprising, considering that an analytic signal is synthesized from *counterclockwise* phasors only.

The quality of the approximation can be judged by computing the magnitude-squared time–frequency coherence $|\bar{\rho}_{xx}(t,f)|^2$ defined in (9.114). For analytic signals, we obtain the simplified expression

$$|\tilde{\rho}_{xx}(t,f)|^2 = \frac{N}{r_{xx}(t,0)(S_{xx}^2(0,f) - |\widetilde{S}_{xx}(2f,f)|^2)}, \tag{9.130}$$

where

$$N = S_{xx}(0,f)(|V_{xx}(t,f)|^2 + |\widetilde{V}_{xx}(t,f)|^2) - 2\,\mathrm{Re}[V_{xx}^*(t,f)\widetilde{V}_{xx}^*(t,f)\widetilde{S}_{xx}(2f,f)e^{j4\pi ft}],$$

if $S_{xx}^2(0,f) \neq |\widetilde{S}_{xx}(2f,f)|^2$. Both in the proper case, which is characterized by $\widetilde{V}_{xx}(t,f) = 0$ and $\widetilde{S}_{xx}(2f,f) = 0$, and in the maximally improper case, characterized by $S_{xx}^2(0,f) = |\widetilde{S}_{xx}(2f,f)|^2$, the magnitude-squared time–frequency coherence simplifies to the magnitude-squared rotational time–frequency coherence (9.93), i.e.,

$$|\tilde{\rho}_{xx}(t,f)|^2 = |\rho_{xx}(t,f)|^2 = \frac{|V_{xx}(t,f)|^2}{r_{xx}(t,0)S_{xx}(0,f)}. \tag{9.131}$$

In these cases $x(t)$ can be estimated from counterclockwise rotating phasors only. In other words, the optimum WLMMSE estimator is the LMMSE estimator, i.e., $W_2(t,-f) = 0$.

9.5 Higher-order statistics

We conclude this chapter with a very brief introduction to the higher-order statistics of a continuous-time signal $x(t)$.[15] We denote the nth-order moment function by

$$m_{x,\diamond}(t,\boldsymbol{\tau}) = E\left[x^{\diamond_n}(t)\prod_{i=1}^{n-1} x^{\diamond_i}(t+\tau_i)\right], \tag{9.132}$$

where, as in Section 8.5, $\boldsymbol{\tau} = [\tau_1,\ldots,\tau_{n-1}]^{\mathrm{T}}$, and $\diamond = [\diamond_n,\diamond_1,\diamond_2,\ldots,\diamond_{n-1}]^{\mathrm{T}}$ contains elements $\diamond_i$ that are either 1 or the conjugating star $*$. This leads to 2^n different nth-order moment functions, depending on which terms are conjugated. As explained in Section 8.5, not all of these functions are required for a complete statistical description.

We assume that $x(t)$ can be represented as in the CL spectral representation (9.47),

$$x(t) = \int_{-\infty}^{\infty} d\xi(f)e^{j2\pi ft}, \tag{9.133}$$

but $\xi(f)$ is now an Nth-order spectral process with moments defined and bounded up to Nth order. That is, the moment functions can be expressed in terms of the increment process $d\xi(f)$ as

$$m_{x,\diamond}(t,\boldsymbol{\tau}) = \int_{\mathbb{R}^n} E\left[d\xi^{\diamond_n}(-\diamond_n f_n)\prod_{i=1}^{n-1} d\xi^{\diamond_i}(\diamond_i f_i)\right] e^{j2\pi[\mathbf{f}^{\mathrm{T}}\boldsymbol{\tau}+(f_1+\cdots+f_{n-1}-f_n)t]}, \tag{9.134}$$

where $n = 1,\ldots,N$, $\mathbf{f} = [f_1,\ldots,f_{n-1}]^{\mathrm{T}}$, and $\diamond_i f_i = -1$ if $\diamond_i$ is the conjugation star $*$. The "$-$" sign for f_n is to ensure that (9.134) complies with (9.50) and (9.51) in the second-order case $n = 2$. Since t is a global time variable and $\boldsymbol{\tau}$ contains local time lags, we call $f_1, \ldots, f_{n-1}$ global frequencies and $\nu = f_1 + \cdots + f_{n-1} - f_n = \mathbf{f}^{\mathrm{T}}\mathbf{1} - f_n$ a local frequency-offset. With this substitution, we obtain

$$m_{x,\diamond}(t,\boldsymbol{\tau}) = \int_{\mathbb{R}^n} E\left[\mathrm{d}\xi^{\diamond_n}(\diamond_n(\nu - \mathbf{f}^{\mathrm{T}}\mathbf{1})) \prod_{i=1}^{n-1} \mathrm{d}\xi^{\diamond_i}(\diamond_i f_i) \right] \mathrm{e}^{\mathrm{j}2\pi(\mathbf{f}^{\mathrm{T}}\boldsymbol{\tau}+\nu t)}. \tag{9.135}$$

We now define

$$S_{x,\diamond}(\nu,\mathbf{f})\mathrm{d}\nu\, \mathrm{d}^{n-1}\mathbf{f} = E\left[\mathrm{d}\xi^{\diamond_n}(\diamond_n(\nu - \mathbf{f}^{\mathrm{T}}\mathbf{1})) \prod_{i=1}^{n-1} \mathrm{d}\xi^{\diamond_i}(\diamond_i f_i) \right]. \tag{9.136}$$

We call $S_{x,\diamond}(\nu,\mathbf{f})$ the nth-order *spectral correlation*, which we recognize as the n-dimensional Fourier transform of the moment function $m_{x,\diamond}(t,\boldsymbol{\tau})$. The spectral correlation may contain δ-functions (e.g., in the stationary case).

Consider now a signal $x(t)$ that is stationary up to order N. Then none of the moment functions $m_{x,\diamond}(t,\boldsymbol{\tau})$, for $n = 1, \ldots, N$ and arbitrary $\diamond$, depends on t: $m_{x,\diamond}(t,\boldsymbol{\tau}) = m_{x,\diamond}(\boldsymbol{\tau})$. This can be true only if $S_{x,\diamond}(\nu,\mathbf{f})$ is zero outside the *stationary manifold* $\nu = 0$. The $(n-1)$-dimensional Fourier transform of $m_{x,\diamond}(\boldsymbol{\tau})$

$$M_{x,\diamond}(\mathbf{f}) = \int_{\mathbb{R}^{n-1}} m_{x,\diamond}(\boldsymbol{\tau})\mathrm{e}^{-\mathrm{j}2\pi\mathbf{f}^{\mathrm{T}}\boldsymbol{\tau}}\, \mathrm{d}^{n-1}\boldsymbol{\tau}, \tag{9.137}$$

is the nth-order moment spectrum introduced in (8.97). The moment spectrum and spectral correlation of a stationary process are related as

$$S_{x,\diamond}(\nu,\mathbf{f}) = M_{x,\diamond}(\mathbf{f})\delta(\nu). \tag{9.138}$$

The moment spectrum can be expressed directly in terms of the increments of the spectral process as

$$M_{x,\diamond}(\mathbf{f})\, \mathrm{d}^{n-1}\mathbf{f} = E\left[\mathrm{d}\xi^{\diamond_n}(-\diamond_n\, \mathbf{f}^{\mathrm{T}}\mathbf{1}) \prod_{i=1}^{n-1} \mathrm{d}\xi^{\diamond_i}(\diamond_i f_i) \right]. \tag{9.139}$$

Notes

1 Section 9.1 closely follows the presentation by Schreier *et al.* (2005). This paper is ©IEEE, and portions are reused with permission.

2 For Grenander's theorem, see for instance, Section VI.B.2 in Poor (1998).

3 Section 9.2 draws on material presented by Picinbono and Bondon (1997) and Schreier and Scharf (2003b).

4 Note that Schreier and Scharf (2003b) use a slightly different way of defining global and local variables: they define $t_1 = t$, $t_2 = t - \tau$ and thus $f_1 = f + \nu$, $f_2 = f$, whereas we define $t_1 = t + \tau$, $t_2 = t$ and thus $f_1 = f$, $f_2 = f - \nu$. While this leads to results that look slightly different, they are fundamentally the same.

5 The Rihaczek distribution for deterministic signals was introduced by Rihaczek (1968). The extension to stochastic signals that is based on harmonizability is due to Martin (1982). Note that we do not follow the (somewhat confusing) convention of differentiating between time–frequency representations for deterministic and stochastic signals by calling the former

"distributions" and the latter "spectra." Time–frequency distributions for stochastic signals in general are discussed by Flandrin (1999), and the Rihaczek distribution in particular by Scharf *et al.* (2005).

6 For a discussion of Wigner's theorem and its consequences for time–frequency analysis, see Flandrin (1999).

7 The Wigner–Ville distribution is the Fourier transform (in τ) of $E[x(t+\tau/2)x^{(*)}(t-\tau/2)]$. That is, the local time lag τ is split symmetrically between $t_1 = t+\tau/2$ and $t_2 = t-\tau/2$, which leads to a symmetric split of the frequency-offset ν between the frequency variables $f_1 = f+\nu/2$ and $f_2 = f-\nu/2$. For a thorough treatment of the Wigner–Ville distribution, see Flandrin (1999).

8 Estimating $x(t)$ from the random variable $\mathrm{d}\xi(f)\mathrm{e}^{\mathrm{j}2\pi ft}$ assumes that $x(t)$ is an energy signal whose Fourier transform $X(f) = \mathrm{d}\xi(f)/df$ exists. If $x(t)$ is not an energy signal, we must estimate the *increment* $x(t)\mathrm{d}f$ instead. This requires only minor changes to the remaining development in Sections 9.3 and 9.4.

9 Following the terminology introduced in Chapter 4, $\rho_{xx}(t, f)$ is a *rotational* correlation coefficient between $x(t)$ and $\mathrm{d}\xi(f)\mathrm{e}^{\mathrm{j}2\pi ft}$, at fixed t and f. However, in this context, the term "coherence" is preferred over "correlation coefficient." The time–frequency coherence should not be confused with the coherence introduced in Definition 8.2, which is a frequency-domain coherence between the rotary components at a given frequency. The time–frequency coherence does *not* simplify to the frequency-domain coherence for a WSS signal.

10 Cohen's class was introduced in quantum mechanics by Cohen (1966). Members of this class can also be regarded as estimators of the Wigner–Ville distribution, see Martin and Flandrin (1985). The estimator of Section 9.3.2 was presented by Scharf *et al.* (2005), building upon prior work by Scharf and Friedlander (2001). The paper Scharf *et al.* (2005) is ©IEEE, and portions are reused with permission.

11 The evaluation of the statistical properties of the HR/CR-TFD estimators is due to Scharf *et al.* (2005), following the lead of Martin and Flandrin (1985). Note again that the results presented in Section 9.3.2 look different at first than those of Scharf *et al.* (2005) even though they are really identical. This is because Scharf *et al.* (2005) define $r_{xx}[k,\kappa] = E(x[k]x^*[k-\kappa])$, whereas this book defines $r_{xx}[k,\kappa] = E(x[k+\kappa]x^*[k])$.

12 Section 9.4 follows Schreier (2008b). The paper is ©IEEE, and portions are reused with permission. Previous extensions of rotary-component and polarization analysis to nonstationary signals have been centered mainly around the wavelet transform, and to a much lesser extent, the STFT. The interested reader is referred to Lilly and Park (1995), Olhede and Walden (2003a, 2003b), Lilly and Gascard (2006), and Roueff *et al.* (2006).

13 In the language of Chapter 4, $\bar{\rho}_{xx}(t, f)$ is a *total* correlation coefficient between $x(t)$ and $[\mathrm{d}\xi(f)\mathrm{e}^{\mathrm{j}2\pi ft}, \mathrm{d}\xi(-f)\mathrm{e}^{-\mathrm{j}2\pi ft}]$.

14 This assumes that $x(t)$ is an energy signal with bounded ESD. If $x(t)$ is a power signal, we have to work with PSDs instead. That is, instead of using $J_{\varepsilon\varepsilon}(t, f)(\mathrm{d}f)^2 = E|\mathrm{d}\zeta(t, f)|^2$, we use $P_{\varepsilon\varepsilon}(t, f)\mathrm{d}f = E|\mathrm{d}\zeta(t, f)|^2$. This requires only minor changes in the remainder of the section.

15 A detailed discussion of higher-order statistics for nonstationary signals (albeit real-valued) is given by Hanssen and Scharf (2003).

10 Cyclostationary processes

Cyclostationary processes are an important class of nonstationary processes that have periodically varying correlation properties. They can model periodic phenomena occurring in science and technology, including communications (modulation, sampling, and multiplexing), meteorology, oceanography, climatology, astronomy (rotation of the Earth and other planets), and economics (seasonality). While cyclostationarity can manifest itself in statistics of arbitrary order, we will restrict our attention to phenomena in which the second-order correlation and complementary correlation functions are periodic in their global time variable.

Our program for this chapter is as follows. In Section 10.1, we discuss the spectral properties of harmonizable cyclostationary processes. We have seen in Chapter 8 that the second-order averages of a WSS process are characterized by the power spectral density (PSD) and complementary power spectral density (C-PSD). These each correspond to a single δ-ridge (the stationary manifold) in the spectral correlation and complementary spectral correlation. Cyclostationary processes have a (possibly countably infinite) number of so-called *cyclic* PSDs and C-PSDs. These correspond to δ-ridges in the spectral correlation and complementary spectral correlation that are parallel to the stationary manifold. In Section 10.2, we derive the cyclic PSDs and C-PSDs of linearly modulated digital communication signals. We will see that there are two types of cyclostationarity: one related to the symbol rate, the other to impropriety and carrier modulation.

Because cyclostationary processes are spectrally correlated between different frequencies, they have spectral redundancy. This redundancy can be exploited in optimum estimation. The widely linear minimum mean-squared error filter is called the *cyclic Wiener filter*, which is discussed in Section 10.3. It can be considered the cyclostationary extension of the WSS Wiener filter. The cyclic Wiener filter consists of a bank of linear shift-invariant filters that are applied to frequency-shifted versions of the signal and its conjugate. In Section 10.4, we develop an efficient causal filter-bank implementation of the cyclic Wiener filter for cyclostationary discrete-time series, which is based on a connection between scalar-valued cyclostationary and vector-valued WSS time series.

We need to emphasize that, due to space constraints, we can barely scratch the surface of the very rich topic of cyclostationarity in this chapter. For further information, we refer the reader to the review article by Gardner *et al.* (2006) and the extensive bibliography by Serpedin *et al.* (2005).

10.1 Characterization and spectral properties

Cyclostationary processes have a correlation function $r_{xx}(t, \tau) = E[x(t + \tau)x^*(t)]$ and complementary correlation function $\tilde{r}_{xx}(t, \tau) = E[x(t + \tau)x(t)]$ that are both periodic in their global time variable t.

Definition 10.1. *A zero-mean process $x(t)$ is* cyclostationary (CS) *if there exists $T > 0$ such that*

$$r_{xx}(t, \tau) = r_{xx}(t + T, \tau), \tag{10.1}$$

$$\tilde{r}_{xx}(t, \tau) = \tilde{r}_{xx}(t + T, \tau) \tag{10.2}$$

for all t and τ.

CS processes are sometimes also called *periodically correlated*. If $x(t)$ does not have zero mean, we require that the mean be T-periodic as well, i.e., $\mu_x(t) = \mu_x(t + T)$ for all t. Note that $r_{xx}(t, \tau)$ might actually be T_1-periodic and $\tilde{r}_{xx}(t, \tau)$ T_2-periodic. The period T would then be the least common multiple of T_1 and T_2. It is important to stress that the periodicity is in the global time variable t, not the local time offset τ. Periodicity in τ may occur *independently* of periodicity in t. It should also be emphasized that periodicity refers to the correlation and complementary correlation functions, not the process itself. That is, in general, $x(t) \neq x(t + T)$.

For WSS processes, (10.1) and (10.2) hold for arbitrary $T > 0$. Thus, WSS processes are a subclass of CS processes. The CS processes we consider are more precisely called *wide-sense* (or second-order) CS, since Definition 10.1 considers only second-order statistics. A process can also be higher-order CS if higher-order correlation functions are periodic in t, and strict-sense CS if the probability density function is periodic in t.

Adding two uncorrelated CS processes with periods T_1 and T_2 produces another CS process whose period is the least common multiple (LCM) of T_1 and T_2. However, if the two periods are incommensurate, e.g., $T_1 = 1$ and $T_2 = \pi$, then the LCM is ∞. Such a process is CS in a generalized sense, which is called *almost CS* because its correlation function is an *almost periodic* function in the sense of Bohr. For fixed τ, the correlation and complementary correlation functions of an almost CS process are each a possibly infinite sum of periodic functions. Many of the results for CS processes generalize to almost CS processes.

10.1.1 Cyclic power spectral density

We now discuss the spectral properties of *harmonizable* CS processes, which possess a Cramér–Loève (CL) spectral representation (cf. Section 9.2). The spectral correlation and complementary correlation are the two-dimensional Fourier transforms of the time correlation and complementary correlation function. However, since $r_{xx}(t, \tau)$ and $\tilde{r}_{xx}(t, \tau)$ are periodic in t, we compute the *Fourier series coefficients* rather than the

Fourier transform in t. These Fourier series coefficients are

$$p_{xx}(n/T,\tau) = \frac{1}{T}\int_0^T r_{xx}(t,\tau)\mathrm{e}^{-\mathrm{j}2\pi nt/T}\,\mathrm{d}t, \tag{10.3}$$

$$\tilde{p}_{xx}(n/T,\tau) = \frac{1}{T}\int_0^T \tilde{r}_{xx}(t,\tau)\mathrm{e}^{-\mathrm{j}2\pi nt/T}\,\mathrm{d}t, \tag{10.4}$$

and they are called the *cyclic correlation function* and *cyclic complementary correlation function*, respectively. There is a subtle difference between the cyclic correlation function and the ambiguity function, which is defined as the Fourier *transform* of the correlation function in t (see Section 9.2.1). The cyclic correlation function and the ambiguity function have different physical units.

The Fourier transform of the cyclic correlation and cyclic complementary correlation functions in τ yields the *cyclic power spectral density* and *cyclic complementary power spectral density*:

$$P_{xx}(n/T,f) = \int_{-\infty}^{\infty} p_{xx}(n/T,\tau)\mathrm{e}^{-\mathrm{j}2\pi f\tau}\,\mathrm{d}\tau, \tag{10.5}$$

$$\widetilde{P}_{xx}(n/T,f) = \int_{-\infty}^{\infty} \tilde{p}_{xx}(n/T,\tau)\mathrm{e}^{-\mathrm{j}2\pi f\tau}\,\mathrm{d}\tau. \tag{10.6}$$

The appropriate generalization for the cyclic correlation and complementary correlation of *almost CS* processes is

$$p_{xx}(\nu_n,\tau) = \lim_{T'\to\infty}\frac{1}{T'}\int_{-T'/2}^{T'/2} r_{xx}(t,\tau)\mathrm{e}^{-\mathrm{j}2\pi\nu_n t}\,\mathrm{d}t, \tag{10.7}$$

$$\tilde{p}_{xx}(\tilde{\nu}_n,\tau) = \lim_{T'\to\infty}\frac{1}{T'}\int_{-T'/2}^{T'/2} \tilde{r}_{xx}(t,\tau)\mathrm{e}^{-\mathrm{j}2\pi\tilde{\nu}_n t}\,\mathrm{d}t. \tag{10.8}$$

The frequency offset variables ν_n and $\tilde{\nu}_n$ are called the *cycle frequencies*. The set of cycle frequencies $\{\nu_n\}$ for which $p_{xx}(\nu_n,\tau)$ is not identically zero and the set of cycle frequencies $\{\tilde{\nu}_n\}$ for which $\tilde{p}_{xx}(\tilde{\nu}_n,\tau)$ is not identically zero are both countable (but possibly countably infinite). These two sets do not have to be identical. CS processes are a subclass of almost CS processes in which the frequencies $\{\nu_n\}$ and $\{\tilde{\nu}_n\}$ are contained in a lattice of the form $\{n/T\}$. Relationships (10.5) and (10.6) remain valid for almost CS processes if n/T is replaced with the cycle frequencies ν_n and $\tilde{\nu}_n$:

$$P_{xx}(\nu_n,f) = \int_{-\infty}^{\infty} p_{xx}(\nu_n,\tau)\mathrm{e}^{-\mathrm{j}2\pi f\tau}\,\mathrm{d}\tau, \tag{10.9}$$

$$\widetilde{P}_{xx}(\tilde{\nu}_n,f) = \int_{-\infty}^{\infty} \tilde{p}_{xx}(\tilde{\nu}_n,\tau)\mathrm{e}^{-\mathrm{j}2\pi f\tau}\,\mathrm{d}\tau. \tag{10.10}$$

Harmonizable signals may be expressed as

$$x(t) = \int_{-\infty}^{\infty} \mathrm{d}\xi(f)\mathrm{e}^{\mathrm{j}2\pi ft}. \tag{10.11}$$

The next result for harmonizable almost CS processes, which follows immediately from the CL spectral representation in Result 9.2, connects the cyclic PSD and C-PSD to the second-order properties of the increments of the spectral process $\xi(f)$.

Result 10.1. *The cyclic PSD* $P_{xx}(\nu_n, f)$ *can be expressed as*

$$P_{xx}(\nu_n, f)\mathrm{d}f = E[\mathrm{d}\xi(f)\mathrm{d}\xi^*(f-\nu_n)], \tag{10.12}$$

and the cyclic C-PSD $\widetilde{P}_{xx}(\tilde{\nu}_n, f)$ *as*

$$\widetilde{P}_{xx}(\tilde{\nu}_n, f)\mathrm{d}f = E[\mathrm{d}\xi(f)\mathrm{d}\xi(\tilde{\nu}_n - f)]. \tag{10.13}$$

Moreover, the spectral correlation $S_{xx}(\nu, f)$ and the complementary spectral correlation $\widetilde{S}_{xx}(\nu, f)$, defined in Result 9.2, contain the cyclic PSD and C-PSD on δ-ridges parallel to the stationary manifold $\nu = 0$:

$$S_{xx}(\nu, f) = \sum_n P_{xx}(\nu_n, f)\delta(\nu - \nu_n), \tag{10.14}$$

$$\widetilde{S}_{xx}(\nu, f) = \sum_n \widetilde{P}_{xx}(\tilde{\nu}_n, f)\delta(\nu - \tilde{\nu}_n). \tag{10.15}$$

On the stationary manifold $\nu = 0$, the cyclic PSD is the usual PSD, i.e., $P_{xx}(0, f) = P_{xx}(f)$, and the cyclic C-PSD the usual C-PSD, i.e., $\widetilde{P}_{xx}(0, f) = \widetilde{P}_{xx}(f)$. It is clear from the bounds in Result 9.3 that it is not possible for the spectral correlation $S_{xx}(\nu, f)$ to have only δ-ridges for $\nu \neq 0$. CS processes must always have a WSS component, i.e., they must have a PSD $P_{xx}(f) \not\equiv 0$. However, it is possible that a process has cyclic PSD $P_{xx}(\nu, f) \equiv 0$ for all $\nu \neq 0$ but cyclic C-PSD $\widetilde{P}_{xx}(\tilde{\nu}_n, f) \not\equiv 0$ for some $\tilde{\nu}_n \neq 0$.

Example 10.1. Let $u(t)$ be a real WSS random process with PSD $P_{uu}(f)$. The complex-modulated process $x(t) = u(t)\mathrm{e}^{\mathrm{j}2\pi f_0 t}$ has PSD $P_{xx}(f) = P_{uu}(f - f_0)$ and cyclic PSD $P_{xx}(\nu, f) \equiv 0$ for all $\nu \neq 0$. However, $x(t)$ is CS rather than WSS because the cyclic C-PSD is nonzero outside the stationary manifold for $\nu = 2f_0$: $\widetilde{P}_{xx}(2f_0, f) = P_{uu}(f - f_0)$.

Unfortunately, there is not universal agreement on the nomenclature for CS processes. It is also quite common to call the cyclic PSD $P_{xx}(\nu_n, f)$ the "spectral correlation," a term that we use for $S_{xx}(\nu, f)$. In the CS literature, one can often find the statement "only (almost) CS processes can exhibit spectral correlation." This can thus be translated as "only (almost) CS processes can have $P_{xx}(\nu_n, f) \not\equiv 0$ for $\nu_n \neq 0$," which means that $S_{xx}(\nu, f)$ has δ-ridges outside the stationary manifold $\nu = 0$. However, there also exist processes whose $S_{xx}(\nu, f)$ have δ-ridges that are curves (not necessarily straight lines) in the (ν, f)-plane. These processes have been termed "generalized almost CS" and were first investigated by Izzo and Napolitano (1998).

10.1.2 Cyclic spectral coherence

The cyclic spectral coherence quantifies how much the frequency-shifted spectral process $\xi(f - \nu_n)$ tells us about the spectral process $\xi(f)$. It is defined as the cyclic

PSD, normalized by the PSDs at f and $f - \nu_n$. There is also a complementary version.

Definition 10.2. *The* rotational cyclic spectral coherence *of an (almost) CS process is defined as*

$$\rho_{xx}(\nu_n, f) = \frac{E[\mathrm{d}\xi(f)\mathrm{d}\xi^*(f-\nu_n)]}{\sqrt{E|\mathrm{d}\xi(f)|^2\, E|\mathrm{d}\xi(f-\nu_n)|^2}} = \frac{P_{xx}(\nu_n, f)}{\sqrt{P_{xx}(f)P_{xx}(f-\nu_n)}} \tag{10.16}$$

and the reflectional cyclic spectral coherence *is*

$$\widetilde{\rho}_{xx}(\tilde{\nu}_n, f) = \frac{E[\mathrm{d}\xi(f)\mathrm{d}\xi(\tilde{\nu}_n - f)]}{\sqrt{E|\mathrm{d}\xi(f)|^2\, E|\mathrm{d}\xi(\tilde{\nu}_n - f)|^2}} = \frac{\widetilde{P}_{xx}(\tilde{\nu}_n, f)}{\sqrt{P_{xx}(f)P_{xx}(\tilde{\nu}_n - f)}}. \tag{10.17}$$

The terms "rotational" and "reflectional" are borrowed from the language of Chapter 4, which discusses rotational and reflectional correlation coefficients. Correlation coefficients in the frequency domain are usually called coherences. It is also possible to define a *total* cyclic spectral coherence, which takes into account both Hermitian and complementary correlations.

The squared magnitudes of the cyclic spectral coherences are bounded as $0 \leq |\rho_{xx}(\nu_n, f)|^2 \leq 1$ and $0 \leq |\widetilde{\rho}_{xx}(\tilde{\nu}_n, f)|^2 \leq 1$. If $|\rho_{xx}(\nu_n, f)|^2 = 1$ for a given (ν_n, f), then $\mathrm{d}\xi(f)$ can be perfectly linearly estimated from $\mathrm{d}\xi(f - \nu_n)$. If $|\widetilde{\rho}_{xx}(\tilde{\nu}_n, f)|^2 = 1$ for a given $(\tilde{\nu}_n, f)$, then $\mathrm{d}\xi(f)$ can be perfectly linearly estimated from $\mathrm{d}\xi^*(\tilde{\nu}_n - f)$. Note that, for $\tilde{\nu}_n = 0$, $\widetilde{\rho}_{xx}(0, f)$ is the coherence $\widetilde{\rho}_{xx}(f)$ in Definition 8.2. It is also clear that $\rho_{xx}(0, f) = 1$.

10.1.3 Estimating the cyclic power-spectral density

In practice, the cyclic correlation function and cyclic PSD must be estimated from realizations of the random process. For ergodic WSS processes, ensemble averages can be replaced by time averages over a single realization. Under certain conditions[1] this is also possible for CS processes. For simplicity, we will consider only the Hermitian cyclic correlation and PSD in this section. Expressions for the complementary quantities follow straightforwardly.

Since the correlation function is T-periodic, we need to perform synchronized T-periodic time-averaging. The correlation function is thus estimated as

$$\hat{r}_{xx}(t, \tau) = \lim_{K\to\infty} \frac{1}{2K+1} \sum_{k=-K}^{K} x(t + kT + \tau)x^*(t + kT). \tag{10.18}$$

In essence, this estimator regards each period of the process as a separate realization. Consequently, an estimate of the cyclic correlation function is obtained by

$$\begin{aligned} \hat{p}_{xx}(n/T, \tau) &= \frac{1}{T}\int_0^T \hat{r}_{xx}(t, \tau)\mathrm{e}^{-\mathrm{j}2\pi nt/T}\,\mathrm{d}t \\ &= \lim_{T'\to\infty} \frac{1}{T'}\int_{-T'/2}^{T'/2} x(t+\tau)x^*(t)\mathrm{e}^{-\mathrm{j}2\pi nt/T}\,\mathrm{d}t. \end{aligned} \tag{10.19}$$

This estimator has the form of (10.7) without the expected value operator. It can also be used for almost CS processes if n/T is replaced with ν_n.

A straightforward estimator of the cyclic PSD is the *cyclic periodogram*

$$\Pi_{xx}(\nu_n, f; t_0, T') = \frac{1}{T'} X(f; t_0, T') X^*(f - \nu_n; t_0, T'), \tag{10.20}$$

where

$$X(f; t_0, T') = \int_{t_0 - T'/2}^{t_0 + T'/2} x(t) \mathrm{e}^{-\mathrm{j}2\pi f t}\, \mathrm{d}t \tag{10.21}$$

is the windowed Fourier transform of $x(t)$ of length T', centered at t_0. This estimator is asymptotically unbiased as $T' \to \infty$ but not consistent. A consistent estimator of the cyclic PSD is the *infinitely time-smoothed cyclic periodogram*

$$\hat{P}^t_{xx}(\nu_n, f) = \lim_{\Delta f \to 0} \lim_{T' \to \infty} \frac{1}{T'} \int_{t_0 - T'/2}^{t_0 + T'/2} \Pi_{xx}(\nu_n, f; t, 1/\Delta f)\, \mathrm{d}t \tag{10.22}$$

or the *infinitely frequency-smoothed cyclic periodogram*

$$\hat{P}^f_{xx}(\nu_n, f) = \lim_{\Delta f \to 0} \lim_{T' \to \infty} \frac{1}{\Delta f} \int_{f - \Delta f/2}^{f + \Delta f/2} \Pi_{xx}(\nu_n, f'; t_0, T')\, \mathrm{d}f'. \tag{10.23}$$

The order of the two limits cannot be interchanged. It can be shown that the two estimators $\hat{P}^t_{xx}(\nu_n, f)$ and $\hat{P}^f_{xx}(\nu_n, f)$ are equivalent.[2]

10.2 Linearly modulated digital communication signals

In this section, we derive the cyclic PSD and C-PSD of linearly modulated digital communication signals such as Pulse Amplitude Modulation (PAM), Phase Shift Keying (PSK), and Quadrature Amplitude Modulation (QAM).[3] We will find that there are two sources of cyclostationarity: one is related to the repetition of pulses at the symbol interval, the other to impropriety and carrier modulation.

10.2.1 Symbol-rate-related cyclostationarity

An expression for a digitally modulated complex baseband signal is

$$x(t) = \sum_{k=-\infty}^{\infty} d_k b(t - kT), \tag{10.24}$$

where $b(t)$ is the deterministic transmit pulse shape, T the symbol interval ($1/T$ is the symbol rate), and d_k a complex-valued WSS random data sequence with correlation $r_{dd}[\kappa] = E[d_{k+\kappa} d_k^*]$ and complementary correlation $\tilde{r}_{dd}[\kappa] = E[d_{k+\kappa} d_k]$. The correlation function of $x(t)$ is

$$r_{xx}(t, \tau) = \sum_{k=-\infty}^{\infty} \sum_{m=-\infty}^{\infty} E[d_k d_m^*] b(t + \tau - kT) b^*(t - mT). \tag{10.25}$$

With the variable substitution $\kappa = k - m$, we obtain

$$r_{xx}(t,\tau) = \sum_{\kappa=-\infty}^{\infty} r_{dd}[\kappa] \sum_{m=-\infty}^{\infty} b(t+\tau-\kappa T - mT)b^*(t-mT). \qquad (10.26)$$

It is easy to see that $r_{xx}(t,\tau) = r_{xx}(t+T,\tau)$, and a similar development for the complementary correlation shows that $\tilde{r}_{xx}(t,\tau) = \tilde{r}_{xx}(t+T,\tau)$. Thus, $x(t)$ is CS with period T. The cyclic correlation function is

$$\begin{aligned} p_{xx}(n/T,\tau) &= \frac{1}{T}\int_0^T r_{xx}(t,\tau)\mathrm{e}^{-\mathrm{j}2\pi nt/T}\,\mathrm{d}t \\ &= \frac{1}{T}\sum_{\kappa=-\infty}^{\infty} r_{dd}[\kappa]\sum_{m=-\infty}^{\infty}\int_0^T b(t+\tau-\kappa T-mT)b^*(t-mT)\mathrm{e}^{-\mathrm{j}2\pi nt/T}\,\mathrm{d}t \\ &= \frac{1}{T}\sum_{\kappa=-\infty}^{\infty} r_{dd}[\kappa]\int_{-\infty}^{\infty} b(t+\tau-\kappa T)b^*(t)\mathrm{e}^{-\mathrm{j}2\pi nt/T}\,\mathrm{d}t. \end{aligned} \qquad (10.27)$$

Now we define

$$\phi(n/T,\tau) = \int_{-\infty}^{\infty} b(t+\tau)b^*(t)\mathrm{e}^{-\mathrm{j}2\pi nt/T}\,\mathrm{d}t, \qquad (10.28)$$

whose Fourier transform $\Phi(n/T, f)$ is the spectral correlation of the deterministic pulse shape $b(t) \longleftrightarrow B(f)$:

$$\Phi(n/T,f) = B(f)B^*(f-n/T). \qquad (10.29)$$

Inserting (10.28) into (10.27) yields

$$p_{xx}(n/T,\tau) = \frac{1}{T}\sum_{\kappa=-\infty}^{\infty} r_{dd}[\kappa]\phi(n/T,\tau-\kappa T). \qquad (10.30)$$

Its Fourier transform in τ is the cyclic PSD

$$P_{xx}(n/T,f) = \frac{1}{T}P_{dd}(f)\Phi(n/T,f) = \frac{1}{T}P_{dd}(f)B(f)B^*(f-n/T), \qquad (10.31)$$

where $P_{dd}(f)$ is the PSD of the data sequence d_k, which is the discrete-time Fourier transform of $r_{dd}[\kappa]$. For $n=0$, $P_{xx}(0,f)$ is the PSD. A completely analogous development leads to the cyclic C-PSD

$$\tilde{P}_{xx}(n/T,f) = \frac{1}{T}\tilde{P}_{dd}(f)B(f)B(n/T-f), \qquad (10.32)$$

where $\tilde{P}_{dd}(f)$ is the C-PSD of the data sequence d_k. For $n=0$, $\tilde{P}_{xx}(0,f)$ is the C-PSD. Applying these findings to the special case of an uncorrelated data sequence leads to the following result.

Result 10.2. *If the data sequence d_k is uncorrelated with zero mean, variance σ_d^2, and complementary variance $\tilde{\sigma}_d^2$, then the cyclic PSD and cyclic C-PSD of the linearly* *modulated baseband signal (10.24) are*

$$P_{xx}(n/T, f) = \frac{\sigma_d^2}{T} B(f) B^*(f - n/T), \tag{10.33}$$

$$\widetilde{P}_{xx}(n/T, f) = \frac{\widetilde{\sigma}_d^2}{T} B(f) B(n/T - f). \tag{10.34}$$

We see that the transmit pulse $b(t)$ determines not only the bandwidth of $x(t)$ but also the number of cycle frequencies for which the cyclic PSD and C-PSD are nonzero.

Example 10.2. In Quaternary Phase Shift Keying (QPSK), the transmitted data are uncorrelated, equally likely $d_k \in \{\pm 1, \pm \mathrm{j}\}$, so that $\sigma_d^2 = 1$ and $\widetilde{\sigma}_d^2 = 0$. This produces a proper baseband signal $x(t)$. We will consider two transmit pulse shapes. The first is the rectangular pulse

$$b_1(t) = \begin{cases} 1, & |t| < T/2, \\ \frac{1}{2}, & |t| = T/2, \\ 0, & |t| > T/2 \end{cases}$$

with Fourier transform $B_1(f) = \sin(\pi f T)/(\pi f)$. Another transmit pulse commonly employed in digital communications is the Nyquist roll-off (or raised-cosine) pulse with roll-off factor $0 \le \alpha \le 1$:

$$b_2(t) = \operatorname{sinc}\left(\frac{t}{T}\right) \frac{\cos(\alpha \pi t/T)}{1 - 4\alpha^2 (t/T)^2}. \tag{10.35}$$

Its Fourier transform is

$$B_2(f) = \begin{cases} T, & |f| \le \dfrac{1-\alpha}{2T}, \\ \dfrac{T}{2}\left[1 - \sin\left(\dfrac{\pi T}{\alpha}\left(|f| - \dfrac{1}{2T}\right)\right)\right], & \dfrac{1-\alpha}{2T} < |f| \le \dfrac{1+\alpha}{2T}, \\ 0, & \text{else.} \end{cases} \tag{10.36}$$

Figure 10.1(a) shows the magnitude of the cyclic PSD of QPSK for the rectangular pulse $b_1(t)$, and Fig. 10.1(b) is for the Nyquist pulse $b_2(t)$ with roll-off factor $\alpha = 1$. The rectangular pulse $b_1(t)$ is not bandlimited, so the cyclic PSD is spread out over the entire (ν, f)-plane. The Nyquist pulse $b_2(t)$, on the other hand, has a limited two-sided bandwidth of $2/T$, and the cyclic PSD is nonzero only for $\{\nu_n\} = \{-1/T, 0, 1/T\}$. Since QPSK signals are proper, the cyclic complementary PSD is identically zero.

In Binary Phase Shift Keying (BPSK), the transmitted data are equally likely $d_k \in \{\pm 1\}$, so that $\sigma_d^2 = \widetilde{\sigma}_d^2 = 1$. The rectangular pulse and the Nyquist roll-off pulse are both real-valued, which means $B(f) = B^*(-f)$. Therefore, the BPSK baseband signal is maximally improper and the cyclic C-PSD equals the cyclic PSD: $\widetilde{P}_{xx}(n/T, f) = P_{xx}(n/T, f)$.

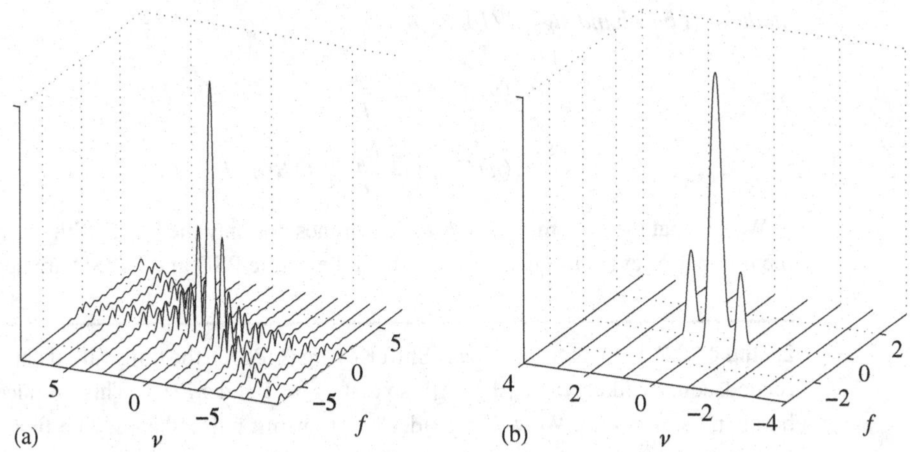

Figure 10.1 The magnitude of the cyclic power spectral density of a QPSK baseband-signal with rectangular transmit pulse (a), and Nyquist roll-off pulse (b), with parameters $T = 1, \alpha = 1$.

10.2.2 Carrier-frequency-related cyclostationarity

Carrier modulation is an additional source of cyclostationarity if the complex baseband signal $x(t)$ is improper.[4] As explained in Section 1.4, the carrier-modulated bandpass signal is

$$p(t) = \operatorname{Re}\{x(t)e^{j2\pi f_0 t}\}. \tag{10.37}$$

We denote the spectral process of $p(t)$ by $\pi(f)$. Its increments are related to the increments of $\xi(f)$ as

$$d\pi(f) = \tfrac{1}{2}[d\xi(f - f_0) + d\xi^*(-f - f_0)], \tag{10.38}$$

and the cyclic PSD of $p(t)$ is therefore

$$\begin{aligned} P_{pp}(\nu, f) = \tfrac{1}{4}[&P_{xx}(\nu, f - f_0) + P^*_{xx}(-\nu, -f - f_0) \\ &+ \widetilde{P}_{xx}(\nu - 2f_0, f - f_0) + \widetilde{P}^*_{xx}(-2f_0 - \nu, -f - f_0)]. \end{aligned} \tag{10.39}$$

We see that the cyclic PSD of the bandpass signal exhibits additional CS components around the cycle frequencies $\nu = \pm 2f_0$ if the baseband signal is improper. Note that, since the modulated signal $p(t)$ is real, its cyclic C-PSD equals the cyclic PSD.

Example 10.3. Figure 10.2 shows the cyclic PSD of a carrier-modulated QPSK signal (a) and BPSK signal (b) using a Nyquist roll-off pulse. Since QPSK has a proper baseband signal, its cyclic PSD has only two peaks, at $\nu = 0$ and $f = \pm f_0$, which correspond to the first two terms in (10.39).

BPSK, on the other hand, has a maximally improper baseband signal with $\widetilde{P}_{xx}(\nu, f) = P_{xx}(\nu, f)$. Its cyclic PSD has two additional peaks located at $\nu = 2f_0, f = f_0$, and $\nu = -2f_0, f = -f_0$, which are due to the last two terms in (10.39).

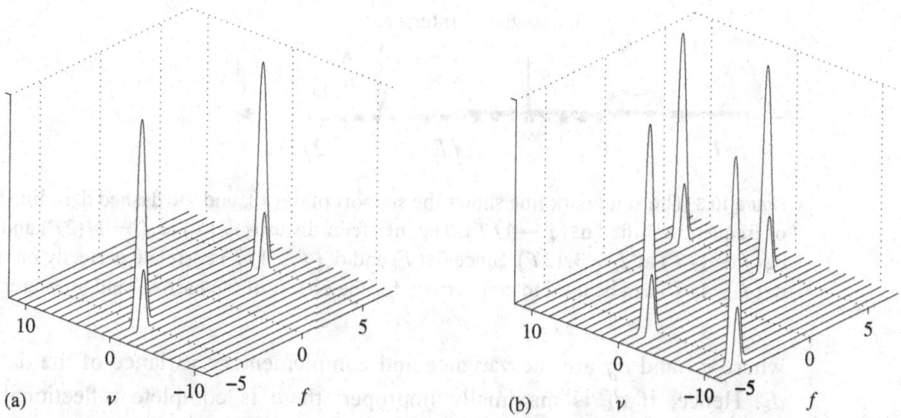

Figure 10.2 The cyclic power spectral density of a QPSK bandpass signal (a) and a BPSK bandpass signal (b) with Nyquist roll-off pulse, with parameters $T = 1$, $\alpha = 1$, $f_0 = 5$.

10.2.3 Cyclostationarity as frequency diversity

Let us compute the rotational cyclic spectral coherence in Definition 10.2 for the baseband signal $x(t)$ in (10.24). Using Result 10.2, we immediately find $|\rho_{xx}(n/T, f)|^2 = 1$ as long as the product $B(f)B^*(f - n/T) \neq 0$. The consequences of this are worth dwelling on. First of all, it means that the baseband signal $x(t)$ is spectrally redundant if its two-sided bandwidth exceeds the Nyquist bandwidth $1/T$. In fact, any nonzero spectral band of width $1/T$ of $d\xi(f)$ can be used to reconstruct the entire process $\xi(f)$ because, for frequencies outside this band, $d\xi(f)$ can be perfectly estimated from frequency-shifted versions $d\xi(f - n/T)$, if $|\rho_{xx}(n/T, f)|^2 = 1$.

Every nonzero spectral band of width $1/T$ can be considered a replica of the signal, which contains the entire information about the transmitted data. Thus, CS processes exhibit a form of *frequency diversity*.[5] Increasing the bandwidth of the transmit pulse $b(t)$ leads to more redundant replicas. This redundancy can be used to combat interference, as illustrated in the following example.

Example 10.4. Consider a baseband QPSK signal, using a Nyquist roll-off pulse with $\alpha = 1$ as described in Example 10.2. This pulse has a two-sided bandwidth of $2/T$ and the spectrum of $x(t)$ thus contains the information about the transmitted data sequence d_k exactly twice. Assume now that there is a narrowband interferer at frequency $f = 1/(2T)$. If this interferer is not CS with period T it will not affect $d\xi(f - 1/T)$ at $f = 1/(2T)$. Hence, $d\xi(f - 1/T)$ can be used to compensate for the effects of the narrowband interferer. This is illustrated in Fig. 10.3.

There is an analogous story for the reflectional cyclic spectral coherence. If we assume a real-valued pulse shape $b(t)$, its Fourier transform satisfies $B(n/T - f) = B^*(f - n/T)$. Thus, as long as the product $B(f)B^*(f - n/T) \neq 0$, we obtain

$$|\tilde{\rho}_{xx}(n/T, f)|^2 = \frac{|\tilde{\sigma}_d^2|^2}{\sigma_d^4}, \tag{10.40}$$

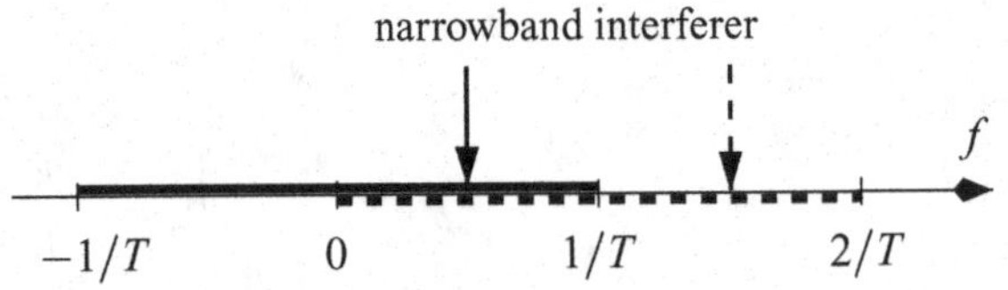

Figure 10.3 The solid dark line shows the support of $d\xi(f)$, and the dashed dark line the support of frequency-shifted $d\xi(f - 1/T)$. The interferer disturbs $d\xi(f)$ at $f = 1/(2T)$ and $d\xi(f - 1/T)$ at $f = 3/(2T)$. Since $d\xi(f)$ and $d\xi(f - 1/T)$ correlate perfectly on $[0, 1/T]$, $d\xi(f - 1/T)$ can be used to compensate for the effects of the narrowband interferer.

where σ_d^2 and $\widetilde{\sigma}_d^2$ are the variance and complementary variance of the data sequence d_k. Hence, if d_k is maximally improper, there is complete reflectional coherence $|\widetilde{\rho}_{xx}(n/T, f)|^2 = 1$ as long as $B(f)B^*(f - n/T) \neq 0$. Thus, $d\xi(f)$ can be perfectly estimated from frequency-shifted conjugated versions $d\xi^*(n/T - f)$. This spectral redundancy comes on top of the spectral redundancy already discussed above. For instance, the spectrum of baseband BPSK, using a Nyquist roll-off pulse with $\alpha = 1$, contains the information about the transmitted real-valued data sequence d_k exactly four times. On the other hand, if d_k is proper (as in QPSK), then $\widetilde{\rho}_{xx}(n/T, f) = 0$ and there is no additional complementary spectral redundancy that can be exploited using conjugate-linear operations.

10.3 Cyclic Wiener filter

The problem discussed in the previous subsection can be generalized to the setting where we observe a random process $y(t)$, with corresponding spectral process $\upsilon(f)$, that is a noisy measurement of a message $x(t)$. Measurement and message are assumed to be individually and jointly (almost) CS. We are interested in constructing a filter that produces an estimate $\hat{x}(t)$ of the message $x(t)$ from the measurement $y(t)$ by exploiting CS properties. In the frequency domain, such a filter adds suitably weighted spectral components of $y(t)$, which have been frequency-shifted by the cyclic frequencies, to produce an estimate of the spectral process:

$$d\hat{\xi}(f) = \sum_{n=1}^{N} G_n(f) d\upsilon(f - \nu_n) + \sum_{n=1}^{\tilde{N}} \widetilde{G}_n(f) d\upsilon^*(\tilde{\nu}_n - f). \tag{10.41}$$

Such a filter is called a (widely linear) *frequency-shift (FRESH) filter*. The corresponding noncausal time-domain estimate is

$$\hat{x}(t) = \sum_{n=1}^{N} g_n(t) * (y(t)e^{j2\pi\nu_n t}) + \sum_{n=1}^{\tilde{N}} \tilde{g}_n(t) * (y^*(t)e^{j2\pi\tilde{\nu}_n t}). \tag{10.42}$$

Since this FRESH filter allows arbitrary frequency shifts $\{\nu_n\}_{n=1}^{N}$ and $\{\tilde{\nu}_n\}_{n=1}^{\tilde{N}}$, with N and $\tilde{N}$ possibly infinite, it can be applied to CS or almost CS signals. Of course, only those frequency shifts that correspond to nonzero cyclic PSD/C-PSD are actually useful.

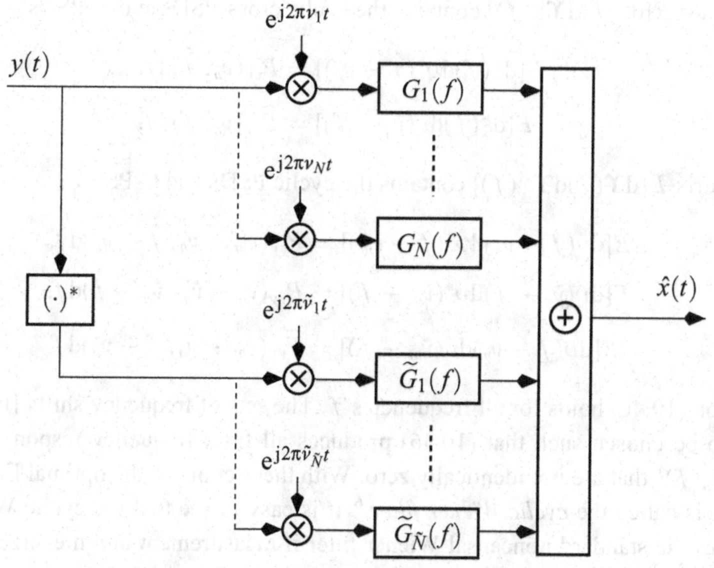

Figure 10.4 Widely linear frequency-shift (FRESH) filter.

For instance, if $y(t)$ is a noisy QPSK baseband signal, using a Nyquist roll-off pulse as described in Example 10.2, then $\{\nu_n\} = \{-1/T, 0, 1/T\}$ and $\{\tilde{\nu}_n\} = \emptyset$ because QPSK is proper. The FRESH filter structure is shown in Fig. 10.4. It consists of a bank of linear shift-invariant filters that are applied to frequency-shifted versions of the signal and its conjugate.

We will now discuss how the optimum FRESH filter responses $\{g_n(t)\} \longleftrightarrow \{G_n(f)\}$ and $\{\tilde{g}_n(t)\} \longleftrightarrow \{\widetilde{G}_n(f)\}$ are determined for minimum mean-squared error estimation. Let us define the $(N+\tilde{N})$-dimensional vectors

$$\mathrm{d}\boldsymbol{\Upsilon}(f) = \left[\mathrm{d}\upsilon(f-\nu_1), \ldots, \mathrm{d}\upsilon(f-\nu_N), \mathrm{d}\upsilon^*(\tilde{\nu}_1 - f), \ldots, \mathrm{d}\upsilon^*(\tilde{\nu}_{\tilde{N}} - f)\right]^{\mathrm{T}}, \quad (10.43)$$

$$\mathbf{F}(f) = \left[G_1(f), \ldots, G_N(f), \widetilde{G}_1(f), \ldots, \widetilde{G}_{\tilde{N}}(f)\right]. \quad (10.44)$$

If $\{\nu_n\} = \{\tilde{\nu}_n\}$, $\mathrm{d}\boldsymbol{\Upsilon}(f)$ and $\mathbf{F}(f)$ both have the structure of augmented vectors. We can now write (10.41) compactly as

$$\mathrm{d}\hat{\xi}(f) = \mathbf{F}(f)\mathrm{d}\boldsymbol{\Upsilon}(f). \quad (10.45)$$

This turns the FRESH filtering problem into a standard Wiener filtering problem. In order to minimize the mean-squared error $E|\mathrm{d}\hat{\xi}(f) - \mathrm{d}\xi(f)|^2$ for any given frequency f, we need to choose

$$\mathbf{F}(f) = E[\mathrm{d}\xi(f)\mathrm{d}\boldsymbol{\Upsilon}^{\mathrm{H}}(f)]E[\mathrm{d}\boldsymbol{\Upsilon}(f)\mathrm{d}\boldsymbol{\Upsilon}^{\mathrm{H}}(f)]^{\dagger}. \quad (10.46)$$

The vector $E[\mathrm{d}\xi(f)\mathrm{d}\boldsymbol{\Upsilon}^{\mathrm{H}}(f)]$ contains the cyclic cross-PSDs and C-PSDs

$$E[\mathrm{d}\xi(f)\mathrm{d}\upsilon^*(f-\nu_n)] = P_{xy}(\nu_n, f)\mathrm{d}f, \tag{10.47}$$

$$E[\mathrm{d}\xi(f)\mathrm{d}\upsilon(\tilde{\nu}_n - f)] = \widetilde{P}_{xy}(\tilde{\nu}_n, f)\mathrm{d}f. \tag{10.48}$$

The matrix $E[\mathrm{d}\boldsymbol{\Upsilon}(f)\mathrm{d}\boldsymbol{\Upsilon}^{\mathrm{H}}(f)]$ contains the cyclic PSDs and C-PSDs

$$E[\mathrm{d}\upsilon(f-\nu_n)\mathrm{d}\upsilon^*(f-\nu_m)] = P_{yy}(\nu_m-\nu_n, f-\nu_n)\mathrm{d}f, \tag{10.49}$$

$$E[\mathrm{d}\upsilon(\tilde{\nu}_n-f)\mathrm{d}\upsilon^*(\tilde{\nu}_m-f)] = P_{yy}(\tilde{\nu}_n-\tilde{\nu}_m, \tilde{\nu}_n-f)\mathrm{d}f, \tag{10.50}$$

$$E[\mathrm{d}\upsilon(f-\nu_n)\mathrm{d}\upsilon(\tilde{\nu}_m-f)] = \widetilde{P}_{yy}(\tilde{\nu}_m-\nu_n, f-\nu_n)\mathrm{d}f. \tag{10.51}$$

Equation (10.46) holds for all frequencies f. The sets of frequency shifts $\{\nu_n\}$ and $\{\tilde{\nu}_n\}$ need to be chosen such that (10.46) produces all filter frequency responses $\{G_n(f)\}$ and $\{\widetilde{G}_n(f)\}$ that are not identically zero. With these choices, the optimal FRESH filter (10.41) is called the *cyclic Wiener filter*.[6] It is easy to see that the cyclic Wiener filter becomes the standard noncausal Wiener filter if measurement and message are jointly WSS.

10.4 Causal filter-bank implementation of the cyclic Wiener filter

We now develop an efficient causal filter-bank implementation of the cyclic Wiener filter for CS discrete-time series. Starting from first principles, we construct a widely linear shift-invariant Multiple-In Multiple-Out (WLSI-MIMO) filter for CS sequences, which is based on a connection between scalar-valued CS and vector-valued WSS processes. This filter suffers from an inherent delay and is thus unsuitable for a causal implementation. This problem is addressed in a modified implementation, which uses a sliding input window and filter branches with staggered execution times. We then show that this implementation is equivalent to FRESH filtering – which we knew it would have to be if it is supposed to optimum. So what have we gained from this detour? By establishing a connection between FRESH filtering and WLSI-MIMO filtering, we are able to enlist the spectral factorization algorithm from Section 8.3 to find the *causal* part of the cyclic Wiener filter. The presentation in this section follows Spurbeck and Schreier (2007).

10.4.1 Connection between scalar CS and vector WSS processes

Let $y[k]$ be a discrete-time CS process whose correlation function $r_{yy}[k,\kappa] = r_{yy}[k+M,\kappa]$ and complementary correlation function $\tilde{r}_{yy}[k,\kappa] = \tilde{r}_{yy}[k+M,\kappa]$ are M-periodic. For simplicity, we assume that the period M is the same *integer* for both correlation and complementary correlation function. If this is not the case, the sampling rate must be converted so that the sampling times are synchronized with the period of correlation. Such a sampling-rate conversion will not be possible for almost CS signals.

Therefore, our results in this section do not apply directly to almost CS signals even though such an extension exists.

We first establish a connection between scalar-valued CS and vector-valued WSS processes.[7] We perform a serial-to-parallel conversion that produces the vector-valued sequence

$$\begin{aligned}\mathbf{y}[k] &= \begin{bmatrix} y[kM] & y[kM-1] & \cdots & y[(k-1)M+1] \end{bmatrix}^{\mathrm{T}} \\ &= \begin{bmatrix} y_0[k] & y_1[k] & \cdots & y_{M-1}[k] \end{bmatrix}^{\mathrm{T}}. \end{aligned} \tag{10.52}$$

For convenience, we index elements in vectors and matrices starting from 0 in this section. The subsequence $y_m[k]$, $m \in \{0, 1, \ldots, M-1\}$, is obtained by delaying the original CS scalar sequence $y[k]$ by m and then decimating it by a factor of M. Each subsequence $y_m[k]$ is WSS and any two subsequences $y_m[k]$ and $y_n[k]$ are also jointly WSS. Hence, the vector-valued time series $\mathbf{y}[k]$ is WSS with matrix correlation

$$\begin{aligned}&\mathbf{R}_{yy}[\kappa] \\ &\quad = E\left[\mathbf{y}[k+\kappa]\mathbf{y}^{\mathrm{H}}[k]\right] \\ &\quad = \begin{bmatrix} r_{yy}[0, \kappa M] & r_{yy}[M-1, \kappa M+1] & \cdots & r_{yy}[1, (\kappa+1)M-1] \\ r^*_{yy}[M-1, -\kappa M+1] & r_{yy}[M-1, \kappa M] & \cdots & r_{yy}[1, (\kappa+1)M-2] \\ \vdots & \vdots & \ddots & \vdots \\ r^*_{yy}[1, (-\kappa+1)M-1] & r^*_{yy}[1, (-\kappa+1)M-2] & \cdots & r_{yy}[1, \kappa M] \end{bmatrix}\end{aligned} \tag{10.53}$$

and matrix complementary correlation

$$\begin{aligned}&\widetilde{\mathbf{R}}_{yy}[\kappa] \\ &\quad = E\left[\mathbf{y}[k+\kappa]\mathbf{y}^{\mathrm{T}}[k]\right] \\ &\quad = \begin{bmatrix} \tilde{r}_{yy}[0, \kappa M] & \tilde{r}_{yy}[M-1, \kappa M+1] & \cdots & \tilde{r}_{yy}[1, (\kappa+1)M-1] \\ \tilde{r}_{yy}[M-1, -\kappa M+1] & \tilde{r}_{yy}[M-1, \kappa M] & \cdots & \tilde{r}_{yy}[1, (\kappa+1)M-2] \\ \vdots & \vdots & \ddots & \vdots \\ \tilde{r}_{yy}[1, (-\kappa+1)M-1] & \tilde{r}_{yy}[1, (-\kappa+1)M-2] & \cdots & \tilde{r}_{yy}[1, \kappa M] \end{bmatrix},\end{aligned} \tag{10.54}$$

both independent of k. As required for a vector-valued WSS time series, the matrix correlation satisfies $\mathbf{R}_{yy}[\kappa] = \mathbf{R}^{\mathrm{H}}_{yy}[-\kappa]$, and the matrix complementary correlation $\widetilde{\mathbf{R}}_{yy}[\kappa] = \widetilde{\mathbf{R}}^{\mathrm{T}}_{yy}[-\kappa]$. The z-transform of $\mathbf{R}_{yy}[\kappa]$ will be denoted by $\mathbb{P}_{yy}(z)$, and the z-transform of $\widetilde{\mathbf{R}}_{yy}[\kappa]$ by $\widetilde{\mathbb{P}}_{yy}(z)$. These can be arranged in the augmented matrix

$$\underline{\mathbb{P}}_{yy}(z) = \begin{bmatrix} \mathbb{P}_{yy}(z) & \widetilde{\mathbb{P}}_{yy}(z) \\ \widetilde{\mathbb{P}}^*_{yy}(z^*) & \mathbb{P}^*_{yy}(z^*) \end{bmatrix}. \tag{10.55}$$

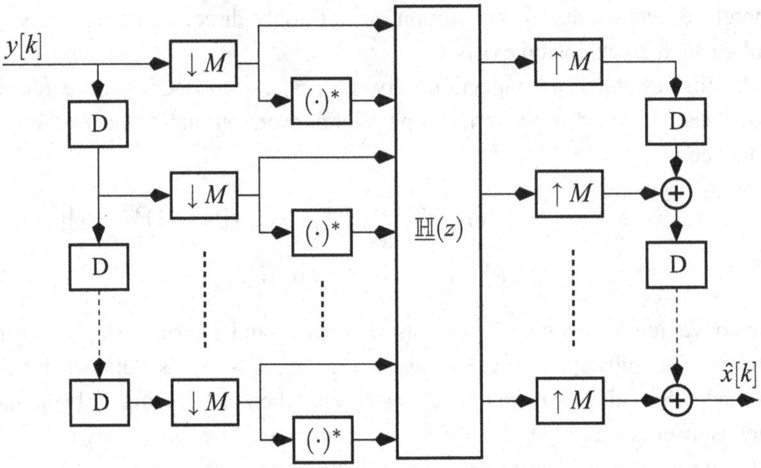

Figure 10.5 A widely linear polyphase filter bank for cyclostationary time series. The filter $\underline{\mathbb{H}}(z)$ operates at a reduced rate that is $1/M$ times the rate of the CS sequence.

10.4.2 Sliding-window filter bank

The correspondence between scalar-valued CS and vector-valued WSS processes allows us to use a WLSI-MIMO filter for CS sequences. As shown in Fig. 10.5, the CS time series is first transformed into a vector-valued WSS process, then MIMO-filtered, and finally transformed back into a CS time series. The MIMO filter operation is

$$\hat{\mathbf{x}}[k] = \sum_{\kappa} \mathbf{H}[\kappa]\mathbf{y}[k-\kappa] + \widetilde{\mathbf{H}}[\kappa]\mathbf{y}^*[k-\kappa]. \tag{10.56}$$

In the z-domain, the filter output is, in augmented notation,

$$\begin{bmatrix} \widehat{\mathbf{X}}(z) \\ \widehat{\mathbf{X}}^*(z^*) \end{bmatrix} = \begin{bmatrix} \mathbb{H}(z) & \widetilde{\mathbb{H}}(z) \\ \widetilde{\mathbb{H}}^*(z^*) & \mathbb{H}^*(z^*) \end{bmatrix} \begin{bmatrix} \mathbf{Y}(z) \\ \mathbf{Y}^*(z^*) \end{bmatrix} \tag{10.57}$$

$$\underline{\widehat{\mathbf{X}}}(z) = \underline{\mathbb{H}}(z)\underline{\mathbf{Y}}(z). \tag{10.58}$$

Therefore,

$$\underline{\mathbb{P}}_{\hat{x}\hat{x}}(z) = \underline{\mathbb{H}}(z)\underline{\mathbb{P}}_{yy}(z)\underline{\mathbb{H}}^{\mathrm{H}}(z^{-*}). \tag{10.59}$$

This filter processes M samples at a time, operating at a reduced rate that is $1/M$ times the rate of the CS sequence. Consequently, the CS output signal $\hat{x}[k]$ is delayed by $M-1$ samples with respect to the input signal $y[k]$. While this delay might not matter in a noncausal filter, it is not acceptable in a causal realization. In order to avoid it, we use a filter bank with a sliding input window such that the output subsequence $\hat{x}_m[k] = \hat{x}[kM-m]$ is produced as soon as the input subsequence $y_m[k] = y[kM-m]$ is available. This necessitates the use of downsamplers and upsamplers that decimate and expand by M with a phase offset corresponding to m sample periods of the incoming CS sequence $y[k]$. These will be denoted by $\downarrow M^{(m)}$ and $\uparrow M^{(m)}$.

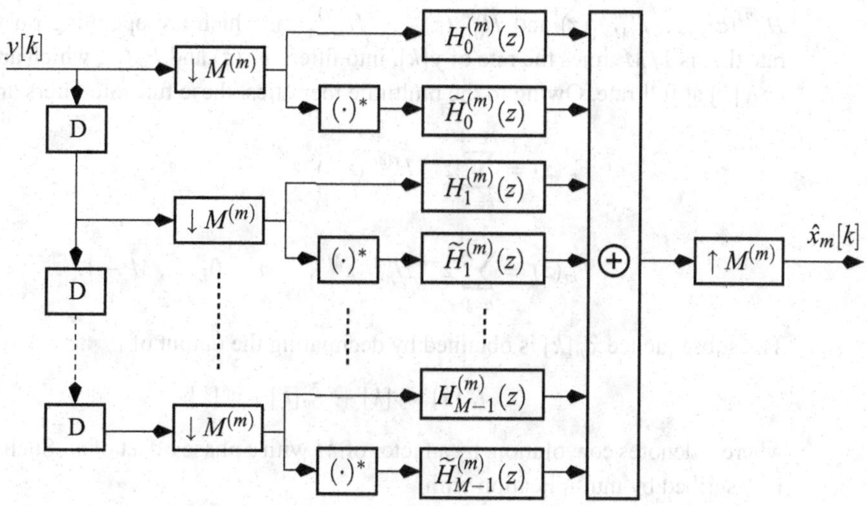

Figure 10.6 Branch m of the widely linear filter bank with sliding input window.

The filter bank consists of M branches, each handling a single output phase. Figure 10.6 shows the mth branch, which produces the subsequence $\hat{x}_m[k]$. It operates on the input vector sequence

$$\mathbf{y}^{(m)}[k] = \begin{bmatrix} y[kM-m] \\ y[kM-m-1] \\ \vdots \\ y[(k-1)M-m+1] \end{bmatrix}, \tag{10.60}$$

which can be more conveniently expressed in the z-domain as

$$\mathbf{Y}^{(m)}(z) = \mathbb{D}^{(m)}(z)\mathbf{Y}(z) \tag{10.61}$$

with

$$\mathbb{D}^{(m)}(z) = \begin{bmatrix} \mathbf{0} & \mathbf{I}_{M-m} \\ z^{-1}\mathbf{I}_m & \mathbf{0} \end{bmatrix}. \tag{10.62}$$

Each branch thus contains M filter pairs, each of which handles a single input phase. Specifically, the filter pair $H_n^{(m)}(z)$ and $\widetilde{H}_n^{(m)}(z)$ determines the contribution from input phase $m+n$ to output phase m. All filter branches operate at a reduced rate that is $1/M$ times the sampling rate of the CS sequence. However, the execution times of the branches are *staggered* by the sampling interval of the CS sequence.

10.4.3 Equivalence to FRESH filtering

The sliding-window filter bank is an efficient equivalent implementation of a FRESH filter. This can be shown along the lines of Ferrara (1985). We first combine the filters

$H_0^{(m)}(z), \ldots, H_{M-1}^{(m)}(z)$ and $\widetilde{H}_0^{(m)}(z), \ldots, \widetilde{H}_{M-1}^{(m)}(z)$, which are operating on $\mathbf{y}^{(m)}[k]$ at a rate that is $1/M$ times the rate of $y[k]$, into filters $F_m(z)$ and $\widetilde{F}_m(z)$, which are operating on $y[k]$ at full rate. Owing to the multirate identities, these full-rate filters are

$$F_m(z) = \sum_{n=0}^{M-1} z^{-n} H_n^{(m)}(z^M), \tag{10.63}$$

$$\widetilde{F}_m(z) = \sum_{n=0}^{M-1} z^{-n} \widetilde{H}_n^{(m)}(z^M), \quad m = 0, \ldots, M-1. \tag{10.64}$$

The subsequence $\hat{x}_m[k]$ is obtained by decimating the output of

$$f_m[k] * y[k] + \tilde{f}_m[k] * y^*[k], \tag{10.65}$$

where $*$ denotes convolution, by a factor of M with a phase offset of m. Such decimation is described by multiplication with

$$\delta[(k-m) \bmod M] = \frac{1}{M} \sum_{n=0}^{M-1} e^{j2\pi n(k-m)/M}, \tag{10.66}$$

where $\delta[\cdot]$ denotes the Kronecker delta. Therefore, the spectral process corresponding to $\hat{x}_m[k]$ has increments

$$d\hat{\xi}_m(\theta) = \frac{1}{M}\left[F_m(\theta)d\upsilon(\theta) + \widetilde{F}_m(\theta)d\upsilon^*(-\theta)\right] * \left[\sum_{n=0}^{M-1} e^{-j2\pi nm/M}\delta(\theta - 2\pi n/M)\right], \tag{10.67}$$

where $\delta(\cdot)$ is the Dirac delta function, and $F_m(\theta) = F_m(z)|_{z=e^{j\theta}}$. The CS output sequence $\hat{x}[k]$ is the sum of its subsequences,

$$\hat{x}[k] = \sum_{m=0}^{M-1} \hat{x}_m[k], \tag{10.68}$$

and the spectral process has increments

$$\begin{aligned} d\hat{\xi}(\theta) = &\sum_{n=0}^{M-1}\left[\frac{1}{M}\sum_{m=0}^{M-1} e^{-j2\pi nm/M} F_m(\theta - 2\pi n/M)\right] d\upsilon(\theta - 2\pi n/M) \\ &+ \sum_{n=0}^{M-1}\left[\frac{1}{M}\sum_{m=0}^{M-1} e^{-j2\pi nm/M} \widetilde{F}_m(\theta - 2\pi n/M)\right] d\upsilon^*(2\pi n/M - \theta). \end{aligned} \tag{10.69}$$

If we define a new set of filter frequency responses by

$$G_n(\theta) = \frac{1}{M}\sum_{m=0}^{M-1} e^{-j2\pi nm/M} F_m(\theta - 2\pi n/M), \tag{10.70}$$

$$\widetilde{G}_n(\theta) = \frac{1}{M}\sum_{m=0}^{M-1} e^{-j2\pi nm/M} \widetilde{F}_m(\theta - 2\pi n/M), \quad n = 0, \ldots, M-1, \tag{10.71}$$

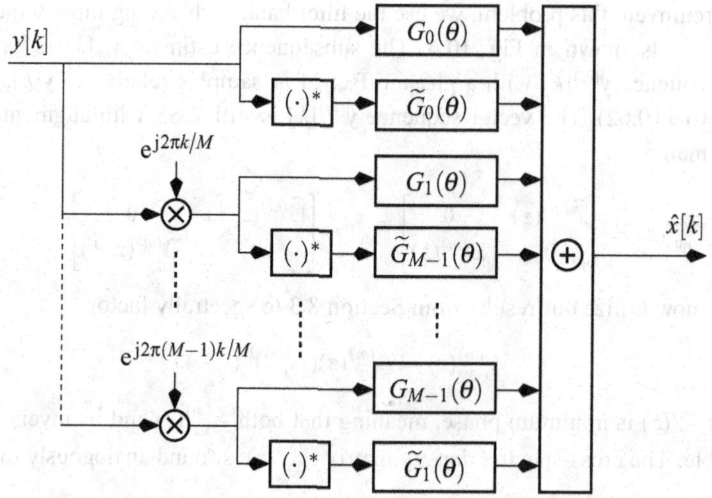

Figure 10.7 A widely linear frequency shift (FRESH) filter.

we can finally write

$$d\hat{\xi}(\theta) = \sum_{n=0}^{M-1} G_n(\theta)d\upsilon(\theta - 2\pi n/M) + \widetilde{G}_n(\theta)d\upsilon^*(2\pi n/M - \theta), \qquad (10.72)$$

which is a FRESH filtering operation. The corresponding time-domain filter operation is

$$\hat{x}[k] = \sum_{n=0}^{M-1} g_n[k] * (y[k]e^{j2\pi n/M}) + \tilde{g}_n[k] * (y^*[k]e^{j2\pi n/M}), \qquad (10.73)$$

which is depicted in Fig. 10.7.

10.4.4 Causal approximation

We can now proceed to the causal approximation of a cyclic Wiener filter. Let $\hat{x}[k]$ be the widely linear estimate of the CS message $x[k]$ based on the noisy observation $y[k]$. Furthermore, let $\hat{\mathbf{x}}[k]$, $\mathbf{x}[k]$, and $\mathbf{y}[k]$ denote the equivalent WSS vector representations. Given the correlation and complementary correlation functions $r_{yy}[k,\kappa]$, $\tilde{r}_{yy}[k,\kappa]$, $r_{xy}[k,\kappa]$, and $\tilde{r}_{xy}[k,\kappa]$, we first compute the augmented spectral density matrices $\underline{\mathbb{P}}_{yy}(z)$ and $\underline{\mathbb{P}}_{xy}(z)$ of their associated WSS sequences, along the lines of Section 10.4.1.

The polyphase filter structure depicted in Fig. 10.5 cannot be used to produce a delay-free, causal estimate $\hat{x}[k]$. The WLSI-MIMO filter in this structure operates at a rate that is $1/M$ times the rate of the incoming CS data sequence. Therefore, the estimated sequence $\hat{x}[k]$ will be delayed by $M-1$ with respect to the input $y[k]$. Moreover, the estimate $\hat{x}[k]$ is not causal because the subsequence $\hat{x}_m[k] = \hat{x}[kM-m]$ depends on the future values $y[kM-m+1], \ldots, y[kM-1]$. The only exception to this is the subsequence $\hat{x}_0[k]$, whose sampling phase matches the phase of the decimators.

To circumvent this problem, we use the filter bank with sliding input window whose mth branch is shown in Fig. 10.6. The subsequence estimate $\hat{x}_m[k]$ is based on the vector sequence $\mathbf{y}^{(m)}[k]$ with a phase offset of m samples relative to $\mathbf{y}[k]$, as defined in (10.60)–(10.62). The vector sequence $\mathbf{y}^{(m)}[k]$ is still WSS with augmented spectral density matrix

$$\underline{\mathbb{P}}_{yy}^{(m)}(z) = \begin{bmatrix} \widehat{\mathbb{D}}^{(m)}(z) & \mathbf{0} \\ \mathbf{0} & \widehat{\mathbb{D}}^{(m)}(z) \end{bmatrix} \underline{\mathbb{P}}_{yy}(z) \begin{bmatrix} \widehat{\mathbb{D}}^{(m)}(z^{-1}) & \mathbf{0} \\ \mathbf{0} & \widehat{\mathbb{D}}^{(m)}(z^{-1}) \end{bmatrix}^{\mathrm{T}}. \tag{10.74}$$

We may now utilize our results from Section 8.3 to spectrally factor

$$\underline{\mathbb{P}}_{yy}^{(m)}(z) = \underline{\mathbb{A}}_{yy}^{(m)}(z)(\underline{\mathbb{A}}_{yy}^{(m)})^{\mathrm{H}}(z^{-*}), \tag{10.75}$$

where $\underline{\mathbb{A}}_{yy}^{(m)}(z)$ is minimum phase, meaning that both $\underline{\mathbb{A}}_{yy}^{(m)}(z)$ and its inverse are causal and stable. The cross-spectral density matrix $\underline{\mathbb{P}}_{xy}^{(m)}(z)$ is found analogously to (10.74),

$$\underline{\mathbb{P}}_{xy}^{(m)}(z) = \begin{bmatrix} \widehat{\mathbb{D}}^{(m)}(z) & \mathbf{0} \\ \mathbf{0} & \widehat{\mathbb{D}}^{(m)}(z) \end{bmatrix} \underline{\mathbb{P}}_{xy}(z) \begin{bmatrix} \widehat{\mathbb{D}}^{(m)}(z^{-1}) & \mathbf{0} \\ \mathbf{0} & \widehat{\mathbb{D}}^{(m)}(z^{-1}) \end{bmatrix}^{\mathrm{T}}. \tag{10.76}$$

The transfer function of a causal WLMMSE filter for the augmented WSS vector sequence $\mathbf{y}^{(m)}[k]$ can now be obtained as

$$\underline{\mathbb{H}}^{(m)}(z) = [\underline{\mathbb{P}}_{xy}^{(m)}(z)(\underline{\mathbb{A}}_{yy}^{(m)})^{-\mathrm{H}}(z^{-*})]_{+}(\underline{\mathbb{A}}_{yy}^{(m)})^{-1}(z), \tag{10.77}$$

where $[\cdot]_{+}$ denotes the causal part. Again, only the subsequence $\hat{x}_m[k]$ produced by this filter is a delay-free, causal MMSE estimate. Hence, for each polyphase matrix $\underline{\mathbb{H}}^{(m)}(z)$, we extract the top row vector, denoted by

$$\left[H_0^{(m)}(z) \quad \cdots \quad H_{M-1}^{(m)}(z) \quad \widetilde{H}_0^{(m)}(z) \quad \cdots \quad \widetilde{H}_{M-1}^{(m)}(z)\right]. \tag{10.78}$$

This vector is the polyphase representation of a causal WLMMSE filter that produces the mth subsequence $\hat{x}_m[k]$ of the CS estimate $\hat{x}[k]$. This filter constitutes the mth branch shown in Fig. 10.6. The causal widely linear MMSE filter for producing $\hat{x}[k]$ from $y[k]$ comprises a parallel collection of M such filters, whose outputs are summed.

Notes

1 CS processes for which ensemble averages can be replaced by time averages are sometimes called *cycloergodic*. An in-depth treatment of cycloergodicity is provided by Boyles and Gardner (1983). Gardner (1988) presents a treatment of cyclostationary processes in the fraction-of-time probabilistic framework, which is entirely based on time averages.

2 A thorough discussion of estimators for the cyclic correlation and PSD, including the proof that the time-smoothed and frequency-smoothed cyclic periodograms are asymptotically equivalent, is provided by Gardner (1986).

3 The cyclic PSDs of several digital communication signals were derived by Gardner *et al.* (1987). A closer look at the cyclic PSD of continuous-phase modulated (CPM) signals is provided by Napolitano and Spooner (2001).

4 Improper digital communication signals include all real-valued baseband signals such as Pulse Amplitude Modulation (PAM) and Binary Phase Shift Keying (BPSK), but also Offset Quaternary Phase Shift Keying (OQPSK), Minimum Shift Keying (MSK), and Gaussian Minimum Shift Keying (GMSK). GMSK is used in the mobile-phone standard GSM.

5 The term "diversity" is commonly used in mobile communications, where the receiver obtains several replicas of the same information-bearing signal. In "frequency diversity," the signal is transmitted on several carriers that are sufficiently separated in frequency.

6 FRESH filtering and the cyclic Wiener filter were introduced and analyzed by Gardner (1988, 1993). The cyclic Wiener filter utilizes the frequency diversity of CS signals in a way that is reminiscent of the RAKE receiver in wireless communications (see Proakis (2001) for background on the RAKE receiver).

7 The connection between scalar-valued CS and vector-valued WSS time series was found by Gladyshev (1961). There is a corresponding result for continuous-time CS processes due to Gladyshev (1963). If $x(t)$ is CS with period T, and θ an independent random variable uniformly distributed over $[0, T]$, then $x_m(t) = x(t-\theta)e^{j2\pi m(t-\theta)/T}$ is WSS, and $x_m(t)$ and $x_n(t)$ are jointly WSS.

Appendix 1
Rudiments of matrix analysis

In this appendix, we present a very concise review of some basic results in matrix analysis that are used in this book. As an excellent reference for further reading we recommend Horn and Johnson (1985).

A1.1 Matrix factorizations

A1.1.1 Partitioned matrices

If **A** and **D** are square matrices and have inverses, there exist the factorizations

$$\mathbb{A} = \begin{bmatrix} \mathbf{A} & \mathbf{B} \\ \mathbf{C} & \mathbf{D} \end{bmatrix} = \begin{bmatrix} \mathbf{I} & \mathbf{BD}^{-1} \\ \mathbf{0} & \mathbf{I} \end{bmatrix} \begin{bmatrix} \mathbf{A} - \mathbf{BD}^{-1}\mathbf{C} & \mathbf{0} \\ \mathbf{0} & \mathbf{D} \end{bmatrix} \begin{bmatrix} \mathbf{I} & \mathbf{0} \\ \mathbf{D}^{-1}\mathbf{C} & \mathbf{I} \end{bmatrix} \tag{A1.1}$$

$$= \begin{bmatrix} \mathbf{I} & \mathbf{0} \\ \mathbf{CA}^{-1} & \mathbf{I} \end{bmatrix} \begin{bmatrix} \mathbf{A} & \mathbf{0} \\ \mathbf{0} & \mathbf{D} - \mathbf{CA}^{-1}\mathbf{B} \end{bmatrix} \begin{bmatrix} \mathbf{I} & \mathbf{A}^{-1}\mathbf{B} \\ \mathbf{0} & \mathbf{I} \end{bmatrix} \tag{A1.2}$$

The factors $\mathbf{A} - \mathbf{BD}^{-1}\mathbf{C}$ and $\mathbf{D} - \mathbf{CA}^{-1}\mathbf{B}$ are referred to as the *Schur complements* of **D** and **A**, respectively, within $\mathbb{A}$. These factorizations establish the determinant formulae

$$\det \mathbb{A} = \det(\mathbf{A} - \mathbf{BD}^{-1}\mathbf{C}) \det \mathbf{D} \tag{A1.3}$$

$$= \det \mathbf{A} \det(\mathbf{D} - \mathbf{CA}^{-1}\mathbf{B}). \tag{A1.4}$$

A1.1.2 Eigenvalue decomposition

A matrix $\mathbf{A} \in \mathbb{C}^{n \times n}$ is *normal* if it commutes with its Hermitian transpose: $\mathbf{AA}^{\mathrm{H}} = \mathbf{A}^{\mathrm{H}}\mathbf{A}$. A matrix is *Hermitian* if it is Hermitian-symmetric: $\mathbf{A}^{\mathrm{H}} = \mathbf{A}$. Any Hermitian matrix is therefore normal.

Every normal matrix has the factorization

$$\mathbf{A} = \mathbf{U} \boldsymbol{\Lambda} \mathbf{U}^{\mathrm{H}}, \tag{A1.5}$$

called an *eigenvalue (or spectral) decomposition* (EVD), where **U** is unitary. The columns $\mathbf{u}_i$ of **U** are called the *eigenvectors*, and the diagonal matrix $\boldsymbol{\Lambda} = \mathbf{Diag}(\lambda_1, \lambda_2, \ldots, \lambda_n)$ contains the *eigenvalues* on its diagonal. We will assume the ordering $|\lambda_1| \geq |\lambda_2| \geq \cdots \geq |\lambda_n|$. The number r of nonzero eigenvalues is the rank of **A**, and the zero eigenvalues are

$\lambda_{r+1} = \lambda_{r+2} = \cdots = \lambda_n = 0$. Thus, **A** may be expressed as

$$\mathbf{A} = \sum_{i=1}^{r} \lambda_i \mathbf{u}_i \mathbf{u}_i^{\mathrm{H}}, \tag{A1.6}$$

and, for any polynomial f,

$$f(\mathbf{A}) = \sum_{i=1}^{r} f(\lambda_i) \mathbf{u}_i \mathbf{u}_i^{\mathrm{H}}. \tag{A1.7}$$

An important fact is that the nonzero eigenvalues of a product of matrices **AB** are the nonzero eigenvalues of **BA**, for arbitrary **A** and **B**.

The following list contains equivalent statements for different types of normal matrices.

$$\mathbf{A} = \mathbf{A}^{\mathrm{H}} \Leftrightarrow \lambda_i \text{ real for all } i, \tag{A1.8}$$

$$\mathbf{A} \text{ unitary} \Leftrightarrow |\lambda_i| = 1 \text{ for all } i, \tag{A1.9}$$

$$\mathbf{A} \text{ positive definite} \Leftrightarrow \lambda_i > 0 \text{ for all } i, \tag{A1.10}$$

$$\mathbf{A} \text{ positive semidefinite} \Leftrightarrow \lambda_i \geq 0 \text{ for all } i. \tag{A1.11}$$

2 × 2 matrix

The two eigenvalues of the 2 × 2 matrix

$$\mathbf{A} = \begin{bmatrix} a & b \\ c & d \end{bmatrix} \tag{A1.12}$$

are

$$\lambda_{1,2} = \tfrac{1}{2}\,\mathrm{tr}\,\mathbf{A} \pm \tfrac{1}{2}\sqrt{\mathrm{tr}^2\mathbf{A} - 4\det\mathbf{A}} \tag{A1.13}$$

and the unit-length eigenvectors are

$$\mathbf{u}_1 = \frac{1}{\|\mathbf{u}_1\|}\begin{bmatrix} b \\ \lambda_1 - a \end{bmatrix}, \tag{A1.14}$$

$$\mathbf{u}_2 = \frac{1}{\|\mathbf{u}_2\|}\begin{bmatrix} b \\ \lambda_2 - a \end{bmatrix}. \tag{A1.15}$$

A1.1.3 Singular value decomposition

Every matrix $\mathbf{A} \in \mathbb{C}^{n \times m}$ has the factorization

$$\mathbf{A} = \mathbf{F}\mathbf{K}\mathbf{G}^{\mathrm{H}}, \tag{A1.16}$$

called a *singular value decomposition* (SVD), where $\mathbf{F} \in \mathbb{C}^{n \times p}$ and $\mathbf{G} \in \mathbb{C}^{m \times p}$, $p = \min(n, m)$, both have unitary columns, $\mathbf{F}^{\mathrm{H}}\mathbf{F} = \mathbf{I}$ and $\mathbf{G}^{\mathrm{H}}\mathbf{G} = \mathbf{I}$. The columns $\mathbf{f}_i$ of **F** are called the *left singular vectors*, and the columns $\mathbf{g}_i$ of **G** are called the *right singular vectors*. The diagonal matrix $\mathbf{K} = \mathbf{Diag}(k_1, k_2, \ldots, k_p)$ contains the *singular values* $k_1 \geq k_2 \geq \cdots \geq k_p \geq 0$ on its diagonal. The number r of nonzero singular values is the rank of **A**.

Since $\mathbf{A}\mathbf{A}^{\mathrm{H}} = \mathbf{F}\mathbf{K}\mathbf{G}^{\mathrm{H}}(\mathbf{G}\mathbf{K}\mathbf{F}^{\mathrm{H}}) = \mathbf{F}\mathbf{K}^2\mathbf{F}^{\mathrm{H}}$ and $\mathbf{A}^{\mathrm{H}}\mathbf{A} = (\mathbf{G}\mathbf{K}\mathbf{F}^{\mathrm{H}})\mathbf{F}\mathbf{K}\mathbf{G}^{\mathrm{H}} = \mathbf{G}\mathbf{K}^2\mathbf{G}^{\mathrm{H}}$, the singular values of $\mathbf{A}$ are the nonnegative roots of the eigenvalues of $\mathbf{A}\mathbf{A}^{\mathrm{H}}$ or $\mathbf{A}^{\mathrm{H}}\mathbf{A}$. So, for *normal* $\mathbf{A}$, how are the EVD $\mathbf{A} = \mathbf{U}\mathbf{\Lambda}\mathbf{U}^{\mathrm{H}}$ and the SVD $\mathbf{A} = \mathbf{F}\mathbf{K}\mathbf{G}^{\mathrm{H}}$ related? If we write $\mathbf{\Lambda} = |\mathbf{\Lambda}|\mathbf{D}$ with $|\mathbf{\Lambda}| = \mathbf{Diag}(|\lambda_1|, |\lambda_2|, \ldots, |\lambda_n|)$ and $\mathbf{D} = \mathbf{Diag}(\exp(\mathrm{j}\measuredangle\lambda_1), \exp(\mathrm{j}\measuredangle\lambda_2), \ldots, \exp(\mathrm{j}\measuredangle\lambda_n))$ then

$$\mathbf{A} = \mathbf{U}\mathbf{\Lambda}\mathbf{U}^{\mathrm{H}} = \mathbf{U}|\mathbf{\Lambda}|(\mathbf{U}\mathbf{D}^{\mathrm{H}})^{\mathrm{H}} = \mathbf{F}\mathbf{K}\mathbf{G}^{\mathrm{H}} \tag{A1.17}$$

is an SVD of $\mathbf{A}$ with $\mathbf{F} = \mathbf{U}$, $\mathbf{K} = |\mathbf{\Lambda}|$, and $\mathbf{G} = \mathbf{U}\mathbf{D}^*$. Therefore, the singular values of $\mathbf{A}$ are the absolute values of the eigenvalues of $\mathbf{A}$. If $\mathbf{A}$ is Hermitian, then the eigenvalues and thus $\mathbf{D}$ are real, and $\mathbf{D} = \mathbf{Diag}(\mathrm{sign}(\lambda_1), \mathrm{sign}(\lambda_2), \ldots, \mathrm{sign}(\lambda_n))$. If $\mathbf{A}$ is Hermitian and positive semidefinite, then $\mathbf{F} = \mathbf{G} = \mathbf{U}$ and $\mathbf{K} = \mathbf{\Lambda}$. That is, the SVD and EVD of a Hermitian positive semidefinite matrix are identical.

A1.2 Positive definite matrices

Positive definite and semidefinite matrices are important because correlation and covariance matrices are positive (semi)definite. A Hermitian matrix $\mathbf{A} \in \mathbb{C}^{n\times n}$ is called positive definite if

$$\mathbf{x}^{\mathrm{H}}\mathbf{A}\mathbf{x} > 0 \text{ for all nonzero } \mathbf{x} \in \mathbb{C}^n \tag{A1.18}$$

and positive semidefinite if the weaker condition $\mathbf{x}^{\mathrm{H}}\mathbf{A}\mathbf{x} \geq 0$ holds. A Hermitian matrix is positive definite if and only if all of its eigenvalues are positive, and positive semidefinite if and only if all eigenvalues are nonnegative.

A1.2.1 Matrix square root and Cholesky decomposition

A matrix $\mathbf{A}$ is positive definite if and only if there exists a nonsingular lower-triangular matrix $\mathbf{L}$ with positive diagonal entries such that $\mathbf{A} = \mathbf{L}\mathbf{L}^{\mathrm{H}}$. This factorization is often called the *Cholesky decomposition* of $\mathbf{A}$. It determines a *square root* $\mathbf{A}^{1/2} = \mathbf{L}$, which is neither positive definite nor Hermitian.

The *unique positive semidefinite* square root of a positive semidefinite matrix $\mathbf{A}$ is obtained via the EVD of $\mathbf{A} = \mathbf{U}\mathbf{\Lambda}\mathbf{U}^{\mathrm{H}}$ as

$$\mathbf{A}^{1/2} = \mathbf{U}\mathbf{\Lambda}^{1/2}\mathbf{U}^{\mathrm{H}} \tag{A1.19}$$

with $\mathbf{\Lambda}^{1/2} = \mathbf{Diag}(\sqrt{\lambda_1}, \sqrt{\lambda_2}, \ldots, \sqrt{\lambda_n})$.

A1.2.2 Updating the Cholesky factors of a Grammian matrix

Consider the data matrix $\mathbf{X}_p = [\mathbf{x}_1, \mathbf{x}_2, \ldots, \mathbf{x}_p] \in \mathbb{C}^{n\times p}$ with $n \geq p$ and positive definite Grammian matrix

$$\mathbf{G}_p = \mathbf{X}_p^{\mathrm{H}}\mathbf{X}_p = \begin{bmatrix} \mathbf{G}_{p-1} & \mathbf{h}_p \\ \mathbf{h}_p^{\mathrm{H}} & h_{pp} \end{bmatrix}, \tag{A1.20}$$

where $\mathbf{G}_{p-1} = \mathbf{X}_{p-1}^{\mathrm{H}}\mathbf{X}_{p-1}$, $\mathbf{h}_p = \mathbf{X}_{p-1}^{\mathrm{H}}\mathbf{x}_p$ and $h_{pp} = \mathbf{x}_p^{\mathrm{H}}\mathbf{x}_p$. The Cholesky decomposition of $\mathbf{G}_{p-1}$ is $\mathbf{G}_{p-1} = \mathbf{L}_{p-1}\mathbf{L}_{p-1}^{\mathrm{H}}$, where $\mathbf{L}_{p-1}$ is lower-triangular. It follows that the Cholesky factors of $\mathbf{G}_p$ may be updated as

$$\mathbf{G}_p = \mathbf{L}_p\mathbf{L}_p^{\mathrm{H}} = \begin{bmatrix} \mathbf{L}_{p-1} & \mathbf{0} \\ \mathbf{m}_p^{\mathrm{T}} & m_{pp} \end{bmatrix} \begin{bmatrix} \mathbf{L}_{p-1}^{\mathrm{H}} & \mathbf{m}_p^* \\ \mathbf{0} & m_{pp}^* \end{bmatrix} = \begin{bmatrix} \mathbf{G}_{p-1} & \mathbf{h}_p \\ \mathbf{h}_p^{\mathrm{H}} & h_{pp} \end{bmatrix}, \tag{A1.21}$$

where $\mathbf{m}_p$ and m_{pp} solve the following consistency equations:

$$\mathbf{L}_{p-1}\mathbf{m}_p^* = \mathbf{h}_p, \tag{A1.22}$$

$$\mathbf{m}_p^{\mathrm{T}}\mathbf{m}_p^* + |m_{pp}|^2 = h_{pp}. \tag{A1.23}$$

This factorization may then be inverted for $\mathbf{G}_p^{-1}$ to achieve the upper-triangular–lower-triangular factorization

$$\mathbf{G}_p^{-1} = \mathbf{L}_p^{-\mathrm{H}}\mathbf{L}_p^{-1} = \begin{bmatrix} \mathbf{L}_{p-1}^{-\mathrm{H}} & \mathbf{k}_p \\ \mathbf{0} & k_{pp} \end{bmatrix} \begin{bmatrix} \mathbf{L}_{p-1}^{-1} & \mathbf{0} \\ \mathbf{k}_p^{\mathrm{H}} & k_{pp}^* \end{bmatrix}. \tag{A1.24}$$

From the constraint that $\mathbf{L}_p^{-1}\mathbf{L}_p = \mathbf{I}$, $\mathbf{k}_p$ and k_{pp} are determined from the consistency equations

$$\mathbf{k}_p^{\mathrm{H}}\mathbf{L}_{p-1} = -k_{pp}^*\mathbf{m}_p^{\mathrm{T}}, \tag{A1.25}$$

$$k_{pp}^* m_{pp} = 1. \tag{A1.26}$$

A1.2.3 Partial ordering

For $\mathbf{A}, \mathbf{B} \in \mathbb{C}^{n\times n}$, we write $\mathbf{A} > \mathbf{B}$ when $\mathbf{A} - \mathbf{B}$ is positive definite, and $\mathbf{A} \geq \mathbf{B}$ when $\mathbf{A} - \mathbf{B}$ is positive semidefinite. This is a partial ordering of the set of $n \times n$ Hermitian matrices. It is partial because we may have $\mathbf{A} \ngeq \mathbf{B}$ and $\mathbf{B} \ngeq \mathbf{A}$. For $\mathbf{A} > \mathbf{0}$, $\mathbf{B} \geq \mathbf{0}$, and nonsingular $\mathbf{C}$, we have

$$\mathbf{C}^{\mathrm{H}}\mathbf{A}\mathbf{C} > \mathbf{0}, \tag{A1.27}$$

$$\mathbf{C}^{\mathrm{H}}\mathbf{B}\mathbf{C} \geq \mathbf{0}, \tag{A1.28}$$

$$\mathbf{A}^{-1} > \mathbf{0}, \tag{A1.29}$$

$$\mathbf{A} + \mathbf{B} \geq \mathbf{A} > \mathbf{0}, \tag{A1.30}$$

$$\mathbf{A} \geq \mathbf{B} \Rightarrow \mathrm{ev}_i(\mathbf{A}) \geq \mathrm{ev}_i(\mathbf{B}). \tag{A1.31}$$

The last statement assumes that the eigenvalues of $\mathbf{A}$ and $\mathbf{B}$ are each arranged in decreasing order. As a particular consequence, $\det\mathbf{A} \geq \det\mathbf{B}$ and $\mathrm{tr}\,\mathbf{A} \geq \mathrm{tr}\,\mathbf{B}$.

The partitioned Hermitian matrix

$$\mathbb{A} = \begin{bmatrix} \mathbf{A} & \mathbf{B} \\ \mathbf{B}^{\mathrm{H}} & \mathbf{D} \end{bmatrix}, \tag{A1.32}$$

with square blocks $\mathbf{A}$ and $\mathbf{D}$, is positive definite if and only if $\mathbf{A} > \mathbf{0}$ and its Schur complement $\mathbf{D} - \mathbf{B}^{\mathrm{H}}\mathbf{A}^{-1}\mathbf{B} > \mathbf{0}$, or $\mathbf{D} > \mathbf{B}^{\mathrm{H}}\mathbf{A}^{-1}\mathbf{B}$.

A1.2.4 Inequalities

There are several important determinant inequalities, which can all be proved from partial-ordering results. Alternative closely related proofs are also possible via majorization, which is discussed in Appendix 3. The *Hadamard determinant inequality* for a positive semidefinite $\mathbf{A} \in \mathbb{C}^{n\times n}$ is

$$\det \mathbf{A} \leq \prod_{i=1}^{n} A_{ii}. \tag{A1.33}$$

For positive definite **A**, equality holds if and only if **A** is diagonal.

The *Fischer determinant inequality* for the partitioned positive definite matrix $\mathbb{A}$ in (A1.32) is

$$\det \mathbb{A} \leq \det \mathbf{A} \det \mathbf{D}. \tag{A1.34}$$

The *Minkowski determinant inequality* for positive definite $\mathbf{A}, \mathbf{B} \in \mathbb{C}^{n\times n}$ is

$$\det^{1/n}(\mathbf{A}+\mathbf{B}) \geq \det^{1/n}\mathbf{A} + \det^{1/n}\mathbf{B}, \tag{A1.35}$$

with equality if and only if $\mathbf{B} = c\mathbf{A}$ for some positive constant c.

A1.3 Matrix inverses

A1.3.1 Partitioned matrices

The inverse of a block-diagonal matrix of nonsingular blocks is obtained by taking the inverses of the individual blocks,

$$[\mathbf{Diag}(\mathbf{A}_1, \mathbf{A}_2, \ldots, \mathbf{A}_k)]^{-1} = \mathbf{Diag}(\mathbf{A}_1^{-1}, \mathbf{A}_2^{-1}, \ldots, \mathbf{A}_k^{-1}). \tag{A1.36}$$

Given the factorizations (A1.1) and (A1.2) for the partitioned matrix $\mathbb{A}$ and the identities

$$\begin{bmatrix} \mathbf{I} & \mathbf{X} \\ \mathbf{0} & \mathbf{I} \end{bmatrix}^{-1} = \begin{bmatrix} \mathbf{I} & -\mathbf{X} \\ \mathbf{0} & \mathbf{I} \end{bmatrix} \quad \text{and} \quad \begin{bmatrix} \mathbf{I} & \mathbf{0} \\ \mathbf{X} & \mathbf{I} \end{bmatrix}^{-1} = \begin{bmatrix} \mathbf{I} & \mathbf{0} \\ -\mathbf{X} & \mathbf{I} \end{bmatrix} \tag{A1.37}$$

there are the following corresponding factorizations for $\mathbb{A}^{-1}$:

$$\mathbb{A}^{-1} = \begin{bmatrix} \mathbf{A} & \mathbf{B} \\ \mathbf{C} & \mathbf{D} \end{bmatrix}^{-1} = \begin{bmatrix} \mathbf{I} & \mathbf{0} \\ -\mathbf{D}^{-1}\mathbf{C} & \mathbf{I} \end{bmatrix} \begin{bmatrix} (\mathbf{A}-\mathbf{B}\mathbf{D}^{-1}\mathbf{C})^{-1} & \mathbf{0} \\ \mathbf{0} & \mathbf{D}^{-1} \end{bmatrix} \begin{bmatrix} \mathbf{I} & -\mathbf{B}\mathbf{D}^{-1} \\ \mathbf{0} & \mathbf{I} \end{bmatrix} \tag{A1.38}$$

$$= \begin{bmatrix} \mathbf{I} & -\mathbf{A}^{-1}\mathbf{B} \\ \mathbf{0} & \mathbf{I} \end{bmatrix} \begin{bmatrix} \mathbf{A}^{-1} & \mathbf{0} \\ \mathbf{0} & (\mathbf{D}-\mathbf{C}\mathbf{A}^{-1}\mathbf{B})^{-1} \end{bmatrix} \begin{bmatrix} \mathbf{I} & \mathbf{0} \\ -\mathbf{C}\mathbf{A}^{-1} & \mathbf{I} \end{bmatrix}. \tag{A1.39}$$

These lead to the determinant formulae

$$\det \mathbb{A}^{-1} = \det(\mathbf{A}-\mathbf{B}\mathbf{D}^{-1}\mathbf{C})^{-1} \det \mathbf{D}^{-1} \tag{A1.40}$$

$$= \det \mathbf{A}^{-1} \det(\mathbf{D}-\mathbf{C}\mathbf{A}^{-1}\mathbf{B})^{-1} \tag{A1.41}$$

and a formula for $\mathbb{A}^{-1}$, which is sometimes called the *matrix-inversion lemma*:

$$\mathbb{A}^{-1} = \begin{bmatrix} \mathbf{A} & \mathbf{B} \\ \mathbf{C} & \mathbf{D} \end{bmatrix}^{-1} = \begin{bmatrix} (\mathbf{A}-\mathbf{B}\mathbf{D}^{-1}\mathbf{C})^{-1} & -(\mathbf{A}-\mathbf{B}\mathbf{D}^{-1}\mathbf{C})^{-1}\mathbf{B}\mathbf{D}^{-1} \\ -(\mathbf{D}-\mathbf{C}\mathbf{A}^{-1}\mathbf{B})^{-1}\mathbf{C}\mathbf{A}^{-1} & (\mathbf{D}-\mathbf{C}\mathbf{A}^{-1}\mathbf{B})^{-1} \end{bmatrix}. \tag{A1.42}$$

Different expressions may be derived by employing the *Woodbury identities*

$$(\mathbf{A}-\mathbf{B}\mathbf{D}^{-1}\mathbf{C})^{-1} = \mathbf{A}^{-1} + \mathbf{A}^{-1}\mathbf{B}(\mathbf{D}-\mathbf{C}\mathbf{A}^{-1}\mathbf{B})^{-1}\mathbf{C}\mathbf{A}^{-1}, \tag{A1.43}$$

$$(\mathbf{A}-\mathbf{B}\mathbf{D}^{-1}\mathbf{C})^{-1}\mathbf{B}\mathbf{D}^{-1} = \mathbf{A}^{-1}\mathbf{B}(\mathbf{D}-\mathbf{C}\mathbf{A}^{-1}\mathbf{B})^{-1}. \tag{A1.44}$$

A simple special case is the inverse of a 2×2 matrix

$$\begin{bmatrix} a & b \\ c & d \end{bmatrix}^{-1} = \frac{1}{ad-bc}\begin{bmatrix} d & -b \\ -c & a \end{bmatrix}. \tag{A1.45}$$

A1.3.2 Moore–Penrose pseudo-inverse

Let $\mathbf{A} \in \mathbb{C}^{n\times m}$, possibly rank-deficient. Its Moore–Penrose pseudo-inverse (or generalized inverse) $\mathbf{A}^\dagger \in \mathbb{C}^{m\times n}$ satisfies the following conditions:

$$\mathbf{A}^\dagger\mathbf{A} = (\mathbf{A}^\dagger\mathbf{A})^{\mathrm{H}}, \tag{A1.46}$$

$$\mathbf{A}\mathbf{A}^\dagger = (\mathbf{A}\mathbf{A}^\dagger)^{\mathrm{H}}, \tag{A1.47}$$

$$\mathbf{A}^\dagger\mathbf{A}\mathbf{A}^\dagger = \mathbf{A}^\dagger, \tag{A1.48}$$

$$\mathbf{A}\mathbf{A}^\dagger\mathbf{A} = \mathbf{A}. \tag{A1.49}$$

The pseudo-inverse always exists and is unique. It may be computed via the SVD of $\mathbf{A} = \mathbf{F}\mathbf{K}\mathbf{G}^{\mathrm{H}}$, as $\mathbf{A}^\dagger = \mathbf{G}\mathbf{K}^\dagger\mathbf{F}^{\mathrm{H}}$. The pseudo-inverse of $\mathbf{K} = \mathbf{Diag}(k_1, k_2, \ldots, k_r, 0, \ldots, 0)$, $k_r > 0$, is $\mathbf{K}^\dagger = \mathbf{Diag}(k_1^{-1}, k_2^{-1}, \ldots, k_r^{-1}, 0, \ldots, 0)$. The SVD is an expensive thing to compute but the pseudo-inverse may be computed inexpensively via a Gram–Schmidt factorization.

Example A1.1. If the 2×2 positive semidefinite Hermitian matrix

$$\mathbf{A} = \begin{bmatrix} a & b \\ b^* & d \end{bmatrix} \tag{A1.50}$$

is singular, i.e., $ad = |b|^2$, then its eigen- and singular values are $\lambda_1 = a + d$ and $\lambda_2 = 0$, thanks to (A1.13). The eigen- and singular vector associated with λ_1 is

$$\mathbf{u}_1 = \frac{1}{\|\mathbf{u}_1\|}\begin{bmatrix} b \\ d \end{bmatrix}. \tag{A1.51}$$

The pseudo-inverse of $\mathbf{A}$ is therefore

$$\mathbf{A}^\dagger = \lambda_1^{-1}\mathbf{u}_1\mathbf{u}_1^{\mathrm{H}} = \frac{1}{(a+d)(|b|^2+d^2)}\begin{bmatrix} |b|^2 & bd \\ b^*d & d^2 \end{bmatrix}. \tag{A1.52}$$

There are special cases when $\mathbf{A}$ has full rank. If $m \leq n$, then $\mathbf{A}^{\dagger} = (\mathbf{A}^{\mathrm{H}}\mathbf{A})^{-1}\mathbf{A}^{\mathrm{H}}$, and $\mathbf{A}^{\dagger}\mathbf{A} = \mathbf{I}_{m \times m}$. If $n \leq m$, then $\mathbf{A}^{\dagger} = \mathbf{A}^{\mathrm{H}}(\mathbf{A}\mathbf{A}^{\mathrm{H}})^{-1}$, and $\mathbf{A}\mathbf{A}^{\dagger} = \mathbf{I}_{n \times n}$. If $\mathbf{A}$ is square and nonsingular, then $\mathbf{A}^{\dagger} = \mathbf{A}^{-1}$.

A1.3.3 Projections

A matrix $\mathbf{P} \in \mathbb{C}^{n \times n}$ that satisfies $\mathbf{P}^2 = \mathbf{P}$ is called a *projection*. If, in addition, $\mathbf{P} = \mathbf{P}^{\mathrm{H}}$ then $\mathbf{P}$ is an *orthogonal projection*. For $\mathbf{H} \in \mathbb{C}^{n \times p}$, the matrix $\mathbf{P}_{\mathbf{H}} = \mathbf{H}\mathbf{H}^{\dagger}$ is an orthogonal projection whose range is the range of $\mathbf{H}$. If $\mathbf{H}$ has full rank and $p \leq n$, $\mathbf{P}_{\mathbf{H}} = \mathbf{H}\mathbf{H}^{\dagger} = \mathbf{H}(\mathbf{H}^{\mathrm{H}}\mathbf{H})^{-1}\mathbf{H}^{\mathrm{H}}$. If $[\mathbf{H}, \mathbf{A}]$ is a full rank $n \times n$ matrix, where $\mathbf{H} \in \mathbb{C}^{n \times p}$ and $\mathbf{A} \in \mathbb{C}^{n \times (n-p)}$, then

$$\mathbf{x} = \mathbf{P}_{\mathbf{H}}\mathbf{x} + \mathbf{P}_{\mathbf{A}}\mathbf{x} \tag{A1.53}$$

decomposes the vector $\mathbf{x}$ into two orthogonal components, one of which lies in $\langle \mathbf{H} \rangle$ and the other of which lies in $\langle \mathbf{A} \rangle$. Note that $\mathbf{P}_{\mathbf{A}} = \mathbf{A}(\mathbf{A}^{\mathrm{H}}\mathbf{A})^{-1}\mathbf{A}^{\mathrm{H}} = \mathbf{I} - \mathbf{P}_{\mathbf{H}}$.

Appendix 2 Complex differential calculus (Wirtinger calculus)

In statistical signal processing, we often deal with a real nonnegative cost function, such as a likelihood function or a quadratic form, which is then either analytically or numerically optimized with respect to a vector or matrix of parameters. This involves taking derivatives with respect to vectors or matrices, leading to gradient vectors and Jacobian and Hessian matrices. What happens when the parameters are complex-valued? That is, how do we differentiate a real-valued function with respect to a complex argument?

What makes this situation confusing is that classical complex analysis tells us that a complex function is differentiable on its entire domain if and only if it is *holomorphic* (which is a synonym for *complex analytic*). A holomorphic function with nonzero derivative is *conformal* because it preserves angles (including their orientations) and the shapes of infinitesimally small figures (but not necessarily their size) in the complex plane. Since nonconstant real-valued functions defined on the complex domain cannot be holomorphic, their classical complex derivatives do not exist.

We can, of course, regard a function f defined on $\mathbb{C}^n$ as a function defined on $\mathbb{R}^{2n}$. If f is differentiable on $\mathbb{R}^{2n}$, it is said to be *real-differentiable*, and if f is differentiable on $\mathbb{C}^n$, it is *complex-differentiable*. A function is complex-differentiable if and only if it is real-differentiable and the *Cauchy–Riemann equations* hold. Is there a way to define *generalized complex derivatives* for functions that are real-differentiable but not complex-differentiable? This would extend complex differential calculus in a way similar to the way that impropriety extends the theory of complex random variables.

It is indeed possible to do this. The theory was developed by the Austrian mathematician Wilhelm Wirtinger (1927), which is why this generalized complex differential calculus is sometimes referred to as *Wirtinger calculus*. In the engineering literature, Wirtinger calculus was rediscovered by Brandwood (1983) and then further developed by van den Bos (1994a). In this appendix, we mainly follow the outline by van den Bos (1994a), with some minor extensions.

The key idea of Wirtinger calculus is to *formally* regard f as a function of two *independent* complex variables x and x^*. A generalized complex derivative is then formally defined as the derivative with respect to x, while treating x^* as a constant. Another generalized derivative is defined as the derivative with respect to x^*, while formally treating x as a constant. The generalized derivatives exist whenever f is real-differentiable. These ideas extend in a straightforward fashion to complex gradients, Jacobians, and Hessians.

A2.1 Complex gradients

We shall begin with a scalar real-valued, totally differentiable function $f(\mathbf{z})$, $\mathbf{z} = [u, v]^{\mathrm{T}} \in \mathbb{R}^2$. Its linear approximation at $\mathbf{z}_0 = [u_0, v_0]^{\mathrm{T}}$ is

$$f(\mathbf{z}) = f(u, v) \approx f(\mathbf{z}_0) + \nabla_z f(\mathbf{z}_0)(\mathbf{z} - \mathbf{z}_0), \tag{A2.1}$$

with the gradient (a row vector)

$$\nabla_z f(\mathbf{z}_0) = \begin{bmatrix} \frac{\partial f}{\partial u}(\mathbf{z}_0) & \frac{\partial f}{\partial v}(\mathbf{z}_0) \end{bmatrix}. \tag{A2.2}$$

How should the approximation (A2.1) work for a real-valued function with complex argument? Letting $x = u + \mathrm{j}v$, we may express f as a function of complex x. In a slight abuse of notation, we write $f(x) = f(\mathbf{z})$. Utilizing the 2×2 real-to-complex matrix

$$\mathbf{T} = \begin{bmatrix} 1 & \mathrm{j} \\ 1 & -\mathrm{j} \end{bmatrix}, \tag{A2.3}$$

which satisfies $\mathbf{T}^{\mathrm{H}}\mathbf{T} = \mathbf{T}\mathbf{T}^{\mathrm{H}} = 2\mathbf{I}$, we now obtain

$$\nabla_z f(\mathbf{z}_0)(\mathbf{z} - \mathbf{z}_0) = \left(\tfrac{1}{2}\nabla_z f(\mathbf{z}_0)\mathbf{T}^{\mathrm{H}}\right)(\mathbf{T}(\mathbf{z} - \mathbf{z}_0)) \tag{A2.4}$$

$$= \nabla_x f(x_0) \begin{bmatrix} x - x_0 \\ x^* - x_0^* \end{bmatrix} \tag{A2.5}$$

$$= \nabla_x f(x_0)(\underline{\mathbf{x}} - \underline{\mathbf{x}}_0). \tag{A2.6}$$

This introduces the augmented vector

$$\underline{\mathbf{x}} = \begin{bmatrix} x \\ x^* \end{bmatrix} = \begin{bmatrix} u + \mathrm{j}v \\ u - \mathrm{j}v \end{bmatrix} = \mathbf{T}\mathbf{z}. \tag{A2.7}$$

Using (A2.4)–(A2.6), we define the *complex gradient*

$$\nabla_x f(x_0) \triangleq \tfrac{1}{2}\nabla_z f(\mathbf{z}_0)\mathbf{T}^{\mathrm{H}} = \begin{bmatrix} \frac{1}{2}\left(\frac{\partial f}{\partial u} - \mathrm{j}\frac{\partial f}{\partial v}\right)(\mathbf{z}_0) & \frac{1}{2}\left(\frac{\partial f}{\partial u} + \mathrm{j}\frac{\partial f}{\partial v}\right)(\mathbf{z}_0) \end{bmatrix}. \tag{A2.8}$$

In (A2.5), the first component of $\nabla_x f(x_0)$ operates on x and the second component operates on x^*. Thus,

$$\nabla_x f(x_0) = \begin{bmatrix} \frac{\partial f}{\partial x}(x_0) & \frac{\partial f}{\partial x^*}(x_0) \end{bmatrix}. \tag{A2.9}$$

By equating this expression with the right-hand side of (A2.8), we obtain the following definition.

Definition A2.1. *The* generalized complex differential operator *is defined as*

$$\frac{\partial}{\partial x} \triangleq \frac{1}{2}\left(\frac{\partial}{\partial u} - \mathrm{j}\frac{\partial}{\partial v}\right) \tag{A2.10}$$

and the conjugate generalized complex differential operator *as*

$$\frac{\partial}{\partial x^*} \triangleq \frac{1}{2}\left(\frac{\partial}{\partial u} + \mathrm{j}\frac{\partial}{\partial v}\right). \tag{A2.11}$$

The derivatives obtained by applying the generalized complex and conjugate complex differential operators are called *Wirtinger derivatives* after the pioneering work of Wirtinger (1927). These generalized differential operators can be formally implemented by treating x and x^* as *independent variables*. That is, when applying $\partial/\partial x$, we take the derivative with respect to x while formally treating x^* as a constant. Similarly, $\partial/\partial x^*$ yields the derivative with respect to x^*, regarding x as a constant. The following example shows how this approach works.

Example A2.1. Consider the function $f(x) = |x|^2 = xx^*$. Treating x and x^* as two independent variables, we find

$$\frac{\partial f(x)}{\partial x} = x^* \quad \text{and} \quad \frac{\partial f(x)}{\partial x^*} = x.$$

This can be checked by writing $f(x) = f(u, v) = u^2 + v^2$ and

$$\frac{1}{2}\left(\frac{\partial}{\partial u} - \mathrm{j}\frac{\partial}{\partial v}\right) f(u, v) = u - \mathrm{j}v = x^*,$$

$$\frac{1}{2}\left(\frac{\partial}{\partial u} + \mathrm{j}\frac{\partial}{\partial v}\right) f(u, v) = u + \mathrm{j}v = x.$$

The linear approximation of f at x_0 is

$$\begin{aligned} f(x) &\approx f(x_0) + \begin{bmatrix} \frac{\partial f}{\partial x}(x_0) & \frac{\partial f}{\partial x^*}(x_0) \end{bmatrix} \begin{bmatrix} x - x_0 \\ x^* - x_0^* \end{bmatrix} \\ &= f(x_0) + x_0^*(x - x_0) + x_0(x^* - x_0^*) \\ &= f(x_0) + 2\,\mathrm{Re}(x_0^* x) - 2|x_0|^2. \end{aligned}$$

There is actually nothing in our development thus far that would prevent us from applying the generalized complex differential operators to complex-valued functions. Therefore, we no longer assume that f is real-valued.

A2.1.1 Holomorphic functions

In classical complex analysis, the derivative of a function $f(x)$ at x_0 is defined as

$$\frac{df}{dx}(x_0) = \lim_{x \to x_0} \frac{f(x) - f(x_0)}{x - x_0}. \tag{A2.12}$$

This limit exists only if it is independent of the direction with which x approaches x_0 in the complex plane. If the limit exists at x_0, f is called complex-differentiable at x_0. A fundamental result in complex analysis is that $f(x)$ is complex-differentiable if and only if $f(u, v)$ is real-differentiable and the *Cauchy–Riemann equations* hold:

$$\frac{\partial(\operatorname{Re} f)}{\partial u} = \frac{\partial(\operatorname{Im} f)}{\partial v} \quad \text{and} \quad \frac{\partial(\operatorname{Re} f)}{\partial v} = -\frac{\partial(\operatorname{Im} f)}{\partial u}. \tag{A2.13}$$

Result A2.1. *The Cauchy–Riemann equations are more simply stated as*

$$\frac{\partial f}{\partial x^*} = 0. \tag{A2.14}$$

Definition A2.2. *If a function f on an open domain $\mathcal{A} \subseteq \mathbb{C}$ is complex-differentiable for every $x \in \mathcal{A}$ it is called* holomorphic *or* analytic.

We now see that a real-differentiable function f is holomorphic if and only if it does *not* depend on x^*. For a holomorphic function f, the Wirtinger derivative $\partial f/\partial x$ is the standard complex derivative. A holomorphic function with nonzero derivative is *conformal* because it only rotates and scales infinitesimally small figures in the complex plane, preserving oriented angles and shapes. The Cauchy–Riemann equations immediately make clear that nonconstant real-valued functions cannot be holomorphic.

A2.1.2 Complex gradients and Jacobians

Wirtinger derivatives may also be computed for functions with vector-valued domain, and vector-valued functions.

Definition A2.3. *Let $f: \mathbb{C}^n \to \mathbb{C}$. Assuming that* $\operatorname{Re} f(\mathbf{z})$ *and* $\operatorname{Im} f(\mathbf{z})$ *are each real-differentiable, the complex gradient is the $1 \times 2n$ row vector*

$$\nabla_x f = \frac{\partial f}{\partial \underline{\mathbf{x}}} = \begin{bmatrix} \frac{\partial f}{\partial \mathbf{x}} & \frac{\partial f}{\partial \mathbf{x}^*} \end{bmatrix} \tag{A2.15}$$

with

$$\frac{\partial f}{\partial \mathbf{x}} = \begin{bmatrix} \frac{\partial f}{\partial x_1} & \frac{\partial f}{\partial x_2} & \cdots & \frac{\partial f}{\partial x_n} \end{bmatrix}, \tag{A2.16}$$

$$\frac{\partial f}{\partial \mathbf{x}^*} = \begin{bmatrix} \frac{\partial f}{\partial x_1^*} & \frac{\partial f}{\partial x_2^*} & \cdots & \frac{\partial f}{\partial x_n^*} \end{bmatrix}. \tag{A2.17}$$

Definition A2.4. *The* complex Jacobian *of a vector-valued function* $\mathbf{f}: \mathbb{C}^n \to \mathbb{C}^m$ *is the $m \times 2n$ matrix*

$$\mathbf{J}_x = \begin{bmatrix} \nabla_x f_1 \\ \nabla_x f_2 \\ \vdots \\ \nabla_x f_m \end{bmatrix}. \tag{A2.18}$$

Result A2.2. *The Cauchy–Riemann equations for a vector-valued function* **f** *are*

$$\frac{\partial \mathbf{f}}{\partial \mathbf{x}^*} = \mathbf{0}. \tag{A2.19}$$

If these hold everywhere on the domain, **f** *is holomorphic. This means that* **f** *must be a function of* **x** *only, and must not depend on* $\mathbf{x}^*$.

A2.1.3 Properties of Wirtinger derivatives

From Definition A2.1, we easily obtain the following rules for working with Wirtinger derivatives:

$$\frac{\partial \mathbf{x}}{\partial \mathbf{x}} = \mathbf{I} \quad \text{and} \quad \frac{\partial \mathbf{x}^*}{\partial \mathbf{x}^*} = \mathbf{I}, \tag{A2.20}$$

$$\frac{\partial \mathbf{x}}{\partial \mathbf{x}^*} = \mathbf{0} \quad \text{and} \quad \frac{\partial \mathbf{x}^*}{\partial \mathbf{x}} = \mathbf{0}, \tag{A2.21}$$

$$\frac{\partial \mathbf{f}^*}{\partial \mathbf{x}^*} = \left(\frac{\partial \mathbf{f}}{\partial \mathbf{x}}\right)^* \quad \text{and} \quad \frac{\partial \mathbf{f}}{\partial \mathbf{x}^*} = \left(\frac{\partial \mathbf{f}^*}{\partial \mathbf{x}}\right)^*, \tag{A2.22}$$

$$\frac{\partial \mathbf{f}(\mathbf{g})}{\partial \mathbf{x}} = \frac{\partial \mathbf{f}}{\partial \mathbf{g}}\frac{\partial \mathbf{g}}{\partial \mathbf{x}} + \frac{\partial \mathbf{f}}{\partial \mathbf{g}^*}\frac{\partial \mathbf{g}^*}{\partial \mathbf{x}}, \tag{A2.23}$$

$$\frac{\partial \mathbf{f}(\mathbf{g})}{\partial \mathbf{x}^*} = \frac{\partial \mathbf{f}}{\partial \mathbf{g}}\frac{\partial \mathbf{g}}{\partial \mathbf{x}^*} + \frac{\partial \mathbf{f}}{\partial \mathbf{g}^*}\frac{\partial \mathbf{g}^*}{\partial \mathbf{x}^*}. \tag{A2.24}$$

The last two identities are the *chain rule* for non-holomorphic functions. If **f** is holomorphic, (A2.23) simplifies to the usual chain rule since the second summand is zero, and (A2.24) vanishes.

If **f** is real-valued, it follows from $\mathbf{f}(\mathbf{x}) = \mathbf{f}^*(\mathbf{x})$ and (A2.22) that

$$\frac{\partial \mathbf{f}}{\partial \mathbf{x}^*} = \left(\frac{\partial \mathbf{f}}{\partial \mathbf{x}}\right)^*. \tag{A2.25}$$

The gradient of a scalar *real*-valued function $f(\mathbf{x})$ is therefore an *augmented* vector

$$\nabla_{\underline{x}} f = \left[\frac{\partial f}{\partial \mathbf{x}} \quad \left(\frac{\partial f}{\partial \mathbf{x}}\right)^*\right], \tag{A2.26}$$

where the last n components are the conjugates of the first n components. The linear approximation of f at $\mathbf{x}_0$ is

$$f(\mathbf{x}) \approx f(\mathbf{x}_0) + \nabla_{\underline{x}} f(\mathbf{x}_0)(\underline{\mathbf{x}} - \underline{\mathbf{x}}_0) = f(\mathbf{x}_0) + 2\,\mathrm{Re}\left[\frac{\partial f}{\partial \mathbf{x}}(\mathbf{x}_0)(\mathbf{x} - \mathbf{x}_0)\right]. \tag{A2.27}$$

When we are looking for local extrema, we search for points with $\nabla_{\underline{x}} f = \mathbf{0}$. The following result shows that local extrema of real-valued functions can be identified using the Wirtinger derivative or the conjugate Wirtinger derivative.

Result A2.3. *For a real-valued function f, the following three conditions are equivalent:*

$$\nabla_x f = \mathbf{0} \Leftrightarrow \frac{\partial f}{\partial \mathbf{x}} = \mathbf{0} \Leftrightarrow \frac{\partial f}{\partial \mathbf{x}^*} = \mathbf{0}. \tag{A2.28}$$

In the engineering literature, the gradient of a real-valued function is frequently defined as $(\partial/\partial\mathbf{u} + \mathrm{j}\,\partial/\partial\mathbf{v})f$, which is equal to $2\partial f/\partial\mathbf{x}^*$. This definition is justified only insofar as it can be used to search for local extrema of f, thanks to Result A2.3. However, the preceding development has made it clear that $(\partial/\partial\mathbf{u} + \mathrm{j}\,\partial/\partial\mathbf{v})f$ is *not* the right definition for the complex gradient of a real (and therefore non-holomorphic) function of a complex vector $\mathbf{x}$.

A2.2 Special cases

In this section, we present some formulae for derivatives of common expressions involving linear transformations, quadratic forms, traces, and determinants. We emphasize that there is no need to develop new differentiation rules for Wirtinger derivatives. All rules for taking derivatives of real functions remain valid. However, care must be taken to properly distinguish between the variables with respect to which differentiation is performed and those that are formally regarded as constants.

Some expressions that involve derivatives with respect to a complex vector $\mathbf{x}$ follow (it is assumed that $\mathbf{a}$ and $\mathbf{A}$ are independent of $\mathbf{x}$ and $\mathbf{x}^*$):

$$\frac{\partial}{\partial \mathbf{x}}\mathbf{a}^{\mathrm{H}}\mathbf{x} = \mathbf{a}^{\mathrm{H}} \quad \text{and} \quad \frac{\partial}{\partial \mathbf{x}^*}\mathbf{a}^{\mathrm{H}}\mathbf{x} = \mathbf{0}, \tag{A2.29}$$

$$\frac{\partial}{\partial \mathbf{x}}\mathbf{x}^{\mathrm{H}}\mathbf{a} = \mathbf{0} \quad \text{and} \quad \frac{\partial}{\partial \mathbf{x}^*}\mathbf{x}^{\mathrm{H}}\mathbf{a} = \mathbf{a}^{\mathrm{T}}, \tag{A2.30}$$

$$\frac{\partial}{\partial \mathbf{x}}\mathbf{x}^{\mathrm{H}}\mathbf{A}\mathbf{x} = \mathbf{x}^{\mathrm{H}}\mathbf{A} \quad \text{and} \quad \frac{\partial}{\partial \mathbf{x}^*}\mathbf{x}^{\mathrm{H}}\mathbf{A}\mathbf{x} = \mathbf{x}^{\mathrm{T}}\mathbf{A}^{\mathrm{T}}, \tag{A2.31}$$

$$\frac{\partial}{\partial \mathbf{x}}\mathbf{x}^{\mathrm{T}}\mathbf{A}\mathbf{x} = \mathbf{x}^{\mathrm{T}}(\mathbf{A} + \mathbf{A}^{\mathrm{T}}) \quad \text{and} \quad \frac{\partial}{\partial \mathbf{x}^*}\mathbf{x}^{\mathrm{T}}\mathbf{A}\mathbf{x} = \mathbf{0}, \tag{A2.32}$$

$$\frac{\partial}{\partial \mathbf{x}}\exp\left(-\tfrac{1}{2}\mathbf{x}^{\mathrm{H}}\mathbf{A}^{-1}\mathbf{x}\right) = -\tfrac{1}{2}\exp\left(-\tfrac{1}{2}\mathbf{x}^{\mathrm{H}}\mathbf{A}^{-1}\mathbf{x}\right)\mathbf{x}^{\mathrm{H}}\mathbf{A}^{-1}, \tag{A2.33}$$

$$\frac{\partial}{\partial \mathbf{x}}\ln\left(\mathbf{x}^{\mathrm{H}}\mathbf{A}\mathbf{x}\right) = \left(\mathbf{x}^{\mathrm{H}}\mathbf{A}\mathbf{x}\right)^{-1}\mathbf{x}^{\mathrm{H}}\mathbf{A}. \tag{A2.34}$$

Sometimes we encounter derivatives of a scalar-valued function $f(\mathbf{X})$ with respect to an $n \times m$ complex matrix $\mathbf{X}$. These derivatives are defined as

$$\frac{\partial f}{\partial \mathbf{X}} = \begin{bmatrix} \dfrac{\partial f}{\partial x_{11}} & \cdots & \dfrac{\partial f}{\partial x_{1m}} \\ \vdots & \ddots & \vdots \\ \dfrac{\partial f}{\partial x_{n1}} & \cdots & \dfrac{\partial f}{\partial x_{nm}} \end{bmatrix} \quad \text{and} \quad \frac{\partial f}{\partial \mathbf{X}^*} = \begin{bmatrix} \dfrac{\partial f}{\partial x_{11}^*} & \cdots & \dfrac{\partial f}{\partial x_{1m}^*} \\ \vdots & \ddots & \vdots \\ \dfrac{\partial f}{\partial x_{n1}^*} & \cdots & \dfrac{\partial f}{\partial x_{nm}^*} \end{bmatrix}. \tag{A2.35}$$

A few important special cases follow:

$$\frac{\partial}{\partial \mathbf{X}}\operatorname{tr}(\mathbf{AX}) = \frac{\partial}{\partial \mathbf{X}}\operatorname{tr}(\mathbf{XA}) = \mathbf{A}^{\mathrm{T}}, \tag{A2.36}$$

$$\frac{\partial}{\partial \mathbf{X}}\operatorname{tr}(\mathbf{AX}^{-1}) = -\mathbf{X}^{-\mathrm{T}}\mathbf{A}^{\mathrm{T}}\mathbf{X}^{-\mathrm{T}}, \tag{A2.37}$$

$$\frac{\partial}{\partial \mathbf{X}}\operatorname{tr}(\mathbf{X}^{\mathrm{H}}\mathbf{AX}) = \mathbf{A}^{\mathrm{T}}\mathbf{X}^{*} \quad \text{and} \quad \frac{\partial}{\partial \mathbf{X}^{*}}\operatorname{tr}(\mathbf{X}^{\mathrm{H}}\mathbf{AX}) = \mathbf{AX}, \tag{A2.38}$$

$$\frac{\partial}{\partial \mathbf{X}}\operatorname{tr}\mathbf{X}^{k} = k(\mathbf{X}^{k-1})^{\mathrm{T}}, \tag{A2.39}$$

$$\frac{\partial}{\partial \mathbf{X}}\det\mathbf{X} = (\det\mathbf{X})\mathbf{X}^{-\mathrm{T}}, \tag{A2.40}$$

$$\frac{\partial}{\partial \mathbf{X}}\ln\det\mathbf{X} = \mathbf{X}^{-\mathrm{T}}, \tag{A2.41}$$

$$\frac{\partial}{\partial \mathbf{X}}\det\mathbf{X}^{k} = k(\det\mathbf{X})^{k}\mathbf{X}^{-\mathrm{T}}. \tag{A2.42}$$

A2.3 Complex Hessians

Consider the second-order approximation of a scalar *real-valued* function $f(\mathbf{z})$, $\mathbf{z} = [\mathbf{u}^{\mathrm{T}}, \mathbf{v}^{\mathrm{T}}]^{\mathrm{T}}$, $\mathbf{u}, \mathbf{v} \in \mathbb{R}^{n}$, at $\mathbf{z}_0 = [\mathbf{u}_0^{\mathrm{T}}, \mathbf{v}_0^{\mathrm{T}}]^{\mathrm{T}}$,

$$f(\mathbf{z}) = f(\mathbf{u},\mathbf{v}) \approx f(\mathbf{z}_0) + \nabla_z f(\mathbf{z}_0)(\mathbf{z}-\mathbf{z}_0) + \tfrac{1}{2}(\mathbf{z}-\mathbf{z}_0)^{\mathrm{T}}\mathbf{H}_{zz}(\mathbf{z}_0)(\mathbf{z}-\mathbf{z}_0), \tag{A2.43}$$

with Hessian matrix

$$\begin{aligned}\mathbf{H}_{zz}(\mathbf{z}_0) &= \frac{\partial}{\partial \mathbf{z}}\left(\frac{\partial f}{\partial \mathbf{z}}(\mathbf{z}_0)\right)^{\mathrm{T}} \\ &= \begin{bmatrix} \dfrac{\partial}{\partial \mathbf{u}}\left(\dfrac{\partial f}{\partial \mathbf{u}}(\mathbf{z}_0)\right)^{\mathrm{T}} & \dfrac{\partial}{\partial \mathbf{v}}\left(\dfrac{\partial f}{\partial \mathbf{u}}(\mathbf{z}_0)\right)^{\mathrm{T}} \\ \dfrac{\partial}{\partial \mathbf{u}}\left(\dfrac{\partial f}{\partial \mathbf{v}}(\mathbf{z}_0)\right)^{\mathrm{T}} & \dfrac{\partial}{\partial \mathbf{v}}\left(\dfrac{\partial f}{\partial \mathbf{v}}(\mathbf{z}_0)\right)^{\mathrm{T}} \end{bmatrix} = \begin{bmatrix} \mathbf{H}_{uu}(\mathbf{z}_0) & \mathbf{H}_{vu}(\mathbf{z}_0) \\ \mathbf{H}_{uv}(\mathbf{z}_0) & \mathbf{H}_{vv}(\mathbf{z}_0)\end{bmatrix}.\end{aligned} \tag{A2.44}$$

Assuming that the second-order partial derivatives are continuous, the Hessian is symmetric, which implies

$$\mathbf{H}_{uu} = \mathbf{H}_{uu}^{\mathrm{T}}, \qquad \mathbf{H}_{vv} = \mathbf{H}_{vv}^{\mathrm{T}}, \qquad \text{and} \qquad \mathbf{H}_{vu} = \mathbf{H}_{uv}^{\mathrm{T}}. \tag{A2.45}$$

However, the Hessian is not generally positive definite or semidefinite.

In Section A2.1, we derived an equivalent description for the linear term in (A2.43) for a function $f(\mathbf{x}) = f(\mathbf{u},\mathbf{v})$ with $\mathbf{x} = \mathbf{u} + \mathrm{j}\mathbf{v}$. We would now like to do the same for the quadratic term. Using the $2n \times 2n$ real-to-complex matrix

$$\mathbf{T} = \begin{bmatrix} \mathbf{I} & \mathrm{j}\mathbf{I} \\ \mathbf{I} & -\mathrm{j}\mathbf{I}\end{bmatrix} \tag{A2.46}$$

with $\mathbf{T}^{\mathrm{H}}\mathbf{T} = \mathbf{T}\mathbf{T}^{\mathrm{H}} = 2\mathbf{I}$, we obtain

$$\begin{aligned}(\mathbf{z}-\mathbf{z}_0)^{\mathrm{T}}\mathbf{H}_{zz}(\mathbf{z}_0)(\mathbf{z}-\mathbf{z}_0) &= \left((\mathbf{z}-\mathbf{z}_0)^{\mathrm{T}}\mathbf{T}^{\mathrm{H}}\right)\left(\tfrac{1}{4}\mathbf{T}\mathbf{H}_{zz}(\mathbf{z}_0)\mathbf{T}^{\mathrm{H}}\right)(\mathbf{T}(\mathbf{z}-\mathbf{z}_0)) \\ &= (\underline{\mathbf{x}}-\underline{\mathbf{x}}_0)^{\mathrm{H}}\underline{\mathbf{H}}_{xx}(\mathbf{x}_0)(\underline{\mathbf{x}}-\underline{\mathbf{x}}_0).\end{aligned} \tag{A2.47}$$

Therefore, we define the *complex augmented Hessian matrix* as

$$\underline{\mathbf{H}}_{xx}(\mathbf{x}_0) \triangleq \tfrac{1}{4}\mathbf{T}\mathbf{H}_{zz}(\mathbf{z}_0)\mathbf{T}^{\mathrm{H}}. \tag{A2.48}$$

This is equivalent to the following more detailed definition.

Definition A2.5. *The* complex augmented Hessian matrix *of a function* $f\colon \mathbb{C}^n \to \mathbb{R}$ *is the* $2n \times 2n$ *matrix*

$$\underline{\mathbf{H}}_{xx} = \frac{\partial f}{\partial \underline{\mathbf{x}}}\left(\frac{\partial f}{\partial \underline{\mathbf{x}}}\right)^{\mathrm{H}} = \begin{bmatrix}\mathbf{H}_{xx} & \widetilde{\mathbf{H}}_{xx} \\ \widetilde{\mathbf{H}}_{xx}^{*} & \mathbf{H}_{xx}^{*}\end{bmatrix}, \tag{A2.49}$$

whose $n \times n$ *blocks are the* complex Hessian matrix

$$\mathbf{H}_{xx} = \frac{\partial}{\partial \mathbf{x}}\left(\frac{\partial f}{\partial \mathbf{x}}\right)^{\mathrm{H}} = \frac{\partial}{\partial \mathbf{x}}\left(\frac{\partial f}{\partial \mathbf{x}^*}\right)^{\mathrm{T}} \tag{A2.50}$$

and the complex complementary Hessian matrix

$$\widetilde{\mathbf{H}}_{xx} = \frac{\partial}{\partial \mathbf{x}^*}\left(\frac{\partial f}{\partial \mathbf{x}}\right)^{\mathrm{H}} = \frac{\partial}{\partial \mathbf{x}^*}\left(\frac{\partial f}{\partial \mathbf{x}^*}\right)^{\mathrm{T}}. \tag{A2.51}$$

Example A2.2. Consider $f(\mathbf{x}) = \mathbf{x}^{\mathrm{H}}\mathbf{A}\mathbf{x} + \mathrm{Re}(\mathbf{a}^{\mathrm{H}}\mathbf{x})$. If f is to be real-valued, we need to have $\mathbf{A} = \mathbf{A}^{\mathrm{H}}$, in which case $\mathbf{x}^{\mathrm{H}}\mathbf{A}\mathbf{x}$ is a Hermitian quadratic form. The complex gradient and Hessians are

$$\begin{aligned}\nabla_x f(\mathbf{x}) &= \left[\mathbf{x}^{\mathrm{H}}\mathbf{A} + \tfrac{1}{2}\mathbf{a}^{\mathrm{H}} \quad \mathbf{x}^{\mathrm{T}}\mathbf{A}^* + \tfrac{1}{2}\mathbf{a}^{\mathrm{T}}\right], \\ \mathbf{H}_{xx} &= \mathbf{A} \quad \text{and} \quad \widetilde{\mathbf{H}}_{xx} = \mathbf{0}.\end{aligned}$$

The gradient is zero at $2\mathbf{A}\mathbf{x} = -\mathbf{a} \Leftrightarrow 2\mathbf{A}^*\mathbf{x}^* = -\mathbf{a}^*$, which again demonstrates the equivalences of Result A2.3.

Now consider $g(\mathbf{x}) = \mathbf{x}^{\mathrm{H}}\mathbf{A}\mathbf{x} + \mathrm{Re}(\mathbf{x}^{\mathrm{H}}\mathbf{B}\mathbf{x}^*) + \mathrm{Re}(\mathbf{a}^{\mathrm{H}}\mathbf{x})$ with $\mathbf{A} = \mathbf{A}^{\mathrm{H}}$ and $\mathbf{B} = \mathbf{B}^{\mathrm{T}}$. We obtain

$$\begin{aligned}\nabla_x g(\mathbf{x}) &= \left[\mathbf{x}^{\mathrm{H}}\mathbf{A} + \mathbf{x}^{\mathrm{T}}\mathbf{B}^* + \tfrac{1}{2}\mathbf{a}^{\mathrm{H}} \quad \mathbf{x}^{\mathrm{T}}\mathbf{A}^* + \mathbf{x}^{\mathrm{H}}\mathbf{B} + \tfrac{1}{2}\mathbf{a}^{\mathrm{T}}\right], \\ \mathbf{H}_{xx} &= \mathbf{A} \quad \text{and} \quad \widetilde{\mathbf{H}}_{xx} = \mathbf{B}.\end{aligned}$$

We note that neither $f(\mathbf{x})$ nor $g(\mathbf{x})$ is holomorphic, yet f somehow seems "better behaved" because $\widetilde{\mathbf{H}}_{xx}$ vanishes.

A2.3.1 Properties

From the connection $\underline{\mathbf{H}}_{xx} = \frac{1}{4}\mathbf{T}\mathbf{H}_{zz}\mathbf{T}^{\mathrm{H}}$ we obtain

$$\mathbf{H}_{xx} = \tfrac{1}{4}[\mathbf{H}_{uu} + \mathbf{H}_{vv} + \mathrm{j}(\mathbf{H}_{uv} - \mathbf{H}_{uv}^{\mathrm{T}})], \tag{A2.52}$$

$$\widetilde{\mathbf{H}}_{xx} = \tfrac{1}{4}[\mathbf{H}_{uu} - \mathbf{H}_{vv} + \mathrm{j}(\mathbf{H}_{uv} + \mathbf{H}_{uv}^{\mathrm{T}})]. \tag{A2.53}$$

The augmented Hessian $\underline{\mathbf{H}}_{xx}$ is Hermitian, so

$$\mathbf{H}_{xx} = \mathbf{H}_{xx}^{\mathrm{H}} \qquad \text{and} \qquad \widetilde{\mathbf{H}}_{xx} = \widetilde{\mathbf{H}}_{xx}^{\mathrm{T}}. \tag{A2.54}$$

The second-order approximation of a scalar real-valued function $f(\mathbf{x})$ at $\mathbf{x}_0$ may therefore be written as

$$\begin{aligned} f(\mathbf{x}) &\approx f(\mathbf{x}_0) + \nabla_{\underline{x}} f(\mathbf{x}_0)(\underline{\mathbf{x}} - \underline{\mathbf{x}}_0) + \tfrac{1}{2}(\underline{\mathbf{x}} - \underline{\mathbf{x}}_0)^{\mathrm{H}}\underline{\mathbf{H}}_{xx}(\mathbf{x}_0)(\underline{\mathbf{x}} - \underline{\mathbf{x}}_0) \\ &= f(\mathbf{x}_0) + 2\,\mathrm{Re}\left[\frac{\partial f}{\partial \mathbf{x}}(\mathbf{x}_0)(\mathbf{x} - \mathbf{x}_0)\right] + (\mathbf{x} - \mathbf{x}_0)^{\mathrm{H}}\mathbf{H}_{xx}(\mathbf{x}_0)(\mathbf{x} - \mathbf{x}_0) \\ &\quad + \mathrm{Re}\left[(\mathbf{x} - \mathbf{x}_0)^{\mathrm{H}}\widetilde{\mathbf{H}}_{xx}(\mathbf{x}_0)(\mathbf{x}^* - \mathbf{x}_0^*)\right]. \end{aligned} \tag{A2.55}$$

While the augmented Hessian matrix $\underline{\mathbf{H}}_{xx}$ (just like any augmented matrix) contains some redundancy, it is often more convenient to work with than considering $\mathbf{H}_{xx}$ and $\widetilde{\mathbf{H}}_{xx}$ separately. In particular, as a simple consequence of $\underline{\mathbf{H}}_{xx} = \frac{1}{4}\mathbf{T}\mathbf{H}_{zz}\mathbf{T}^{\mathrm{H}}$, the following holds.

Result A2.4. *The eigenvalues of $\underline{\mathbf{H}}_{xx}$ are the eigenvalues of $\mathbf{H}_{zz}$ multiplied by $1/2$.*

This means that $\underline{\mathbf{H}}_{xx}$ shares the definiteness properties of $\mathbf{H}_{zz}$. That is, $\underline{\mathbf{H}}_{xx}$ is positive or negative (semi)definite if and only if $\mathbf{H}_{zz}$ is respectively positive or negative (semi) definite. Moreover, the condition numbers of $\underline{\mathbf{H}}_{xx}$ and $\mathbf{H}_{zz}$ are identical. These properties are important for numerical optimization methods.

A2.3.2 Extension to complex-valued functions

We can also define the Hessian of a *complex-valued* function f. The key difference is that the Hessian

$$\frac{\partial f}{\partial \underline{\mathbf{x}}}\left(\frac{\partial f}{\partial \underline{\mathbf{x}}}\right)^{\mathrm{H}} \tag{A2.56}$$

no longer satisfies the block pattern of an augmented matrix because

$$\mathbf{H}'_{xx} \triangleq \frac{\partial}{\partial \mathbf{x}}\left(\frac{\partial f}{\partial \mathbf{x}}\right)^{\mathrm{T}} \neq \left[\frac{\partial}{\partial \mathbf{x}^*}\left(\frac{\partial f}{\partial \mathbf{x}}\right)^{\mathrm{H}}\right]^* = \widetilde{\mathbf{H}}_{xx}^*. \tag{A2.57}$$

On the other hand,

$$\frac{\partial}{\partial \mathbf{x}^*}\left(\frac{\partial f}{\partial \mathbf{x}}\right)^{\mathrm{T}} = \left[\frac{\partial}{\partial \mathbf{x}}\left(\frac{\partial f}{\partial \mathbf{x}}\right)^{\mathrm{H}}\right]^{\mathrm{T}} = \mathbf{H}_{xx}^{\mathrm{T}}, \tag{A2.58}$$

but $\mathbf{H}_{xx} \neq \mathbf{H}_{xx}^{\mathrm{H}}$. The following example illustrates that, for complex-valued f, we need to consider three $n \times n$ Hessian matrices: $\mathbf{H}_{xx}$, $\widetilde{\mathbf{H}}_{xx}$, and $\mathbf{H}'_{xx}$ defined in (A2.57).

Example A2.3. Let $f(\mathbf{x}) = \mathbf{x}^{\mathrm{H}}\mathbf{A}\mathbf{x} + \frac{1}{2}\mathbf{x}^{\mathrm{H}}\mathbf{B}\mathbf{x}^* + \frac{1}{2}\mathbf{x}^{\mathrm{T}}\mathbf{C}\mathbf{x}$ with $\mathbf{B} = \mathbf{B}^{\mathrm{T}}$ and $\mathbf{C} = \mathbf{C}^{\mathrm{T}}$. However, we may have $\mathbf{A} \neq \mathbf{A}^{\mathrm{H}}$ so that $\mathbf{x}^{\mathrm{H}}\mathbf{A}\mathbf{x}$ is generally complex. We obtain

$$\mathbf{H}_{xx} = \frac{\partial}{\partial \mathbf{x}}\left(\frac{\partial f}{\partial \mathbf{x}}\right)^{\mathrm{H}} = \mathbf{A}^{\mathrm{H}},$$

$$\widetilde{\mathbf{H}}_{xx} = \frac{\partial}{\partial \mathbf{x}^*}\left(\frac{\partial f}{\partial \mathbf{x}^*}\right)^{\mathrm{T}} = \mathbf{B}.$$

However, in order to fully characterize the quadratic function f, we also need to consider the third Hessian matrix

$$\mathbf{H}'_{xx} = \frac{\partial}{\partial \mathbf{x}}\left(\frac{\partial f}{\partial \mathbf{x}}\right)^{\mathrm{T}} = \mathbf{C}.$$

For complex-valued f, the second-order approximation at $\mathbf{x}_0$ is therefore

$$\begin{aligned} f(\mathbf{x}) &\approx f(\mathbf{x}_0) + \nabla_{\underline{x}} f(\underline{\mathbf{x}} - \underline{\mathbf{x}}_0) + \tfrac{1}{2}(\underline{\mathbf{x}} - \underline{\mathbf{x}}_0)^{\mathrm{H}}\underline{\mathbf{H}}_{xx}(\underline{\mathbf{x}} - \underline{\mathbf{x}}_0) \\ &= f(\mathbf{x}_0) + \frac{\partial f}{\partial \mathbf{x}}(\mathbf{x} - \mathbf{x}_0) + \frac{\partial f}{\partial \mathbf{x}^*}(\mathbf{x}^* - \mathbf{x}_0^*) \\ &\quad + \tfrac{1}{2}(\mathbf{x} - \mathbf{x}_0)^{\mathrm{H}}\mathbf{H}_{xx}(\mathbf{x} - \mathbf{x}_0) + \tfrac{1}{2}(\mathbf{x} - \mathbf{x}_0)^{\mathrm{T}}\mathbf{H}_{xx}^{\mathrm{T}}(\mathbf{x}^* - \mathbf{x}_0^*) \\ &\quad + \tfrac{1}{2}(\mathbf{x} - \mathbf{x}_0)^{\mathrm{H}}\widetilde{\mathbf{H}}_{xx}(\mathbf{x}^* - \mathbf{x}_0^*) + \tfrac{1}{2}(\mathbf{x} - \mathbf{x}_0)^{\mathrm{T}}\mathbf{H}'_{xx}(\mathbf{x} - \mathbf{x}_0), \end{aligned} \tag{A2.59}$$

where all gradients and Hessians are evaluated at $\mathbf{x}_0$. For holomorphic f, the only nonzero Hessian is $\mathbf{H}'_{xx}$.

Appendix 3 Introduction to majorization

The origins of majorization theory can be traced to Schur (1923), who studied the relationship between the diagonal elements of a positive semidefinite Hermitian matrix **H** and its eigenvalues as a means to illuminate Hadamard's determinant inequality. Schur found that the diagonal elements $H_{11}, \ldots, H_{nn}$ are majorized by the eigenvalues $\lambda_1, \ldots, \lambda_n$, written as

$$[H_{11}, \ldots, H_{nn}]^{\mathrm{T}} \prec [\lambda_1, \ldots, \lambda_n]^{\mathrm{T}},$$

which we shall precisely define in due course. Intuitively, this majorization relation means that the eigenvalues are more spread out than the diagonal elements. Importantly, majorization defines an ordering. Schur identified all functions g that preserve this ordering, i.e.,

$$\mathbf{x} \prec \mathbf{y} \Rightarrow g(\mathbf{x}) \leq g(\mathbf{y}).$$

These functions are now called Schur-convex. In doing so, Schur implicitly characterized all possible inequalities that relate a function of the diagonal elements of a Hermitian matrix to the same function of the eigenvalues. (Schur originally worked with positive semidefinite matrices, but this restriction subsequently turned out to be unnecessary.) For us, the main implication of this result is the following. In order to maximize *any* Schur-convex function (or minimize *any* Schur-concave function) of the diagonal elements of a Hermitian matrix **H** with given eigenvalues $\lambda_1, \ldots, \lambda_n$, we must unitarily diagonalize **H**. For instance, from the fact that the product $x_1 x_2 \cdots x_n$ is Schur-concave, we obtain Hadamard's determinant inequality

$$\prod_{i=1}^{n} H_{ii} \geq \prod_{i=1}^{n} \lambda_i = \det \mathbf{H}.$$

Hence, the product of the diagonal elements of **H** with prescribed eigenvalues is minimized if **H** is diagonal. It is possible to generalize these results to apply to the singular values of nonsquare non-Hermitian matrices.

Majorization is not as well known in the signal-processing community as it perhaps should be. In this book, we consider many problems from a majorization point of view, even though they can also be solved using other, better-known but also more cumbersome, algebraic tools. The concise introduction presented in this appendix barely scratches the surface of the rich theory of majorization. For an excellent treatment of this topic, including proofs and references for all unreferenced results in this chapter, we refer the reader to the book by Marshall and Olkin (1979).

A3.1 Basic definitions

A3.1.1 Majorization

Definition A3.1. *A vector* $\mathbf{x} \in \mathbb{R}^n$ *is said to be* majorized *by a vector* $\mathbf{y} \in \mathbb{R}^n$, *written as* $\mathbf{x} \prec \mathbf{y}$, *if*

$$\sum_{i=1}^{r} x_{[i]} \leq \sum_{i=1}^{r} y_{[i]}, \quad r = 1, \ldots, n-1, \tag{A3.1}$$

$$\sum_{i=1}^{n} x_{[i]} = \sum_{i=1}^{n} y_{[i]}, \tag{A3.2}$$

where $[\cdot]$ *is a permutation such that* $x_{[1]} \geq x_{[2]} \geq \cdots \geq x_{[n]}$.

That is, the sum of the r largest components of $\mathbf{x}$ is less than or equal to the sum of the r largest components of $\mathbf{y}$, with equality required for the total sum. Intuitively, if $\mathbf{x} \prec \mathbf{y}$, the components of $\mathbf{x}$ are *less spread out* or "more equal" than the components of $\mathbf{y}$.

Example A3.1. Under the constraints

$$\sum_{i=1}^{n} x_i = N, \quad x_i \geq 0, i = 1, \ldots, n,$$

the vector with the *least spread-out* components in the sense of majorization is $[N/n, N/n, \ldots, N/n]^{\mathrm{T}}$, and the vector with the *most spread-out* components is $[N, 0, \ldots, 0]^{\mathrm{T}}$ (or any permutation thereof). Thus, for any vector $\mathbf{x}$,

$$\frac{1}{n}[N, N, \ldots, N]^{\mathrm{T}} \prec \mathbf{x} \prec [N, 0, \ldots, 0]^{\mathrm{T}}.$$

Definition A3.2. *If the weaker condition*

$$\sum_{i=1}^{r} x_{[i]} \leq \sum_{i=1}^{r} y_{[i]}, \quad r = 1, \ldots, n \tag{A3.3}$$

holds, without equality for $r = n$, $\mathbf{x}$ *is said to be* weakly majorized *by* $\mathbf{y}$, *written as* $\mathbf{x} \prec_{\mathrm{w}} \mathbf{y}$.

The inequality (A3.3) is sometimes referred to as weak submajorization in order to distinguish it from another form of weak majorization, which is called weak supermajorization. For our purposes, we will not require the concept of weak supermajorization.

Because the components of $\mathbf{x}$ and $\mathbf{y}$ are reordered in the definitions of majorization and weak majorization, their original order is irrelevant. Note, however, that majorization is sometimes also defined with respect to *increasing* rather than *decreasing* order. In that case, a vector $\mathbf{x}$ is majorized by a vector $\mathbf{y}$ if the components of $\mathbf{x}$ are *more* spread out than the components of $\mathbf{y}$.

Majorization defines a *preordering* on a set $\mathcal{A}$ since it is *reflexive,*

$$\mathbf{x} \prec \mathbf{x}, \quad \forall \mathbf{x} \in \mathcal{A}, \tag{A3.4}$$

and *transitive,*

$$(\mathbf{x} \prec \mathbf{y} \text{ and } \mathbf{y} \prec \mathbf{z}) \Rightarrow \mathbf{x} \prec \mathbf{z}, \quad \forall (\mathbf{x}, \mathbf{y}, \mathbf{z}) \in \mathcal{A}^3. \tag{A3.5}$$

Yet, in a strict sense, majorization does not constitute a *partial ordering* because it is not *antisymmetric,*

$$\mathbf{x} \prec \mathbf{y} \prec \mathbf{x} \not\Rightarrow \mathbf{x} = \mathbf{y}. \tag{A3.6}$$

However, if $\mathbf{x} \prec \mathbf{y}$ and $\mathbf{y} \prec \mathbf{x}$, then $\mathbf{x}$ is simply a permutation of $\mathbf{y}$. Therefore, majorization does determine a partial ordering if it is restricted to the set of ordered n-tuples $\mathcal{D} = \{\mathbf{x}\colon x_1 \geq x_2 \geq \cdots \geq x_n\}$. It is straightforward to verify that all statements in this paragraph regarding preordering and partial ordering also apply to weak majorization.

A3.1.2 Schur-convex functions

Functions that preserve the preordering of majorization are called *Schur-convex.*

Definition A3.3. *A real-valued function g defined on a set* $\mathcal{A} \subset \mathbb{R}^n$ *is Schur-convex on* $\mathcal{A}$ *if*

$$\mathbf{x} \prec \mathbf{y} \text{ on } \mathcal{A} \Rightarrow g(\mathbf{x}) \leq g(\mathbf{y}). \tag{A3.7}$$

If strict inequality holds when $\mathbf{x}$ *is not a permutation of* $\mathbf{y}$*, then g is called* strictly Schur-convex. *A function g is called* (strictly) Schur-concave *if* $-g$ *is (strictly) Schur-convex.*

Functions that preserve the preordering of *weak* majorization must be Schur-convex *and* increasing.

Definition A3.4. *A real-valued function g defined on* $\mathcal{A} \subset \mathbb{R}^n$ *is called* increasing *if*

$$(x_i \leq y_i, \quad i = 1, \ldots, n) \Rightarrow g(\mathbf{x}) \leq g(\mathbf{y}). \tag{A3.8}$$

It is called strictly increasing *if the right-hand inequality in (A3.8) is strict when* $\mathbf{x} \neq \mathbf{y}$, *and* (strictly) decreasing *if* $-g$ *is (strictly) increasing.*

Result A3.1. *We have*

$$\mathbf{x} \prec_w \mathbf{y} \text{ on } \mathcal{A} \Rightarrow g(\mathbf{x}) \leq g(\mathbf{y}) \tag{A3.9}$$

if and only if g is Schur-convex and increasing on $\mathcal{A}$.

The strict version of this result goes as follows:

$$(\mathbf{x} \prec_w \mathbf{y} \text{ on } \mathcal{A} \text{ and } \mathbf{x} \text{ not a permutation of } \mathbf{y}) \Rightarrow g(\mathbf{x}) < g(\mathbf{y}) \tag{A3.10}$$

if and only if g is strictly Schur-convex and strictly increasing on $\mathcal{A}$.

Testing whether or not a function g is Schur-convex is usually straightforward but possibly tedious. We will look at tests for Schur-convexity in the next section. If g is differentiable, it is a simple matter to check whether it is increasing.

Result A3.2. *If $\mathcal{I} \subset \mathbb{R}$ is an open interval, then g is increasing on $\mathcal{I}$ if and only if the derivative $g'(x) \geq 0$ for all $x \in \mathcal{I}$. If $\mathcal{A} \subset \mathbb{R}^n$ is a convex set with nonempty interior and g is differentiable on the interior of $\mathcal{A}$ and continuous on the boundary of $\mathcal{A}$, then g is increasing on $\mathcal{A}$ if and only if all partial derivatives*

$$\frac{\partial}{\partial x_i} g(\mathbf{x}) \geq 0, \quad i = 1, \ldots, n,$$

for all $\mathbf{x}$ in the interior of $\mathcal{A}$.

A3.2 Tests for Schur-convexity

We will need the following definition.

Definition A3.5. *A function g is called* symmetric *if it is invariant under permutations of the arguments, i.e., $g(\mathbf{x}) = g(\mathbf{\Pi x})$ for all $\mathbf{x}$ and all permutation matrices $\mathbf{\Pi}$.*

One of the most important characterizations of Schur-convex functions is the following.

Result A3.3. *Let $\mathcal{I} \subset \mathbb{R}$ be an open interval and $g: \mathcal{I}^n \to \mathbb{R}$ be continuously differentiable. Then g is Schur-convex on $\mathcal{I}^n$ if and only if*

$$g \text{ is symmetric on } \mathcal{I}^n \tag{A3.11}$$

and

$$(x_1 - x_2)\left[\frac{\partial}{\partial x_1} g(\mathbf{x}) - \frac{\partial}{\partial x_2} g(\mathbf{x})\right] \geq 0, \quad \forall \mathbf{x} \in \mathcal{I}^n. \tag{A3.12}$$

The fact that (A3.12) can be evaluated for x_1 and x_2 rather than two general components x_i and x_j is a consequence of the symmetry of g. The conditions (A3.11) and (A3.12) are also necessary and sufficient if g has domain $\mathcal{A} \subset \mathbb{R}^n$, which is not a Cartesian product, provided that

(1) $\mathbf{x} \in \mathcal{A} \Rightarrow \mathbf{\Pi x} \in \mathcal{A}$, $\forall$ permutation matrices $\mathbf{\Pi}$,
(2) $\mathcal{A}$ is convex and has nonempty interior, and
(3) g is continuously differentiable on the interior of $\mathcal{A}$ and continuous on $\mathcal{A}$.

We will not consider tests for *strict* Schur-convexity in this book.

Example A3.2. Consider the function

$$g(\mathbf{x}) = \sum_{i=1}^{n} \frac{x_i^2}{1 - x_i^2}$$

defined on $\mathcal{I}^n = (0, 1)^n$. This function is obviously symmetric on $\mathcal{I}^n$. It is also increasing on $\mathcal{I}^n$ since, for all $\mathbf{x} \in \mathcal{I}^n$,

$$\frac{\partial}{\partial x_i} g(\mathbf{x}) = \frac{2x_i}{(1 - x_i^2)^2} \geq 0, \quad i = 1, \ldots, n.$$

The condition (A3.12) becomes

$$(x_1 - x_2)\left(\frac{2x_1}{(1 - x_1^2)^2} - \frac{2x_2}{(1 - x_2^2)^2}\right) \geq 0, \quad \forall \mathbf{x} \in \mathcal{I}^n.$$

This can be verified by computing the derivative of

$$f(x) = \frac{2x}{(1 - x^2)^2},$$

which can be seen to be positive for all $x \in \mathcal{I}$,

$$f'(x) = \frac{2(1 - x^2)(1 + 3x^2)}{(1 - x^2)^4} > 0.$$

Thus, g is Schur-convex.

A3.2.1 Specialized tests

It is possible to explicitly determine the form of condition (A3.12) for special cases. The following list contains but a small sample of these results.

1. If $\mathcal{I}$ is an interval and $h\colon \mathcal{I} \to \mathbb{R}$ is convex, then
$$g(\mathbf{x}) = \sum_{i=1}^{n} h(x_i) \tag{A3.13}$$
is Schur-convex on $\mathcal{I}^n$.
2. Let h be a continuous nonnegative function on $\mathcal{I} \subset \mathbb{R}$. The function
$$g(\mathbf{x}) = \prod_{i=1}^{n} h(x_i) \tag{A3.14}$$
is Schur-convex on $\mathcal{I}^n$ if and only if $\log h(x)$ is convex on $\mathcal{I}$.
3. A more general version, which subsumes the first two results, considers the composition of functions of the form
$$g(\mathbf{x}) = f(h(x_1), h(x_2), \ldots, h(x_n)), \tag{A3.15}$$
where $f\colon \mathbb{R}^n \to \mathbb{R}$ and $h\colon \mathbb{R} \to \mathbb{R}$. If f is increasing and Schur-convex and h is convex (increasing and convex), then g is Schur-convex (increasing and Schur-convex). Alternatively, if f is decreasing and Schur-convex and h is concave (decreasing and concave), then g is also Schur-convex (increasing and Schur-convex).

These results establish that there is a connection between convexity (in the conventional sense) and Schur-convexity. In addition to the cases listed above, if g is symmetric and convex, then it is also Schur-convex. The reverse conclusion is not valid.

Example A3.3. Consider again the function $g(\mathbf{x})$ from Example A3.2, which is of the form (A3.13) with

$$h(x_i) = \frac{x_i^2}{1 - x_i^2}.$$

It can be checked that $h(x_i)$ is convex by showing that its second derivative is nonnegative. Therefore, $g(\mathbf{x})$ is Schur-convex. This is exactly what we have done in Example A3.2, thus implicitly deriving the special rule for functions of the form (A3.13).

Example A3.4. Consider a discrete random variable X that takes on n values with probabilities $p_1, p_2, \ldots, p_n$, which we collect in the vector $\mathbf{p} = [p_1, p_2, \ldots, p_n]^{\mathrm{T}}$. The entropy of X is given by

$$H(X) = -\sum_{i=1}^{n} p_i \log p_i.$$

Since $h(x) = x \log x$ is strictly convex, entropy is strictly Schur-concave. Using the result from Example A3.1, it follows that minimum entropy is achieved when probabilities are "most unequal", i.e., $\mathbf{p} = [1, 0, \ldots, 0]^{\mathrm{T}}$ or any permutation thereof. Maximum entropy is achieved when all probabilities are equal, i.e., $\mathbf{p} = [1/n, 1/n, \ldots, 1/n]^{\mathrm{T}}$.

A3.2.2 Functions defined on $\mathcal{D}$

We know from Result A3.3 that Schur-convex functions are necessarily symmetric if they are defined on $\mathbb{R}^n$. However, it is not required that functions defined on the set of ordered n-tuples $\mathcal{D} = \{\mathbf{x}\colon x_1 \geq x_2 \geq \cdots \geq x_n\}$ be symmetric in order to be Schur-convex.

Result A3.4. *Let $g\colon \mathcal{D} \to \mathbb{R}$ be continuous on $\mathcal{D}$ and continuously differentiable on the interior of $\mathcal{D}$. Then g is Schur-convex if and only if*

$$\frac{\partial}{\partial x_i} g(\mathbf{x}) \geq \frac{\partial}{\partial x_{i+1}} g(\mathbf{x}), \quad i = 1, \ldots, n-1, \tag{A3.16}$$

for all $\mathbf{x}$ in the interior of $\mathcal{D}$. For functions of the form

$$g(\mathbf{x}) = \sum_{i=1}^{n} h_i(x_i), \tag{A3.17}$$

where each $h_i\colon \mathbb{R} \to \mathbb{R}$ is differentiable, this condition simplifies to

$$h_i'(x_i) \geq h_{i+1}'(x_{i+1}), \quad \forall x_i \geq x_{i+1},\ i = 1, \ldots, n-1. \tag{A3.18}$$

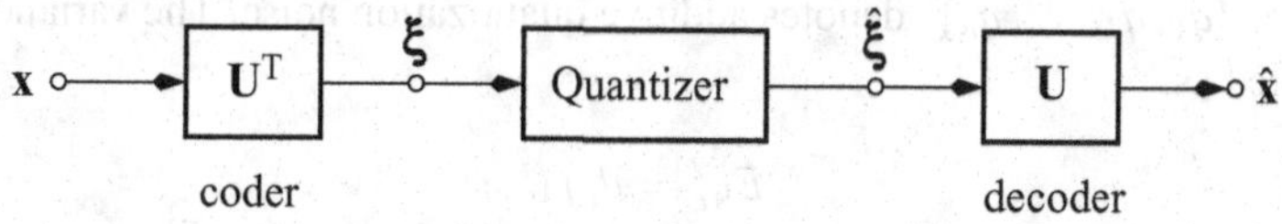

Figure A3.1 An orthogonal transform coder.

In some situations, it is possible to assume without loss of generality that the components of vectors are always arranged in decreasing order. This applies, for instance, when there is no a-priori ordering of components, as is the case with eigenvalues and singular values. Under this assumption, majorization is indeed a partial ordering since, on $\mathcal{D}$, $\mathbf{x} \prec \mathbf{y} \prec \mathbf{x}$ implies that $\mathbf{x} = \mathbf{y}$.

In the context of Schur-convex functions, the fact that majorization only defines a preordering rather than a partial ordering on $\mathbb{R}^n$ is actually irrelevant. On $\mathbb{R}^n$, $\mathbf{x} \prec \mathbf{y} \prec \mathbf{x}$ means that $\mathbf{x}$ is a permutation of $\mathbf{y}$. However, since Schur-convex functions defined on $\mathbb{R}^n$ are necessarily symmetric (i.e., invariant with respect to permutations), $\mathbf{x} \prec \mathbf{y} \prec \mathbf{x}$ implies $g(\mathbf{x}) = g(\mathbf{y})$ for all Schur-convex functions g.

A3.3 Eigenvalues and singular values

A3.3.1 Diagonal elements and eigenvalues

Result A3.5. *Let* $\mathbf{H}$ *be a complex* $n \times n$ *Hermitian matrix with diagonal elements* $\mathbf{diag}(\mathbf{H}) = [H_{11}, H_{22}, \ldots, H_{nn}]^{\mathrm{T}}$ *and eigenvalues* $\mathbf{ev}(\mathbf{H}) = [\lambda_1, \lambda_2, \ldots, \lambda_n]^{\mathrm{T}}$. *Then,*

$$\mathbf{diag}(\mathbf{H}) \prec \mathbf{ev}(\mathbf{H}). \tag{A3.19}$$

Combined with the concept of Schur-convexity, this result can be used to establish a number of well-known inequalities (e.g., Hadamard's determinant inequality) and the optimality of the eigenvalue decomposition for many applications.

Example A3.5. Consider the transform coder shown in Fig. A3.1. The real random vector $\mathbf{x} = [x_1, x_2, \ldots, x_n]^{\mathrm{T}}$, which is assumed to be zero-mean Gaussian, is passed through an orthogonal $n \times n$ coder $\mathbf{U}^{\mathrm{T}}$. The output of the coder is $\boldsymbol{\xi} = \mathbf{U}^{\mathrm{T}}\mathbf{x}$, which is subsequently processed by a bank of n scalar quantizers. That is, each component of $\boldsymbol{\xi}$ is independently quantized. The quantizer output $\hat{\boldsymbol{\xi}}$ is then decoded as $\hat{\mathbf{x}} = \mathbf{U}\hat{\boldsymbol{\xi}}$. The transformation $\mathbf{U}$ therefore determines the internal coordinate system in which quantization takes place.

It is well known that the mean-squared error $E\,\|\mathbf{x} - \hat{\mathbf{x}}\|^2$ is minimized by choosing the coder $\mathbf{U}^{\mathrm{T}}$ to contain the n eigenvectors of the covariance matrix $\mathbf{R}_{xx}$, irrespective of how the total bit budget for quantization is distributed over individual components. In this example, we use majorization to prove this result, following ideas by Goyal *et al.* (2000) and also Schreier and Scharf (2006a). We first write the quantizer output as $\hat{\boldsymbol{\xi}} = \boldsymbol{\xi} + \mathbf{q}$, where $\mathbf{q} = [q_1, q_2, \ldots, q_n]^{\mathrm{T}}$ denotes additive quantization noise. The variance of q_i can be modeled as

$$Eq_i^2 = d_i f(b_i),$$

where $d_i \geq 0$ is the variance of component ξ_i, and $f(b_i)$ is a decreasing function of the number of bits b_i spent on quantizing ξ_i. The function f characterizes the quantizer. We may now express the MSE as

$$E\|\mathbf{x} - \hat{\mathbf{x}}\|^2 = E\|\mathbf{U}(\boldsymbol{\xi} - \hat{\boldsymbol{\xi}})\|^2 = E\|\mathbf{q}\|^2 = \sum_{i=1}^{n} d_i f(b_i).$$

We can assume without loss of generality that the components of $\boldsymbol{\xi}$ are arranged such that

$$d_i \geq d_{i+1}, \quad i = 1, \ldots, n-1.$$

Then, the minimum MSE solution requires that $b_i \geq b_{i+1}$, $i = 1, \ldots, n-1$, because $f(b_i)$ is a decreasing function. With these assumptions, $\mathbf{diag}(\mathbf{R}_{\xi\xi}) = [d_1, d_2, \ldots, d_n]^{\mathrm{T}}$ and $\mathbf{b} = [b_1, b_2, \ldots, b_n]^{\mathrm{T}}$ are both ordered n-tuples.

For a fixed but arbitrary, ordered, bit assignment vector $\mathbf{b}$, the MSE is a Schur-concave function of the diagonal elements of $\mathbf{R}_{\xi\xi} = \mathbf{U}^{\mathrm{T}}\mathbf{R}_{xx}\mathbf{U}$. In order to show this, we note that the MSE is of the form (A3.17) and $f(b_i) \leq f(b_{i+1})$, $i = 1, \ldots, n-1$, because f is decreasing and $b_i \geq b_{i+1}$.

With this in mind, the majorization

$$\mathbf{diag}(\mathbf{U}^{\mathrm{T}}\mathbf{R}_{xx}\mathbf{U}) \prec \mathbf{ev}(\mathbf{R}_{xx})$$

establishes that the MSE is minimized if $\mathbf{U}^{\mathrm{T}}\mathbf{R}_{xx}\mathbf{U}$ is diagonal. Therefore, $\mathbf{U}$ is determined by the eigenvalue decomposition of $\mathbf{R}_{xx}$.

A3.3.2 Diagonal elements and singular values

For an arbitrary nonsquare, non-Hermitian matrix $\mathbf{A} \in \mathbb{C}^{m\times n}$, the diagonal elements and eigenvalues are generally complex, so that majorization would be possible only for absolute values or real parts. Unfortunately, no such relationship exists. However, there is the following comparison, which involves the *singular values* of $\mathbf{A}$.

Result A3.6. *With* $p = \min(m, n)$ *and* $|\mathbf{diag}(\mathbf{A})| \triangleq [|A_{11}|, |A_{22}|, \ldots, |A_{pp}|]^{\mathrm{T}}$, *we have the* weak *majorization*

$$|\mathbf{diag}(\mathbf{A})| \prec_{\mathrm{w}} \mathbf{sv}(\mathbf{A}). \tag{A3.20}$$

This generalizes Result A3.5 to arbitrary matrices. Just as Result A3.5 establishes many optimality results for the eigenvalue decomposition, Result A3.6 does so for the singular value decomposition.

A3.3.3 Partitioned matrices

In Result A3.5, eigenvalues are compared with diagonal elements. We shall now compare the eigenvalues of the $n \times n$ Hermitian matrix

$$\mathbf{H} = \begin{bmatrix} \mathbf{H}_{11} & \mathbf{H}_{12} \\ \mathbf{H}_{12}^{\mathrm{H}} & \mathbf{H}_{22} \end{bmatrix} \tag{A3.21}$$

with the eigenvalues of the corresponding *block-diagonal* matrix

$$\mathbf{H}_0 = \begin{bmatrix} \mathbf{H}_{11} & \mathbf{0} \\ \mathbf{0} & \mathbf{H}_{22} \end{bmatrix}. \tag{A3.22}$$

The submatrices are $\mathbf{H}_{11} \in \mathbb{C}^{n_1 \times n_1}$, $\mathbf{H}_{12} \in \mathbb{C}^{n_1 \times n_2}$, and $\mathbf{H}_{22} \in \mathbb{C}^{n_2 \times n_2}$, where $n = n_1 + n_2$. Obviously, $\mathbf{ev}(\mathbf{H}_0) = (\mathbf{ev}(\mathbf{H}_{11}), \mathbf{ev}(\mathbf{H}_{22}))$, and $\operatorname{tr}\mathbf{H} = \operatorname{tr}\mathbf{H}_0$.

Result A3.7. *There exists the majorization*

$$\mathbf{ev}(\mathbf{H}_0) \prec \boldsymbol{\lambda} = \mathbf{ev}(\mathbf{H}). \tag{A3.23}$$

More generally, let

$$\mathbf{H}_\alpha = \begin{bmatrix} \mathbf{H}_{11} & \alpha\mathbf{H}_{12} \\ \alpha\mathbf{H}_{12}^{\mathrm{H}} & \mathbf{H}_{22} \end{bmatrix}. \tag{A3.24}$$

Then,

$$\mathbf{ev}(\mathbf{H}_{\alpha_1}) \prec \mathbf{ev}(\mathbf{H}_{\alpha_2}), \quad 0 \le \alpha_1 < \alpha_2 \le 1. \tag{A3.25}$$

This shows that the "stronger" the off-diagonal blocks are, the more spread out the eigenvalues become. An immediate consequence is the "block Hadamard inequality" $\det \mathbf{H}_0 \ge \det \mathbf{H}$, or more generally, $\det \mathbf{H}_{\alpha_1} \ge \det \mathbf{H}_{\alpha_2}$.

There is another result for block matrices of the form (A3.21) with $n_1 = n_2 = n/2$, due to Thompson and Therianos (1972).

Result A3.8. *Assuming that $n_1 = n_2$ and the eigenvalues of* $\mathbf{H}$, $\mathbf{H}_{11}$, *and* $\mathbf{H}_{22}$ *are each ordered decreasingly, we have*

$$\sum_{i=1}^{k} \lambda_i + \lambda_{n-k+i} \le \sum_{i=1}^{k} \mathrm{ev}_i(\mathbf{H}_{11}) + \sum_{i=1}^{k} \mathrm{ev}_i(\mathbf{H}_{22}), \quad k = 1, \ldots, n_1. \tag{A3.26}$$

These inequalities are reminiscent of majorization. It is not necessarily majorization because the partial sums on the left-hand side of (A3.26) need not be maximized for some k. However, if

$$\lambda_i + \lambda_{n-i+1} \ge \lambda_{i+1} + \lambda_{n-i}, \quad i = 1, \ldots, n_1 - 1, \tag{A3.27}$$

then (A3.26) becomes

$$[\lambda_1 + \lambda_n, \lambda_2 + \lambda_{n-1}, \ldots, \lambda_{n_1} + \lambda_{n-n_1+1}]^{\mathrm{T}} \prec \mathbf{ev}(\mathbf{H}_{11}) + \mathbf{ev}(\mathbf{H}_{22}), \tag{A3.28}$$

which is an actual majorization relation.

References

Adali, T. and Calhoun, V. D. (2007). Complex ICA of brain imaging data. *IEEE Signal Processing Mag.*, 24:136–139.

Adali, T., Li, H., Novey, M., and Cardoso, J.-F. (2008). Complex ICA using nonlinear functions. *IEEE Trans. Signal Processing*, 56:4536–4544.

Amblard, P. O., Gaeta, M., and Lacoume, J. L. (1996a). Statistics for complex variables and signals – part I: variables. *Signal Processing*, 53:1–13.

(1996b). Statistics for complex variables and signals – part II: signals. *Signal Processing*, 53:15–25.

Andersson, S. A. and Perlman, M. D. (1984). Two testing problems relating the real and complex multivariate normal distribution. *J. Multivariate Analysis*, 15:21–51.

Anttila, L., Valkama, M., and Renfors, M. (2008). Circularity-based I/Q imbalance compensation in wideband direct-conversion receivers. *IEEE Trans. Vehicular Techn.*, 57:2099–2113.

Bangs, W. J. (1971). *Array Processing with Generalized Beamformers*. Ph.D. thesis, Yale University.

Bartmann, F. C. and Bloomfield, P. (1981). Inefficiency and correlation. *Biometrika*, 68:67–71.

Besson, O., Scharf, L. L., and Vincent, F. (2005). Matched direction detectors and estimators for array processing with subspace steering vector uncertainties. *IEEE Trans. Signal Processing*, 53:4453–4463.

Blahut, R. E. (1985). *Fast Algorithms for Digital Signal Processing*. Reading, MA: Addison-Wesley.

Bloomfield, P. and Watson, G. S. (1975). The inefficiency of least squares. *Biometrika*, 62:121–128.

Born, M. and Wolf, E. (1999). *Principles of Optics*. Cambridge: Cambridge University Press.

Boyles, R. A. and Gardner, W. A. (1983). Cycloergodic properties of discrete parameter non-stationary stochastic processes. *IEEE Trans. Inform. Theory*, 29:105–114.

Brandwood, D. H. (1983). A complex gradient operator and its application in adaptive array theory. *IEE Proceedings H*, 130:11–16.

Brown, W. M. and Crane, R. B. (1969). Conjugate linear filtering. *IEEE Trans. Inform. Theory*, 15:462–465.

Burt, W., Cummings, T., and Paulson, C. A. (1974). Mesoscale wind field over ocean. *J. Geophys. Res.*, 79:5625–5632.

Buzzi, S., Lops, M., and Sardellitti, S. (2006). Widely linear reception strategies for layered space-time wireless communications. *IEEE Trans. Signal Processing*, 54:2252–2262.

Cacciapuoti, A. S., Gelli, G., and Verde, F. (2007). FIR zero-forcing multiuser detection and code designs for downlink MC-CDMA. *IEEE Trans. Signal Processing*, 55:4737–4751.

Calman, J. (1978). On the interpretation of ocean current spectra. *J. Physical Oceanography*, 8:627–652.

Charge, P., Wang, Y., and Saillard, J. (2001). A non-circular sources direction finding method using polynomial rooting. *Signal Processing*, 81:1765–1770.

Chen, M., Chen, Z., and Chen, G. (1997). *Approximate Solutions of Operator Equations*. Singapore: World Scientific.

Chevalier, P. and Blin, A. (2007). Widely linear MVDR beamformers for the reception of an unknown signal corrupted by noncircular interferences. *IEEE Trans. Signal Processing*, 55:5323–5336.

Chevalier, P. and Picinbono, B. (1996). Complex linear-quadratic systems for detection and array processing. *IEEE Trans. Signal Processing*, 44:2631–2634.

Chevalier, P. and Pipon, F. (2006). New insights into optimal widely linear array receivers for the demodulation of BPSK, MSK, and GMSK interferences – application to SAIC. *IEEE Trans. Signal Processing*, 54:870–883.

Cohen, L. (1966). Generalized phase-space distribution functions. *J. Math. Phys.*, 7:781–786.

Comon, P. (1994). Independent component analysis, a new concept? *Signal Processing*, 36:287–314.

Conte, E., De Maio, A., and Ricci, G. (2001). GLRT-based adaptive detection algorithms for range-spread targets. *IEEE Trans. Signal Processing*, 49:1336–1348.

Coxhead, P. (1974). Measuring the relationship between two sets of variables. *Brit. J. Math. Statist. Psych.*, 27:205–212.

Cramer, E. M. and Nicewander, W. A. (1979). Some symmetric, invariant measures of multivariate association. *Psychometrika*, 44:43–54.

Davis, M. C. (1963). Factoring the spectral matrix. *IEEE Trans. Automatic Control*, 8:296–305.

DeLathauwer, L. and DeMoor, B. (2002). On the blind separation of non-circular sources, in *Proc. EUSIPCO*, pp. 99–102.

Delmas, J. P. (2004). Asymptotically minimum variance second-order estimation for noncircular signals with application to DOA estimation. *IEEE Trans. Signal Processing*, 52: 1235–1241.

Delmas, J. P. and Abeida, H. (2004). Stochastic Cramér–Rao bound for noncircular signals with application to DOA estimation. *IEEE Trans. Signal Processing*, 52:3192–3199.

Dietl, G., Zoltowski, M. D., and Joham, M. (2001). Recursive reduced-rank adaptive equalization for wireless communication, *Proc. SPIE*, vol. 4395.

Drury, S. W. (2002). The canonical correlations of a 2×2 block matrix with given eigenvalues. *Lin. Algebra Appl.*, 354:103–117.

Drury, S. W., Liu, S., Lu, C.-Y., Puntanen, S., and Styan, G. P. H. (2002). Some comments on several matrix inequalities with applications to canonical correlations: historical background and recent development. *Sankhya A*, 64:453–507.

Eaton, M. L. (1983). *Multivariate Statistics: A Vector Space Approach*. New York: Wiley.

Eriksson, J. and Koivunen, V. (2006). Complex random vectors and ICA models: identifiability, uniqueness, and separability. *IEEE Trans. Inform. Theory*, 52:1017–1029.

Erkmen, B. I. (2008). *Phase-Sensitive Light: Coherence Theory and Applications to Optical Imaging*. Ph.D. thesis, Massachusetts Institute of Technology.

Erkmen, B. I. and Shapiro, J. H. (2006). Optical coherence theory for phase-sensitive light, in *Proc. SPIE*, volume 6305. Bellingham, WA: The International Society for Optical Engineering.

Fang, K.-T., Kotz, S., and Ng, K. W. (1990). *Symmetric Multivariate and Related Distributions*. London: Chapman and Hall.

Ferguson, T. S. (1967). *Mathematical Statistics: A Decision Theoretic Approach*. New York: Academic Press.

Ferrara, E. R. (1985). Frequency-domain implementations of periodically time-varying filters. *IEEE Trans. Acoustics, Speech, Signal Processing*, 33:883–892.

Flandrin, P. (1999). *Time–Frequency/Time–Scale Analysis*. San Diego: Academic.

Gardner, W. A. (1986). Measurement of spectral correlation. *IEEE Trans. Acoust. Speech Signal Processing*, 34:1111–1123.

(1988). *Statistical Spectral Analysis*. Englewood Cliffs, NJ: Prentice Hall.

(1993). Cyclic Wiener filtering: theory and method. *IEEE Trans. Commun.*, 41:151–163.

Gardner, W. A., Brown, W. A., and Chen, C.-K. (1987). Spectral correlation of modulated signals, part II: digital modulation. *IEEE Trans. Commun.*, 35:595–601.

Gardner, W. A., Napolitano, A., and Paura, L. (2006). Cyclostationarity: half a century of research. *Signal Processing*, 86:639–697.

Gelli, G., Paura, L., and Ragozini, A. R. P. (2000). Blind widely linear multiuser detection. *IEEE Commun. Lett.*, 4:187–189.

Gersho, A. and Gray, R. M. (1992). *Vector Quantization and Signal Compression*. Boston, MA: Kluwer.

Gerstacker, H., Schober, R., and Lampe, A. (2003). Receivers with widely linear processing for frequency-selective channels. *IEEE Trans. Commun.*, 51:1512–1523.

Gladyshev, E. (1963). Periodically and almost periodically correlated random processes with continuous time parameter. *Theory Probab. Applic.*, 8:173–177.

Gladyshev, E. D. (1961). Periodically correlated random sequences. *Soviet Math. Dokl.*, 2:385–388.

Gleason, T. C. (1976). On redundancy in canonical analysis. *Psych. Bull.*, 83:1004–1006.

Goh, S. L. and Mandic, D. P. (2007a). An augmented ACRTRL for complex valued recurrent neural networks. *Neural Networks*, 20:1061–1066.

(2007b). An augmented extended Kalman filter algorithm for complex-valued recurrent neural networks. *Neural Computation*, 19:1039–1055.

Goldstein, J. S., Reed, I. S., and Scharf, L. L. (1998). A multistage representation of the Wiener filter based on orthogonal projections. *IEEE Trans. Inform. Theory*, 44:2943–2959.

Gonella, J. (1972). A rotary-component method for analysing meteorological and oceanographic vector time series. *Deep-Sea Res.*, 19:833–846.

Goodman, N. R. (1963). Statistical analysis based on a certain multivariate complex Gaussian distribution (an introduction). *Ann. Math. Statist.*, 34:152–177.

Gorman, J. D. and Hero, A. (1990). Lower bounds for parametric estimation with constraints. *IEEE Trans. Inform. Theory*, 26:1285–1301.

Goyal, V. K., Zhuang, J., and Vetterli, M. (2000). Transform coding with backward adaptive updates. *IEEE Trans. Inform. Theory*, 46:1623–1633.

Grettenberg, T. L. (1965). A representation theorem for complex normal processes. *IEEE Trans. Inform. Theory*, 11:305–306.

Haardt, M. and Roemer, F. (2004). Enhancements of unitary ESPRIT for non-circular sources, in *Proc. ICASSP*, pp. 101–104.

Hanson, B., Klink, K., Matsuura, K., Robeson, S. M., and Willmott, C. J. (1992). Vector correlation: review, exposition, and geographic application. *Ann. Association Am. Geographers*, 82:103–116.

Hanssen, A. and Scharf, L. (2003). A theory of polyspectra for nonstationary stochastic processes. *IEEE Trans. Signal Processing*, 51:1243–1252.

Hindberg, H., Birkelund, Y., Øigård, T. A., and Hanssen, A. (2006). Kernel-based estimators for the Kirkwood–Rihaczek time-frequency spectrum, in *Proc. European Signal Processing Conference*.

Horn, R. A. and Johnson, C. R. (1985). *Matrix Analysis*. Cambridge: Cambridge University Press.

Hotelling, H. (1936). Relations between two sets of variates. *Biometrika*, 28:321–377.

Hua, Y., Nikpour, M., and Stoica, P. (2001). Optimal reduced-rank estimation and filtering. *IEEE Trans. Signal Processing*, 49:457–469.

Izzo, L. and Napolitano, A. (1997). Higher-order statistics for Rice's representation of cyclostationary signals. *Signal Processing*, 56:179–292.

(1998). Multirate processing of time-series exhibiting higher-order cyclostationarity. *IEEE Trans. Signal Processing*, 46:429–439.

Jeon, J. J., Andrews, J. G., and Sung, K. M. (2006). The blind widely linear minimum output energy algorithm for DS-CDMA systems. *IEEE Trans. Signal Processing*, 54:1926–1931.

Jezek, J. and Kucera, V. (1985). Efficient algorithm for matrix spectral factorization. *Automatica*, 21:663–669.

Jones, A. G. (1979). On the difference between polarisation and coherence. *J. Geophys.*, 45:223–229.

Jones, R. C. (1941). New calculus for the treatment of optical systems. *J. Opt. Soc. Am.*, 31:488–493.

Jouny, I. I. and Moses, R. L. (1992). The bispectrum of complex signals: definitions and properties. *IEEE Trans. Signal Processing*, 40:2833–2836.

Jupp, P. E. and Mardia, K. V. (1980). A general correlation coefficient for directional data and related regression problems. *Biometrika*, 67:163–173.

Kelly, E. J. (1986). An adaptive detection algorithm. *IEEE Trans. Aerosp. Electron. Syst.*, 22:115–127.

Kelly, E. J. and Root, W. L. (1960). A representation of vector-valued random processes. *Group Rept. MIT Lincoln Laboratory*, no. 55–21.

Kraut, S., Scharf, L. L., and McWhorter, L. T. (2001). Adaptive subspace detectors. *IEEE Trans. Signal Processing*, 49:1–16.

Lampe, A., Schober, R., and Gerstacker, W. (2002). A novel iterative multiuser detector for complex modulation schemes. *IEEE J. Sel. Areas Commun.*, 20:339–350.

Lee, E. A. and Messerschmitt, D. G. (1994). *Digital Communication*. Boston, MA: Kluwer.

Lehmann, E. L. and Romano, J. P. (2005). *Testing Statistical Hypotheses*, 2nd edn. New York: Springer.

Li, H. and Adali, T. (2008). A class of complex ICA algorithms based on the kurtosis cost function. *IEEE Trans. Neural Networks*, 19:408–420.

Lilly, J. M. and Gascard, J.-C. (2006). Wavelet ridge diagnosis of time-varying elliptical signals with application to an oceanic eddy. *Nonlin. Processes Geophys.*, 13:467–483.

Lilly, J. M. and Park, J. (1995). Multiwavelet spectral and polarization analysis. *Geophys. J. Int.*, 122:1001–1021.

Loève, M. (1978). *Probability Theory II*, 4th edn. New York: Springer.

Mandic, D. P. and Goh, V. S. L. (2009). *Complex Valued Nonlinear Adaptive Filters*. New York: Wiley.

Mardia, K. V., Kent, J. T., and Bibby, J. M. (1979). *Multivariate Analysis*. New York: Academic.

Marshall, A. W. and Olkin, I. (1979). *Inequalities: Theory of Majorization and Its Applications*. New York: Academic.

Martin, W. (1982). Time–frequency analysis of random signals, in *Proc. Int. Conf. Acoustics., Speech, Signal Processing*, pp. 1325–1328.

Martin, W. and Flandrin, P. (1985). Wigner–Ville spectral analysis of nonstationary processes. *IEEE Trans. Acoustics, Speech, Signal Processing*, 33:1461–1470.

Marzetta, T. L. (1993). A simple derivation of the constrained multiple parameter Cramer–Rao bound. *IEEE Trans. Signal Processing*, 41:2247–2249.

McWhorter, L. T. and Scharf, L. (1993a). Geometry of the Cramer–Rao bound. *Signal Processing*, 31:301–311.

(1993b). Cramér–Rao bounds for deterministic modal analysis. *IEEE Trans. Signal Processing*, 41:1847–1865.

(1993c). Properties of quadratic covariance bounds, in *Proc. 27th Asilomar Conf. Signals, Systems, Computers*, pp. 1176–1180.

McWhorter, T. and Schreier, P. (2003). Widely-linear beamforming, in *Proc. 37th Asilomar Conf. Signals, Systems, Comput.*, pp. 753–759.

Mirbagheri, A., Plataniotis, N., and Pasupathy, S. (2006). An enhanced widely linear CDMA receiver with OQPSK modulation. *IEEE Trans. Commun.*, 54:261–272.

Mitra, S. K. (2006). *Digital Signal Processing*, 3rd edn. Boston, MA: McGraw-Hill.

Molle, J. W. D. and Hinich, M. J. (1995). Trispectral analysis of stationary random time series. *IEEE Trans. Signal Processing*, 97:2963–2978.

Mooers, C. N. K. (1973). A technique for the cross spectrum analysis of pairs of complex-valued time series, components and rotational invariants. *Deep-Sea Res.*, 20:1129–1141.

Morgan, D. R. (2006). Variance and correlation of square-law detected allpass channels with bandpass harmonic signals in Gaussian noise. *IEEE Trans. Signal Processing*, 54:2964–2975.

Morgan, D. R. and Madsen, C. K. (2006). Wide-band system identification using multiple tones with allpass filters and square-law detectors. *IEEE Trans. Circuits Systems I*, 53:1151–1165.

Mueller, H. (1948). The foundation of optics. *J. Opt. Soc. Am.*, 38:661–672.

Mullis, C. T. and Scharf, L. L. (1996). *Applied Probability*. Course notes for ECE5612, Univ. of Colorado, Boulder, CO.

Napolitano, A. and Spooner, C. M. (2001). Cyclic spectral analysis of continuous-phase modulated signals. *IEEE Trans. Signal Processing*, 49:30–44.

Napolitano, A. and Tanda, M. (2004). Doppler-channel blind identification for noncircular transmissions in multiple-access systems. *IEEE Trans. Commun.*, 52:2073–2078.

Navarro-Moreno, J., Ruiz-Molina, J. C., and Fernández-Alcalá, R. M. (2006). Approximate series representations of linear operations on second-order stochastic processes: application to simulation. *IEEE Trans. Inform. Theory*, 52:1789–1794.

Neeser, F. D. and Massey, J. L. (1993). Proper complex random processes with applications to information theory. *IEEE Trans. Inform. Theory*, 39:1293–1302.

Nilsson, R., Sjoberg, F., and LeBlanc, J. P. (2003). A rank-reduced LMMSE canceller for narrowband interference suppression in OFDM-based systems. *IEEE Trans. Commun.*, 51:2126–2140.

Novey, M. and Adali, T. (2008a). Complex ICA by negentropy maximization. *IEEE Trans. Neural Networks*, 19:596–609.

(2008b). On extending the complex FastICA algorithm to noncircular sources. *IEEE Trans. Signal Processing*, 56:2148–2154.

Olhede, S. and Walden, A. T. (2003a). Polarization phase relationships via multiple Morse wavelets. I. Fundamentals. *Proc. Roy. Soc. Lond. A Mat.*, 459:413–444.

(2003b). Polarization phase relationships via multiple Morse wavelets. II. Data analysis. *Proc. Roy. Soc. Lond. A Mat.*, 459:641–657.

Ollila, E. (2008). On the circularity of a complex random variable. *IEEE Signal Processing Lett.*, 15:841–844.

Ollila, E. and Koivunen, V. (2004). Generalized complex elliptical distributions, in *Proc. SAM Workshop*, pp. 460–464.

(2009). Complex ICA using generalized uncorrelating transform. *Signal Processing*, 89:365–377.

Pflug, L. A., Ioup, G. E., Ioup, J. W., and Field, R. L. (1992). Properties of higher-order correlations and spectra for bandlimited, deterministic transients. *J. Acoust. Soc. Am.*, 91:975–988.

Picinbono, B. (1994). On circularity. *IEEE Trans. Signal Processing*, 42:3473–3482.

(1996). Second-order complex random vectors and normal distributions. *IEEE Trans. Signal Processing*, 44:2637–2640.

Picinbono, B. and Bondon, P. (1997). Second-order statistics of complex signals. *IEEE Trans. Signal Processing*, 45:411–419.

Picinbono, B. and Chevalier, P. (1995). Widely linear estimation with complex data. *IEEE Trans. Signal Processing*, 43:2030–2033.

Picinbono, B. and Duvaut, P. (1988). Optimal linear-quadratic systems for detection and estimation. *IEEE Trans. Inform. Theory*, 34:304–311.

Poor, H. V. (1998). *An Introduction to Signal Detection and Estimation*. New York: Springer.

Proakis, J. G. (2001). *Digital Communications*, 4th edn. Boston, MA: McGraw-Hill.

Ramsay, J. O., ten Berge, J., and Styan, G. P. H. (1984). Matrix correlation. *Psychometrika*, 49:402–423.

Reed, I. S., Mallett, J. D., and Brennan, L. E. (1974). Rapid convergence rate in adaptive arrays. *IEEE Trans. Aerosp. Electron. Syst.*, 10:853–863.

Renyi, A. (1959). On measures of dependence. *Acta Math. Acad. Sci. Hungary*, 10:441–451.

Richmond, C. D. (2006). Mean-squared error and threshold SNR prediction of maximum likelihood signal parameter estimation with estimated colored noise covariances. *IEEE Trans. Inform. Theory*, 52:2146–2164.

Rihaczek, A. W. (1968). Signal energy distribution in time and frequency. *IEEE Trans. Inform. Theory*, 14:369–374.

Rivet, B., Girin, L., and Jutten, C. (2007). Log-Rayleigh distribution: a simple and efficient statistical representation of log-spectral coefficients. *IEEE Trans. Audio, Speech, Language Processing*, 15:796–802.

Robert, P. and Escoufier, Y. (1976). A unifying tool for linear multivariate statistical methods: the RV-coefficient. *Appl. Statist.*, 25:257–265.

Robey, F. C., Fuhrmann, D. R., Kelly, E. J., and Nitzberg, R. A. (1992). A CFAR adaptive matched filter detector. *IEEE Trans. Aerosp. Electron. Syst.*, 28:208–216.

Römer, F. and Haardt, M. (2007). Deterministic Cramér–Rao bounds for strict sense non-circular sources, in *Proc. ITG/IEEE Workshop on Smart Antennas*.

(2009). Multidimensional unitary tensor-esprit for non-circular sources, in *Proc. ICASSP*.

Roueff, A., Chanussot, J., and Mars, J. I. (2006). Estimation of polarization parameters using time-frequency representations and its application to waves separation. *Signal Processing*, 86:3714–3731.

Rozanov, Y. A. (1963). *Stationary Random Processes*. San Francisco, CA: Holden-Day.

Rozeboom, W. W. (1965). Linear correlation between sets of variables. *Psychometrika*, 30: 57–71.

Rubin-Delanchy, P. and Walden, A. T. (2007). Simulation of improper complex-valued sequences. *IEEE Trans. Signal Processing*, 55:5517–5521.

(2008). Kinematics of complex-valued time series. *IEEE Trans. Signal Processing*, 56:4189–4198.

Rykaczewski, P., Valkama, M., and Renfors, M. (2008). On the connection of I/Q imbalance and channel equalization in direct-conversion transceivers. *IEEE Trans. Vehicular Techn.*, 57:1630–1636.

Sampson, P. D., Streissguth, A. P., Barr, H. M., and Bookstein, F. L. (1989). Neurobehavioral effects of prenatal alcohol: part II. Partial least squares analysis. *Neurotoxicology Teratology*, 11:477–491.

Samson, J. C. (1980). Comments on polarization and coherence. *J. Geophys.*, 48:195–198.

Scharf, L. L. (1991). *Statistical Signal Processing*. Reading, MA: Addison-Wesley.

Scharf, L. L., Chong, E. K. P., Zoltowski, M. D., Goldstein, J. S., and Reed, I. S. (2008). Subspace expansion and the equivalence of conjugate direction and multistage Wiener filters. *IEEE Trans. Signal Processing*, 56:5013–5019.

Scharf, L. L. and Friedlander, B. (1994). Matched subspace detectors. *IEEE Trans. Signal Processing*, 42:2146–2157.

(2001). Toeplitz and Hankel kernels for estimating time-varying spectra of discrete-time random processes. *IEEE Trans. Signal Processing*, 49:179–189.

Scharf, L. L., Schreier, P. J., and Hanssen, A. (2005). The Hilbert space geometry of the Rihaczek distribution for stochastic analytic signals. *IEEE Signal Processing Lett.*, 12:297–300.

Schreier, P. J. (2008a). Bounds on the degree of impropriety of complex random vectors. *IEEE Signal Processing Lett.*, 15:190–193.

(2008b). Polarization ellipse analysis of nonstationary random signals. *IEEE Trans. Signal Processing*, 56:4330–4339.

(2008c). A unifying discussion of correlation analysis for complex random vectors. *IEEE Trans. Signal Processing*, 56:1327–1336.

Schreier, P. J., Adalı, T., and Scharf, L. L. (2009). On ICA of improper and noncircular sources, in *Proc. Int. Conf. Acoustics, Speech, Signal Processing*.

Schreier, P. J. and Scharf, L. L. (2003a). Second-order analysis of improper complex random vectors and processes. *IEEE Trans. Signal Processing*, 51:714–725.

(2003b). Stochastic time-frequency analysis using the analytic signal: why the complementary distribution matters. *IEEE Trans. Signal Processing*, 51:3071–3079.

(2006a). Canonical coordinates for transform coding of noisy sources. *IEEE Trans. Signal Processing*, 54:235–243.

(2006b). Higher-order spectral analysis of complex signals. *Signal Processing*, 86:3321–3333.

Schreier, P. J., Scharf, L. L., and Hanssen, A. (2006). A generalized likelihood ratio test for impropriety of complex signals. *IEEE Signal Processing Lett.*, 13:433–436.

Schreier, P. J., Scharf, L. L., and Mullis, C. T. (2005). Detection and estimation of improper complex random signals. *IEEE Trans. Inform. Theory*, 51:306–312.

Schur, I. (1923). Über eine Klasse von Mittelbildungen mit Anwendungen auf die Determinantentheorie. *Sitzungsber. Berliner Math. Ges.*, 22:9–20.

Serpedin, E., Panduru, F., Sari, I., and Giannakis, G. B. (2005). Bibliography on cyclostationarity. *Signal Processing*, 85:2233–2303.

Shapiro, J. H. and Erkmen, B. I. (2007). Imaging with phase-sensitive light, in *Int. Conf. on Quantum Information*. New York: Optical Society of America.

Slepian, D. (1954). Estimation of signal parameters in the presence of noise. *Trans. IRE Prof. Group Inform. Theory*, 3:68–89.

Spurbeck, M. S. and Mullis, C. T. (1998). Least squares approximation of perfect reconstruction filter banks. *IEEE Trans. Signal Processing*, 46:968–978.

Spurbeck, M. S. and Schreier, P. J. (2007). Causal Wiener filter banks for periodically correlated time series. *Signal Processing*, 87:1179–1187.

Srivastava, M. S. (1965). On the complex Wishart distribution. *Ann. Math. Statist.*, 36:313–315.

Stewart, D. and Love, W. (1968). A general canonical correlation index. *Psych. Bull.*, 70:160–163.

Stoica, P. and Ng, B. C. (1998). On the Cramer–Rao bound under parametric constraints. *IEEE Signal Processing Lett.*, 5:177–179.

Stokes, G. G. (1852). On the composition and resolution of streams of polarized light from different sources. *Trans. Cambr. Phil. Soc.*, 9:399–423.

Tauböck, G. (2007). Complex noise analysis of DMT. *IEEE Trans. Signal Processing*, 55:5739–5754.

Thompson, R. C. and Therianos, S. (1972). Inequalities connecting the eigenvalues of a Hermitian matrix with the eigenvalues of complementary principal submatrices. *Bull. Austral. Math. Soc.*, 6:117–132.

Tishler, A. and Lipovetsky, S. (2000). Modelling and forecasting with robust canonical analysis: method and application. *Comput. Op. Res.*, 27:217–232.

van den Bos, A. (1994a). Complex gradient and Hessian. *IEE Proc. Vision, Image, Signal Processing*, 141:380–383.

(1994b). A Cramér–Rao lower bound for complex parameters. *IEEE Trans. Signal Processing*, 42:2859.

(1995). The multivariate complex normal distribution – a generalization. *IEEE Trans. Inform. Theory*, 41:537–539.

Van Trees, H. L. (2001). *Detection, Estimation, and Modulation Theory: Part I.* New York: Wiley.

Van Trees, H. L. and Bell, K. L., editors (2007). *Bayesian Bounds for Parameter Estimation and Nonlinear Filtering/Tracking*. New York: IEEE and Wiley-Interscience.

Wahlberg, P. and Schreier, P. J. (2008). Spectral relations for multidimensional complex improper stationary and (almost) cyclostationary processes. *IEEE Trans. Inform. Theory*, 54:1670–1682.

Walden, A. T. and Rubin-Delanchy, P. (2009). On testing for impropriety of complex-valued Gaussian vectors. *IEEE Trans. Signal Processing*, 57:825–834.

Weinstein, E. and Weiss, A. J. (1988). A general class of lower bounds in parameter estimation. *IEEE Trans. Inform. Theory*, 34:338–342.

Weippert, M. E., Hiemstra, J. D., Goldstein, J. S., and Zoltowski, M. D. (2002). Insights from the relationship between the multistage Wiener filter and the method of conjugated gradients, *Proc. IEEE Workshop on Sensor Array Multichannel Signal Processing*, pp. 388–392.

Weiss, A. J. and Weinstein, E. (1985). A lower bound on the mean squared error in random parameter estimation. *IEEE Trans. Inform. Theory*, 31:680–682.

Wiener, N. and Masani, P. (1957). The prediction theory of multivariate stochastic processes, part I. *Acta Math.*, 98:111–150.

(1958). The prediction theory of multivariate stochastic processes, part II. *Acta Math.*, 98:93–137.

Wilson, G. T. (1972). The factorization of matricial spectral densities. *SIAM J. Appl. Math.*, 23:420–426.

Wirtinger, W. (1927). Zur formalen Theorie der Funktionen von mehr komplexen Veränderlichen. *Math. Ann.*, 97:357–375.

Witzke, M. (2005). Linear and widely linear filtering applied to iterative detection of generalized MIMO signals. *Ann. Telecommun.*, 60:147–168.

Wold, H. (1975). Path models with latent variables: the NIPALS approach, in Blalock, H. M., editor, *Quantitative Sociology: International Perspectives on Mathematical and Statistical Modeling*. New York: Academic, pp. 307–357.

(1985). Partial least squares, in Kotz, S. and Johnson, N. L., editors, *Encyclopedia of the Statistical Sciences*, New York: Wiley, pp. 581–591.

Wolf, E. (1959). Coherence properties of partially polarized electromagnetic radiation. *Nuovo Cimento*, 13:1165–1181.

Wooding, R. A. (1956). The multivariate distribution of complex normal variables. *Biometrika*, 43:212–215.

Yanai, H. (1974). Unification of various techniques of multivariate analysis by means of generalized coefficient of determination (G.C.D.). *J. Behaviormetrics*, 1:45–54.

Yoon, Y. C. and Leib, H. (1997). Maximizing SNR in improper complex noise and applications to CDMA. *IEEE Commun. Letters*, 1:5–8.

Youla, D. C. (1961). On the factorization of rational matrices. *IRE Trans. Inform. Theory*, 7:172–189.

Zou, Y., Valkama, M., and Renfors, M. (2008). Digital compensation of I/Q imbalance effects in space-time coded transmit diversity systems. *IEEE Trans. Signal Processing*, 56:2496–2508.

Index

Printed in the United States
By Bookmasters